Tetrahedrally-Bonded Amorphous Semiconductors

Institute for Amorphous Studies Series

Series editors

David Adler
Massachusetts Institute of Technology
Cambridge, Massachusetts

and

Brian B. Schwartz
Institute for Amorphous Studies
Bloomfield Hills, Michigan
and Brooklyn College of the City University of New York
Brooklyn, New York

PHYSICAL PROPERTIES OF AMORPHOUS MATERIALS
Edited by David Adler, Brian B. Schwartz, and Martin C. Steele

PHYSICS OF DISORDERED MATERIALS
Edited by David Adler, Hellmut Fritzsche, and Stanford R. Ovshinsky

TETRAHEDRALLY-BONDED AMORPHOUS SEMICONDUCTORS
Edited by David Adler and Hellmut Fritzsche

LOCALIZATION AND METAL–INSULATOR TRANSITIONS
Edited by Hellmut Fritzsche and David Adler

A Continuation Order Plan is available for this series. A continuation order will bring delivery of each new volume immediately upon publication. Volumes are billed only upon actual shipment. For further information please contact the publisher.

Tetrahedrally-Bonded Amorphous Semiconductors

Edited by
David Adler
Massachusetts Institute of Technology
Cambridge, Massachusetts

and
Hellmut Fritzsche
University of Chicago
Chicago, Illinois

Springer Science+Business Media, LLC

Library of Congress Cataloging in Publication Data

Main entry under title:

Tetrahedrally-bonded amorphous semiconductors.

(Institute for Amorphous Studies series)
Bibliography: p.
Includes index.
1. Amorphous semiconductors—Addresses, essays, lectures. 2. Mott, N. F. (Nevill
Francis), Sir, 1905–   . I. Adler, David. II. Fritzsche, Hellmut. III. Series.
QC611.8.A5T4698 1985   530.4'1   85-12140
ISBN 978-1-4899-5363-6   ISBN 978-1-4899-5361-2 (eBook)
DOI 10.1007/978-1-4899-5361-2

To
Sir Nevill Mott
with respect and affection on the occasion
of his eightieth birthday

PREFACE

This volume and its two companion volumes, entitled Physics of Disordered Materials and Localization and Metal-Insulator Transitions, are our way of paying special tribute to Sir Nevill Mott and to express our heartfelt wishes to him on the occasion of his eightieth birthday. Sir Nevill has set the highest standards as a physicist, teacher, and scientific leader. Our feelings for him include not only the respect and admiration due a great scientist, but also a deep affection for a great human being, who possesses a rare combination of outstanding personal qualities. We thank him for enriching our lives, and we shall forever carry cherished memories of this noble man.

Scientists best express their thanks by contributing their thoughts and observations to a Festschrift. This one honoring Sir Nevill fills three volumes, with literally hundreds of authors meeting a strict deadline. The fact that contributions poured in from all parts of the world attests to the international cohesion of our scientific community. It is a tribute to Sir Nevill's stand for peace and understanding, transcending national borders.

The editors wish to express their gratitude to Ghazaleh Koefod for her diligence and expertise in deciphering and typing many of the papers, as well as helping in numerous other ways. The blame for the errors that remain belongs to the editors.

David Adler
Massachusetts Institute
of Technology
Cambridge, Massachusetts

Hellmut Fritzsche
The University of Chicago
Chicago, Illinois

Stanford R. Ovshinsky
Energy Conversion Devices
Troy, Michigan

CONTENTS

PART ONE: GROWTH AND STRUCTURE

PART TWO: HYDROGEN INCORPORATION, IMPURITIES, AND DEFECTS

PART THREE: OPTICAL PROPERTIES, GAP STATES,
EQUILIBRIUM, AND NONEQUILIBRIUM TRANSPORT

PART FOUR: HETEROSTRUCTURES AND DEVICES

Contents of Companion Volumes:

INTRODUCTION

David Adler and Hellmut Fritzsche

This volume, one of three in this series presented as a Festschrift in honor of the eightieth birthday of Sir Nevill Mott, is concerned with the subject of tetrahedrally bonded amorphous semiconductors in general, although the vast majority of contributions are restricted to hydrogenated amorphous silicon in particular. Over the past 10 years, interest in this system and its related alloys has mushroomed, concomitant with their potential applications in the areas of solar cells, thin-film transistors, electrophotographic copiers, laser printers, computer memories, and a myriad of other devices. The present volume contains 44 papers by 87 authors from 10 different countries, and is divided into four parts.

Part one consists of eight papers principally dealing with growth and structure. It begins with a historical review of the origins of hydrogenated amorphous silicon (a-Si:H) by its discoverers, Chittick and Sterling. In the grand tradition of materials science, the original discovery of this important semiconductor was an accident, an undesired byproduct of the race to produce ultrapure crystalline Si (c-Si). In the succeeding paper, the heterogeneity of a-Si:H is discussed in detail by Chenevas-Paule and Bellissent, followed by two contributions relating to amorphous germanium (a-Ge), a description of its shock crystallization by Kikuchi and a review of recent structural studies by Paesler and Sayers. Finally, Grigorovici and Gartner analyze the chemical origin of short-range order in tetrahedral solids, Collins et al. detail the results of their investigation of growth and microstructure in a-Si:H, Von Roedern and Madan discuss the effects of disorder, and Moustakas details the results of his study of the growth of rf-sputtered microcrystalline Si. These papers, as a group, suggest some of the future directions of the field, viz. coming to grips with the heterogeneity, improving the quality of amorphous germanium alloys, and understanding the mechanism of crystallization.

Part Two of the volume contains 13 papers on the incorporation of hydrogen, selected impurities, and intrinsic defects in disordered Si and Ge alloys and III-V semiconductors. It begins with a review of the relationship between hydrogen bonding and structure in high-quality a-Si:H by P. John and J.I.B. Wilson and an analogous discussion of hydrogen passivation of defects in c-Si by Pankove. Beyer then analyzes the mechanisms of hydrogen incorporation and release in a-Si:H and Persans et al. demonstrate the lack of hydrogen passivation in a-Ge:H relative to a-Si:H. The effects of several different types of impurities on the properties of a-Si:H are then reviewed, both by Zavetova and Arkimchenko and by Carlson. These are followed by three theoretical studies, in which Robertson analyzes the doping mechanism and defects, T. Shimizu et al. suggest the importance of weak bonds, and Lin and Lucovsky calculate the energies of dangling bond centers in a-Si-Ge alloys. Gheorghiu and Theye investigate disorder in a-InP films, followed by three papers dealing with defect states in a-Si:H, one by Morigaki et al. on trapped-hole centers, one by Guha on deep centers, and one by Tanaka et al. on defect states in phosphorus-doped films.

The 13 papers of Part Three are primarily concerned with the density of states, transport, and carrier trapping and recombination, especially in a-Si:H. It begins with the presentation of a new technique for calculating the density of states, $g(E)$, of a disordered system by Hayes and Beeby, followed by a discussion of the meaning of the optical gap in a-Si:H by Frova and Selloni. Overhof then analyzes the relationship between $g(E)$ and transport in a-Ge:H, Cohen and Lang present arguments that $g(E)$ in the bulk of a-Si:H films is accurately determined by deep level transient spectroscopy, and Crandall summarizes data on photostructural effects in a-Si:H. Butcher et al. describe a unified theory for dc and ac hopping transport and then apply it to experiments on a-Si:H and a-Ge:H. The remainder of the section is concerned with time-resolved spectroscopy of a-Si:H and related alloys. Tauc reviews photoinduced absorption experiments in the 10^{-12}-10^{-2}s range, Schiff analyzes transient photoconductivity measurements, and Carius et al. report the results of a dual-beam photoconductivity study of a-Si:H films. Oda et al. then describe time-of-flight experiments on a high-quality small band gap alloy, a-Si-Ge:H:F. Finally, the section concludes with three papers on a-Si:H, a discussion of the effects of diffusion-limited kinetics by Silver and Cannella, a review of recombination processes by B.A. Wilson, and an analysis of the picosecond high-energy luminescence by Scher.

The concluding section of the volume is devoted to the exciting areas of heterostructures and device applications. It begins with a series of papers on periodic multilayered films, including the presentations of new methods for determining $g(E)$ by Döhler and for analyzing optical absorption by Ugur et al., a calculation of the mean free path by Tsu, a discussion of luminescence by Hirose et al., and a photoemission analysis by Evangelisti. Amorphous-crystalline heterojunctions are investigated theoretically by Herman and Lambin and experimentally by Smid et al.; Nitta et al. discuss the structural and optical properties of primarily polymeric, $(SiH_2)_n$ films, and the volume concludes with reviews of technological applications by Hamakawa and by Mort and Jansen.

As a whole, this volume clearly demonstrates how sophisticated the science and technology of amorphous silicon alloys have become in such a short time. We now routinely refer to sub-picosecond spectroscopy, periodic structures with layers less than 10Å thick, and efficient foot-square multijunction solar cells, without realizing the scorn with which predictions of these would have been greeted only five years ago. Truly a new age has dawned.

GLOW DISCHARGE DEPOSITION OF AMORPHOUS SEMICONDUCTORS:

THE EARLY YEARS

R.C. Chittick and H.F. Sterling

Standard Telecommunication Laboratories

London Road, Harlow, Essex. UK.

INTRODUCTION

During the past ten years, we have seen an unprecedented interest in the electrodeless r.f. glow discharge technique as a method for depositing thin films of non-crystalline semiconductors. This is, in the main, due to the great advances in the understanding of the behaviour of glow discharge (GD) deposited amorphous silicon (a-Si) that followed the successful n and p-type doping announced by Spear and Le Comber in 1975[1] and the subsequent demonstration of practical a-Si solar cells.[2,3]

The first accidental discovery, however, that led to this modern development of the glow discharge technique was made over thirty years ago during the race to produce high purity silicon in single crystal form. J.M. Wilson and his group at Standard Telecommunication Laboratories,[4] using thermal decomposition of silane by r.f. heating, noticed that unheated parts of their reaction tube had a deposit of a non-crystalline form of silicon that was characterized by a very high resistivity. This was found to be associated with unwanted glow and/or corona discharges produced by the r.f. field.

These findings, intriguing though they were, could not at that time be allowed to detract from the main task of producing high purity single crystals. It was not until after the successful completion of that project that further investigation of glow discharge deposition could be undertaken.

In this paper we would like to follow the evolution at STL of the electrodeless r.f. glow discharge method from this early stage in 1954 to the development of a practical equipment in the 1960s, capable of producing from gaseous covalent hydrides, high quality amorphous films including silicon, germanium, silicon oxides and silicon nitride.

Some of the properties of our early GD amorphous silicon and germanium[5,6] are re-assessed in the light of more recent developments and the long term (18 years) stability of our original n-type doped GD a-Si is demonstrated.

In the early fifties many groups were involved in the search for a satisfactory route leading to the production of silicon in single crystal form and of sufficiently high purity for device fabrication.

At Standard Telecommunication Laboratories, J.M. Wilson and his team elected to use silane gas (SiH_4) as the chemical starting material for conversion to massive silicon. Since silane gas was not available commercially, it had to be made and purified in the laboratory and problems concerning its flammability and reputed explosive nature had to be overcome. Little was known about any physiological effects – after all, other covalent hydrides are extremely poisonous. Not only was it necessary for the silane to be ultra-pure itself, but a decomposition method was needed that would preserve this semiconductor purity during silicon production. A radio frequency heating method was chosen and developed, so that a temperature of around 1100°C was maintained in the growing silicon rod while other parts of the equipment remained unheated.

Operating pressures of around 10 Torr were found to be satisfactory so that homogeneous reactions were suppressed in favour of the heterogeneous surface reaction with a reasonable silicon deposition rate.

It was during the development of such equipment that unwanted glow and/or corona discharges sometimes took place. In areas away from the main growing zone, deposits of silicon were noticed although the temperatures there were less than 100°C.

A cursory examination of these deposits, some millimetres thick in places, showed them to have a high electrical resistivity. Broken, they showed a conchoidal fracture but retained the colour and lustre of crystalline silicon.

Silane decomposition methods

During our experimental work many silane decomposition methods were tried and rejected. Here it is relevant to detail only the particular construction and operating conditions which resulted in the deposition of amorphous silicon.

Figure 1 shows the equipment concerned. A Pyrex glass tube some 4 inches diameter and 1/8 inch wall thickness was mounted between neoprene seals set in brass end-plates and clamped with insulating fibre rods. Inside the chamber a radio frequency current-concentrator transformer, made of copper, was mounted through the base plate by means of its water cooling pipes. This concentrator was split vertically to promote a high current flow around its centre hole.

Silane gas was admitted by a jet above. Below this hole was mounted the silicon 'boule' or 'candle', and its tip was heated to 1100°C by the induced radio frequency current. Since the thermal gradient was very sharp below the heated zone, vertical growth was preserved as the boule was automatically withdrawn. In practice, a photocell focused on the boule tip was used to control a servo-loop fed back to the induction heating unit. Hydrogen from the thermal reaction was exhausted by means of a pump which was throttled by a valve in order to maintain the desired gas velocity and low pressure profile at the growing interface.

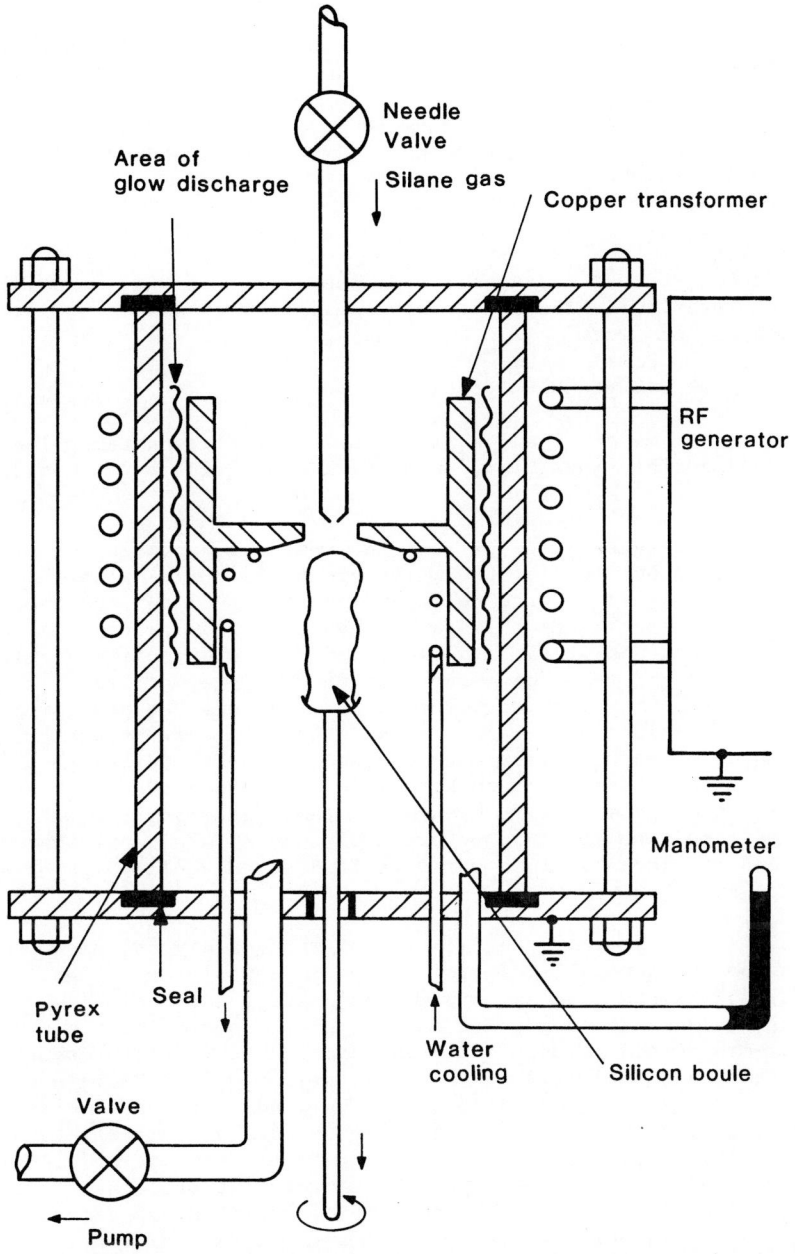

Figure 1. Equipment for deposition of high purity massive silicon by
thermal decomposition of silane gas; showing unwanted glow
discharge.

3

Energy from the 20 kW generator was supplied at 450 kHz to the primary coil which was close coupled to the concentrator through the Pyrex tube wall. It was in this area that an unwanted electrical discharge promoted the anomalous deposition of silicon between the glass wall and the copper concentrator, as shown in the diagram. A normal run continued for some four hours, during which time a six inch rod of silicon was built up. Concurrently, the amorphous deposition on the chamber walls continued and would grow to a millimeter or so thick. Sometimes it would crack off and bring pieces of the glass surface with it, like small scallops. Preliminary physical measurements showed this material to be a form of elemental silicon with an extremely high resistivity.

Speculation at that time centred on the nature of this material. Was its high resistivity due to microcrystallinity or was it a truly disordered amorphous form?

EXPERIMENTAL GLOW DISCHARGE DEPOSITION

Work on other material and device aspects of silicon technology prevented an immediate investigation of film deposition by r.f. glow discharge.

An opportunity arose when current interest was centering on silicon device passivation by means of thermally grown silicon dioxide. Speculation and discussion with colleagues at STL produced the concept that perhaps silicon nitride films would have superior properties. As silicon nitride had not previously been produced in thin film form, a deposition method had to be devised. Fortuitously, a concurrent investigation was being carried out at that time into the problem of 'auto-doping' of epitaxial silicon films on low resistivity substrates (i.e. diffusion of dopants from the substrate into the epitaxial film during deposition). The approach being taken was low pressure thermal decomposition of silane by r.f. heating at temperatures below 900°C. Since the gas pressure was in the region of 1 Torr, a glow discharge was initiated and amorphous silicon was deposited on the cold walls of the silica reaction tube.

It required only minor modification of the epitaxy equipment to realize an apparatus capable of glow discharge deposition of thin silicon nitride films from the covalent hydrides of nitrogen and silicon. Figure 2 is a schematic diagram of the equipment used. Ammonia and silane gas were pumped through a silica reaction tube at a pressure of about 0.1 Torr, and subjected to an inductively coupled r.f. glow discharge. The substrate could be heated by placing it on a molybdenum or carbon susceptor. In view of the geometry of the small reaction chamber and electrical matching considerations, a frequency of 1 MHz was chosen. From the start of the experimental work, it was clear that continuous 'glassy' layers of silicon nitride could be deposited easily, even at temperatures below ambient [7,8]. This material proved to be useful in the passivation of p–n junctions in silicon, and in consequence a patent application was filed in 1963 assigned to Standard Telephones & Cables [9]. It also showed promise as a capacitor dielectric, having a dielectric constant of 8 and low loss characteristics similar to mica [10,11].

Further to these experiments, our attention now turned to the possible preparation of silicon oxides [7-13]. In order to produce the silicon oxides under controlled conditions, it was necessary to search for a suitable gaseous oxidizing agent bearing in mind that silane is spontaneously flammable in oxygen. The agent would also need to be easily decomposed chemically in the radio frequency discharge. Further,

4

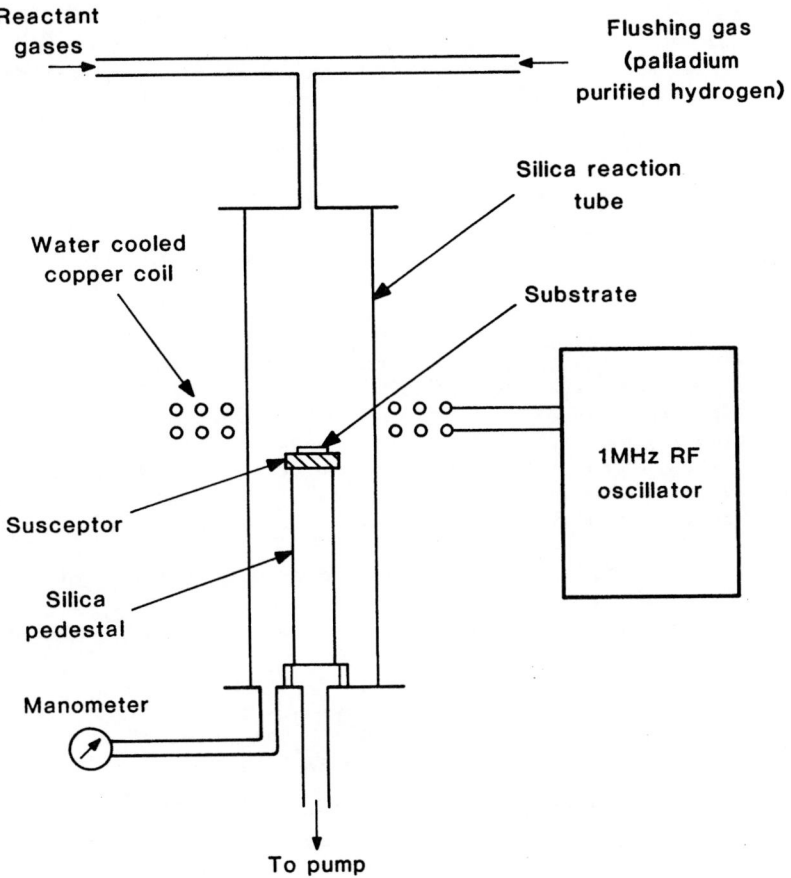

Figure 2. Early electrodeless r.f. glow discharge deposition system
derived from silicon epitaxy equipment.

its complementary element must be itself gaseous and not contribute to
the reaction. Both nitrous oxide and carbon dioxide proved to have the
necessary properties.

An important feature of the glow discharge method was the ability to
continuously vary the composition of films by varying the ratios of
reacting gases. For example, a complete range of amorphous non-
stoichiometric films could be produced between Si and SiO_2 [10,11,13].

The properties of oxide and nitride films were also very dependent
on substrate temperature, partly due to excessive hydrogen incorporation
in films deposited below 300°C.[7,10,11,13].

Temperature control of the substrate in the earliest equipment was
found to be difficult in view of the fact that the power to heat the
susceptor was derived from the same r.f. source as that sustaining the
discharge. Temperatures were varied by simply moving the r.f. coil in
relation to the susceptor. This, however, had the effect of moving the
discharge. A greater degree of independence between temperature control

and discharge conditions was achieved by the use of pulsed r.f. power.
The heating of the susceptor could then be varied by changing the pulse
width and the pulse repetition frequency, while maintaining constant
discharge parameters, provided that the pulse off period was less than
the lifetime of the active species produced in the discharge. Additional
control of deposition could be effected by the use of magnetic fields
using permanent or electromagnets to vary the uniformity of the
films.[12] This was not a simple focusing effect. The deposition rate
of the films was affected by the magnetic field, resulting in
characteristic interference patterns, e.g. an axial field would produce
concentric rings.

We were now in a position to examine a range of materials and carry
out depositions under varying conditions and onto a variety of
substrates. We could also sequentially combine this method of deposition
with that of silicon epitaxy. Among the other films deposited were
silicon carbide from silane and ethylene, titanium dioxide from carbon
dioxide and titanium tetrachloride and boron nitride from boron
tribromide and ammonia. Metallic films were also produced from the
volatile carbonyls of nickel, molybdenum and tungsten.

A demonstration of experimental glow discharge deposition was given
at The Physics Exhibition held in Manchester, England in 1965 [14].
Considerable interest was provoked and reports in the British Press
followed [15].

Following this work and our success with the oxides and nitrides of
silicon, our attention again turned to the examination of the silicon
previously deposited in the electrodeless glow discharge.

AMORPHOUS SILICON AND GERMANIUM

In April 1967, a short exploratory study of glow discharge deposited
silicon was undertaken,[5,6] followed in 1968 by a similar exercise on
germanium.[6] Initially the objective was to assess the feasibility of
using these materials for the fabrication of thin film devices such as
resistors and thermistors, but soon after the start of the project we
became aware of the growing interest in amorphous semiconductors,
encouraged largely by Mott and his group at the Cavendish Laboratory.

The first silicon films were deposited at room temperature using an
inductively coupled 500 W r.f. generator at 1 MHz and a dynamic gas
pressure of between 0.1 and 1 Torr of silane. Higher substrate
temperatures were achieved by placing the substrate on a molybdenum
susceptor, the temperature of which was monitored by an infrared
pyrometer. Although the accuracy of temperature control and the ability
to alter the discharge parameters were limited using this technique, we
believed that this was preferable to having electrical connections in the
reaction tube that could alter the discharge conditions. Deposition
rates were 2-12 Angstroms/second.

The films were shown to be amorphous by Coleman and Thomas [16] who
favoured a structural model based on the presence of both the normal
diamond cubic form and pentagonal dodecahedra.

Initial measurements on the films revealed very high room
temperature resistivities and activation energies, far in excess of the
corresponding values for amorphous silicon deposited by other methods. A
high oxygen content was ruled out since great care was taken to keep the
system clean and free from leaks. Clearly the GD a-Si had a much lower
defect density than other types. It is now, of course, recognized that

6

this is due to the saturation of dangling bonds by hydrogen but at that time we believed that the glow discharge method resulted in films with an inherently low number of dangling bonds.

Both resistivity and activation energy were found to be strongly dependent on deposition temperature changing from 10^{10} ohm cm and 0.83 eV for 21°C depositions to 5 x 10^4 and 0.25 eV at 550°C, as was the unexpectedly large photoconductive effect that was most prominent in films deposited between 250°C and 300°C.

It was the discovery of this photoconductive effect and the consequent promise of cheap large area amorphous photo-devices that enabled the project to be continued longer than the time originally proposed.

Since the resistivity of the material was higher than we would have liked for the devices we had in mind, which by this time included photo-elements, we set out to explore the possibility of doping. We managed to obtain phosphine but diborane was not available to us at that time. Phosphine/silane concentrations of 40 ppm to 4% were achieved by introducing phosphine diluted in hydrogen into the discharge tube, and n-type films with a wide range of resistivity were produced. Room temperature deposited films had a resistivity of about 10^4 ohm cm for the highest doping levels, while that of films deposited between 250°C and 300°C was up to an order of magnitude lower.

Instability of activation energy and resistivity with both time and temperature was also reported for the doped and undoped material deposited at substrate temperatures of 200°C – 500°C. It has been suggested more recently by Staebler and Wronski[17] that these effects

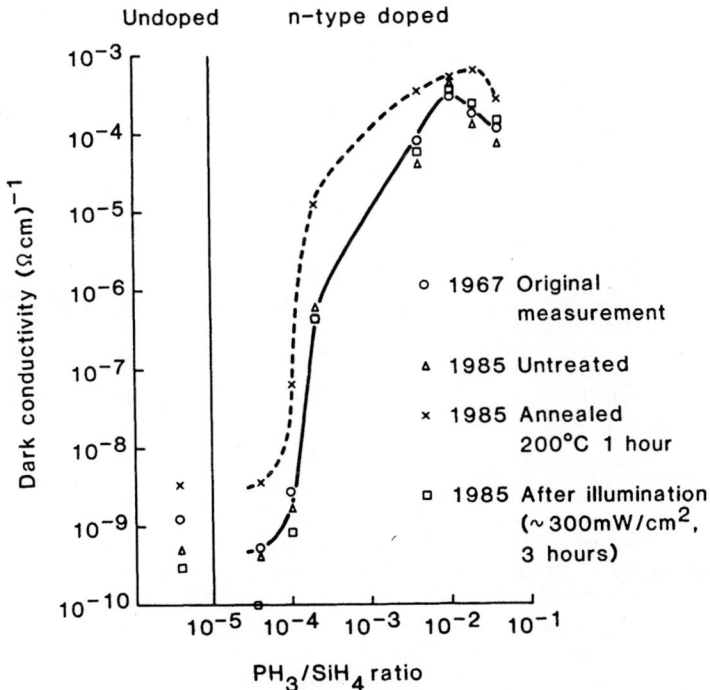

Figure 3. Room temperature dark conductivity as a function of phosphine in silane concentration for eight GD a-Si samples made in 1967.

may have the same origin as the photo-induced degradation of dark current observed after illumination with an intense light source (Staebler-Wronski effect). This does appear to be the case. Figure 3 shows the dark conductivity as a function of the PH_3/SiH_4 concentration for films deposited at between 250°C and 280°C that we still have after 18 years. The films have only been exposed to low ambient light during this period and have had no voltage bias. Therefore the results cannot be considered as from a true life test. However, it is of some interest to note that very little change has occurred in the dark conductivity during the period. Recent measurements show that after annealing at 200°C, the dark conductivity increases and exposure to light (about 300 mW/cm, tungsten-halogen) brings it back towards the original level in accordance with the Staebler-Wronski effect.

The relatively low sensitivity of our early material to doping, compared with the later results of Spear and Le Comber[1,18], particularly at the low PH_3/SiH_4 levels, may be related to the high degree of dilution of phosphine in hydrogen (0.03% for PH_3/SiH_4 up to 4×10^{-4} and 3% for PH_3/SiH_4 greater than 10^{-3}).

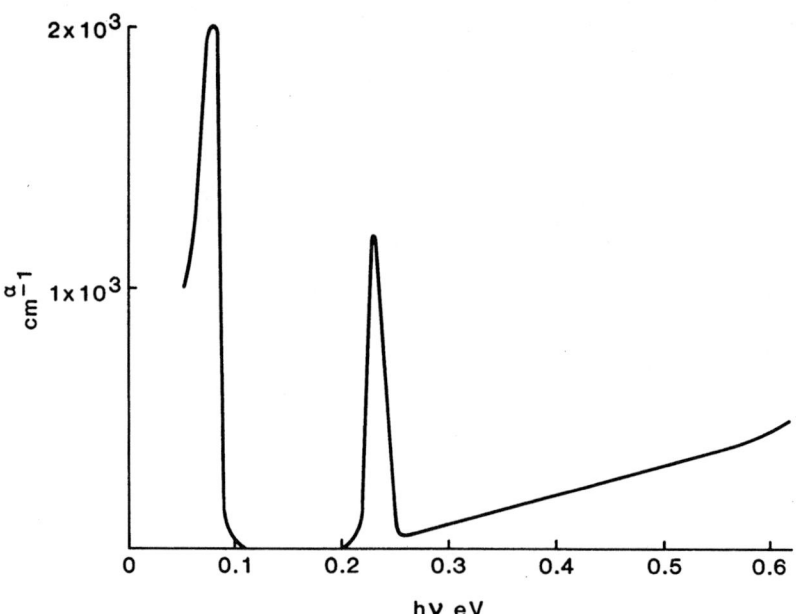

Figure 4. Infrared absorption of GD a-Ge, showing 0.23 eV and 0.07 eV bands.

Towards the end of our exploratory study on GD a-Si, some work was carried out on GD a-Ge deposited from germane[6]. We believed that the resistivity of this material would be considerably lower than that of GD a-Si (and as such might be more useful in device fabrication either by itself or in combination with GD a-Si as an alloy). This proved to be the case. Room temperature resistivities were 5 or 6 orders of magnitude lower than those of GD a-Si with activation energies of 0.3 eV to 0.45 eV depending on deposition temperature. No measurable photoconductivity was found.

The infrared spectra of GD a-Ge (Figure 4) published by Chittick in 1970[6] and that of electrolytic a-Ge by Tauc et al[19] at the same time, show prominent bands at 0.23 eV and 0.07 eV. At that time these were ascribed to defect acceptor levels formed by dangling bonds. They are now recognized as the hydrogen vibration modes. It is, however, as yet unexplained why no hydrogen vibration modes were found for our GD a-Si. We were, in fact well aware of the possibility of hydrogen incorporation from our earlier work on silicon nitride and oxides in which Si-H and Si-OH vibration modes were present in films deposited at room temperature[7,10,11,13]. The intensity of these bands, however, diminished as deposition temperatures increased until at about 300°C they were extremely weak or absent.

EPILOGUE

Due to pressure of other work, our investigations on amorphous semiconductors ceased at the end of 1968. However, the r.f. glow discharge technique had by then become an established method for the deposition of non-crystalline materials and it was not long before other workers were beginning their own studies, in particular on GD a-Si. Our direct successors in this field were Spear and his group at Dundee with whom we had close contact. Their early drift mobility studies on our films[20] lent much support to the Mott-CFO type model of amorphous semiconductors[21-23] and their further development of the glow discharge technique resulted in their successful n and p-type substitutional doping of GD a-Si which opened the door for a wide variety of amorphous devices.

REFERENCES

1. W.E. Spear and P.G. Le Comber, Solid State Commun. 17:1193 (1975).

2. D.E. Carlson and C.R. Wronski, Appl. Phys. Lett. 28:671 (1976).

3. W.E. Spear, P.G. Le Comber, S. Kinmond and M.H. Brodsky, Appl. Phys. Lett. 28:105 (1976).

4. J.M. Wilson, Research, 12:91 (1959).

5. R.C. Chittick, J.H. Alexander and H.F. Sterling, J. Electrochem. Soc. 116:77 (1969).

6. R.C. Chittick, J. Non-Cryst. Solids 3:255 (1970).

7. H.F. Sterling, R.C.G. Swann, Solid State Electronics 8:653 (1965).

8. H.F. Sterling, R.C.G. Swann, Physics and Chemistry of Glasses 6:108 (1965).

9. H.F. Sterling, C.F. Drake, BP 1006803 May 10th 1963.

10. H.F. Sterling, J.H. Alexander, R.J. Joyce, Le Vide, No. Special, AVI SEM, October 1966.

11. R.J. Joyce, H.F. Sterling and J.H. Alexander, Thin Solid Films 1:481 (1967/68).

12. H.F. Sterling, R.C.G. Swann, BP1104935, 8th May 1964.

13. H.F. Sterling, J.H. Alexander and R.J. Joyce, Special Ceramics 4:139, British Ceramic Research Assoc. (1968),

14. 49th Exhibition of the Institute of Physics and the Physical Society, Manchester, England, April 5th-8th 1965.

15. The Guardian, "Matchbox computer nearer" April 6th 1965,

16. M.V. Coleman and D.J.D. Thomas, Phys. Stat. Sol., 24:K111 (1967).

17. D.L. Staebler and C.R. Wronski, J. Appl. Phys. 51:3262 (1980).

18. W.E. Spear and P.G. Le Comber, Phil. Mag. 33:935 (1976).

19. J. Tauc, A. Abraham, R. Zallen and M. Slade, J. Non-Cryst. Solids 4:279 (1970).

20. P.G. Le Comber and W.E. Spear, Phys. Rev. Letters 25:509 (1970).

21. N.F. Mott, Phil. Mag. 19:835.(1969).

22. M.H. Cohen, H. Fritzche and S.R. Ovshinsky, Phys. Rev. Lett. 22:1065 (1969).

23. N.F. Mott and E.A. Davis, "Electronic Processes in Non-Crystalline Materials," Clarendon Press, Oxford, (1979).

MEDIUM RANGE ORDER AND

MICRO STRUCTURE OF a-Si:H, NEW TRENDS

A. Chenevas-Paule   and   R. Bellissent

LETI C.E.S. CEN Grenoble   L.L.B. CEN Saclay
85X 38041 Grenoble   91191 Gif-sur-Yvette, cedex
FRANCE   FRANCE

INTRODUCTION

It is now well accepted that a-Si:H does not form a continuous random network (CRN) and its structure is still poorly understood especially concerning its medium range order.
The amount of covalently bonded hydrogen as silicon hydride or dihydride reaches 10 to 15 atomic percent in the conventional material prepared by plasma decomposition of silane or by sputtering in Ar/H_2. As is well known this element plays a predominent role in the electronic properties of the material as well as in its cohesiveness. Hydrogen being a monovalent atom is obviously a terminal one in the a-Si network, and has two main effects on the material properties : (i) it removes deep gap states and (ii) it decreases cohesiveness (cohesive energy) cutting weak bonds.

Consequently it is of fundamental importance that we know the spatial distribution of this element in a-Si:H. This amount the determination of the topology of defects (i.e. topological defects) assuming that H decorates pre-existing defects in the a-Si network such as dangling bonds or weak bonds. This is conceptually a good approach if we consider that the material obtained by plasma post-hydrogenation of CVD a-Si[1] is of high quality, and very close to the best obtained by glow discharge. According to proton NMR[2] experiments on various samples included high quality ones, two kinds types of bonded hydrogen exist : (i) a clustered one (60 to 80% of the total hydrogen), and (ii) a diluted one (40 to 15% of the total hydrogen).

Recently a small amount of molecular hydrogen (about 1% of total hydrogen) filling small voids ($\Phi \simeq 10$ Å) under high pressure have been observed in high quality samples by low temperature calorimetry[3,4], infrared spectroscopy[5], and NMR measurements[6,7]. It is widely accepted that the first step in the condensed matter physics is the characterisation of atomic structures. Theoretical evaluations of various physical quantities in a material and in particular elementary excitations are made possible only when the atomic structure in that material is characterized properly.

The information obtained from diffraction data or from EXAFS studies is limited by the inherent one-dimensional nature of the resulting correlation function.

A result we might obtain concerning topological disorder without long range order is that the coordination number is practically unaltered. No information on the medium range order is available from diffraction data.

It is impossible to determine the total structure of any amorphous solid and it is for this reason that modelling plays such an important part in the study of glass structures. Fundamental differences between amorphous materials and crystals are :

- in crystals local order is compatible with translation symmetry, which gives (microscopic homogeneity) to these materials

- in amorphous materials the homogeneity is not compatible with translation symmetry (homogeneity is not a constructive symmetry for these materials)[8].

Hence <u>spatial propagation</u> in the three dimensional space of the local order induces a fundamental heterogeneity.

Amorphous materials are consequently fundamentally <u>heterogeneous but hide a subtle order</u>.

In the case of hydrogenated amorphous silicon, it is fundamental to know whether hydrogen is distributed homogeneously (alloy model)[9] or clustered in voids[10]. Hydrogen may also be considered as a probe, if we assume that it decorates irreducible defects. As we shall see later, the problem is complex since very often the growth structure (the so-called columnar structure) is embedded in an intrinsic microstructure. It is vital to determine the homogeneity of the material since the aggregation of hydrogen may be very important on a scale of the order of 10 Å. Beyond this scale one has to consider a multigap or a quantum wells system as proposed by Brodsky[10]. The difficulty is that conventional tools for structural investigation (x-rays, electrons) cannot directly "see" hydrogen.

Phillips[11] was the first to take into account the mismatch between the bonding constraints the number of degrees of freedom in three dimensions the flexibility required to accomodate the mismatch, and the need to treat the problem quantitatively. For the a-Si:H case some authors point out that much more hydrogen is present than is needed to quench the spin resonance signals[1]. Phillips[11] proposed a topological model with internal microvoids on the inner surfaces of which dangling bonds reconstruct as they do on crystalline surfaces, giving a global spin density of about 10^{19} cm^{-3}.

Hydrogenation occurs by chemisorption on the surface of these voids. The two dimensional nature of voids is now contested since, as we shall see later, it is energetically infavourable. Indeed, many authors have speculated about the existence of voids but their shape and topology have never been experimentally determined. New concepts are now needed to describe a-Si and a-Si:H in order to obtain a comprehensive structural model which fits data.

New theoretical approaches have been recently developed.

The Rivier[8] approach is purely geometrical (or topological) for it proposes that odd-numbered rings of bonds (five, seven...), in CRN that are <u>indices of the local curvature</u>, are aligned in such a manner that they surround the core of a disclination. Odd-numbered rings can all be stepped through by single lines that avoid any numbered rings and form closed loops or terminate at the surface of the materials. This model takes into account only one kind of disclination.

Disclinations are also defined as rotation defects[12], Rivier demonstrated that this kind of defect is topologically stable and is suitable case for homotopy[8]. According to Rivier these defects can perhaps also account for some physical properties such as low temperature anomalous specific heat due to two-level tunneling modes, new expected elementary excitations characteristic of the non-crystalline state. Rivier demonstrates also that disclinations are the only possible defects in glass[8].

The Sadoc-Kleman approach starts from curved spaces in order to have the best space filling. It searches for the best way to map a regular arrangement of curved spaces into the three-dimensional Euclidian space[13]. Thus the amorphous disorder appears as a consequence of the mapping of regular structures defined in a three-dimensional curved space. Mapping may induce different kinds of defects : the two-dimensional one works but may be unrealistic with regard to the material connectivity. The disclination created by cutting the structure and adding (or removing) a wedge of material between the two lips of the cut seems to be the best mapping. The final result is similar to those obtained by Rivier concerning the odd-numbered rings of bonds and leads to a three-dimensional disclination network. Kleman[14] has proposed for the best final structure two conjugated irregular disclination networks with respectively opposite signs for associated distorsions.

More recently Gaspard et al[15] have used a multi-curved space (corrugated space) instead of a single curved space which had lead also to a similar description of amorphous materials.

According to these new approaches amorphous or vitreous materials may be described as consisting of highly ordered regions connected each other by line defects. These lines, the density of which is function of radius cuvature, form a three dimensional random network of disclinations in Euclidian space. It should be noted that some crystalline materials with complex structure may be described in the same manner. Sadoc[16] has shown for example that polytetrahedral structures reticulated by a periodic network of disclination lines may describe very well complex structures like A_{15} compounds, Mn, W for exemple.

The a-Si/hydrogen system is useful for testing the ideal structures foreseen by these kinds of structural models. In this paper, using both short and medium range order investigation methods *, we show that hydrogen seems to be concentrated along linear defects that it decorates in both sputtered and glow discharge samples.

NEUTRON SCATTERING AND EXAFS

We have seen that the role played by hydrogen is essential for a good understanding of the structural and electronic properties of this material. However conventional means of structural investigation are very unsensitive to hydrogen. Therefore neutron scattering appears to be the most efficient way to study the structure of an hydrogenated substance for the following reasons : (i) hydrogen cross section for neutrons is of the same order of magnitude as the silicon cross section, and (ii) hydrogen may be substituted by deuterium whose cross section is quite different.

The last point allows us contrast variation without structural change.
Neutron scattering has been used for structural investigation on a-Si:H in two ways (i.e. correlations distances ranging from 20 to 200 Å). Firstly the medium range order has been investigated by small angle Neutron scattering (SANS). This medium range ordering typically describes fluctuation densities and clustering. Secondly short range order (SRO) has been investigated by Neutron scattering and EXAFS. It leads to a description of atomic order up to about 10 Å (i.e. about 3 or 4 coordination shells for an amorphous system).

Medium range order by SANS

Experimental details. The aim of the medium range order investigation of a-Si:H by SANS was first to determine if hydrogen was randomly distributed

---

*Neutron diffraction, EXAFS and Small Angle Neutron Scattering (SANS) and High Resolution Electron Microscopy respectively (HREM).

or clustered in hydrogen rich regions. Owing to the fact that neutrons interact with the nucleus of the atom it is possible to change the inter-action by isotopic substitution without changing the chemical bond. This method which has been extensively used on hydrogenated substances and espe-cially chemical and biological system is known as the contrast variation method[17].

Hydrogen is a very convenient element for the contrast variation method since its coherent scattering length is negative compared to those of deuterium and silicon as shown in table I.

Table I.

	$b \times 10^{-12}$ cm	σ_{coh} (barns)	σ_{inc} (barns)	
H	$- 0.374$	1.8	81.67	
D	0.667	5.6	2.	$\sigma_{coh} = 4\pi b^2$
Si	0.414	2.15	0.1	

If we consider the low angle coherently scattered intensity due to dilute clusters of mean scattering length b_c in a matrix of mean scattering length b_m this intensity may be expressed by :

$$I_{coh}(Q) = \alpha \rho_c V_c^2 \ (<b_m> - <b_c>)^2 \qquad (1)$$
$$Q \to 0$$

where ρ_c and V_c are the cluster density and the cluster mean volume respec-tively.

$$<b_m> = \sum_{i(m)} x_{im} b_i \qquad ; \qquad <b_c> = \sum_{j(c)} x_{cm} b_j \qquad (2)$$

x_{im} and x_{jc} are the concentrations of the component of which the coherent scattering length are bi and bj in the matrix and in the clusters respectively.

From equation (1) and (2) one can see that if hydrogen rich regions do exist in a SiH it will be quite easy to observe strong variation on the small angle scattered intensity by changing the H/D ratio of the sample. Five isotopic ratio ranging from pure hydrogen to pure deuterium in steps of 25% in concentration have been used to prepare our samples. A pure a-Si sample was also been studied as a reference sample.

Our experiments were performed on D17 (Grenoble)[18] and D11 at ILL[19,20]. The momentum transfer range was from $5 \ 10^{-3}$ to $.28 \ \text{Å}^{-1}$ corresponding in direct space to distances ranging from 10 to 1000 Å.

Experimental results. The scattered intensities for the higher value of the momentum transfer have been represented on the Figure 1 for the larger values of the momentum transfer. It should be noted at this point that the scattered intensities we observed on both pure amorphous silicon and a-Si:H are much smaller than those reported by Leadbetter[21,22]. These smaller values are consistent with the high temperature deposition of our sputtered samples (400°C). The a-Si or a-Si:H obtained in such a way on to fused sili-con is a compact materiel of which the bulk density is very close to those of the crystalline material. Thus we do not observe the columnar voids refe-red by Leadbetter and his coworkers. The thickness of the amorphous material was around 30 μm and four stacked samples were used to obtain an adequate signal-to-noise ratio. We have compared this kind of sample with those sput-tered onto crystalline silicon at room temperature. Figure 2 shows the scan-ning electron micrographs of fractured surfaces of these films parallel to the growth direction for room and high-temperature materials. Similar results were obtained by Knights and Lujant[23] using the glow-discharge technique. The low temperature sample which exhibits the typical columnar structure, is similar to those used by Leadbetter.

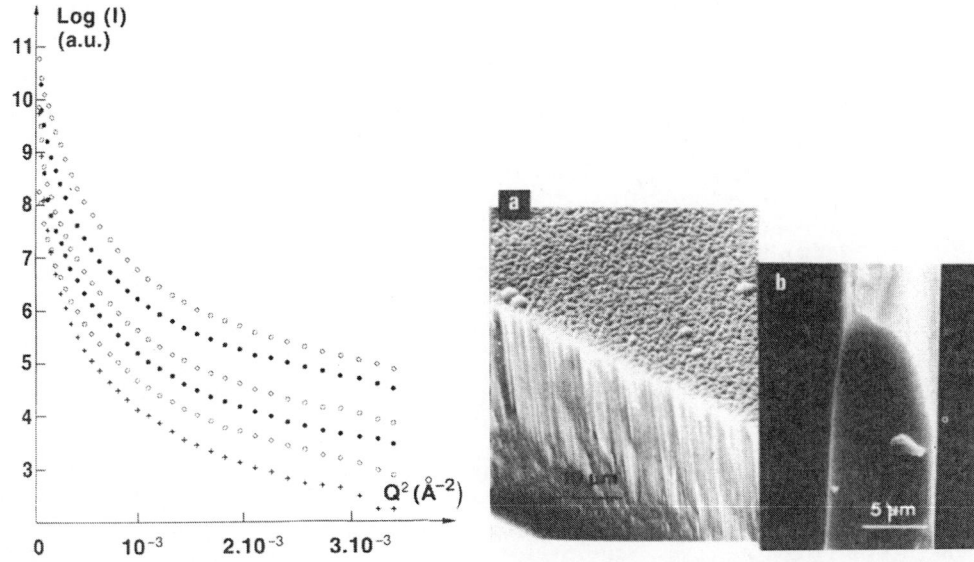

Fig. 1. Guinier plots of a-Si:H$_{1-x}$D$_x$ and a-Si

Fig. 2. SEM micrographs of a fractured edge of (a) a-Si:H sample sputtered onto a low-temperature substrate of SiO$_2$, (b) a-Si:H sputtered onto a high-temperature (400°C) SiO$_2$ substrate.

As shown by Figure 3 a small bump in our scattered intensity curve, slightly enhanced in the hydrogenated sample and located at about 22 Å^{-1} suggests the existence of a correlation length of about 30 Å. Such a distance has already been observed by SANS by Postol[24].

However the existence of medium range oder is more clearly indicated by the strong increase of the scattered intensity for the smallest values of the momentum transfer.

We have seen that the limiting value of the scattered intensity, for a zero value of the momentum transfer, by clusters in a given matrix was related to the contrast between the matrix and the cluster scattering length. The size of the cluster should be related to the scattered intensity for non zero small values of the momentum transfer Q. This size can be represented by the radius of gyration R$_g$ wich takes the form :

$$R_g^2 = \sum_i R_i^2 \qquad (3)$$

Ri are the distances of particles of the cluster to a given origin inside the cluster.
Rg is a global parameter and needs no assumption on the shape of the clusters. An approximate relation between the scattered intensity and the radius of gyrations is the Guinier law which can be written as[25].

$$I(Q) = Io \exp\left(-\frac{1}{3} Q^2 R_g^2\right) \qquad (4)$$

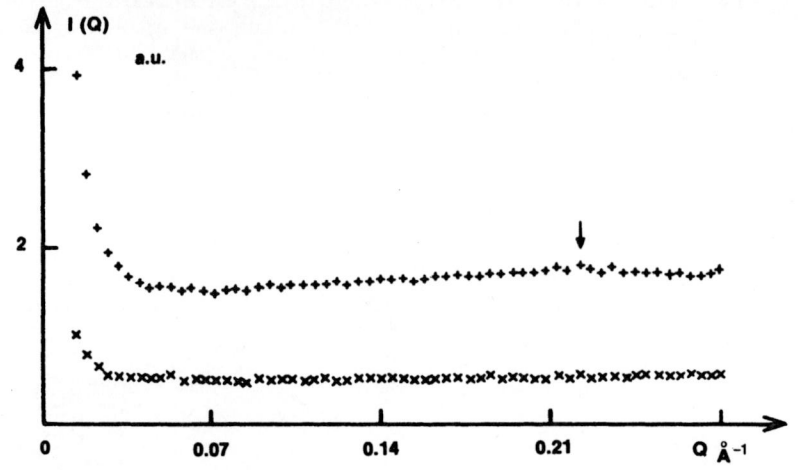

Fig. 3. Scattered intensity for a-Si:H(+) and a-Si(X) versus |Q|.

where Io, the limiting value of the intensity has been defined by equation (1).

The Guinier law is valid only for describing the scattering from dilute non-interacting clusters. More over the Q range in which this law may be used is defined using the condition : $QR_y \simeq 1$

Figure 3 represents the Guinier plots $\left(\text{Log}\,(I) = f(Q^2)\right)$ for a-Si and a-Si:H$_{1-x}$D$_x$ for the five isotopic compositions used.

All these curves exhibit two linear parts. The first is located in the range $2.10^{-3} < Q^2 < 3.10^{-3}$ Å^{-2}. In this region our curves exhibit similar slopes from which we are able to define a radius of gyration of about 30 Å for both pure and hydrogenated amorphous silicon. Such a value is consistent with the condition $QR_y \simeq 1$. The second linear part is located in the range $5\,10^{-5} < Q^2 < 5\,10^{-4}$ Å^{-2} and corresponds to a radius of gyration of approximatively 120 Å which is also consistent with the above condition in the Q range used.

These two values of the radius of gyration have already been observed in a previous SANS measurement[24]. The larger corresponds to the value observed by Leadbetter[21,22]. The smaller one has been observed by electron diffraction[26] and small angle X-Ray diffraction[27].

Discussion

The SANS experiment clearly demonstrates the heterogeneous nature of pure and hydrogenated amorphous silicon. The presence of a strong small angle scattered intensity forbids a completely random distribution of hydrogen. This point has already been underlined by NMR measurement[28,29,30] but it is more clearly indicated by SANS. Equation 1 shows that the scattered intensity for Q = 0 is proportional to the contrast between the coherent cross sections of the matrix and of the cluster. We have tried to explain the large variation of the zero Q scattered intensity by assuming hydrogen cluster scattering in a silicon matrix. Thus we have used equation (1) with:

$$<b_m> = <b_{Si}> \qquad , \qquad <b_c> = x_H <b_H> + x_D <b_D> \qquad (5)$$

x_H and x_D representing H and D concentrations in the various samples*.

$\overline{\quad b_H \quad , \quad b_D \quad , \quad b_{Si}}$ are the coherent scattering length for H, D and Si respectively.

16

Such a crude representation enabled us to fit the experimental zero-Q scattered intensities within 10% for all samples. Moreover this fit has also been made for pure amorphous silicon by assuming $<bc> = 0$, which corresponds to voids.

Thus it can be deducted from SANS that : (i) a-Si is heterogeneous, with voids or lower density regions, and (ii) in a-Si:H, H is mainly located in the pre-existency low density regions of a-Si.

Due to the difficulty in obtaining absolute values of the scattered intensity in the relation (1), the proportion of hydrogen entering the cluster and those entering the matrix has not been directly determined. However the fact that the zero-Q behaviour may be represented using the contrast between a silicon matrix and H clusters did not allow the presence of much hydrogen in the matrix since in this case the contrast should be much lower. Thus, the zero-Q intensity values indicate a mainly heterogeneous distribution of hydrogen.

Therefore a-Si:H may be described as an heterogeneous medium which consists of hydrogen rich clusters in a silicon matrix containing no more than 1 or 2% hydrogen.

At this point it is essential to make a distinction between materials showing large defects, columnar structures for instance[21] and the materials that we have studied of which the bulk density is quite similar to the density of crystalline silicon. Our material for instance appears to be homogeneous regarding electron-microscopy[31].

Thus the 30 Å and 120 Å radius of gyration that we have observed certainly does not correspond to three dimensional clusters but more likely to plane or linear array filled by most of the hydrogen.

Such a picture is consistent with disinclination models, or the topological models developed by Rivier[8] Sadoc and Mosseri[32] and Kleman[14]. A linear model is also necessary to support the connectivity between both silicon an hydrogenated regions.

A more precise fit of SANS data may enable us to refine our model. However to clarify the local order at the atomic scale and particularly by Si-H bonding we have undertaken both a neutron diffraction study, using again isotopic substitution of hydrogen by deuterium, and an EXAFS study near the silicon K. edge of a-Si:H.

Neutron diffraction and EXAFS

Generalities. The aim of short range order (SRO) investigation of a-Si:H was to describe its structure over interatomic distances (from 1 to 10 A typically).

Previous investigations of SRO have been done on a-Si:H during the past five years by X-Ray[33,34] electrons[35,36,26].

However these experiments were unsensitive to hydrogen and so were unable to elucidate out hydrogen ordering in a-Si:H.

Following Faber and Ziman[37] the coherently scattered intensity from a binary mixture of components A and B may be expressed as :

$$I_{coh}(Q) = N[<b^2> - <b>^2 + c_A^2 b_A^2 S_{AA}(Q) + c_B^2 b_B^2 S_{BB}(Q) + 2 c_A c_B S_{AB}(Q)] \quad (6)$$

Where

$$<b^2> - <b>^2 = c_A c_B (b_A - b_B)^2 \quad (7)$$

17

c_A, c_B, bA, bB are respectively the concentrations and the coherent scattering length of A and B components.
S_{AA}, sAB and sBB are the Faber Ziman partial structure factors and may be defined as the Fourier transforms of the pair partial correlation functions g_{ij} (r) as follows :

$$S_{ij}(Q) = 1 + \frac{4\pi\rho_o}{Q} \int_0^\infty (g(r) - 1) \sin(Qr)dr \qquad (8)$$

$$g_{ij}(r) = 1 + \frac{(c_i c_j)^{\frac{1}{2}}}{2\pi^2 \rho_o r} \int_0^\infty (S_{ij}(Q) - 1) \sin(Qr)dQ \qquad (9)$$

where   is the bulk number density of the system.
Since the nature of neutron-atom interaction in nuclear isotopic substitution did not affect chemical bonding. Therefore, by isotopic substitution the scattered intensity given by the relation 6 may be changed without any change of SRO. Thus, using 3 different isotopic compositions, one may determine the 3 partial structure factors of a binary mixture and the 3 partial pair correlation functions by Fourier transform following the equation (9).

This method has been widely used during the last 10 years to determine local order in amorphous or liquid binary systems.

Experimental details. Table 1 gives the values of the coherent scattering length of H, D and Si for thermal neutrons.
As these scattering lengths were of opposit sign for hydrogen and deuterium, it was possible to prepare 3 samples for which the total scattered intensities were fairly different. Our experiments were carried out on the 7C2[38,39] spectrometer at the "Laboratoire Leon Brillouin" at Saclay and on the D4B spectrometer on the High Flux Reactor at ILL, Grenoble[40]. Both spectrometers are built on reactor hot sources in order to provide a high flux for small wave lengths. Therefore the obtained momentum transfer range was : $0.2 < Q < 17 \text{ Å}^{-1}$ using a wave length of 0.7 Å. This maximum value of 17 Å for the momentum transfer corresponds to a resolution   r definied as

$$\Delta r = \frac{2\pi}{Q_{max}} \simeq 0.3 \text{ Å}$$

in real space. The use of position sensitive detectors has provided us with high counting rates.

Data reduction has been performed using the conventional corrections[41] for transmission and multiple scattering[42]. The main difficulty in obtaining the three total structure factors was the correction for the incoherent inelastic scattering of hydrogen. It can be seen from table 1 that incoherent scattering due to hydrogen is about 80 barns (compared to 2 barns for coherent scattering). Moreover owing to the identity of neutron and proton mass the recoil effect is very important for the interaction hydrogen-neutron. Therefore it was impossible to use conventional Placzek correction for the incoherent inelastic contribution[43,44].
Therefore we have used a method developped by Chieux at ILL for scattering by light elements[40].

Experimental results. The 3 total structure factors are represented in Figure 4. The spreading of the dots gives an idea of the statistical spreading of the measured intensities (about .5%). The statistical error is

18

a little larger for the hydrogenated sample due to the subtraction of a important incoherent inelastic term.

The partial structure factors (PSF) $sH–H$, $S_{Si–Si}$ and $S_{Si–H}$ have been calculated by using the relation 6 for each total structure factor and solving the system of equation. They are represented in Figure 5.

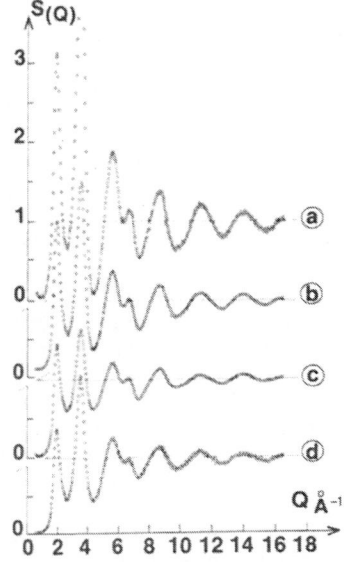 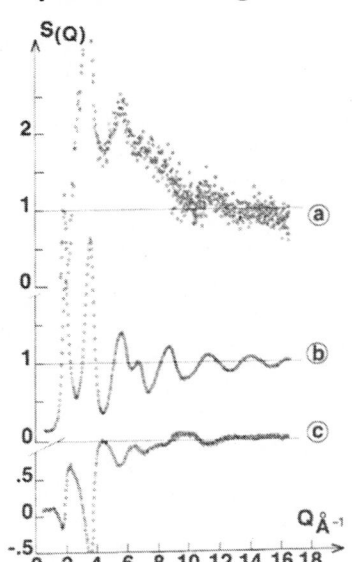

Fig. 4. a–Si:H total structure
factors for a–Si:$H_{1–X}D_X$.

	H	D
a –	15	0
b –	9	6
c –	0	15
d –	a–Si	

Fig. 5. Partial structure factors
    a   H–H
    b   Si–Si
    c   Si–H

Due to the importance of the applied corrections and particularly of the incoherent inelastic term due to scattering by hydrogen, these PSF, are not yet sufficiently reliable to calculate reasonable pair partial correlation functions by Fourier transform . However we have tried to obtain some preliminary results at least from $g_{Si–Si}$ and $g_{Si–H}$.
The $g_{Si–H}$ function exhibits a first neighbours distance of about 1.4 A which corresponds to silicon-hydrogen distance in Si–H in SiH_2 units[45,46,47].
This has already been assumed by IR and Raman spectroscopy. However neutron scattering provides us with direct evidence and measurement of this distance.

Discussion. Until now from neutron scattering experiment it has only be possible to identify in the Si–H radial distribution function a distance of 1.40 Å corresponding to one silicon hydrogen distance. However we have demonstrated that it is possible to perform a neutron scattering determination of the partial structure factors of a–Si:H. We are presently working on the correction for the incoherent inelastic term in order to improve the accuracy of our data. We have undertaken a new determination of the structure factor by using several incident energies for the incident neutron beam in order to determine experimentally the hydrogen effective mass in a–Si:H. Such an improvement of the inelasticity corrections will lead to partial structure factors accurate enough to determine the hydrogen-hydrogen pair partial distribution functions. This function will provide us with the various H-H nearest neighbours distances and therefore allows us to estimate the proportion of H_2 for instance compared to those of Si-H or $Si-H_2$ or even the $Si-H/Si-H_2$ ratio. Thus we will obtain a very precise knowledge of a-Si:H local order at the atomic scale.

EXAFS study of a-Si:H. An additional study of the short range order of a-Si:H has been undertaken by EXAFS at the silicon K-edge in order to compare the first coordination shell of the pure and hydrogenated amorphous silicon to that of the crystalline form[48]. The experiments have been performed on the ACO machine at the LURE Laboratory, Orsay, FRANCE. Our samples consisted of thin films of pure or hydrogenate amorphous silicon 1.5   m in thickness obtained by sputtering. Another a-Si:H sample has been prepared by glow discharge in the same way and a crystalline probe has been also prepared for calibration. The obtained EXAFS spectra have been represented in Figure 6 within an energy range of 450 eV above the absorption[49].

K edge of silicon. The analys is of the EXAFS data has been made in the conventional manner by substracting a pre-edge Victoreen fit of the absorption, and applying before Fourier transforming Hanning window which extends from 70 to 450 eV, in order to remove the effect of the XANES region (0 to 70 eV). The corresponding amplitudes of the Fourier transforms have been represented in Figure 7 for all the samples, non corrected for the phase shifts. As expected for amorphous covalent systems, the first neighbour distance does not change going from crystal to the glass, and the differences between the samples are just a matter of apparent coordination number and width of the distribution. The two hydrogenated samples exhibit an intermediate behaviour between amorphous and crystalline silicon with regard to local order. The nearest neighbours number as well as the width of the nearest number distribution of the hydrogenated sample are intermediate between crystalline and amorphous silicon. Assuming that amorphous silicon has an heterogeneous structure with voids associated in the broken bonds we can expect quite large distorsions of the tetrahedral units compared to the crystalline ones. This static disorder is well known to generate an apparent decrease in coordination number as seen by EXAFS[50].

In hydrogenated amorphous silicon the dangling bonds will be saturated by hydrogen giving rise to a less distorted structure which is consistent with the higher value of the nearest neighbour number and with the narrowing of the distribution that we have observed.

HIGH-RESOLUTION ELECTRON MICROSCOPY (HREM)

High-resolution electron microscopes are now capable of providing unambiguous structural information on crystalline solids at the atomic level (resolution r   2 A). Thus it is tempting to use this characterization tool to obtain structural information about one-, two-, or three-dimensional defects in a-Si and a-Si:H as was done for crystalline silicon. Nevertheless, it should be noted that the electron scattering coefficient for hydrogen is very low and only important local atomic density fluctuations result in a noticeable contrast. The phase contrast is not well understood, especially since specific defects in amorphous materials are not yet well known[51]. We have thus used this technique essentially to show the existence of certain defects[52].

A bright-field electron micrograph provides a representation of the atomic density of the sample projected onto a plane perpendicular to the electron beam. The image contrast corresponding to a Fourier component of the sample (with period d) is the product of the scattered amplitude, repre-

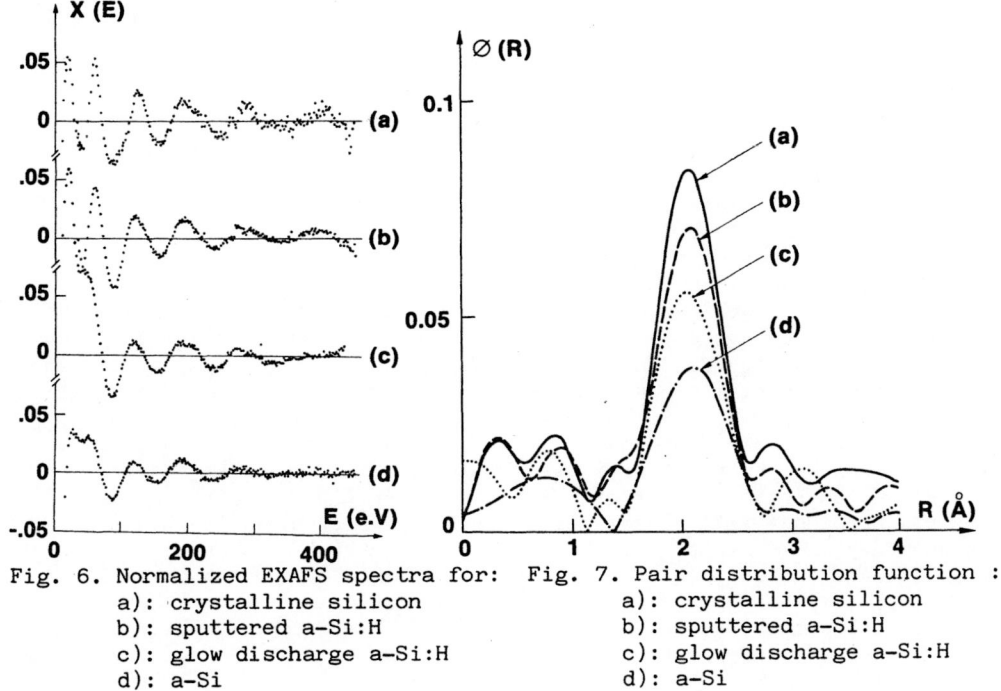

Fig. 6. Normalized EXAFS spectra for:
    a): crystalline silicon
    b): sputtered a-Si:H
    c): glow discharge a-Si:H
    d): a-Si

Fig. 7. Pair distribution function :
    a): crystalline silicon
    b): sputtered a-Si:H
    c): glow discharge a-Si:H
    d): a-Si

sented by the square root of the structure factor and transfer function of the microscope.

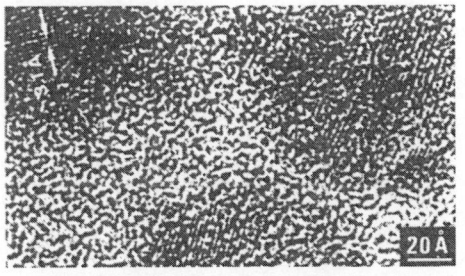

Fig. 8. HREM ABF micrograph of 100 Å a-Si film partly crystallized.

In the case of a completely random distribution, the output frequency of such an object is identical to that of an isolated atom with the intensities being multiplied by the number N of atoms. One can check this on thin samples (thicknesses lower than 100 Å) : the microscope acts as a spatial frequency filter, the band pass of which can be varied by varying the defocusing value ΔZ. Thereby, a large spatial frequency range can be continuously visualized. It has been shown[51] that the use of different spatial frequencies does not give more usable structural information for thicknesses in the 50-100 Å range. But if we want to work in a <u>differential manner</u>, for example comparing a-Si and a-Si:H (obtained by post hydrogenation of a-Si), it is necessary to work at a very well defined ΔZ.

Figure 8 gives a good illustration of the resolution of the microscope we used (JEOL CX 200 ; $r \simeq 2.2$ Å). Showing that the interplanar spacing for (111) silicon plane, may be measured directly. It is possible to identify in a straightforward manner crystallites as small as 15 Å in diameter in amorphous silicon matrix for this sample partly crystallized.

<u>Preparation of samples and observations</u> :

Samples were prepared by sputtering in a pure Argon atmosphere

$(P_{Ar} \simeq 10^{-2}$ TORR) onto fused SiO2 substrate at 450°C. A thin film of 100 A thickness of a-Si was deposited after a long pre-sputtering time ($\simeq 1$ hour) in order to benefit from a getter effect and to obtain a very pure material, the cathode being made of "FZ" single crystal silicon.

Some of these films were converted into a-Si:H by plasma post hydrogenation ($P_{H2} \simeq 10^{-2}$ TORR) at 400°C over 30 minutes. The a-Si and a-Si:H films were treated in a HF/H2O mixture to strip them from their SiO$_2$ substrates and then place on special microscope grids prepared to receive very thin films (100 Å or less). Figure 9 shows typical micrographs : 9-a is for a-Si without treatment, 9-b for a-Si post hydrogenated at 400°C. These micrographs were obtained under the same defocusing conditions with a JEOL CX 200. Micrograph 9-a exhibits the classical contrast (fringes) seen for every amorphous system in ABF mode, while 9-b shows contrast associated with important density fluctuations of the silicon atoms. This latter kind of contrast has sometimes been interpreted as being due to grain boundaries. In order to make a realistic interpretation of this contrast we have developed stereoscopie techniques. Two micrographs (type 9-b) were taken tilting the sample by $\pm 10°$ with respect to the electron beam axis. This stereoscopic analysis shows that the contrast in Figure 9-b is due to cylindrical voids ($\Phi \simeq 10$ Å) which are distributed as shown in Figure 10. Complementary experiments have allowed us to demonstrate that photo-etching phenomena have to be taken into account which would * produce etch pits during the strip-off treatment with HF under exposure to light. Due to their poor photo conductivity, (pure a-Si samples) as we shall see later, do not show any contrast and seem to be non or homogeneously etched and never show the kind of contrast seen in hydrogenated samples.

We have also prepared pure a-Si films as thin as 50 A at 450°C : they are also homogeneous and exhibit neither a nuclation pattern nor a crystalline phase. Our samples are similar to the non-density-deficient ones studied by Donovan et al[53] on a-Ge and different from those observed by many experimentalists which, owing to incomplete coalescence, exhibit a super-network of grain boundaries to be associated with the so-called "columnar structure". In addition, micrographs taken after the a-Si thin films (50-100 A thick) on these special HREM grids were submitted to the plasma hydrogenation treatment do not exhibit any density fluctuation or any modification compared to non-hydrogenated initial state.

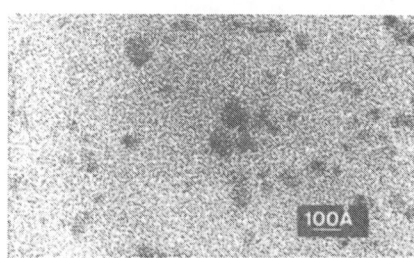

Fig. 9-a. Fig. 9-b
HREM ABF micrograph for self supporting samples :
a-Si a-Si:H plasma post hydrogenated at 400°C.

This seems to demonstrate the etched nature of the contrast exhibited by the 9-b micrograph but in this case the temperature was not as well control-led as for the plasma hydrogenated substrate.

In order to interpret these experimental results, we propose the fol-lowing explanations : (i) Photocorrosion scheme

During the plasma hydrogenation treatment, hydrogen diffuses into the material through linear defects which it decorates.

The same etching has been carried out in the dark.

This inhomogeneous distribution of bonded hydrogen in the material induces large local fluctuations of the chemical potential which are revealed by HF/H2O solution acting as a specific etchant. These etch pits do not form in pure a-Si because chemical potential fluctuations are not so large when undecorated, and the life time of photogenerated carriers is very short in this material in comparaison with a-Si:H. Indeed a-Si:H corrosion in a HF/H2O mixture is light dependent and occurs probably through photoxydation and subsequent SiO2 etching by HF of pure a-Si regions.

Owing to their high optical gap, near by hydrogen-dressed regions remains unetched. Photogenerated carriers accumulate in low optical gap regions (pure a-Si regions) where localized chemical reactions such as photoxydation in aqueous media may occur. The 2-b micrograph and its three-dimensional schematic representation may be considered as a negative picture of interconnected quantum wells.
This microcorrosion phenomenon is not specific to amorphous materials, the HREM experimentalists always observed pinholes in crystalline thin films along decorated dislocations.

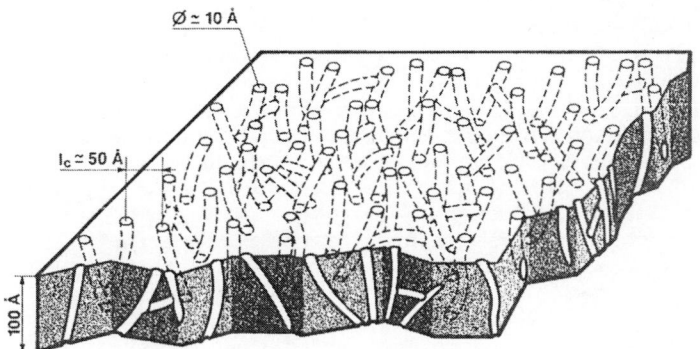

Fig. 10. Three-dimensional representation of voids in a 100 Å hydrogenated a-Si sample.

Figure 10 deserves another comment since according to this diagram, it seems that a certain axial anisotropy exists in the distribution of the filaments. This is not true and is due to the fact that only defects which emerge at the surfaces may be etched : (ii) Filled molecular hydrogen voids formation

Contrast in Fig. 9-b may also be explained by the formation of microvoids due to molecular hydrogen precipitation in hydrogen rich regions. The temperature of the substrate (400°C) and the time duration of the plasma treatment are close to these at which molecular hydrogen appears in thick sample as we observed by NMR[54,55]. This microvoid ($\Phi \simeq 10$ Å) may coalesce along defects and form thread-like voids as shown in Fig. 3. It should be noted that the two above explanations, although quite different, Lead to the same scheme.Systematic experiments including striping treatment in the dark to avoid any photo corrosion phenomena, are in progress in our laboratory to indicate the more likely explanation.

In conclusion (HREM) is not yet able to give straightforward evidence for structural defects in amorphous materials. But by combining HREM and decoration by hydrogen one shoud be able to obtain valuable information about defects in a-Si topologically consistent with decorated disclinations theoretically forsseen by Kleman, Sadoc and Rivier. Using new electron microscopy and spectroscopy techniques such as Extended Electron Energy Loss fine structure we may hope to see this kind of defects directly in the future[56].

CONCLUSION

In conclusion we have shown that it is possible to prepare a-Si:H where strutural defects due to incomplete coalescence (columnar structure) are absent as shown by electron microscopy.

The structure of a-Si:H has been investigated within a range of correlation distances from 30 to 200 Å by means of SANS and at the atomic scale by neutron diffraction and EXAFS. SANS provided us with the experimental evidence that a-Si and a-Si:H have an heterogeneous structure, even for the high density sputtered material that we have studied. Amorphous silicon behaves as a very compact silicon matrix with low density regions of which the radius of gyration could take discrete values around 30 Å and 120 Å. For the hydrogenated material the observed contrast for zero values of the momentum transfer reveals that the hydrogen is located in the low density regions. Owing to the high value of the mean bulk density of these materials it seems to be quite impossible than low density regions corresponding to a radius of gyration of 120 Å are three dimensional. Therefore the low density regions should be considered to be 2 or 1 dimensional regions. The large amount of hydrogen (12% to 15%) suggests that these regions are strongly correlated and may constitute an interconnected random network as well as the high density regions the cohesiveness of which is near perfect. Such a structure would be consistent with the theoretical models proposed by Rivier, Kleman, Sadoc and Mosseri.

On the atomic scale we have measured by neutron scattering the presence of a-Si-H bond lengths of 1.40 Å. This result suggests that an important proportion of the hydrogen is involved in Si-H bonds and saturates the dangling bonds of a-Si. Such an interpretation is consistent with EXAFS results on the radial distribution functions from which the widths of the first neighbours distribution in a-Si:H is intermediate between a-Si and crystalline Si.

The a-Si structure seems to be less distorted due to relaxation effects when hydrogen saturates dangling bonds. To obtain a more precise description of a-Si:H it would be useful to have a precise knowledge of the H-H radial distribution functions. To this end neutron diffraction experiments and improvement in data reduction are presently under study.

One of the consequences of hydrogen clustering is likely at the origin of small filled molecular hydrogen voids formation the H-H bond being the more stable in the Si/H system.

From an electronic point of view the schematic model we obtain does not introduce any discontinuity and allows the propagation of orientational order along several atomic distances. This is consistent with the recent result obtained by several teams which have used superlattices to probe localisation and long range interactions in amorphous semiconductors. Exxon team in particular has shown that the coherent length of the electronic wave function is at least 30 Å in a-Si:H.

The origin of tail states which is not well established has to be reviewed with this structural model which can take into account short and medium range order contributions as well as electrostatic charge fluctuations.

Other consequences of such structural model may be summarized as follows : (i) hydrogen decorates preexisting linear defects, so the electronic structure of a-Si:H may be better described by a Brodsky-like interconnected quantum wells system containing pure but non-crystalline silicon regions without hydrogen rather than by an alloy or a two-phase system. Indeed hydrogen, due to SiH energy bond, builds walls of the quantum wells, preventing recombination in partly passivate defects rich regions, and (ii) optical gap depends both of hydrogen concentration and quantum wells dimensions, and (iii) the interaction between impurities, doping atoms, point defects in general, and linear defects (sources of elastic perturbations) must be considered. They are probably the origin of the doping ineffeciency, and (iii) one must expect collective manifestations of these hydrogen dressed defects such as 214 cm^{-1} vibrational mode, and perhaps the existence of two level tunneling states.

REFERENCES

1.   D. Kaplan, N. Sol, and G. Velasco, Appl. Phys. Lett. 33:440 (1978).
2. W. E. Carlos, and P. C. Taylor, Phys. Rev. B 16:3605 (1982).
3. J. E. Graebner, B. Golding, L. C. Allen and D. K. Bielgelsen and
      M. Stutzmann, Phys. Rev. Lett. 52:553 (1984).

4. H. V. Löhneysen, H. J. Schink, and W. Beyer, Phys. Rev. Lett. 52:549
      (1984).
5. Y. J. Chabal and C. K. N. Patel, Phys. Rev. Lett. 53:210 (1984).
6.   B. Lamotte, Phys. Rev. Lett. 53:576 (1984).
7. J. B. Boyce and M. Stuzman, Phys. Rev. Lett. 54:562 (1984)
8.   N. Rivier, Philo. Mag. (8), 40:859 (1979).
             Philo. Mag. (8), 45:1081 (1982).
      and Rivier, Dand dufy, D.M. J. Physique 43:293 (1982).
9.   I. Solomon, J. Perrin, and B. Bourdon, Proc. Int. Conf. Phys.
      Semicond. 14th, (1978).
10. M. H. Brodsky, Solid Stte Commun. 36;55 (1980).
11. J. C. Philips, Phys. Rev. Lett. 42:1151 (1979).
12. W. F. Harris, Sci. Am. 237:130 (1977).
13.   M. Kleman, and J. F. Sadoc, J. Phys. Lett. 40:L569 (1979).
14.   M. Kleman, J. Phys. Lett. 44:L295 (1983).
15. J. P. Gaspard, R. Mosseri and J. F. Sadoc, Philo. Mag. 550:557 (1984).
16. J. F. Sadoc, J. Phys. Lett. 44:L707 (1983).
17.   B. Jacrot, Rep. Prog. Phys. 34:911 (1976).
18.   R. Bellissent, A. chenevas-Paule, and M. Roth, Physica 117-118B:941
      (1983a).
19.   R. Bellissent, A. chenevas-Paule, and M. Roth, J. Now Cryst. Solids,
      59-60, 229-32 (1983b).
20.   B. Maier, "Neutron Beam Facilities Available for users", ILL, 156X,
      Grenoble cedex, FRANCE (1981).
21. A. J. Leadbetter, A. A. M. Raschid, N. Colenutt, A. F. Wright, and
      J. C. Knights, Solid State comm. 33:973-7 (1980).
22. A. J. Leadbetter, A. A. M. Raschid, N. Colenutt, A. F. Wright, and
      J. C. Knights, Solid State comm. 38:957 (1981).
23. J. C. Knights, and R. A. Lujan, Appl. Phys. Lett. 35:244 (1979).
24. T. A. Postol, C. M. Falco, R. T. Kampwirth, I. K. Schuller and
      W. B. Yelon, Phys. Rev. Lett. (1980).
25.   G. Kostorz "Treatise ou Materials Science and Technology", Vol. 15,
      P. 227 Academic Press, New York (1979)..
26.   R. Vanderhagen, B. Chaurand and B. Drevillon, Poly-Micro-Crystalline
      and Amorphous Semiconductors MRS Europe ed. de physique (1984).
27. P. H. D'antonio, J. H. Konnert, Phys. Rev. lett. 43, 16:1161 (1979).
28. J. C. Knights, Solar Cells, 2, 409:19 (1980).
29.   B. Lamotte, A. Rousseau, and A. Chenevas-Paule, J. Physique Suppl.
      42:839 (1981).

30. J. A. Reimer, R. W. Vaughan, and J. C. Knights, Phys. Rev. Lett. 44:183.
31. A. Chenevas-Paule, "Semiconductors and Semi-metals" Vol. 21 "Hydrogenated Amorphous Silicon", ed. J. Pankove part A.
32. J. F. Sadoc, R. Mosseri, Phil. Mag. (8), 45:467 (1982).
33. R. Mosseri, C. Sella, and J. Diximier, Phys. Stat. Solidi, A52, 475:9 (1979).
34. W. Schulke, Pjil. Mag. 43, 3, 451:68 (1981).
35. A. Barna, P. B. Barna, G. Radnoczi, L. Toht, and P. Thomas, Phys. Stat. Solidi (a), 41, 48:4 (1977).
36. J. F. Graczyk, Phys. Stat. Solidi A55:231 (1979).
37. R. E. Faber, and J. F. Ziman, Philo. Mag. 11:153 (1965).
38. J. P. Ambroise, M. C. Bellissent-Funnel, R. Bellissent, Rev. Phys. Appl. 19, 731:34 (1984).
39. Bellissent 80.
40. P. Chieux, R. De Kouchkovsky, and B. Boucher, J. Phys. F., 14, 2239:57 (1984).
41. P. Chieux, Topics in Current Physics, Neutron diffraction, vol. 6, ed., H. DACHS (Berlin : Springer) P. 87 (1978).
42. I. A. Blech, B. L. Averbach, Phys. Rev. A137, 1113-1 (1965).
43. G. Placezk, Phys. Rev. 86:377, (1952).
44. J. L. Yarnell, M. J. Katz, R. G. Wenzel, H. S. Koenig, Phys. Rev. A7, 2130:2144, (1973).
45. G. Lucovsky, R. J. Nemanich, and J. C. Knights, Phys. Rev. B, 19,4, 2064:73, (1979).
46. G. Lucovsky, Solar Cells, 2, 431:32 (1983).
47. M. H. Brodsky, M. Cardona, J. Now Cryst. Solids, 31, 81:108, (1978).
48. R. Bellissent, A. Chenevas-Paule, P. Lagarde, D. Bazin, D. Raoux, J. Now Cryst; Solids, 59-60, 237:40 (1983c).
49. A. Fontaine et al, J. Phys. F 9 2143 (1979).
50. P. Eisenberger, G. S. Brown, Solid State Comm. 29:481 (1979).
51. D. J. Smith, W. O. Saxton, J. R. A. Cleaver, and C. J. D. Catto, J. Microsc., Oxford, 119:19 (1980).
52. A. Chenevas-Paule, and A. Bourret, J. of Non-Cryst. Solids, 59 & 60, 233 (1983).
53. T. M. Donovan, and K. Heinemann, Phys. Rev. Lett. 27:1794 (1971).
54. J. B. Boyce, and M. Stuzman, Phys. Rev. Lett. 54:562 (1985).
55. B. Lamotte, and A. Chenevas-Paule to be published.
56. J. C. H. Spence, Ultramicroscopy 7:59 (1981).

A DETECTIVE STORY "SHOCK CRYSTALLIZATION

OF SPUTTERED AMORPHOUS GERMANIUM FILMS"

Makoto Kikuchi

Sony Research Center
174 Fujitsuka-cho Hodogaya-ku
Yokohama   240   Japan

INTRODUCTION

This paper is our cordial gift for the 80th birthday of Prof. Nevill
Mott. We believe it is worth a birthday gift for the big scientist because
this work is one of the most enjoyable research projects of ours in the field
of amorphous semiconductors performed in these 20 years. The project really
made us smell and feel the hidden sources of interests in amorphous materials.
The course of events was like a detective story. We still keep vivid
memories of the work, by which we came to tightly believe that the amorphous
semiconductor research was really enjoyable. I have selected this topic also
because the nature of the sputtered film is not yet clearly understood even
now. Although the work mentioned here is not new as will be mentioned below,
I think it may be useful to refresh the memory of old research on the fundamental nature of sputtered amorphous semiconductor films.

START OF THE EVENTS

In August 1971, we had International Conference on Amorphous and Liquid
Semiconductors in the University of Michigan, Ann Arbor, Michigan. I was
still working in the government laboratory, Electrotechnical Laboratory, and
so I could not get any financial support. Fortunately, however, I was one of
the steering committee members of the Conference and finally I could receive
financial aid to join it. It was the Conference with unforgettable event.
We had a panel discussion with many stars. I remember that Prof. Mott and
Dr. Grigorovici joined it as panel members and the discussion was chaired by
Prof. Fritzsche. On the third day of the Conference, I finished my job to
chair one of the afternoon sessions, and I wanted to take rest joining
another session after the coffee break. During the purposeless listening to
a speaker in that session, I was hit by some interesting words, "I dropped my
specimen on the floor, and when I picked it up again, I found everything
crystallized". The specimen in that presentation was a sputtered amorphous
germanium film on a glass substrate. Although I was tired the words were
kept hanging in the corner of my brain. On the way back to Japan, I was
sometimes reminded of this interesting happening on the amorphous films, and
finally the words motivated me to check the phenomenon myself. When I
returned to Japan, I called up T. Kurosu and A. Mineo to my office. Several
groups were working with me then in the Electrotechnical Laboratory on
amorphous semiconductors. Switching, memory phenomena, photon-induced

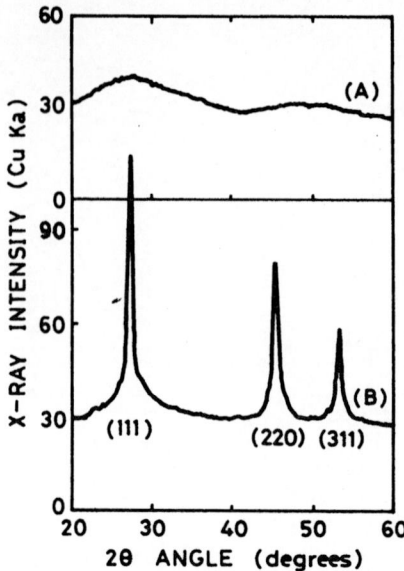

Fig. 1. X-ray diffraction patterns
for (A) as-sputtered film
and (B) the film after
pricking by a needle,
respectively.

effects etc. were being studied. Kurosu and Mineo were temporary staffs from
Tokai University. I told them about my interest on the phenomenon I had
found in the Conference, and I suggested them to trace the effect in the
possible nearest future. It took some time to make up necessary instruments
for preparing sputtered germanium films.

FIRST OBSERVATION

We prepared germanium film on the substrate, micro slide galss, by r-f
sputtering in argon atmosphere of 5×10^{-3}Torr. Single crystal of germanium
(77mmϕ, 5mm^t) was used as the target. Distance between the target and the
substrate glass was 29mm, power density was 3.3W/cm^2. For making a film of
5 microns in thickness it took about 4 hours.

Around the end of 1971, we could prepare good specimens under reasonable
controllability. One morning, we gathered near a table. A senior staff
A. Matsuda joined us. We first tried to hit a corner of the specimen by the
tip of a sharp knife. Instant and explosive crystallization did occur there!
It was a pleasant shock, so we tentatively named it shock crystallization.
For our naked eyes, the phenomenon looked like an explosion. The film
jumped out of the substrate in small fragments. Colour changed showing the
change in nature. We had to check if it was really "crystallization". Fig.1
shows the result of X-ray diffraction on (A)as-sputtered film and (B)after
the pricking, respectively. Because the film scattered around after the
pricking we had to carefully collect the fragments. The result clearly
showed the diamond-type crystalline structure for the case (B). (2)

Now the phenomenon was traced and we were sure that "crystallization had
taken place in the sputtered amorphous germanium film by indentation with a
knife edge".

28

PROPAGATION OBSERVATION

Even if the effect looked like an explosion for our naked eyes, there should have been "cause and effect" relationship. Cause was the indentation by the tip of a knife into the specimen at the corner. Effect was the sudden change of the film. The pricking was given at a corner but whole film revealed crystallization. Therefore, there should have been some "propagation" of the events. I wanted to <u>observe</u> the propagation. Kurosu and Mineo gave me independent ideas respectively, and I gave them mine. After the discussions, we made up our mind to make a "slow-motion picture".

Now, what kind of velocity would be involved in the propagation phenomenon? It might be extremely fast and it might be rather slow as well. There had been a previous work early in 1972 by Takamori et al(1), in which authors had mentioned the possibility of acoustic wave propagation. If this was really the case, we needed very high-speed camera to make useful slow-motion movie. Concerning this case, I tried to borrow a professional high speed camera from Japan Broadcasting Corporation (NHK) but I failed it because of the bureaucratic nature of the Corporation. Finally, Mineo found a very nice channel of his own to rent 16mm semi-professional high speed camera and we could prepare a studio for this movie-experiment in our laboratory.

BETTING 200 YEN

Mineo prepared specimens of sputtered germanium films, Matsuda was the camera operator, Kurosu checked the system and I was the director doing nothing. We had to take special care to make good "timing" for this experiment because the film ran at very high speed. Mis-driving of camera would cause a big loss of film so good coincidence in time was extremely important. Our minds kept very good friendship for doing this task.

At this time, George M. Hatoyama visited us. Hatoyama had been our boss in the Electrotechnical Laboratory before, and he was the Managing Director of Sony then. He was the president of Japan Physical Society sometime before that. We talked about the outcome of the slow-motion picture and discussed on the nature of propagation. I proposed "betting".

Many years later, I came to know that W. Shockley had often let his staff scientists bet on the results of experiments in Bell Labs in 1940's and very frequently (almost everytime) lost his dollars. Shockley gave talk on this on 19th of December 1972 at the dinner of 25 year Anniversary of Transistor Invention. W. Brattain disclosed the reason why he had proposed the closing of betting. First reason was that he had been always the winner which made no sense of "betting", and second reason was that he had become more reluctant to skim money from his supervisor.

Hatoyama bet 200 yen on "acoustic wave propagation" and I bet on "non-acoustic and slower".

FIRST VIVID OBSERVATION

Figure 2 is a part of 16mm slow motion picture taken in this way. Speed of the film is corresponding to 2msec per frame. As you can see in the pictures, crystallization is gradually spreading from the point of indentation by a knife. As a movie we could observe beautiful movement of the wavefront just as we had expected. We could also observe the segments of crystallized germanium film flying out of the substrate.

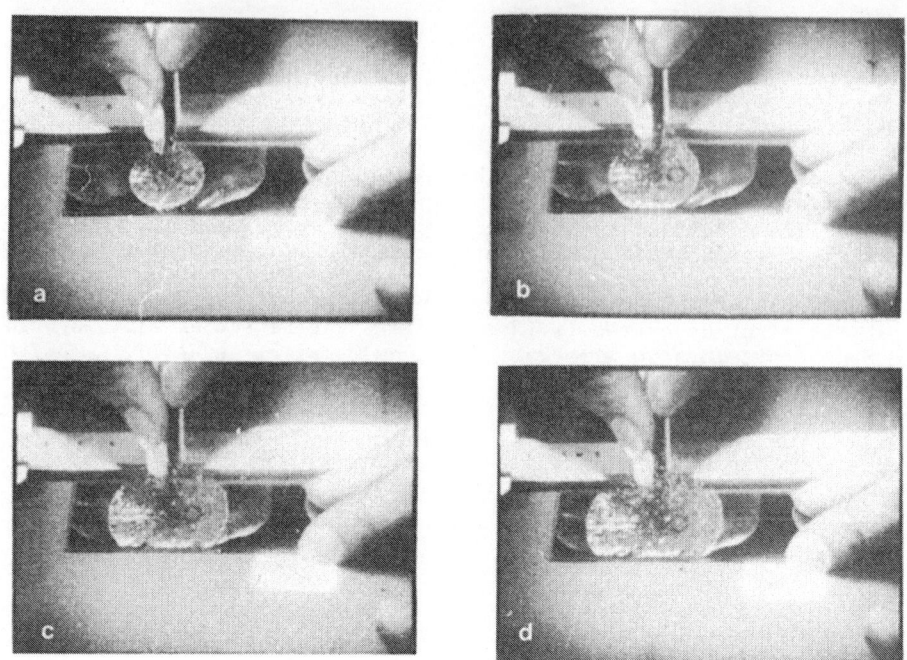

Fig. 2. Slow-motion movie of the shock crystallization.
(Speed; 500 frames/second)

The propagation of crystallized region was carefully measured from frame to frame and plotted as shown in Fig. 3. It is shown that the propagation velocity of the wavefront is around 100cm/sec and almost constant. This value is much lower than the sound velocity of 5×10^5cm/sec. I received 200 yen from Hatoyama in this way.

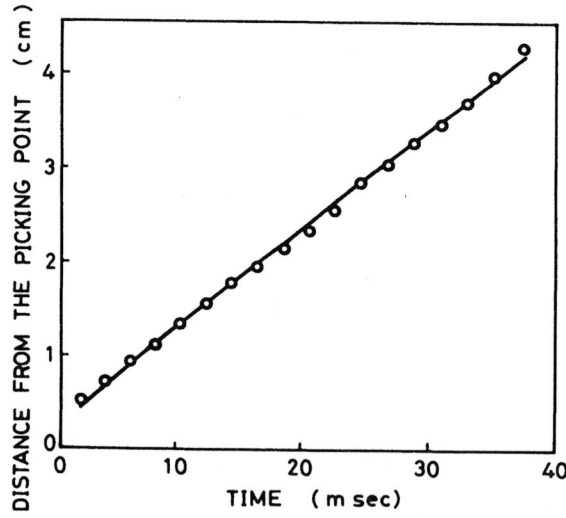

Fig. 3. Position of the crystallization
wave-front as a function of time.

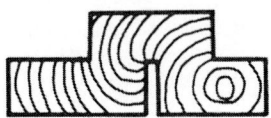

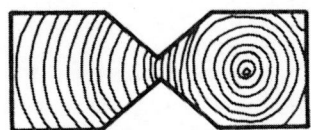

Fig. 4. Schematic representation of the move-
ment of crystalli-
zation front at
every 2m-sec in
differently shaped
specimens.

My interest was on the nature of propatation. We tried to observe if
the shape of the specimen would give any effect on the propagation. Mineo
and Matsuda made a variety of different shapes of specimens as shown in Fig. 4.
It was found out that the propagation is retarted near the narrow portion of
the film(4). Contour of the wavefront is corresponding at each 2msec. Wave
propagation reminded us of Heygens' theorem. In other wards, it looked like
"domino".

FINGER TIP

We tried to make more slow-motion pictures changing several factors of
sputtered germanium films, such as thickness, shape etc. One morning, we
were in our studio. After taking one shot, Mineo tried to change the
specimen. By mistake, he dropped the specimen onto another new specimen.
The edge of the specimen must have hit the new one, and the latter film
instantly revealed the shock crystallization. Mineo cried out "ouch!". I
asked him what had happened on him. He said, "I felt very hot on my hand".

This was the first time for us to feel <u>heat</u> in this experiment. "O.K.
Our next movie should be taken in the way like this. Mineo, you put your
finger tip at the center of the new specimen and hit the corner of the film
by a knife. We can take the movie and Mineo can feel the heat", I said.
Mineo strongly resisted against my instruction. "Why should <u>I</u> burn my finger
tip?" I changed my mind then and suggested an experiment as shown in Fig 5.

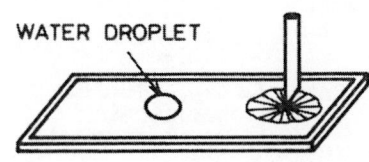

WATER DROPLET

Fig. 5. Sketch of the experiment with a water drop-
let on the surface.

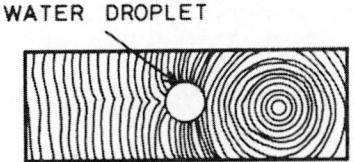

WATER DROPLET

Fig. 6. Movement of Crystallization front at every
2msec in the specimen
shown in Fig. 5.

"Carefully putting a big drop of water at the center of the specimen and take a movie of crystallization[4]". Instead of putting human fingertip, a water drop would be used to check thermal process.

The result is shown schematically in Fig. 6. As shown there, crystallization wave front proceeded along the periphery of the water drop like Heygens's law, showing re-entrant wave in the back side of the drop. It would be noted that propagation along the periphery of the water drop was slower. It was our wonderful surprise to find "original amorphous film" beneath the water drop free of crystallization. What Mineo had felt at the dropping accident turned out to be really important. Thermal process was essentially involved in the shock crystallization phenomenon. Water drop acted as a sufficiently good heat sink which prevented the propagation of crystallization in the film under it.

TEMPERATURE MEASUREMENT

At this stage, we came to talk about the simulation of this phenomenon. "Domino" was one of good ones. Taking the similarity of the heat-transfer, Matsuda made another simulation by spontaneous combustion in a filter paper treated by suitable chemical solution. If the chemical treatment was done in a similar shape as shown in Fig. 4, the burning process was almost exactly same as observed by the slow-motion movie of real experiment. Then I wanted to measure the temperature rise directly. My purpose was to check the validity of the thermal model, and I wanted to know if the temperature rise was reasonably high.

Very fortunately, my position in the laboratory then was a special one directly reporting to the Director of the Laboratory so I could negotiage him personally to get financial support free from the fixed annual budget. We could purchase Barnes RM-50 with InSb detector by which we could measure the temperature of a region 25 microns in diameter[3].

After setting specimen as shown in the figure, we gave pricking at the corner of the film. When crystallization propagated to the point of observation, signal triggered the sweep of the synchroscope. The result is shown in Fig. 8. We could find out in this way that the temperature rise was about 500°C. The shape of the curve in Fig. 8 should be explained as follows: the temperature rose rapidly when crystallization wavefront reached to the point and then started to decay rather slowly. However, as mentioned previously, the film started to fly out from the substrate in fragments, so the IR detector lost images of hot fragments in shorter time than the natural cooling process.

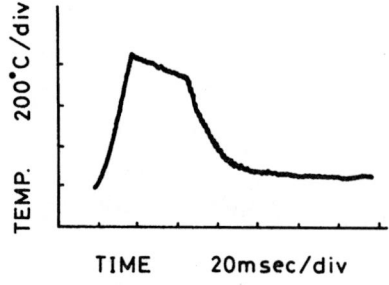

Fig. 7. Set up of the experiment for heat radiation measurement.

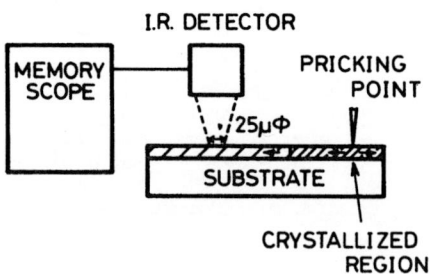

Fig. 8. Typical result of the I-R signal in the experiment shown in Fig. 7.

From these results, we could believe that sputtered amorphous germanium film had stored considerably large amount of internal energy which was emitted as heat on conversion to crystalline state.

ARGON ATOMS

In the early stage of experiments, sputtered germanium films could be classified by the success and failure of shock crystallization. Matsuda concentrated his interest on this classification and tried extensive characterization study of films. He changed factors for preparing sputtered films and made measurements of differential scanning calorimetric analyses (DSC), mass spectroscopy, electron spin resonance (ESR) and X-ray diffraction. He found interesting results[5]. First, he obtained the crystallization temperature(Tc) of "as-sputtered" film and "sputtered and peeled-off" film to be 400 and 510°C, respectively. Therefore, as-sputtered amorphous germanium film had Tc which was about 100°C lower than that of the "sputtered and peeled of" film. He measured the amount of argon atoms in the film by the mass-spectroscopy upon heating-up the specimen. He found that the as-sputtered film contained large amount of argon, and that the "sputtered and peeled-off" film had much smaller mount of argon. The amount of argon turned out to have intimate correlation to the success and failure of the shock crystallization. The fact that a film which contained large amount of argon revealed lower crystallization temperature gave a good hint for the consideration of fundamental mechanisms underlying the phenomenon.

A SISTER JOINED THE GROUP

In 1972, another temporary staff joined us. K. J. Callanan, a Sister of an American religious activity in Tokyo, visited me telling that she wanted to work with us on amorphous semiconductors. She held teacher's licence of physics in Hawaii and had experience in a laboratory of American company before. There was some bureaucratic friction in the government laboratory but finally I could accept her in our group.

Sometime later, when Prof. Fritzsche of Chicago University visited me in Tokyo, I said to him, "I don't really understand why a Sister was interested in amorphous semiconductors". Prof. Fritzsche answered me almost instantly, "May be because there are a lot of dis-orders in the society".

Callanan performed the experiment on the sputtered germanium films, especially to check the triggering of shock crystallization by thermal spikes generated by electrical process. She made many electrode pairs on specimens and measured current voltage characteristics. Then, increasing the voltage untill she observed the triggering of crystallization. The results were correlated to the characterization of films.(7)

Interesting results came out from Callanan's experiments. There were classifications of amorphous germanium films pertaining to the easiness of triggering the crystallization. For the films which showed beautiful and complete shock crystallization, the energy Ps, the thermal spike necessary to trigger the effect, was very low. Ps was found to be high for the specimens which showed only partial crystallization.

MORE DIRECT EVIDENCE

We came to imagine energy states of amorphous germanium films before and after the shock crystallization. An example is shown in Fig 9. Amorphous film in the state A surmounts a barrier and comes down to C after the

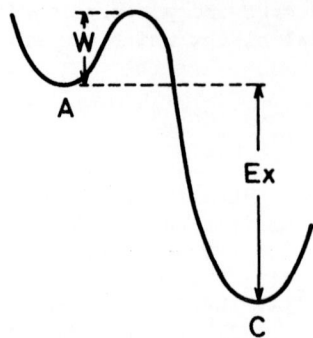

Fig. 9. Relative energy states of (A) amorphous and (C) crystallized germanium films

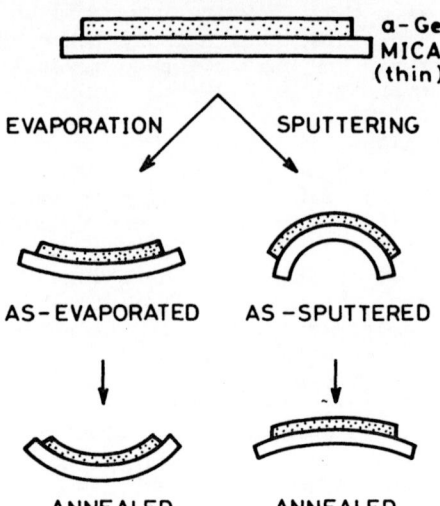

Fig. 10. Schematic representation of deformation in evaporated and sputtered specimens with thin mica substrates.

crystallization. It gives off energy during this transition. If the original state is high, it can give off energy which is sufficient to raise the temperature of neighbouring region so that the latter can surmount the barrier. In this case, crystallization can propagate. Therefore, we considered the position A was the main factor which influenced the possiblity of shock crystallization.

Now we came to a key question. "Why we had never observed shock crystallization in evaporated films?" A key answer to this key question seemed to be related again to argon atoms which caused internal stress in the film.

One night, a hint hit me, i.e., an experiment with mica sheet. Next morning, I told Mineo to try new experiments[6]. First experiment is shown in Fig. 10. I imagined that the existence of argon in germanium should have drastically increased the stress, so I wanted to show it. We used thin mica plates as for the substrate and made specimens by sputtering and evaporation, respectively.

After the preparation, we carefully observed the shape of specimens and found the results as shown in Fig. 10. Evaporated and sputtered films bent in concave and convex, respectively. After the annealing at 100°C for 10 minutes we checked again. Evaporated specimen showed increased bending, whereas the sputtered specimen decreased bending. More interesting results were obtained by another experiment shown in Fig. 11.

We prepared thick mica plate. Cutting the plate into two halves and germanium was sputtered on these substrates. After the sputtering, we observed shock crystallization by the pricking in the first half. Now, the other half was carefully peeled from the bottom side of the mica. Using a sharp razor-blade, mica substrate was reduced in thickness. Then, gradually the specimen bent and the curvature increased. In this situation, we tried pricking but nothing occurred.

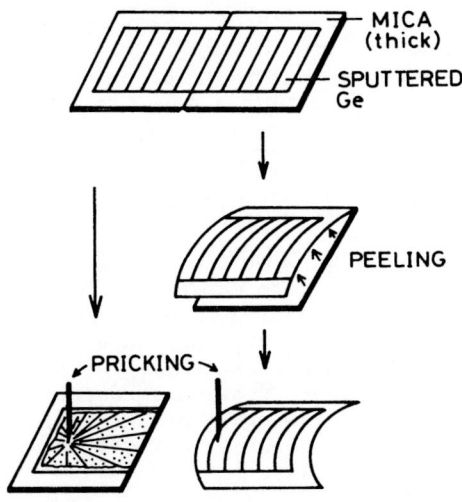

Fig. 11. Effect of stress relief
on shock crystallization.

Mental picture became clear by this experiment.
(1) Argon in the sputtered film induced internal stress and this was observed
by the bending of specimen.
(2) The internal stress caused higher energy as amorphous state. This
increase in energy caused the triggering of neighbouring region of amorphous
film when crystallization took place at a corner.
(3) In the evaporation process, even in the case where argon gas was introduced, amount of argon included in the evaporated germanium film was too low
to cause strong stress for raising energy of amorphous state. This was the
reason why no shock crystallization had ever been observed in evaporated
germanium films.
(4) Degree of easiness for triggering shock crystallization was directly
related to the amount of stress, which in turn was related to the amount of
argon atoms. Annealing effect, or peel-off should have essentially reduced
these factors.
(5) The reasons why crystalline germanium film jumped off from the substrate
were as follows. Firstly, because of heating through crystallization process,
film could not keep original size because of thermal expansion, and also
because of the phase change. Secondly, argon atoms must leave the film upon
crystallization. Argon must then form a gas phase which should occupy much
larger volume than before, which should have given the event an explosive
behavior.

EVIDENCE FOR THE TRIGGERING ACTION

Why was the shock crystallization triggered by pricking? Why did the
phenomenon start just by hitting the film by the tip of a knife? Our last
question was this. We once again tried to observe the pricking by the use
of Barnes IR radiometer. We first observed the temperature rise when a sharp
metal scriber was pushed down on a glass plate and moved on it. Temperature
rise about 120∿150°C was reproducibly observed.

In the next experiment, tip of a knife was pushed into the surface of a
glass, just in the same way as performed for the pricking in the experiment.
In this case, pulse of 350∿450°C was the result of temperature rise measurement.

It was concluded that when the tip of a knife broke into the surface of the glass, it induced intensive local heating. This reminded us of the facts in previous series of experiments that indentation by a knife into the substrate penetrating the sputtered germanium film was necessary to cause the crystallization.

PRESENTATION AT NAGOYA MEETING

In September 1972, we had annual meeting of Japan Society of Applied Physics in Nagoya University. All five members of our group on this shock crystallization research project prepared five papers, each for each contribution. (2-8). Matsuda gave Callanan extremely hard training of Japanese language for her presentation before this meeting. Callanan gave her paper as the fourth of the series of papers from our group. She gave it in perfect Japanese, which was really better than that of Matsuda! When she closed her talk, big audience gave her long applause, the first time a speaker received it in physical society meetings in Japan.

EPILOG

Shock crystallization was one of our unforgettable memories in the research of amorphous semiconductors. It was one of typical attacks for unveiling secrets of nature. The group of people engaged in this work has been disintegrated. Callanan left Japan and returned to the United States, Mineo is in Nippon Electric Co. and Kurosu is a professor of Tokai University. Matsuda is still in Electrotechnical Laboratory which is now located in Tsukuba. I guess he is still with the ETL because he liked to be with wide playground. He was an excellent socker player.

Amorphous semiconducturs were and still are, full of hidden sources of new knowledges which excite our "will to think".

ACKNOWLEDGMENTS

The author wishes to express his sincere thanks to A. Matsuda, T. Kurosu, A. Mineo and K. J. Callanan for providing him of their valuable and continuous contributions in the research of amorphous semiconductors, and also for permitting him to prepare this manuscript by his name. He also wishes to thank K. Tanaka and S. Iizima for joining the discussions throughout this work, who were working on different research projects in his group in Electrotechnical Laboratory then. He is quite sure that all of the persons named here join him for presenting cordial Congraturations to Prof. Mott for the happiest birthday.

REFERENCES

1. T. Takamori, et al, Appl. Phys. Lett. 20:201 (1972)
2. A. Mineo, et al, Solid State Comm. 13:329 (1973)
3. A. Matsuda, et al, Solid State Comm. 13:1165 (1973)
4. A. Mineo, et al, Solid State Comm. 13:1307 (1973)
5. A. Matsuda, et al, Solid State Comm. 13:1685 (1973)
6. A. Mineo, et al, Solid State Comm. 13:1945 (1973)
7. K. J. Callanan, et al, Solid State Comm. 13:119 (1974)
8. K. Kikuchi, et al, Solid State Comm. 13:731 (1974)

ORDERING IN AMORPHOUS GERMANIUM

M.A. Paesler and D.E. Sayers

Department of Physics
North Carolina State University
Raleigh, NC 27695-8202 USA

ABSTRACT

The preparation and characterization of fully relaxed amorphous germanium (a-Ge) is discussed. Deposition and annealing conditions are presented for samples of a-Ge which represent a free energy minimum. That is, further relaxation in this amorphous phase is not possible and further annealing results in the onset of crystallization. Using a light scattering technique, these samples are shown to have an atomic fraction of crystallinity less than 10^{-3}. EXAFS measurements indicate that for such samples the disorder induced spread in the $109°$ tetrahedral bond angle is $7°$. XANES experiments on the L_{III} edge reveal two features in the electronic density of conduction band states separated by 3 eV. The higher energy feature is apparently associated with the relaxed amorphous phase while the lower energy feature--at the band edge--appears only in crystallized samples. We discuss the importance of the incorporation of these spectroscopic results on any modelling of a-Ge.

INTRODUCTION

One of the principal issues addressed in the era of accelerating interest in amorphous semiconductors that began in the late 1960's was the determination of the nature of the "purely disordered" state. That is, investigators were interested in characterizing the lowest free energy amorphous state in an elemental system where only positional disorder is possible. Theorists conceived of models of continuous random networks (crn's) that in one way or another preserved local atomic order while disallowing long-range order. Experimentalists meanwhile measured the structure of amorphous silicon (Si) and germanium in attempts to characterize the order in actual films. These investigations--though initially the focal point of much of the research--represented a smaller and smaller share of the fundamental research in amorphous semiconductors until the mid-1970's when such studies rarely appeared in the literature. The diminished interest among theorists was due in part to the fact that the predicted measures of the disorder of the various theoretical crn's (given by the radial distribution functions, or rdf's) were essentially identical. Experimentally, decreasing interest may have been due to an inability of various laboratories to produce identical materials as well as the insensitivity of the results of the x-ray experiments--the rdf's-- to characterize the intermediate range order. Not inconsequential was the concurrent building competitive interest in studies of hydrogenated Si and Ge that presented many technological and fundamental challenges.

In the last several years a renewed interest has been taking shape as techniques are becoming available that make it possible to look at the nature of positional disorder with new precision. Thus, researchers have been able to define and investigate "ideal" amorphous semiconductors-- thermally relaxed, low free energy samples of Si and Ge. Below we discuss the results of some of these investigations, and identify samples that may represent a fundamental lower limit of disorder in the amorphous state.

The identification of the lowest lying amorphous state cannot be understood too heavily when considering the importance of recent work. For although work from earlier years involved samples prepared at different substrate temperatures or sometimes annealed to various temperatures, the samples were indeed different and could not be unambiguously identified as representing lowest lying amorphous state. The so-called "relaxed amorphous" samples used in the more recent investigations discussed below represent a more well-defined species: an amorphous elemental semiconductor at a configurational free energy minumum for which further structural relaxation can be made only through crystallization.

In the following section we outline the technique we used to fabricate samples for our studies as well as the light scattering experiments employed to insure that our samples were not polycrystalline. Next, we present a section on experiments designed to measure the atomic ordering of our sample followed by a section on measurements of the electronic ordering. In the subsequent discussion section, we draw on our results as well as the results of other researchers to consider pre-crystallization ordering in group IV amorphous semiconductors.

PREPARATION AND CHARACTERIZATION OF FULLY RELAXED a-Ge

Two elements that make the current investigations particularly timely are the ability to prepare and identify a true free energy minimum amorphous state and the ability to investigate this state with more powerful experimental tools. The identification of a fully annealed amorphous state with minimal crystallinity has been made possible by using a light scattering technique developed by Tsu and co-workers (Hernandez and Tsu, 1983). In most of the work described below four kinds of samples are investigated (as-deposited, fully-relaxed, mixed phase a-c, and poly-crystalline). Because the identification of the fully relaxed sample is central to the thrust of much of the work, we describe in some detail the light-scattering technique used to identify the relaxed amorphous state.

Amorphous germanium (a-Ge) samples 5-8μm thick were sputtered in a magnetron Ar sputtering system (base pressure $\approx 10^{-7}$ torr) onto substrates held at 20°C. Substrate material was quartz for the EXAFS and Raman experiments and 50μm aluminum foil for the x-ray scattering experiments. One-hour tube furnace annealing in flowing nitrogen was performed at pressures slightly in excess of one atmosphere.

The relaxed samples were annealed to a state where the volume fraction of microcrystallinity was determined to be below 0.1% yet the annealing temperature was as high as possible to allow annealing without the appearance of microcrystallinity. The mixed-phase samples were annealed at temperatures high enough to cause the onset of crystallization. Raman experiments were used to determine the volume fraction of crystallinity in a method described by Tsu and co-workers (Tsu et al., 1982; Hernandez and Tsu, 1983).

For a-Ge the TO-phonon frequency of the as-deposited sample is located at 267 cm^{-1} and increases to 274 cm^{-1} after annealing to 460°C. The Γ-point TO phonon begins to appear at 460°C directly, indicating the presence of microcrystallinity. The observation of the Γ-point phonon gives a lower limit to the establishment of the existence of the volume fraction of microcrystallinity x on the order of 3 vol %. Further annealing makes it possible to track the growth and change in wavenumber of the Γ-point peak so that by extrapolation one can estimate x for samples annealed at lower temperatures.

After determining relatively large values of x, it is then possible to relay on the Avrami expression (Burke and Turnbull, 1952) to determine the relationship between the volume fraction of crystallinity before and after annealing. Rearrangement of the Avrami expression (Hernandez and Tsu, 1983) results in an Arrhenius expression which clearly indicates the difference between growth and nucleation plus growth behavior. Relating annealing temperatures to these two very different behaviors makes it possible to infer the value of x well below the lower limit of detectability of the Γ-point phonon. Using this technique we investigated samples having x < 0.1% (Paesler et al., 1983).

EXAFS RESULTS

Extended x-ray absorption fine structure (EXAFS) experiments were performed on samples of a-Ge, making it possible to determine the mean-square deviation spread in the second nearest neighbor bond distance $(\Delta\sigma_2)^2$. From this we calculated the spread in bond angle $\Delta\theta$ for each sample. Measurements were made at the Stanford Synchrotron Radiation Laboratory (SSRL) on the 1-5 beam line monochromator using a Si(220) cut channel crystal. Beam energy was 3 GeV and beam currents were typically 40-60 mA. The EXAFS data runs are identified using five symbols such as GeNTO. The Ge identifies the sample as germanium. The 3rd and 4th characters indicate the measurement temperature which may be either liquid nitrogen (NT) or room temperature (RT) while the last character refers to annealing history. This character may indicate either unannealed (0), 400°C annealed fully relaxed amorphous (4), mixed-phase amorphous-crystalline (5), and polycrystalline (6) samples. We tabulate the sample name and character and list the figures in which spectra of each appear in Table 1.

The Fourier transforms of the EXAFS of GeNTO and GeNT4 are shown in Fig. 1. All data were multiplied by k^3 before transforming, and transforms were done over a k-space range of 3.0 - 13.1Å^{-1}. An indication of the noise level can be obtained by viewing the large r-value (6-10Å^{-1})

features where a perfectly noise-free spectrum would be flat. In Fig. 2 the transforms of runs taken at room temperature are shown.

Previously reported EXAFS data taken on a-Ge show no evidence of ordering past the first nearest neighbor peak (Hayes, 1979). However, our data--in particular Figs. 1b and 2b--show evidence of ordering at $r \approx 3.7$Å where the second neighbor feature would be expected to appear. Clouding the identification of this feature are the maxima at 2.8 and 3.1Å which are satellites of the first nearest neighbor peak and should not be confused with higher shell ordering. This conclusion is given further weight when the choice of transform integration range is varied and the 3.7Å feature is seen to persist in all transforms (Paesler et al., 1984).

Table 1   Sample designation, character, measurement
temperature and the figures in which each appears.

Sample Name	Character	Meas. Temp.(K)	Figure
GeNTO	as-deposited	80	1
GeRTO	as-deposited	300	2,4a
GeNT4	relaxed amorphous	80	1,3
GeRT4	relaxed amorphous	300	2,4b,5
GeRT5	mixed phase a-c	300	4c,6
GeRT6	poly-crystalline	300	4d

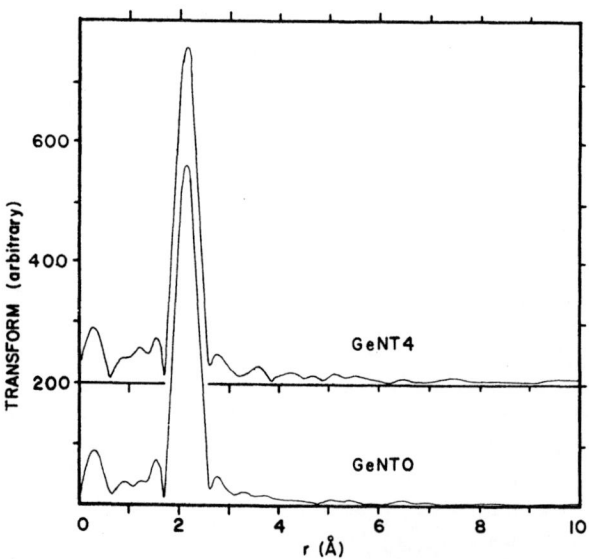

Fig. 1   The transforms of the EXAFS measured at 80K for unannealed (GeNTO, figure 1a) and 400°C annealed (GeNT4, figure 1b) samples. The data were multiplied by k^3 and transformed over the range of 3.5-13.1Å. The feature at $r \approx 3.7$Å, which rises above the noise only for GeNT4, results from the second nearest neighbor shell.

Also lending credence to this conclusion is the absence of the 3.7Å feature in as-deposited samples where disorder is at a maximum and a second nearest neighbor feature is unexpected.

The bond angle deviation $\Delta\theta$ can be found from the mean square deviation of the second nearest neighbor distance, $(\Delta\sigma_2)^2$. To determine $(\Delta\sigma_2)^2$ values which best fit our data, we model the first three shells of c-Ge and then introduce disorder by increasing $(\Delta\sigma_2)^2$ and $(\Delta\sigma_3)^2$. For successively larger disorder, we show the resultant transforms in Fig. 3. At the top of the figure, GeNT4 is shown for comparison. From this figure it is clear that the sample used in GeNT4 has a bond angle deviation such that $4^\circ < \Delta\theta < 10^\circ$. The features in the range between 3 and 5Å are best

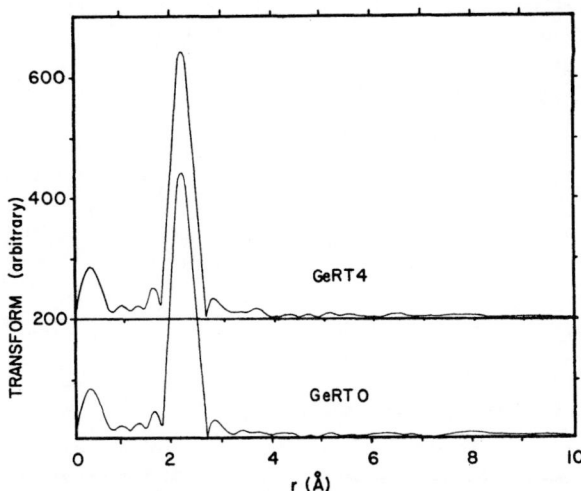

Fig. 2   The transforms of the EXAFS measured at room temperature for un-annealed (GeRTO, figure 2a) and 400°C annealed (GeRT4, figure 2b) samples. The feature at $r \approx 3.7$Å, which rises above the noise only for GeRT4, results from the second nearest neighbor shell.

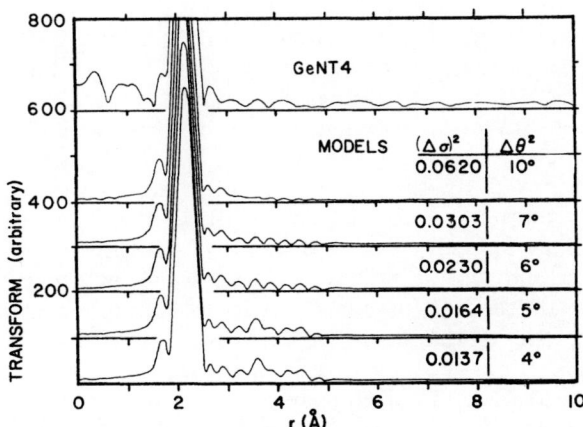

Fig. 3   The transforms of the EXAFS of sample GeNT4 (top) and models of c-Ge incorporating decreasing amounts of disorder.

modeled by a total $(\Delta\sigma_2)^2$ value of 0.0303Å^2. Since the crystalline sample had $(\Delta\sigma_2)^2 = 0.0014\text{Å}^2$, we find an additional amount of disorder relative to the crystal of 0.0289Å^2. The mean square deviation in second nearest neighbor distance is related to the spread in bond angle by

$$(\Delta\sigma_2)^2 = (r_1 \cos(\theta/2)d\theta)^2.$$

Using $r_1 = 2.45\text{Å}$ and $\theta = 109°28'$, we calculate $\Delta\theta = 7°$ for the relaxed amorphous sample GeNT4.

Further confirmation of the determination of the value of $\Delta\theta$ that best fits our data can also be made by comparing the peak height of the second nearest neighbor feature with model calculations. Such modelling shows a temperature dependence consistent with theory (Chou et al., 1985). In heating from 80 to 300K, our samples exhibit an increase in mean-square spreads $(\Delta\sigma_1)^2$ equal to $1.4 \times 10^{-5}\text{Å}^2$ while the value quoted by Chou and co-workers (1985) is $1.5 \times 10^{-5}\text{Å}^2$.

Our data also show that the first shell becomes slightly more ordered when annealed at 400°C. For example $(\Delta\sigma_1)^2$ decreases from 0 to $(-2.3 \pm 2) \times 10^{-5}\text{Å}^2$ from GeNT4, where the minus sign indicates that the sample is more ordered than the model compound. Thus, the bond angle variation decreases with annealing and there is a small decrease in the first bond length distortion as well.

XANES RESULTS

The L_{III} edge x-ray absorption near edge structure (XANES) of our samples was measured at SSRL on beam line 111-3 using the JUMBO monochromator with beryl crystals. Samples were sputtered cleaned prior to XANES measurements, and spectra were taken at room temperature in the total yield mode. Four spectra on each sample were taken and summed to improve statistics. For the XANES runs an additional sample (GeRT5) was investigated. This sample had been annealed at 500°C for one hour and thus was a partially crystallized sample.

Figure 4 shows a comparison of the L_{III} edges of GeRTO, GeRT4, GeRT5, and GeRT6. The energy scale has been shifted somewhat to align the inflection points of the edges. The relative shifts are small and not significant to our analysis. In addition, their vertical scales have been normalized to conserve total oscillator strength. The figure shows that there are significant differences between the as-deposited sample (GeRTO, curve a), the relaxed amorphous sample (GeRT4, curve b), the partially crystallized sample (GeRT5, curve c) and the crystallized sample (GeRT6, curve d). The as-deposited sample (curve a) is essentially featureless, indicating sufficient disorder to smear out any distinct electronic states. The spectrum of the relaxed amorphous sample, curve b, shows a very definite feature at 1221 eV indicating that a significant change has taken place in the average electronic environment of the material relative to the as-deposited sample. The spectrum of the crystallized sample (curved) shows a dominant feature at 1218 eV with a suggestion at 1221 eV of a remnant of the feature present in the amorphous sample. The spectrum of the sample which may be a mixed-phase amorphous-crystalline specimen, shows both features at 1218 and 1221 eV.

It is possible to determine to what extent the annealed samples are mixed phases consisting of as-deposited and crystalline Ge by attempting to express the spectra of GeRT3 and GeRT4 as linear combinations of GeRTO and GeRT5. One can see that the spectrum of GeRT3 is not such a combination by viewing Fig. 5. In this figure the solid line is the spectrum of the relaxed amorphous phase and the dotted line is a multi-regressive fit

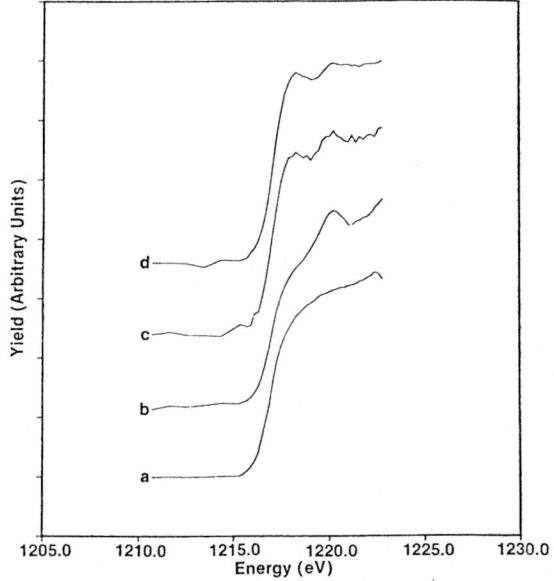

Fig. 4 L_{III} absorption edges of GeRTO, GeRT4, GeRT5, and GeRT6 (from bottom to top, a–d respectively).

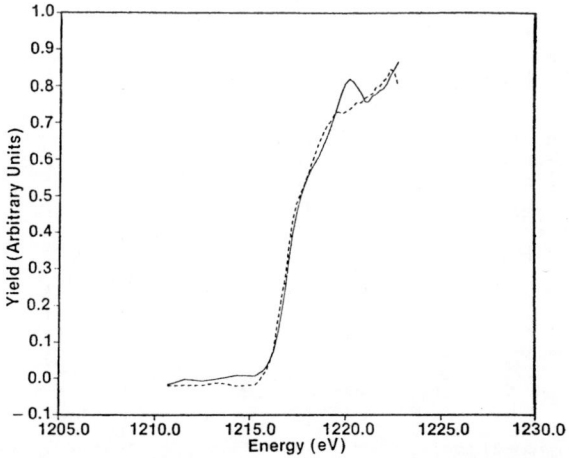

Fig. 5   Multi-regressive fit of GeRTO and GeRT6 to the spectrum of GeRT4. From the mismatch of these curves, we conclude that the relaxed amorphous phase cannot be considered a two-phase system with c-Ge embedded in an as-deposited a-network.

of the GeRT4 spectrum using a linear combination of spectra GeRTO and GeRT6. Clearly, the relaxed phase cannot be considered a two-phase system with c-Ge embedded in an a-network.

The nature of GeRT5 can be deduced in a similar way. Figure 6a shows a comparison of GeRT5 (solid) and GeRT6 (dashed). Some differences in the spectra are apparent, and it is most reasonable to assume that we can count for these differences by attributing them to a mix of a- and c-phases. Figure 6b shows a comparison of a combination of GeRTO and GeRT6 spectra (dashed) with the GeRT5 spectrum (dashed). Some improvement can be seen in adding the spectra of GeRTO, GeRT4 and GeRT5 (Fig. 6c), but the coefficient of the GeRTO spectrum is small and negative, which is an unphysical result. The best fit is found by combining 0.15 GeRT4 and 0.85 GeRT6 (Fig. 6d). Thus, one might expect that GeRT5 is indeed a combination of c- and relaxed a-phases.

DISCUSSION

In both our EXAFS and XANES experiments we are observing the approach to the low free energy metastable state from which the abrupt transition to the lowest free energy state--the crystal--takes place. This latter transition has been studied by many researchers. In a model of the a-c interface in Ge and Si, Spaepen (1978) modeled the a-c interface and determined the bond angle distortion across this interface. His model incorporated a four monolayer transition with a first and second amorphous layer and a first and second crystalline layer as one moved away from the interface in the a- and c-directions, respectively. Beyond the second crystalline layer,

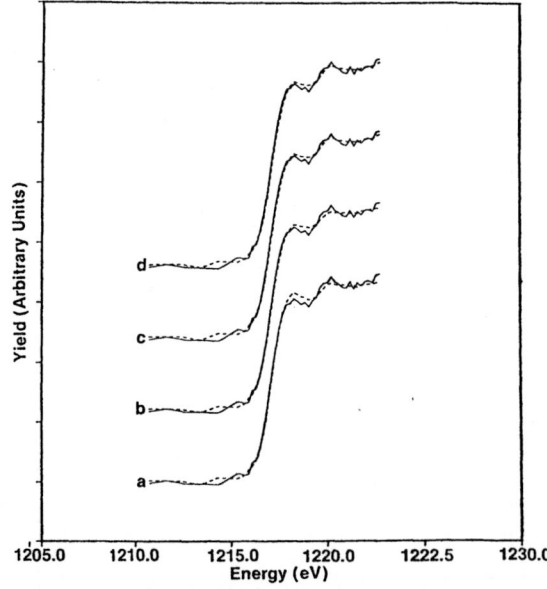

Fig. 6   Multi-regressive fits of the spectrum of the mixed-phase a-c sample (GeRT5) to combinations of: a) GeRT6, b) GeRTO and GeRT6, c) GeRTO, GeRT4, and GeRT6, d) GeRT4 and GeRT6. From the goodness of the fit of curve d, we conclude that GeRT5 is indeed a combination of the c- and a-phases and represents a system with crystallites embedded in a relaxed a-network.

material indistinguishable from bulk c-Ge or c-Si exists according to Spaepen. Within the second amorphous layer, he calculates a bond angle distortion of 10°. This calculation--based on a stick and ball model-- makes no attempt to identify a minimum free energy amorphous state, but does provide a benchmark calculation of bond angle deviations. An early well-known model of the structure of a-Ge was the Polk model (Polk, 1971) which was based on several criteria. Among these was the requirement that $\Delta\theta$ not increase as the model is made larger. Polk's model involved 5, 6, and 7-membered rings and had $\Delta\theta = 9°$. Refinements made by Steinhart et al. (1974) resulted in $\Delta\theta = 7°$ in a model designed to minimize the elastic potential energy.

The ordering--or relaxation-- of the amorphous network involves not only a decrease in the bond angle distortion, but also an alignment of the dihedral angles. Dihedral angle ordering is not directly observable in EXAFS experiments since it involves no change in first and second nearest neighbor distances. This can be visualized by considering two tetrahedra joined by a common bond in an eight atom model. Changes in the dihedral angle with no changes in bond angle distortion for such a construct result in no changes in the second nearest neighbor distance. On the other hand, changes in bond angle distortion result in increased spread in the second nearest neighbor distance. Relaxation might be thought of as involving either dihedral angle ordering only, bond angle ordering only, or a mix of dihedral and bond angle ordering. The first case can be ruled out from our EXAFS results since if the ordering were purely dihedral, the second near- est neighbor peak in deposited samples would be much stronger than it is seen to be in Fig. 1. This lack of second nearest neighbor effects implies that the disorder involves a substantial bond angle component.

Neutron diffraction experiments can be used to examine the disorder in a-Ge, and work by Etherington et al. (1982) has been used to evaluate various models of a-Ge. Models are most successful when the strain energy has been minimized using a Keating potential. Even for these models, however, the second nearest neighbor peaks are much sharper than experiment, thus showing the distribution of bond angles in the models to be too small. Etherington and co-workers conclude that this is due to the fact that not enough strain energy is built into models rather than any consequence of model topology. For deposited films, these workers find $\Delta\theta = 9.7°$, a value in agreement with the lower limit of 10° seen in our as-deposited samples. Models compared by Etherington include those of Polk and Boudreaux (1973), Connell and Temkin (1974) and Beeman and Bobbs (1975). X-ray scattering experiments have not been performed on amorphous samples shown to be fully relaxed, and such experiments are now particularly timely. The resulting rdf's could be compared to neutron diffraction results, models of crn's, and EXAFS data. Such experiments are currently in progress in our laboratory.

Perhaps the strongest confirmation of our conclusions about the spread in tetrahedral bond angle comes from Raman scattering results (Beeman et al., 1985; Tsu and Hernandez, and Pollack, 1984). These authors compute the Raman scattering from various structures for a-Si and a-Ge. By relating the width of the optic peak to bond angle distortion, Beeman and co-workers find that $\Delta\theta$ lies between 7.6 and 11° for a-Si with the smaller angle corresponding to networks that had been annealed to higher temperatures. In comparing their results with modelling calculations, Beeman and co-workers (1985) conclude that their results are consistent with model building experience which shows that it has so far been impossible to construct a fully bonded amorphous network with $\Delta\theta$ less than about 6.5°. Tsu and colleagues (1984) find that for a-Ge the anneal-stabilized material has $\Delta\theta \approx 6.1^\circ$.

Stern and co-workers have performed EXAFS experiments on a-Ge with a different thrust than that of the present work. In one investigation (Stern et al., 1983) the incipient a-c transition is studied. For samples prepared at different substrate temperatures, EXAFS evidence of micro-crystallinity in an otherwise amorphous network is presented. In the other work (Bouldin et al., 1984) the role of hydrogen in hydrogenated a-Ge is probed. In neither case is the bond angle deviation calculated, nor is any attempt made to identify and characterize a lowest lying free energy amorphous state. Impacting on our work, however, is their conclusion that the amorphous state is a continuous random network.

Our EXAFS measurements on annealed a-Ge have shown a decrease in the deviation of the first and second nearest neighbor distances from the crystalline values with increasing annealing temperature. We find a decrease of the bond angle distortion to 7° in an anneal stabilized sample and find a slight tightening in the first nearest neighbor distance. These EXAFS results cannot by themselves, however, rule out an increase in order due to an increasing fraction of microcrystals in an amorphous network. The coupling of the Raman results, however, makes it possible to set an upper limit on volume fraction of crystallinity in our fully relaxed sample. This limit is 0.001 so that the EXAFS measurements on this sample represent evidence of the bond angle deviation in anneal stabilized yet still amorphous Ge. Our Raman-EXAFS experiment has shown that an anneal stabilized a-Ge film exhibits a minimum bond angle deviation of 7°.

In calorimetric investigations of the crystallization of a-Si and a-Ge prepared by ion implanation of single crystals, Donovan et al. (1985) found distinct differences between the behavior of Si and Ge. Amorphous Ge was found to relax continuously to an apparently lowest lying free energy state with a total enthalpy of relaxation of 6.0 kJ/mole, while a-Si showed no evidence of heat release due to relaxation prior to crystallization. The enthalpy of crystallization for both a-Ge and a-Si were found to be approximately 12 kJ/mole. This striking difference between a-Si and a-Ge underscores the importance of taking care in drawing any conclusions about the behavior of one material from observations on the other. Likewise, since these measurements were made on samples amorphized by implantation, any conclusions about deposited films must be viewed with caution. Keeping these caveats in mind, we feel that the 6.0 kJ/mole value for the enthalpy of relaxation for a-Ge represents a process that involves a decrease in $\Delta\theta$ from a value somewhat in excess of 10° to a value near 7°.

Electronic investigations of ordering in Ge have been performed at both SSRL (Morrison et al., 1985) and Frascati (Proietti et al., 1984). In the latter work, the principal thrust was to illucidate the relationship between fluctuations in bond distances and edge shape. Samples investigated were a-Ge deposited on substrates held at different temperatures, and of particular interest was the effect of crystalline grains in an amorphous network. Thus, the work of Proietti et al. (1984) has little bearing on the present conclusions.

The study by Morrison and co-workers (1985) does impact on our results in that it gives information about changes in the electronic structure of a-Ge as one approaches the lowest lying disordered state. Investigations of electronic ordering in germanium begin with studies of the crystalline series FC-2, 2H-4, BC-8, and ST-12. Joannopoulos and Cohen (1974) contend that observation of trends in the electronic density of states (EDOS), as one proceeds through this sequence, makes it possible to glean information about short-range order. Although the series involves only crystalline polymorphs, it represents a progression in successively larger amounts of short-range disorder. Thus, the sequence FC-2, 2H-4, BC-8, ST-12, and a-Ge represents an increasingly disordered list of polymorphs of Ge.

Calculations of the EDOS through this sequence of structures results in a family of EDOS spectra that exhibit shifts from feature to feature in the spectra. These features derive from considerations of the ring statistics involved in the various structures and are not of interest here. What does pertain to the current discussion is the general smoothing of features as the disorder is increased. Indeed, the amorphous EDOS shows a spectrum so broadened as to reduce the multitude of distinct features to two broad featureless bumps. These broadened peaks correspond to clearly identifiable features from the crystalline series that have been disorder broadened.

The calculations of the EDOS of a-Ge generally involve no discernible structure in the conduction band. This is understandable since the delocalization of conduction band electrons allows a larger sampling volume and increased sensitivity to disorder. The valence band, on the other hand, shows two broad features that are generally identified as the S band and the M band (indicating s-like and mixed s&p-like character, respectively). Experimental photoemission data (Ley et al., 1972) support these findings and show spectra with essentially the same features.

Our XANES results also show a smeared out spectrum for the most disordered state GeRTO, but the feature at 1221 eV in sample GeRT4 represents a distinct electronic signature of the relaxed amorphous state. From our Raman results we are assured of the amorphicity of sample GeRT4, yet our XANES results show an electronic character distinct from the as-deposited sample GeRTO. With the onset of crystallinity in sample GeRT5, and additional XANES feature near the band edge (at 1218 eV) indicates the presence of longer range correlations.

The band structure of c-Ge suggests that the two features separated by 3 eV may be associated with conduction band minima (Phillips et al., 1962). A direct gap minimum at the Γ-point (Γ_2) and a minimum at the same k-value but 3 eV higher (Γ_{15}) may represent the 1218 and 1221 eV features, respectively. The higher minimum is more indicative of the Penn-gap (or average energy gap) of the material and thus could result in a strong feature in the amorphous spectrum. In contrast, the direct gap Γ minimum is more dependent upon symmetry considerations and thus might be expected to be associated with the crystalline spectrum. It may be the case that the XANES results provide the most sensitive measure of the accuracy of any structural model if the nature of conduction band states is particularly sensitive to model-dependent structures.

The only study of the effect of d-states on the EDOS of Ge, of which we are aware, involves the work by Louie (1983) on c-Ge. This work implies that consideration of d-orbitals on the Γ-point energy eigenvalues is quite small. This is encouraging in that it suggests that the 3 eV difference between the two XANES features we observe might very well correspond to the $\Gamma_{2'}$ and Γ_{15} transitions. These calculations, however, deal only with c-Ge and yield only energy eigenvalues. They are thus of limited use in the present discussion. In the past, most effort at using EDOS calculations to determine the efficacy of given structural models has centered on the valence band where discernible features in the photoemission spectra could be used to determine the viability of various models. It can hardly be said that these efforts have resulted in an unambiguous model of the structure of a-Ge or a-Si. With XANES we now have a new handle on one particular feature of the EDOS of a-Ge.

tions centered on the conduction band EDOS will unlock secrets about the structure of a-Ge.

The short-range order of the crystalline state is to a large extent preserved in a-Ge. The evolution of features in the electronic density of states comes instead from changes in long-range correlations in atomic structure. For example, investigations of the ordering of dihedral angles (Lucovsky et al., 1984) has led to a better understanding of the relationship between atomic structural ordering and changes in the electronic density of states of amorphous chalcogenides.

In our studies of ordering of Ge we are led to the conclusion that the relaxed amorphous state has a bond angle spread of $\Delta\theta = 7^\circ$ and that associated with this state is a distinct electronic environment manifesting itself in the XANES feature at 1221 eV. The electronic environment in the relaxed state is different from that in the as-deposited state and any accurate model of relaxed a-Ge must exhibit the XANES feature at 1221 eV.

ACKNOWLEDGEMENTS

We gratefully acknowledge the many far-reaching contributions to the study of disordered materials made by Sir Nevill Mott. We are honored to be invited to this Festschrift in his honor. The work on which the present paper was based was done with colleagues T.I. Morrison, R. Tsu, and J.G. Hernandez. Discussions with J. Rehr and E.A. Stern are likewise acknowledged. Work at SSRL is supported by the NSF (MRL) and the NIH through the Biotechnology Resources Program in the Division of Research Resources (in cooperation with the DOE).

REFERENCES

Beeman, D., Bobbs, B.L., 1975, "Computer Restructuring of Continuous Random Tetrahedral Networks," Phys. Rev. B 12:1399.
Beeman, D., Tsu, R., and Thrope, M.F., 1985, "Structural Information from the Raman Spectrum of a-Si," Submitted to Phys. Rev. B.
Bouldin, C.E., Stern, E.A., and von Roedern, B., 1984, "A Structural Study of Hydrogenated a-Ge Using EXAFS," Phys. Rev. B 30:4462.
Chou, S., Bunker, B.A., and Rehr, J.J., Unpublished.
Connell, G.A.N., and Temkin, R.J., 1974, "Modelling the Structure of

Amorphous Tetrahedrally Coordinated Semiconductors," Phys. Rev. B 9:5323.

Cook, J.W., and Sayers, D.E., 1981, "Criteria for Automatic X-ray Absorption Fine Structure Background Removal," J. Appl. Phys. 52: 5024.

Donovan, E.P., Spaepen, F., Turnbull, D., Poate; J.M., and Jacobson, D.C., 1985, "Calorimetric Studies of Crystallization and Relaxation of Amorphous Si and Ge Prepared by Ion Implantation," J. Appl. Phys., in press.

Etherington, G., Wright, A.C., Wenzel, J.T., Dore, J.C., Clarke, J.H., Sinclair, R.N., 1982, "A Neutron Diffraction Study of the Structure of Evaporated a-Ge," J. Non-Cryst. Solids 48:265.

Hayes, T., 1979, <u>Atomic Scale Structure of Amorphous Solids</u>, edited by G.S. Cargill and P. Chaudhari, North Holland, New York.

Hernandez, J.G., and Tsu, R., 1983, "Nucleation and Growth Rate of a-Si Alloys," Appl. Phys. Lett. 42:90.

Joannopolous, J.D., and Cohen, M.L., 1974, "Electronic Properties of Complex Crystalline and Amorphous Phases of Ge and Si," Phys. Rev. B 31:71.

Ley, L., Kowalczyk, S., Pollack, R., and Shirley, D.A., 1972, "X-ray Photoemission Spectra of Crystalline and Amorphous Si and Ge Valence Bands," Phys. Rev. Lett. 29:1088.

Louie, S.G., 1980, "New Localized-Orbital Method for Calculating the Electronic Structure of Molecules and Solids: Covalent Semiconductors," Phys. Rev. B 22:1933.

Lucovsky, G., Wong, C.K., and Pollard, W., 1984, "Vibrational Properties of Glasses: Intermediate Range Order," J. Non-Cryst. Solids 59:839.

Morrison, T.I., Paesler, M.A., Sayers, D.E., Tsu, R., and Hernandez, J.G., 1985, "Electronic Characterization of Ordering of Amorphous Germanium Using XANES," Phys. Rev. B, in press.

Paesler, M.A., Sayers, D.E., Tsu, R., and Hernandez, J.G., 1984, "Ordering of Amorphous Germanium Prior to Crystallization," Phys. Rev. 28:4550.

Phillips, J.C., Brust, D., and Bassani, F., 1962, "Energy Bands and Lineshape of UV Reflectivity of Ge and Si," <u>Proc. Int. Conf. on the Physics of Semiconductors</u>, Bartholomew, England, p. 564.

Polk, D.E., 1971, "Structural Model for a-Si and a-Ge," J. Non-Cryst. Solids 5:365.

Polk, D.E., and Boudreaux, D.S., 1973, "Tetrahedrally Coordinated Random-Network Structure," Phys. Rev. Lett. 31:92.

Proietti, M.G., Incoccia, L., Mobilio, S., Gargano, A., and Evangelisti, F., 1985, "Effect of Intermediate Range Order on the Lineshape of the Ge K Edge," in <u>EXAFS and Near Edge Structure III</u>, Springer-Verlag, New York.

Spaepen, F., 1978, "A Structural Model for the Interface Between Amorphous and Crystalline Si or Ge," Acta. Metall. 26:1167.

Stern, E.A., Bouldin, C.E., von Roedern, B., and Azoulay, J., 1983, "Incipient Amorphous to Crystalline Transition in Ge," Phys. Rev. B 27:6557.

Steinhardt, P., Alben, R., and Weaire, D., 1974, "Relaxed Continuous Random Network Models," J. Non-Cryst. Solids 15:199.

Tsu, R., Hernandez, J.G., Chao, S.S., Lee, S.C., and Tanaka, K., 1982, "Critical Volume Fraction of Crystallinity for Conductivity Percolation in Phosphorous-Doped Si:F:H Alloys," Appl. Phys. Lett. 40:534.

Tsu, R., and Hernandez, J.G., 1984, "Determination of the Energy Barrier for Structural Relaxation in a-Si and a-Ge by Raman Scattering," J. Non-Cryst. Solids 66:109.

BONDING IN DISTORTED TETRAHEDRA BY S-P-D HYBRID BONDS

R. Grigorovici and P. Gartner

Institute of Physics and Technology of Materials

Bucharest-Magurele, P.O.B. M7, Romania

ABSTRACT

It is shown that the short-range order in tetrahedrally bonded meta-stable crystals and random networks can be described quantitatively by non-equivalent s-p-d hybrid bonds.

1. INTRODUCTION

In the evolution of our knowledge about non-crystalline materials, Mott and Davis's book on electronic processes in these materials [1] was a landmark. Not only did it offer a review of this knowledge at the beginning of the 70's, but it also clarified the ways in which the outstanding problems were tackled later.

The essential question is formulated on page 1: "...we have to ask first which of the concepts appropriate to crystalline solids can be used in non-crystalline materials." In accordance with the title of the book, this question is analyzed mainly from the point of view of electronic processes.

However, exactly the same question can be put in connection with chemical bonding: how far can the concepts of chemical bonding in crystalline solids be used in non-crystalline materials?

In the early reviews of one of the present authors on the structure of amorphous semiconductors [2-4] and in his early contributions to the structural modelling of tetrahedrally bonded (TB) amorphous semiconductors [5-9], chemical bonding in chemically related crystals and amorphous solids was supposed to be identical.

Objections against this concept were raised from different quarters. Ovshinsky expressed his conviction [10] that the chemical bonds in disordered materials cannot "be separated from their surrounding environment, for it is their interaction with it and the resulting feedback which create a combined new entity" which he calls "the total interactive environment (TIE)."

In fact, Anderson [11] and Mosseri and Gaspard [12] had already approached the problem of chemical bonding in TB atom clusters by the formalism of molecular orbitals. Agreement with experiments was not entirely satisfactory. The addition of d contributions to the usual s and p type contributions seemed to improve the situation. However, in this approach structural model builders lose all contact with their intuitive tetrapods and, we think, theoretical physicists do not feel at home either.

In a series of previous papers one of us tried a step-by-step approach whose goal was to formally describe the interatomic bonds in TB amorphous elements by hybrid atomic orbitals, without losing the feedback from the TIE [13].

It is generally admitted that the energy stored in the metastable phases of TB elements, be they crystalline or amorphous, results from the local bond distortions [5] and is thus of elastic origin. This is implied, e.g., in all relaxation procedures applied to structural network models. Still, in these networks every atom had a different energy stored by elastic deformation [14].

While the stored energy equalled more or less the crystallization enthalpy of a-Ge, there were striking discrepancies between these two values in the high-pressure polymorphs of a-Ge [8] indicating that the elastic approach is unsatisfactory and that there must have taken important changes in the character of chemical bonding when the increase in density imposed angular and radial deformations on the formerly regular tetrahedron. Implicitly these changes would have to be rather slight in the amorphous TB elements, whose density remains very little changed in spite of the topological disorder.

It followed from a detailed analysis of the ring statistics about individual atoms in a relaxed RNM [15] that the angular distortions at the central atom are determined primarily by this statistics, though certainly other features of the TIE also contributed significantly to the local angular and radial disorder. Thus, the TIE is indeed finding its expression in the distortion of the central tetrahedron.

An attempt to apply Coulson's concept of non-equivalent s-p hybridization [16] to the description of angular disorder in TB RNMs [15] was mostly a failure, even as an approximation, because of the lack of an appropriate symmetry. This was particularly obvious when two bond angles facing each other across the central atom were both bigger than 109.5°. So the hope was expressed that the addition of d contributions might avoid these difficulties.

Indeed, on the one hand the very complexity of the d orbitals seemed to enhance the chances of success for this approach. On the other hand, the addition of d contributions takes to some extent into account the existence of weaker interactions, i.e., the influence of the TIE.

The present paper deals with the first attempt of the above approach as applied to both metastable, high-pressure TB polymorphs of Si and Ge (Section 3) and to amorphous Si and Ge (Section 4).

2. THE PROCEDURE

Hitherto, many attempts have been made to describe by appropriate s-p, s-p-d, and p-d hybrids local structural configurations in crystalline solids (see, e.g., refs. 17 and 18). Only the angular bond distortions were taken into account. Changes in bond lengths were not considered, probably because the calculation of superposition integrals is a rather complicated matter.

We shall use Pauling's "bond strength" S [17] as a simple and thus appropriate approximation for the evaluation of changes in superposition integrals. It is defined by the value of the wave function of the hybrid orbital in the direction of its maximum. Its value is equal 2 for the pure sp^3 bond, smaller for any other s-p hybrid and changes drastically under the influence of d-type contributions.

We postulate that the bond length is inverse proportional to the bond strength. Thus, a weaker bond is correspondingly elongated and vice versa.

Energy storage is achieved by the redistribution of the available valence electrons over another set of orbitals. In sp^x hybrids with $x \neq 3$ the energy stems from the fact that, after promotion of an s electron into a p orbital, the gain achieved by the formation of a hybrid bond is less than if x would have been equal 3. In the case of s-p-d hybrids, more energy may be stored due to the promotion of a fraction of all electrons into the higher lying d states.

We shall now consider in more detail s-p-d hybrids. Obviously, the bonds must be normalized and orthogonal, must point in the right directions and must have the strength imposed by the actual structure. In high-pressure polymorphs the number of bonds may exceed the number of valence electrons, thus creating an unusual situation.

If the distorted tetrahedra and the supplementary bonds display a high degree of symmetry, the above conditions are few and the type of d orbitals to be added are easily guessed.

When, as in the case of some high-pressure polymorphs and in all amorphous networks, symmetry is lost completely, a procedure must be devised in order to find the "best" solution, i.e., non-equivalent s-p-d hybrids satisfying not only orthonormalization, direction and bond strength, but also achieving minimal d contribution.

An analysis of a particular Ge RNM [19] showed that, to the difference with high-pressure polymers, higher-order neighbours do not grasp the opportunity to come much nearer the central atom by using the free space created by an increase in a bond angle. So normalization is always confined to the four covalent bonds.

Preliminary numerical evidence confirmed the intuitive picture that s-p orbitals, while unable to satisfy both the orthonormality conditions and correct spatial orientation, can however be made to point in the right directions by adding a very small amount of d contributions. The d character is so weak that it barely affects the orthonormality conditions. Therefore, as a first approximation, it is reasonable to ask only for the s-p part of the orbitals to be mutually orthogonal.

This is the starting point of our iterative procedure in which one computes in turn the s-p part and the d part of the orbitals. When self-consistency is achieved, the hybrids satisfy the orthonormation and spatial constraints with the desired precision. At each step, given the s-p part, the minimal d part is found out among those which ensure the right direction and strength of the bonds. The criterion of minimality we used was simply the minimization of the squares of all the d coefficients from all the four hybrid orbitals.

On the other hand, when calculating the s-p part, one takes into account in the orthonormality conditions the d components previously computed. Otherwise, the s-p parts are subject only to the requirement of minimum departure (in the sense of minimum mean square deviation) from some prescribed s-p hybrids.

It was felt that a good choice of these "reference" hybrids would be provided by the s-p combinations which point in the direction of the bonds, are normalized (though not orthogonal), and have the strengths as close as possible (exactly equal, if the given strength is not greater than 2) to the real bond strengths. This way one is from the very beginning pretty close to the solution and, since the iteration scheme proceeds in steps which try to keep the d contribution at the lowest level, one may say that the final result is not far from the exact solution having the absolute minimum d admixture.

The scheme converges very quickly (3 to 4 iterations are sufficient for a precision of 10^{-7}) and, when tested against the special cases which can be described by pure s-p orbitals, correctly yields a vanishing d contribution.

4. HIGH-PRESSURE POLYMORPHS

Si and Ge acquire under high pressure the white tin structure (Si II and Ge II). Every atom is surrounded by four first neighbours in the configuration of a flattened tetrahedron (Fig. 1). Other two bonds are slightly longer and bisect the two big bond angles of 128.39^o. All bond lengths are increased compared with the equilibrium Si I and Ge I crystals.

Krebs [18] has succeeded long ago to describe the angular aspects of the flattened tetrahedron configuration by p-d$_{3z^2-1}$ hybrids. He has, however, not tried to describe also the radial aspects, nor has he taken into account the presence of the other two longer bonds.

The symmetry of the short-range order immediately suggests an s-p-d$_{3z^2-1}$ hybridization, where the main lobes of the d orbital are pointing towards the two weaker bonded neighbours (D-bond). We include in our calculation the values of the bond strengths S_T and S_D, supposed to be inversely proportional to the bond length. S being equal 2 for the ideal sp^3 orbital, we get in our case $S_T = 1.8568$ and $S_D = 1.7638$. Obviously, the tetrahedral bonds must also point in the right directions.

The corresponding wave functions have the following contributions:

	$\underline{s}$	$\underline{p}_x$	$\underline{p}_y$	$\underline{p}_z$	$\underline{d}_{3z^2-1}$
ψ_1	a	λ	0	μ	f
ψ_2	a	0	λ	$-\mu$	f
ψ_3	a	$-\lambda$	0	μ	f
ψ_4	a	0	$-\lambda$	$-\mu$	f
ψ_z	α	0	0	0	δ

The above-mentioned conditions are contained in the following equations:

Orthogonality:

$$a^2 - \mu^2 + f^2 = 0 \tag{1}$$

$$a^2 - \lambda^2 + \mu^2 + f^2 = 0 \tag{2}$$

$$a\alpha + f\delta = 0 \tag{3}$$

Tetrahedral bond directions:

$$\sqrt{3}\lambda - \Lambda x = 0 \tag{4}$$

$$\sqrt{3}\mu + 3\sqrt{5}\,f - \Lambda z = 0 \tag{5}$$

Bond strength,
  T-bonds:

$$a + \sqrt{3}(\lambda x + \mu z) + \frac{\sqrt{5}}{2} f(3z^2 - 1) = S_T \tag{6}$$

  D-bonds:

$$\alpha + \sqrt{5}\,\delta = S_D \tag{7}$$

where $x = \cos\theta$; $z = -\sin\theta$ (see Fig. 1) and Λ is a Lagrange multiplier. The system is sufficient for the calculation of the seven unknown quantities: a, λ, μ, f, α, δ, and Λ. One gets

$$\alpha = 0.17494 \qquad \delta = 0.71055 \qquad a = 0.42433$$
$$\lambda = 0.64263 \qquad \mu = -0.45441 \qquad f = -0.16252$$

The normalization condition is not met by each individual orbitals, but, surprisingly enough, the sum of their norm is

$$N = N_D + 4 N_T = 0.5355 + 4 \cdot 0.8259 = 3.8562$$

i.e., only 3.7 per cent short of 4, which is the number of bonding electrons. If one may interpret the normalization constant of each orbital as a

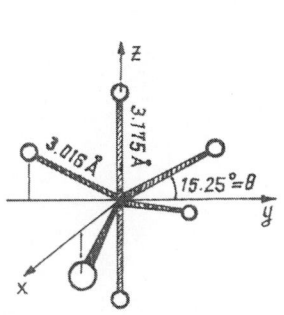

Fig. 1   The white-tin structure.

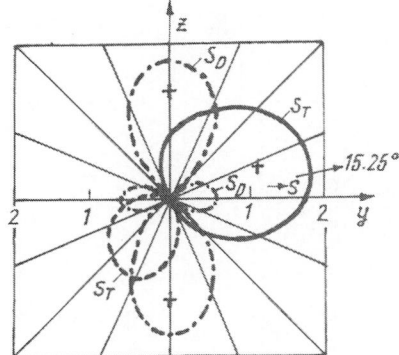

Fig. 2   Angular distribution of $\underline{S}_D$ and $\underline{S}_T$ in Si II and Ge II.

rough estimate of the degree of the electron to the corresponding bond, then the result turns out to be satisfactory.

The angular distribution of the bond strengths in the s-p-d and s-d orbitals are represented in Fig. 2. The incomplete filling of the orbitals and their strong overlapping explains the metallic properties of Si II, Ge II and white tin. Normalized to D =1, the D-bond is a $S^{0.0606}$ hybrid; normalized to S = 1, the T-bonds are $SP^{3.44}D^{0.147}$ hybrids. The total d contribution amounts to 15.8 per cent.

Si III and Ge IV are metastable high-pressure polymorphs resulting at moderate pressures by a slow reconstructive transformation. All atoms are equivalent. The short-range order is depicted in Fig. 3. One of the bonds (1) lies on a trigonal symmetry axis. In Si III its length is 2.306Å; the corresponding bond strength is S_1 = 2.0381. The other three equivalent bonds are forming a very flat pyramid, the bond angle being 117.9°. Their length of 2.392Å corresponds to a bond strength S_2 = 1.9648. There is also a fifth, not bonded neighbour lying at 3.441Å on the trigonal axis, i.e., nearer than the nearest second order neighbours. Were it bonded to the central atom, its bond strength would be S_3 = 1.3658.

If one considers only the bond angle distribution, s-p hybrids are capable of describing it exactly. The p contribution to bond (1) would be high, equal 4.6783. Normalizing bond (1) to S = 1 gives it a $SP^{21.89}$ character, i.e., of a nearly pure p bond. In contrast, the p contribution to the other three bonds is only 1.462 each; so they are $SP^{2.14}$ hybrids which fits their near coplanarity and trigonal symmetry.

If one calculates, however, the corresponding bond strengths, one finds S_1 = 1.9028 and S_2 = 1.9942. These values not only do not tally the real values, but bond (1) should be longer, not shorter than the other three bonds.

This discrepancy can be avoided by adding to bond (1) a small d_{3z^2-1} contribution oriented with its main lobes after the trigonal axis. The wave functions are thus given by

	s	p_x	p_y	p_z	d_{3z^2-1}
ψ_1	a	0	0	ξ	δ
ψ_2	b	0	0	$-\rho$	f
ψ_3	b	$\frac{\sqrt{3}}{2}\mu$	$\frac{1}{2}\mu$	$-\rho$	f
ψ_4	b	$-\frac{\sqrt{3}}{2}\mu$	$\frac{1}{2}\mu$	$-\rho$	f

with the condition for

Normalization: $a^2 + \xi^2 + \delta^2 = 1$ (1)

$b^2 + \mu^2 + \rho^2 + f^2 = 1$ (2)

Orthogonality: $a b - \rho\xi + \delta f = 0$ (3)

$b^2 - \frac{1}{2}\mu^2 + \rho^2 + f^2 = 0$ (4)

Directions: $\sqrt{3}\rho + 3\sqrt{5}\, f \sin\theta - 2\, tg\,\theta = 0$ (5)

Bond strengths:
$$a + \sqrt{3}\xi + \sqrt{5}\delta = S_1 \tag{6}$$
$$b + \sqrt{3}\rho\sin\theta - \frac{\sqrt{5}}{2}f(1 - 3\sin^2\theta) + \sqrt{2}\cos\theta = \underline{S}_2 \tag{7}$$

The unknown quantities being seven, the system yields

$$a = 0.17937 \qquad \xi = 0.98121 \qquad \delta = 0.07121$$

$$b = 0.56697 \qquad \mu = \sqrt{2/3} = 0.81650 \qquad \rho = 0.10558 \qquad f = 0.02669$$

The bonds (1) and (2) to (4) have $SP^{29.93}$ and $SP^{2.109}D^{0.0022}$ character, respectively. So, except for the small $\underline{d}$ contributions, the character of the bonds is like before, but now the bond strengths agree with reality.

Even more satisfactory is the following surprising result. Towards negative $\underline{z}$'s, the $\underline{p}$ and $\underline{d}$ orbitals are of opposite sign so that the remaining, mainly p-type, negative lob (Fig. 4) is able to bind a fifth atom with a calculated bond strength $S_3 = a - \sqrt{3}\xi - \sqrt{5}\delta = -1.3609$ that tallies exactly the above predicted value for a fifth bond. This is a remarkable result. The total $\underline{d}$ contributions amounts to only 0.721 per cent.

It is worthwhile mentioning that by applying the iteration procedure described in Section 2 to the same case, one obtains the correct bond strengths and directions with a lower $\underline{d}$ contribution of 0.430 per cent and similar bond characters (namely, $SP^{8.33}D^{0.0106}$ and $SP^{2.36}D^{0.0035}$), but the bond strength to the fifth neighbour is 1.23, i.e., lower than required. Thus, in order to obtain the right result, the presence of an allegedly not bonded, but near enough lying neighbour, must be taken into consideration from the very beginning.

4.   AMORPHOUS Si and Ge

A relatively small (155 atoms) relaxed a-Ge random network model [19] was used to select 32 atoms whose environment comprised at least six--mostly many more--complete rings which included each of the six bond angles centred on the chosen atom.

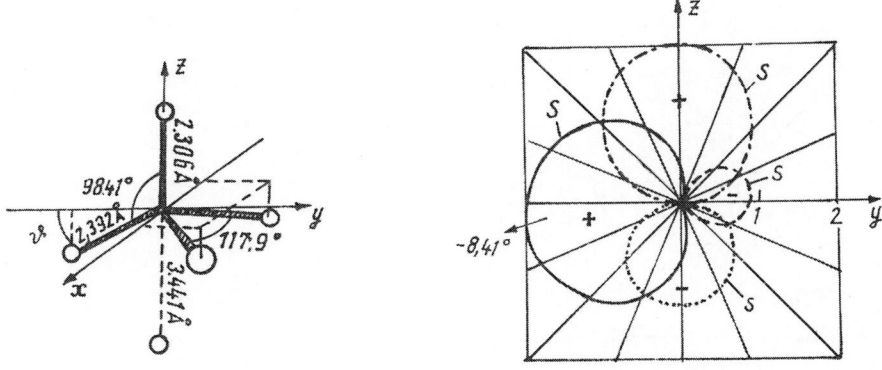

Fig. 3   Bonding of Si atom in Si III.   Fig. 4   Angular distribution of $\underline{S}$ in Si III.

The radial and angular distribution describing the short-range order of ten of these atoms are displayed in Table 1. The bond angles were arranged in the following order: A_{12} is the biggest one; A_{34} is its opposite angle. Follow the pair of opposite angles A_{13}/A_{24}, A_{24} being the next biggest angle after A_{12}, and then A_{14}/A_{23}. Radial distances are expressed by the corresponding bond strengths S_1 to S_4.

The calculations whose goal it was to establish the appropriate hybrid bond asking for the smallest total d contribution $\emptyset$ and described in Section 2 were performed on a Sharp PC-1500 computer. The results obtained for the atoms listed in Table 1 are given in Table 2. The atoms are arranged in order of increasing $\emptyset$ and selected from the whole list of atoms as follows: the first three have the lowest $\emptyset$ values; the last three the highest ones. The other four are taken from the middle of the whole list.

The results show that even if the resulting hybrid bonds do not represent the very best solutions from the point of view of minimizing the supplementary energy necessary for achieving the angular and radial distortions present in our RNM, the main contributions still come from the s and p parts which vary rather much as shown by the scatter of the exponents of P in the hybrids normalized to S = 1 (see Table 2).

The d contributions necessary to overcome the relatively important remaining discrepancies in bond angles and bond strengths between the pure s-p hybrids and the actual values are surprisingly low, mostly below 1 per cent. Big d contributions are often, but not always, associated with big deviations of the p contribution from the standard value of 3.

Obviously, the angular bond strength distribution must vary markedly with the relative values of the s, p and d contributions. Figures 5 and 6 offer two examples. Both correspond to atom No. 30 which has the highest d contribution. In Fig. 5 the plane of the drawing coincides with that of the angle A_{12} and the distribution of S to bond (2) which has a very low standardized p contribution of only 1.169 (see Table 2).

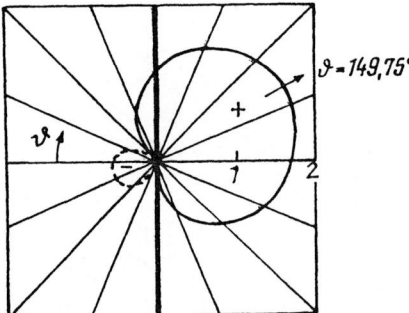

Fig. 5 Angular bond strength distribution for bond (2) of atom No. 30.

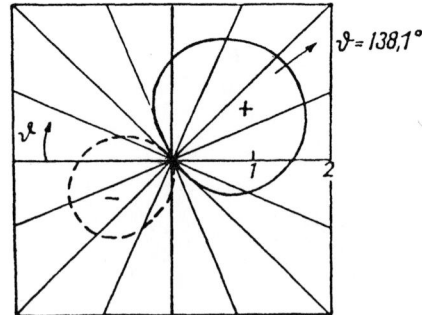

Fig. 6 Angular bond strength distribution for bond (4) of atom No. 30.

Table 1   Angles $\underline{A}$ and bond strengths $\underline{S}$ in tetrahedrally bonded RNM [19].

Atom No.	A_{12}/A_{34}	A_{13}/A_{24}	A_{14}/A_{23}	S_1	S_2	S_3	S_4
27	119.7/103.0	100.1/119.2	105.3/106.6	2.000	1.962	1.983	1.976
25	112.7/112.6	102.6/111.6	109.5/107.5	2.001	2.016	1.993	2.064
2	117.9/107.4	109.2/111.9	105.3/104.8	2.004	1.994	2.041	2.042
5	116.6/109.8	107.7/114.1	107.8/100.9	2.002	1.993	2.059	1.956
4	116.5/115.1	106.4/113.4	99.3/106.2	2.044	1.986	2.012	2.044
14	117.0/109.1	109.3/113.1	95.6/111.6	2.067	1.983	2.044	1.958
23	116.9/101.0	109.3/113.2	112.0/102.7	1.921	1.968	2.029	2.093
31	116.7/114.0	106.1/110.0	107.3/102.9	1.950	1.991	2.117	2.110
15	117.6/113.4	107.5/107.5	105.2/105.9	1.875	2.081	1.993	2.071
30	119.5/114.9	113.7/117.3	96.2/ 96.6	1.930	1.875	1.921	1.909

Table 2   Total $\underline{d}$ contribution $\emptyset$ and $\underline{s}$-$\underline{p}$-$\underline{d}$ hybrid bonds

Atom No.	$\emptyset$ (%)	Normalized Bond Characters											
		\multicolumn 1			2			3			4		
		S	P	D	S	P	D	S	P	D	S	P	D
27	.119	1	3.23	.0007	1	1.19	.0014	1	6.47	.0024	1	4.77	.0005
25	.177	1	2.98	.0007	1	2.67	.0002	1	4.06	.0021	1	2.60	.0041
2	.192	1	2.00	.0011	1	2.27	.0020	1	3.62	.0029	1	3.41	.0018
5	.365	1	3.61	.0037	1	2.83	.0046	1	4.54	.0078	1	1.93	.0007
4	.387	1	3.65	.0051	1	2.08	.0012	1	3.38	.0036	1	3.33	.0068
14	.410	1	3.21	.0052	1	1.63	.0017	1	2.92	.0047	1	6.89	.0083
23	.429	1	1.68	.0021	1	2.44	.0025	1	5.59	.0051	1	4.46	.0126
31	.960	1	1.94	.0013	1	2.87	.0009	1	4.27	.0262	1	3.77	.0190
15	1.020	1	1.28	.0042	1	3.82	.1092	1	5.30	.0115	1	4.18	.0241
30	2.052	1	1.70	.0100	1	1.17	.0100	1	11.03	.0789	1	11.06	.0698

Figure 6 lies in the plane of the angle A_{14} and the bond strength distribution corresponds to that of bond (4) which has an exceptionally high p contribution of 11.06. This difference is clearly reflected in the shapes of the positive and negative lobes.

Figures 5 and 6 are based on the data contained in Table 3 which gives a detailed example of the contributions of the s, p, and d orbitals to the wave functions corresponding to the four bonds of atom No. 30. The sum of the squares of all contributions is 0.999997 for each of the bonds after three iterations, showing the good quality of the normalization achieved by the procedure used in our calculations.

Finally, Table 4 correlates the value of the global d contribution, $\emptyset$, with the determining quantities:

- $a = \sqrt{\overline{C^2}}$ is the rms deviations of the three products between the cosinuses of opposing angles, thus characterizing the deviation from Coulson's condition of pure s-p hybridixation;

- $b = \sqrt{\overline{S^2}}$ is the rms deviation of the four bond strengths from their average, thus characterizing the radial distortion of the tetrahedron;

- $c = (\overline{S} - 2)^2$ is the square of the deviation of the average bond strength $\overline{S}$ from the ideal value of 2, thus characterizing local compression or dilation.

An analysis of the data of Table 4 shows that all these quantities have their influence on $\emptyset$ and the empiric formula

$$\emptyset = 0.0388\, a + 1.136\, b^2 + 235\, c^2$$

gives, if not the exact values of $\emptyset$, at least the general trend of its dependence on a, b, and c. The calculated values of $\emptyset$ are given in the last column of Table 4.

5. CONCLUSIONS

Clearly the attempt of describing the short-range order in elemental TB high-pressure crystals and amorphous networks containing distorted tetrahedra by non-equivalent s-p-d hybrid bonds has succeeded. Even increases in coordination leading to metallization could be described, when during a phase transition a supplementary neighbour or two approached the central atom enough to lie at a distance comprised between that of the first- and second-order neighbours.

We intend to extend this approach beyond this limit in order to include as much as possible of the total interactive environment into the description.

Obviously, this approach establishes a close link between elastic properties and electronic phenomena, describing elastic deformations by processes of electronic excitations.

The thermodynamics of phase transformations may also profit from this approach.

Finally, results like those presented in Section 4 may constitute an

Table 3 s, p, and d contributions to bonds (1) to (4) of atom No. 30

	ψ_1	ψ_2	ψ_3	ψ_4
s	0.60712	0.67745	0.28738	0.28715
p_x	0.08569	-0.09735	0.72273	-0.67740
p_y	-0.72548	0.67383	0.06293	-0.11960
p_z	0.30670	0.27009	-0.62016	-0.66228
d_{yz}	-0.01189	0.00759	0.01162	-0.00524
d_{xz}	-0.01931	0.02193	0.03374	-0.03087
d_{xy}	0.03310	0.03761	0.04101	0.03166
$d_{x^2-y^2}$	0.02835	0.03419	0.02092	0.01940
d_{3z^2-1}	0.03551	0.03828	-0.05586	-0.05821

Table 4 Quantitites which determine the total d contribution ∅.

Atom No.	∅ (%)	$\sqrt{\overline{C^2}} = a$	$\sqrt{\overline{S^2}} = b$	$(\bar{S}-2)^2 = c$	∅ calc.
27	.119	0.0262	0.013	0.0004	.122
25	.117	0.0285	0.028	0.0003	.199
2	.192	0.0309	0.022	0.0004	.177
5	.365	0.0408	0.044	0.0000	.377
4	.387	0.0589	0.024	0.0005	.301
14	.410	0.0490	0.044	0.0002	.411
23	.429	0.0216	0.065	0.0000	.544
31	.960	0.0495	0.073	0.0017	.865
15	1.020	0.0491	0.083	0.0000	.973
30	2.052	0.0868	0.021	0.0084	2.047

improved starting point for density of states calculations based on local bonding like those initiated by Weaire and Thorpe [20] and developed later by other authors (see e.g., [21]).

REFERENCES

1. N.F. Mott and E.A. Davis, Electronic Processes in Non-Crystalline Materials, Clarendon Press, Oxford (1971).
2. R. Grigorovici, J. Non-Cryst. Solids 1, 303 (1969).
3. R. Grigorovici, "The Structure of Amorphous Semiconductors," in Electronic and Structural Properties of Amorphous Semiconductors, P. LeComber and J. Mort, eds., Academic Press, London (1973) p. 192.
4. R. Grigorovici, "The Structure of Amorphous Semiconductors," in Amorphous and Liquid Semiconductors, J. Tauc, ed., Plenum Press, London (1973) p. 45.
5. R. Grigorovici and R. Manaila, Thin Solid Films 1, 343 (1967).
6. R. Grigorovici and R. Manaila, J. Non-Cryst. Solids 1, 371 (1969).
7. R. Grigorovici and R. Manaila, Nature 226, 143 (1970).
8. R. Grigorovici, R. Manaila, M. Popescu and C. Stanescu, in Proc. of Conf. in Tetrah. Bonded Amorphous Semiconductors, Yorktown Heights (1974) p. 200.
9. R. Grigorovici, M. Popescu and R. Manaila, J. Non-Cryst. Solids 23, 229 (1977).
10. S.R. Ovshinsky, Rev. Roum . Phys. 26, 893 (1981). (See also Disordered Materials: Science and Technology, Selected Papers by S.R. Ovshinsky, D. Adler, ed., Amorphous Institute Press, Bloomfield Hills (1982) p. 282.
11. R. Grigorovici, Solar Energy Mats. 8, 177 (1982).
12. A.B. Anderson, J. Chem. Phys. 63, 4430 (1975).
13. R. Mosseri and J.P. Gaspard, J. de Phys. 42, C4-245 (1981).
14. R. Grigorovici, J. Non-Cryst. Solids 35-36, 1167 (1980).
15. R. Grigorovici, J. Non-Cryst. Solids 59-60, 221 (1983).
16. C.A. Coulson, Valence, Oxford Univ. Press, Oxford (1961) p. 203.
17. L. Pauling, The Nature of the Chemical Bond, Cornell Univ. Press, Ithaca (1960).
18. H. Krebs, Inorganic Crystal Chemistry, McGraw-Hill, London (1968).
19. M. Popescu, Ph.D. Thesis, Bucharest (1975).
20. D. Weaire and M.F. Thorpe, Phys. Rev. B 4, 2508 (1971).
21. J. Ziman, Models of Disorder, Cambridge Univ. Press, Cambridge (1979) Ch. 11.3.

SPECTROSCOPIC ELLIPSOMETRY STUDIES OF THE GROWTH
AND MICROSTRUCTURE OF HYDROGENATED AMORPHOUS SILICON

R.W. Collins, A.H. Clark, and C.-Y. Huang

The Standard Oil Company (OHIO)
Research Center
4440 Warrensville Center Road
Cleveland OH 44128

INTRODUCTION

Today's understanding of the electrical and optical properties of bulk semiconductor glasses has been achieved in good part as a result of the seminal research contributions of Sir Nevill Mott, to whom this Festschrift is dedicated. The models to describe the corresponding properties of hydrogenated amorphous silicon (a-Si:H), prepared in thin film form, have been developed along a similar theoretical framework as the bulk glasses, and primarily without regard to the microstructure established by the nucleation and growth process. Defect states detected through a broadened absorption edge or a reduced luminescence efficiency, for example, are seemingly well attributed to uniform intrinsic disorder and randomly distributed bulk dangling bonds, respectively. Earlier morphological studies, however, have shown that non-optimally prepared glow discharge a-Si:H exhibits direct evidence of columnar growth which is established in the initial stages of deposition when nucleation occurs at preferred sites.[1] Only under optimal plasma conditions do the thin film nuclei merge to invisibility. As a result, defects such as strained and dangling bonds and polymeric $(SiH_2)_n$ chains would tend to congregate along the internal surfaces of the incompletely merged nuclei, not unlike grain boundaries in poly-crystalline materials.

In more recent work, a-Si:H prepared under conditions that give high efficiency solar cells was shown to exhibit surface microstructure thought to arise from initial nucleation followed by shadowing effects, causing only the most prominent nuclei to dominate the microstructural scene.[2] For these

63

films, however, the strength of bonding along the merged surfaces was just as great as internal to the surfaces. Thus, the interesting surface structure masked a rather homogeneous bulk in comparison to the earlier results on the non-optimally prepared films. This work also stressed the importance of substrate material in its influence on the surface morphology of the films. A similar conclusion was drawn earlier from the observation that the low temperature photoluminescence intensity of low defect density a-Si:H deposited on a Ni coated region of a fused silica substrate was significantly reduced relative to a bare region.[1] This effect could be attributed to differences in the microscopic surface structure of the a-Si:H over the two regions which would cause differences in the reflectivity of the interface for excitation and emission.

There are still many unresolved questions pertaining to the growth of a-Si:H. Since thin (<200Å) amorphous layers are required in most device oriented a-Si:H applications, these questions are particularly urgent. For these thin layers, the surface structure may have significant influence on the photoelectronic properties of the layer. Can one reliably assess the behavior of the thin layers by studying the properties of thicker (>0.3μm) layers? Correspondingly, because of substrate dependent effects, measurements on films deposited other than in the device configuration may not be justifiably applied to an understanding of the device. Clearly sensitive non-destructive surface and interface spectroscopies are required to begin answering such questions.

In this work we will discuss the applications of spectroscopic ellipsometry in characterizing the near surface microstructure of thin and thick amorphous and microcrystalline silicon films of practical importance. Earlier studies have clearly demonstrated the usefulness of both in situ growth monitoring of (CVD[3] and dc multipole plasma[4] deposited) a-Si:H by single wavelength ellipsometry and post-deposition characterization of thick a-Si by spectroscopic ellipsometry.[4,5] For the studies performed in situ, the lack of spectroscopic data limits the extent of the analysis; whereas for the post-deposition studies oxidation is a complicating factor to be included in the data analysis. In both cases, evidence for film growth via three-dimensional nucleation was presented. The studies performed on CVD a-Si:H in situ were consistent with an average nucleation center spacing of approximately 100Å in the initial stages of growth.[3]

The data presented in our study include: (1) the results of a post-deposition analysis of photovoltaic quality glow discharge a-Si:H deposited

on glass and c-Si substrates to intended thicknesses of 20-1000A, and
(2) an optical analysis of thin phosphorus-doped microcrystalline Si:H:F
deposited on intrinsic a-Si:H (in the n-i-p configuration) in comparison to
identically prepared thick Si:H:F on a quartz substrate. In these
applications the power of spectroscopic ellipsometry for the study of
imperfect, non-crystalline thin film surfaces is evident. The derivable
information may include the index of refraction and extinction coefficient
throughout the optical range for an unknown component of a multilayer
structure, thus making the technique particularly well suited for the study
of photovoltaic materials.

EXPERIMENTAL TECHNIQUE AND DATA ANALYSIS

     In an ellipsometry experiment on an unknown surface, one determines the
change experienced by the polarization state of a collimated monochromatic
light beam upon reflection from the surface.[6] From this measurement, the
complex reflectance ratio,

$$\rho = r_p/r_s = \tan\psi \exp(i\Delta) \tag{1}$$

can be calculated. Here r_p and r_s are the complex reflectances for light
polarized parallel (p) and perpendiular (s) to the plane of incidence. The
ellipsometric angles ψ and Δ represent the amplitude ratio and phase difference between the reflectances for the two linearly polarized states. The
quantity ρ (or equivalently ψ and Δ) is determined by the nature of the
specular interface whether it be a perfect dielectric discontinuity between
two homogeneous media or (as is normally the case in thin film applications)
a complicated stack of heterogeneous layers. In the rare cases where a
single perfect dielectric discontinuity between the surface and its ambient
(air) can be assumed, then the dielectric function, ϵ, of the unknown bulk
is uniquely determined:

$$\epsilon = \sin^2\theta \left(1 + \left[\frac{1-\rho}{1+\rho} \right]^2 \tan^2\theta \right) \tag{2}$$

Here θ is the angle of incidence. In thin film applications where overlayers
(ie. oxides or surface roughness) can be neglected this formula can be
applied over the spectral range where the film is opaque. For the three
phase system: ambient/overlayer/bulk in the opaque regime (or
ambient/film/substrate for an ideal film surface for photon energies where
the film transmits), a corresponding equation can be written which expresses

65

ρ as a function of the overlayer (or film) thickness, the angle of incidence and the dielectric functions of the three phases. In general, even if two of the three dielectric functions are known, the three phase formula cannot be inverted to obtain the thickness and the third dielectric function.

In the case of the ambient/film/substrate system, various ingenious schemes have been developed using spectroscopic data to obtain the dielectric function and thickness of the film (and in some cases even correct for overlayers such as surface roughness).[7,8,9] For example,[8] if the substrate exhibits sharp structure in its (known) dielectric function over the accessible spectral range, one can make a guess for the thickness of the film. The three phase expression for ρ can then be numerically inverted to obtain the dielectric function of the film. If the guess was incorrect, erroneous structure would appear in the film dielectric function at the spectral position of the structure in the substrate dielectric function. The thickness is chosen to eliminate, or at least minimize, the erroneous structure.

In certain cases of practical interest such as thin a-Si:H on c-Si a five phase system might be expected:

ambient/overlayer(1)/a-Si:H/overlayer(2)/c-Si,

where overlayer (1) may be due to a combination oxide (unless the sample surface has been chemically etched) and surface modulation and overlayer (2) is the corresponding layer on the substrate. It is clearly not reasonable to seek a complete solution to this problem to obtain three unknown dielectric functions and thicknesses -- especially since the overlayers are surely not homogeneous. A more realistic approach to this problem has been developed by Aspnes.[10] Since this analysis forms the basis for the sample structure determinations, it will be described in further detail. In such complicated situations, one assumes that each layer is composed of a material or microscopic mixture of materials of known dielectric function. The unknowns in the problem are then the thicknesses of each of the layers d_j ($j=2,\ldots m-1$; m = no. of phases) and the independent component volume fractions (f_{ij} $i=1,\ldots L(j); j=2,\ldots,m$; L(j) is the number of components of layer j) for each inhomogeneous layer. Since all unknowns are spectrally invariant, the analysis is considerably simplified. This analysis proceeds as follows:

(1) An educated guess is made for the structure of the sample: the number of layers, which layers are inhomogeneous, and the material components of each layer.

(2) Starting values are set for all unknown parameters f_{ij} and d_j.

(3) For the mixed layers, an effective medium approximation (EMA) can be used to determine an effective dielectric function. Although various techniques of wide ranging complexity have been developed to calculate the dielectric function of a heterogeneous material from those of its components, the Bruggeman EMA has been particularly successful in dealing with semiconductor problems.[5] The Bruggeman EMA is derived under the same general assumptions as the Maxwell Garnett EMA with the added condition that the host dielectric function be equal to the unknown effective dielectric function, a self-consistent approach taken when the detailed microstructure is unknown. The Bruggeman EMA is expressed by

$$0 = f_{1j} \frac{\epsilon_{1j} - \epsilon_j}{\epsilon_{1j} + 2\epsilon_j} + \cdots + f_{Lj} \frac{\epsilon_{Lj} - \epsilon_j}{\epsilon_{Lj} + 2\epsilon_j} \qquad (3)$$

$$1 = f_{1j} + \cdots + f_{Lj}$$

where ϵ_{ij} is the dielectric function of the ith component of the jth layer and ϵ_j is the effective dielectric function. This equation is solved for ϵ_j.

(4) At this point, $\rho_{calc}(h\nu)$ can be calculated for the proposed structure and the initial guesses for f_{ij} and d_j. The calculated and experimental functions are then compared and a linear regression analysis (LRA) provides corrections to f_{ij} and d_j to eliminate the discrepancy.

(5) Steps (3)-(4) are repeated until these corrections converge to zero. The resulting f_{ij} and d_j parameters represent the solution for the proposed structure. It is necessary to determine the unbiased estimator of the mean square deviation between the experimental and calculated ρ spectra as an assessment of the quality of the fit for comparisons with other proposed structures. It is also necessary to calculate confidence limits which can be used as a guide to prevent excessive use of free parameters.

(6) A second structure is proposed and steps (1) through (5) are repeated in an effort to minimize the unbiased estimator.

The main problem that remains, therefore, is to establish bulk dielectric functions to represent the known components. For materials such as SiO_2 and c-Si, typical substrate and overlayer components, up to date data from the literature can be used.[11,12] Considering the wide range of preparation techniques used to deposit a-Si:H, and the wide range of H-contents that can be obtained, it is impossible to identify a unique dielectric function for a-Si:H. However, the main differences among high photon energy ($>2.5eV$) dielectric data of a-Si:H samples can be attributed to the effects of overlayers (surface structure) and/or bulk density deficit. Overlayer and density deficit effects tend to reduce the peak value of ϵ_2 ($\epsilon=\epsilon_1-i\epsilon_2$) observed near 3.75eV. It would then follow that the a-Si(:H) dielectric function spectrum exhibiting the largest ϵ_2 value at its peak is most representative of a smooth sample of completely coordinated a-Si. An apparently limiting value of ϵ_2 of 29.6 has been suggested[9] and values very close to this have been obtained from glow discharge[13] and multipole plasma[4] techniques. The effects of H incorporation on the dielectric function are expected to be minimal in the dangling bond compensation regime and stronger at high H incorporation levels (>2 at%) where bulk density deficits and H-induced microstructure may be exhibited.[14] Obviously these arguments do not apply to the optical constants near the band edge and below where the effects of H incorporation are dramatic.

Figures 1a and b exhibit pseudo-dielectric functions (dielectric functions determined under the assumptions of Eq. 1) calculated to simulate the effects of both a bulk and surface density deficit. These data are presented to show the decrease in the peak value of ϵ_2 caused by a bulk density deficit or overlayers. The latter exhibits what might be expected in the case of surface roughness of increasing modulation. In these calcul- ations as with the fitting analyses described below, we employed the dielectric function for low pressure CVD a-Si corrected for the effects of overlayers given elsewhere.[9]

The ellipsometer used for the collection of data reported below was of

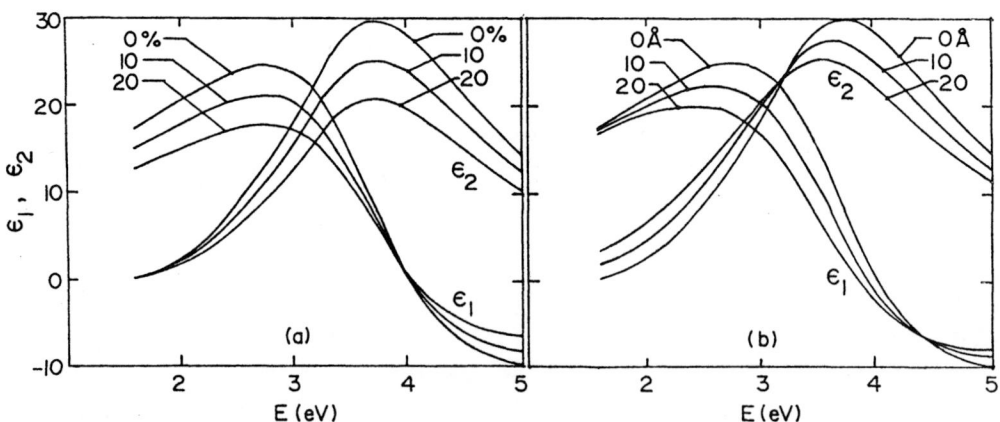

Fig. 1. Calculated pseudo-dielectric functions (see text) simulating the effects of (a) bulk density deficit due to voids and (b) surface roughness (modelled by a layer at the film-ambient interface composed of a 50-50 mixture of a-Si and voids).

Table 1. The structural parameters, 90% confidence limits, and unbiased estimator (δ) for the best fit models to the structure of thin a-Si:H deposited on c-Si and glass. In cases where two overlayers are listed, the first is at the top of the sample.

Time of Dep. (s)	Overlayer(s) Thickness/Comp.		Bulk Thickness/Composition		δ
ON C-SI:					
8	38±1A:	oxide	6±1A:	0.85 a-Si 0.15 void	0.0071
20	20±3A:	0.70 oxide 0.30 a-Si	27±3A:	0.89±0.05 a-Si 0.11±0.05 void	0.0049
40	22±3A:	0.50 oxide 0.50 a-Si	92±4A:	0.88±0.02 a-Si 0.12±0.02 void	0.0078
80	9±5A: 21±7A:	oxide 0.40 oxide 0.60 a-Si	217±19A:	0.97±0.02 a-Si 0.03±0.02 void	0.0102
400	23±3A: 10±3A:	oxide 0.50 oxide 0.50 a-Si	---	0.96±0.01 a-Si 0.04±0.01 void	0.0065
ON GLASS:					
20	61±4A:	0.28±0.07 a-Si 0.72±0.07 oxide	25±1A:	a-Si	0.020
40	105±10A:	0.54±0.02 a-Si 0.46±0.02 void	55±6A:	a-Si	0.017

the automatic rotating analyzer (RAE) type operating over the 1.6 to 5.0eV spectral range. A compensator was used only for the thin a-Si:H samples prepared on glass. In all other cases the spectral range of analysis was limited from 3.0 to 5.0eV where both film and substrate were opaque. It is in this region that the RAE (without compensator) is most accurate. Since quartz optics were used, the data discussed here were corrected for optical activity effects of the polarizer, compensator, and analyzer.[15,16] During measurement, samples were placed under a continuous flow of filtered N_2 gas to maintain stable surfaces. No surface treatments other than methanol were used to prepare samples previous to measurement; thus, any oxide overlayers were treated explicitly in the multilayer analysis.

The silicon and glass substrates used for the deposition of the very thin films were studied separately to determine the surface structure in the former case and the optical constants in the latter. An excellent fit to our data for the c-Si substrates was obtained using dielectric function data in the literature and an SiO_2 thickness of 26±1A. (Here the ± value indicates the 90% confidence limits in the LRA-EMA fit.)

RESULTS AND DISCUSSION

A. The Thickness Dependence of the Structure of Thin a-Si:H

Table 1 summarizes the results of the best fit solutions to the structure of glow discharge a-Si:H deposited under conditions that give efficient solar cells. The solutions tabulated give the lowest unbiased estimator of the mean square deviation (δ) between the experimental and calculated ρ values among a large number of possible structures for which a fit was attempted. For the films on c-Si, values of the overlayer compositions were held constant to obtain the lowest value of δ. This approach was necessary

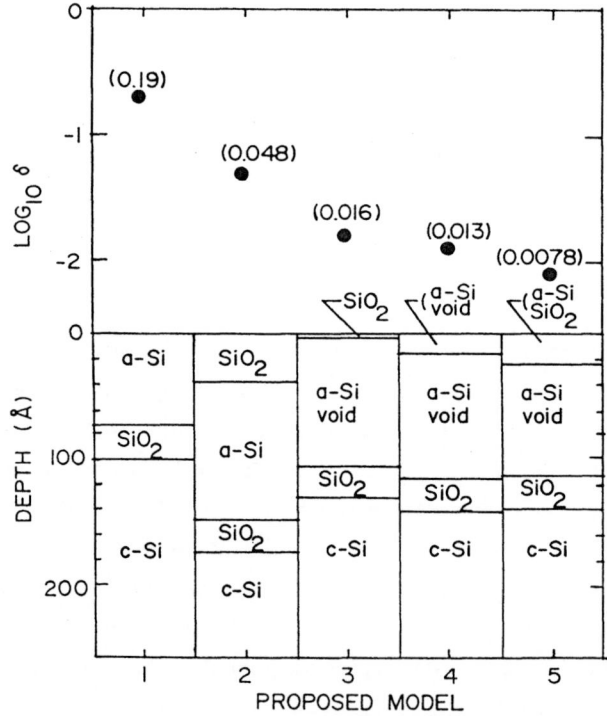

Fig. 2. A series of structures proposed to model the 40s a-Si:H deposition. Plotted at the top is the unbiased estimator for the model structure depicted below it. Depth into the sample is plotted downward at lower left.

Table 2.　The best fit parameters and 90% confidence
limits for the series of models used to fit
the structure of the 40s a-Si:H deposition.
The models are shown schematically in Fig. 2.

Bulk Thickness/comp.			Overlayer Thickness/composition		Column in Fig. 2
74±48A:	a-Si		---		1
110±33A:	a-Si		39±5A:	oxide	2
103±7A:	0.78±0.03 a-Si 0.22±0.03 void		3±6A:	oxide	3
100±6A:	0.85±0.02 a-Si 0.15±0.02 void		16±4A:	0.5 a-Si 0.5 void	4
92±4A:	0.88±0.02 a-Si 0.12±0.02 void		22±3A:	0.5 a-Si 0.5 oxide	5

because the small thickness of the overlayer prevented the determination of
a sharply defined composition by LRA techniques. As a result, the 90% confidence limits could not be determined. Similarly for the thin a-Si:H on
glass, equally good fits were obtained for fixed bulk void volumes in the
range of 0-20% with only very minor variations in the other structural
parameters. In this case, the large thickness of the overlayer prevented an
accurate determination of the bulk composition.

Figure 2 provides a schematic of the solutions to five models proposed
for the structure of the 40s a-Si:H deposition. This sequence of solutions
shows by example how we arrived at the best fit models of Table 1. Scaled at
the top of the figure is the unbiased estimator of the mean square deviation
for the solution to the model structure depicted directly below it. In Table
2, the structural parameters and 90% confidence limits are given for the
five solutions. The model in column 3 of Fig. 2 would be expected to provide
the best fit if the film structure was ideal, ie. sharp interfaces with no
surface roughness. The void component in the bulk would account for any bulk
density deficit caused, for example, by hydrogenation. A factor of two
improvement in the fit was obtained by the incorporation of a mixed composition surface layer as shown in the final column of Fig. 2. Additional
attempts at complicating this best fit model, such as the incorporation of a
mixed composition layer at the film-substrate interface, resulted in an
increase in δ. Figure 3 shows the experimental data, expressed as tanψ, cosΔ
(solid), and the calculated fits based on the models of columns 3 (ideal
structure) and 5 (best fit structure) of Fig. 2. Clearly a significant

improvement in the agreement is achieved assuming a mixed composition top layer.

The observation that boundary layers of <u>mixed composition</u> between the ambient and the film bulk were necessary in each case to obtain the best fit to the data is suggestive that surface microstructure is present which influences the ellipsometry-derived optical constants of these films. For the films on c-Si with thicknesses 50-250A, a surface roughness of about 20A is suggested. This thickness-independent structure for the best fit solutions gives us confidence that the reference dielectric function used for bulk a-Si:H is appropriate. Otherwise spurious overlayers would show up in our best fit solutions to compensate for an inappropriate reference--an effect which would be different for different a-Si:H thicknesses. The data for a-Si:H on glass point out a significantly thicker mixed composition overlayer: 50-100A.

Figure 4 depicts the type of structure that these data appear to suggest. The mixture of a-Si and SiO_2 compositions in the Table provided the best two-component fit. The three-component (a-Si+SiO_2+void) system would be expected for surface roughness coated with an oxide such that the modulation is on the order of the oxide thickness. However, the data cannot sensitively distinguish between the two- and three-component models. One should keep in

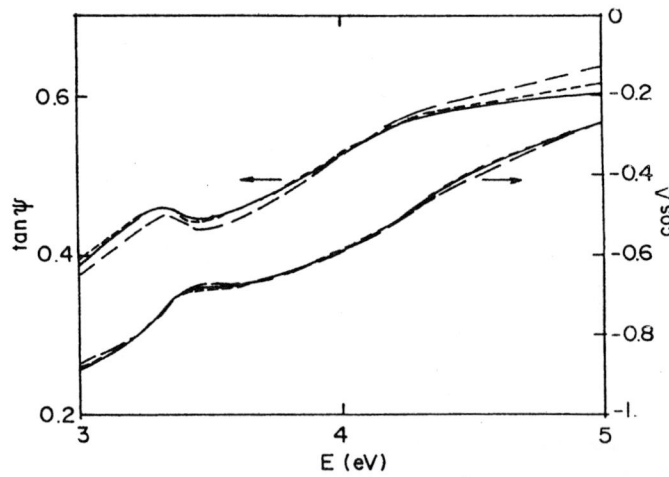

Fig. 3. Experimental (solid) and calculated tanψ and cosΔ data for the 40s a-Si:H deposition. The calculated results are based on the model of column 3, Fig. 2 (long dash) and the best fit model of column 5 (short dash).

mind that the thicknesses and volume fractions of the layers determined ellipsometrically should be used as guides since the actual structure is sure to be graded.

The results for the thinnest film on c-Si indicate that, except for a 6A layer of density deficit a-Si:H at the substrate interface, the film has been completely oxidized, possibly because it may have a more porous structure than the thicker layers. The 6A layer may provide a measure of the extent of merging of thin film growth nuclei. For the thickest film measured (400s of deposition; about 1000A) the surface structure is less pronounced and a 20A continuous oxide layer can be incorporated into the best fit model. The films on c-Si thicker than 200A gave a bulk density close to that of the reference, whereas those less than 100A exhibited at least 10% voids. This may be a reflection of the fact that in the thinner films, growth nuclei have not merged sufficiently to obtain the corresponding thick-film density.

The data summarized in Table 1, therefore, are consistent with the picture of nucleation and growth deduced from single wavelength ellipsometry studies during the deposition of CVD a-Si on Si_3N_4 substrates.[3] Apparently photovoltaic quality glow discharge a-Si:H is subject to verifiable non-uniform growth processes. The 100A average nucleation center spacing suggested in the work on CVD a-Si appears to be closer to the results for a-Si:H on glass. Table 1 indicates more uniform growth on oxidized c-Si, possibly a result of the microscopic substrate texturing which allows more closely spaced growth nuclei. Such substrate dependent effects have also been observed for ion-beam prepared a-Si:H as well.[17] For codeposited ion-beam prepared samples of thickness 0.5μm, those on oxidized c-Si were found to be consistent with <5A surface modulation, whereas those on glass were

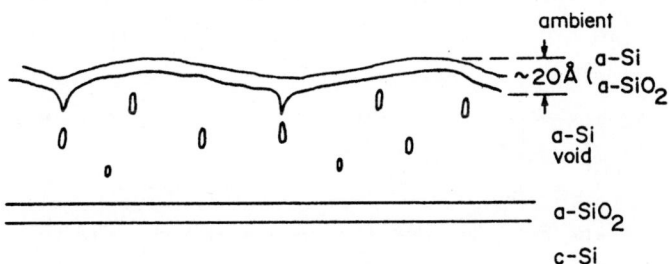

Fig. 4. Schematic surface structure on thin a-Si:H suggested by spectroscopic ellipsometry measurements. The horizontal scale cannot be determined from the measurements.

Table 3. The structural parameters, 90% confidence limits, and unbiased estimator for the best fit models to the structure of a thick n-layer deposited on quartz and a thin n-layer deposited on a-Si:H. In the latter case, the LRA established the void density in the substrate (i material) as zero, within the confidence limits.

Overlayer(s) Thickness/comp.		Bulk Thickness/composition		δ
(a) THICK N-LAYER ON QUARTZ:				
38±1A:	0.13±0.02 a-Si 0.87±0.02 oxide	0.74±0.01 a-Si 0.09±0.02 c-Si 0.17±0.03 void		0.0056
(b) THIN N-LAYER ON a-Si:H:				
95±20A:	0.50 a-Si 0.50 oxide	0.76±0.01 a-Si 0.24±0.01 void		0.012

consistent with a 25A rough layer. The results show that nucleation imposed microstructure is present on the surface of a-Si throughout its growth history.

B. The a-Si:H Based n-i Structure

A second problem of more practical interest is the study of the microstructure of the (top-most) n-layer of an a-Si:H n-i-p cell. If the n-layer was deposited under identical conditions as the i-layer by the introduction of a small flow of phosphine without interruption of the plasma, the spectroscopic ellipsometry measurements would be of more limited value. In such a case, since even heavily doped a-Si:H exhibits a dielectric function very close to that of a-Si:H at high photon energies,[9] there would be no detectable interface between the two layers. However, in cases where the top-layer is significantly different in composition or microstructure than the intrinsic layer, spectroscopic ellipsometry can be used to study its structure. The particular case where the n-layer is microcrystalline Si:H:F will be dicussed in more detail.[18]

Table 3 describes the best fit models for (a) the top surface of a 0.3μm layer of Si:H:F and (b) an n-i structure composed of an intended 150A thick, identically prepared Si:H:F atop 0.3μm conventional i-type material. In both cases, the substrate material was quartz. Figure 5 shows the pseudo-dielectric functions of the n-type material (solid) compared with that

calculated from the structure in the Table (short dash). The third
dielectric function in Fig. 5 is that calculated for the thin n-layer on
intrinsic a-Si:H using the structural parameters of Table 3. (The dielectric
function of a single unknown component of a multilayer structure can be
numerically determined as long as all other structural parameters are
known.) Two principal observations can be made on the basis of the results
in Table 3 and Fig. 5:

(1) The bulk of the thick n-layer can be well described as an effective
medium mixture of a-Si (0.74), c-Si (0.09), and void (0.17). The crystalline
silicon component improves the fit considerably (δ decreased from 0.010 to
0.0056 upon addition of c-Si). Note the increase in ϵ_1 extending to high
photon energies for the data of Fig. 5 which is absent in a-Si (see Figs.
1). The sharper feature near 3.35eV predicted from the EMA calculation is
not matched by the data, however (see Figure 5). Apparently this may
represent a breakdown in the assumption implicit in the EMA, that the
components of the mixture maintain their bulk dielectric identities. Such a
breakdown may be the case for Si crystallites 50-100Å in size.

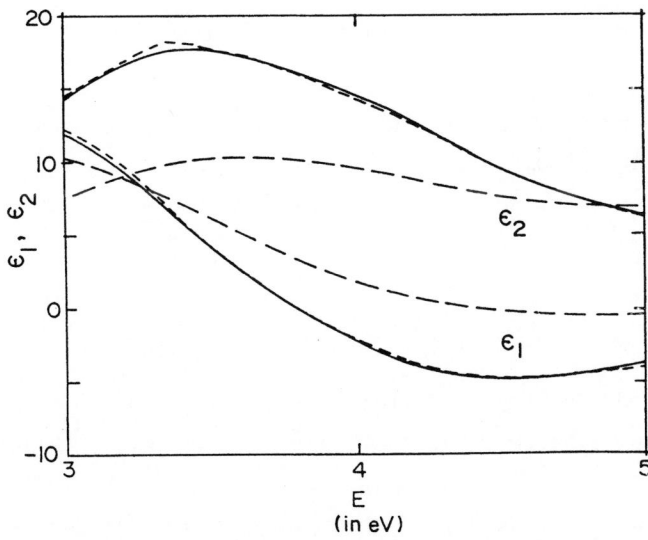

Fig. 5. The pseudo-dielectric functions calculated from
the experimental data for the thick n-layer
(solid) and from the best fit model of the first
entry in Table 3 (short dashes). Also included is
the pseudo-dielectric function for the n on i
determined numerically from the experimental data
using the thickness and i-layer dielectric
function from the best fit model of the second
entry of Table 3 (long dashes).

(2) The thin n on i shows a near surface density deficit region which is much thicker than that on the 0.3μm n-layer. This region occupies a significant fraction of the total thickness of the n-layer. As shown by the low value of the ϵ_2 peak for the thin n-layer in Fig. 5, the density deficit region clearly dominates the effective optical constants of the layer. The addition of a volume fraction of c-Si in the n-layer for the best fit in Table 3 to the structure of the n-i was attempted. Although a solution was obtained, no significant decrease in δ was obtained, and the 90% confidence limits were too large to verify the presence of a crystalline component. (In the LRA, the solution for only a limited number of free parameters can be established with confidence.)

This example demonstrates the importance of characterizing thin layer components of an a-Si:H based device in the device configuration. This is particularly crucial when the thickness of the component is of the same order as the scale of nucleation and growth imposed surface structure. Under these circumstances, the optical and electrical characteristics of the material as determined from studies of thick films should be applied to the thinner layers with care.

ACKNOWLEDGMENTS

We would like to thank Dr. S. Guha and J. Kulman of Energy Conversion Devices, Inc. for the preparation of the n-i structure and C. Nickoson for assistance in the preparation of the thin a-Si:H samples.

REFERENCES

1. J.C. Knights, J. Non-Cryst. Solids 35&36, 180 (1980).
2. R.C. Ross, A.G. Johncock, and A.R. Chan, J. Non-Cryst. Solids 66, 81 (1984).
3. F. Hottier and J.B. Theeten, J. of Crystal Growth, 48, 644 (1980).
4. B. Drevillon, J. Perrin, J.M. Siefert, J. Huc, A. Lloret, G. de Rosny, and J.P.M. Schmitt, Appl. Phys. Lett. 42, 801 (1983).
5. D.E. Aspnes, J.B. Theeten, and F. Hottier, Phys. Rev. B 20, 3292 (1979).
6. R.M.A. Azzam and N.M. Bashara, Ellipsometry and Polarized Light (North-Holland, Amsterdam, 1977).

7.   B.D. Cahan, Surf. Sci. <u>56</u>, 354 (1976).

8.   H. Arwin and D.E. Aspnes, Thin Solid Films <u>113</u>, 101 (1984).

9.   D.E. Aspnes, A.A. Studna, and E. Kinsbron, Phys. Rev. B <u>29</u>, 768 (1984).

10. D.E. Aspnes in SPIE Proc. <u>276</u>, 188 (1981).

11. I.H. Malitson, J. Opt. Soc. Am. <u>55</u>, 1205 (1965).

12. D.E. Aspnes and A.A. Studna, Phys. Rev. B <u>27</u>, 985 (1983).

13. D. Ewald, M. Milleville, and G. Weiser, Philos. Mag. B <u>40</u>, 291 (1979).

14. See, for example, W. Paul and D.A. Anderson, Solar Energy Mater. <u>5</u>, 229 (1981).

15. D.A. Radman and B.D. Cahan, J. Opt. Soc. Am. <u>71</u>, 1546 (1981).

16. D.E. Aspnes, J. Opt. Soc. Am. <u>64</u>, 812 (1975).

17. R.W. Collins and H. Windischmann, unpublished.

18. R. Tsu, S.S. Chao, M. Izu, S.R. Ovshinsky, G.J. Jan, and F.H. Pollak, J. Phys. (Paris) <u>42</u>, C4-269 (1981).

THE INFLUENCE OF DISORDER ON THE PROPERTIES OF HYDROGENATED AMORPHOUS SILICON AND RELATED ALLOYS

B. Von Roedern and A. Madan

Glasstech Solar Inc.
P.O. Box 52
Wheat Ridge, CO 80034

INTRODUCTION

Amorphous Silicon (a-Si) based alloys have attracted a considerable amount of attention because of their potential application for electophotography,[1] image pickup systems,[2] and as field effect transistors for large area display applications.[3] However, the major effort has been towards their potential use as inexpensive solar cells. Impressive progress in this field has been made since the group at University of Dundee demonstrated that a low defect, device quality hydrogenated amorphous silicon (a-Si:H) material could be produced using the radio frequency (r.f.) glow discharge in SiH_4 gas[4,5] and that the material could be doped n- and p-type[6]. These results spurred a worldwide interest in a-Si based alloys, especially for photovoltaic devices which has resulted in a conversion efficiency approaching 12%.[7] There is now a quest for even higher conversion efficiencies by using the multijunction cell approach. This necessitates the synthesis of new materials of differing bandgaps which in principle amorphous semiconductors can achieve. In this article, we review some of this work and consider from a device and a materials point of view the hurdles which have to be overcome before this type of concept can be realized.

DEFECTS IN AMORPHOUS SEMICONDUCTORS

The major distinguishing feature of amorphous semiconductors is the continuum of defect states, g(E), within the mobility gap. Tetrahedrally coordinated amorphous semiconductors, such as a-Si:H, are well described by a model which separates the distribution of states into extended, tail and deep localized states, as shown in Fig. 1. The tail states arise from deformed but intact bonds whereas the deep states are due to incorrect coordination. Upon hydrogenation, all three regimes are affected but the most pronounced effect occurs for the extended and for the deep localized states. Hydrogen tends to remove states from the top of the valence band[8] which leads to an increase in the optical bandgap, E_g, with increasing H content, C_H. In most cases the density of states, $g(E)$, around the Fermi level, E_f, is reduced by a factor of 10^2 to 10^3 from the values observed in its unhydrogenated counterpart ($10^{20} cm^{-3} ev^{-1}$); this is attributed to the passivation of dangling bonds.[9] Low defect state density $g(E_f) = 2 \cdot 10^{15} cm^{-3} eV^{-1}$ has been achieved only in the hydrogenated (glow discharge or sputtered[10]) and fluorinated[11] amorphous silicon type materials; the low values of $g(E_f)$ thus represent a reduction of defects to 1 per 10^7 atoms.

There are now three different deposition methods for the preparation of hydrogenated amorphous silicon (glow discharge (gd), sputtering (sp) or CVD of Si_2H_6), which have yielded materials with at least "fair" electronic properties and hence are viable for device applications.

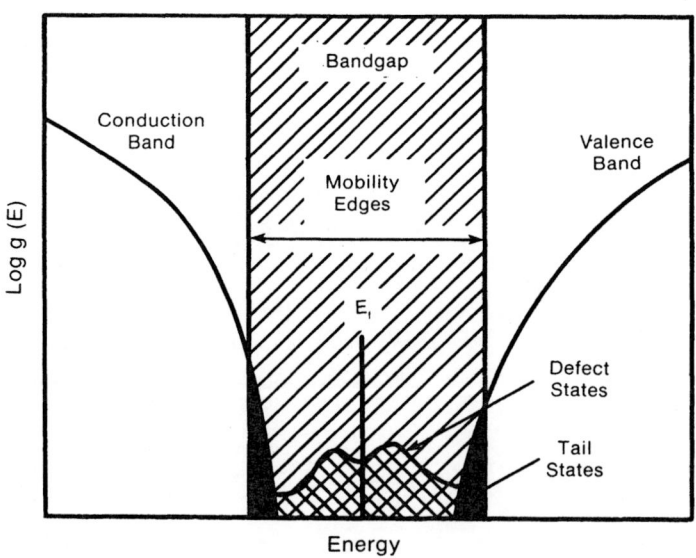

Fig. 1. Distribution of density of states, $g(E)$, in amorphous semiconductors.

Distinct differences between these materials exist and it is perhaps appropriate to reconcile these. A better understanding of the nature of defects may be helpful for the development of new amorphous semiconductor alloys, such as a-SiY:H (Y=C, Ge, Sn), since alloying adds a further independent parameter in addition to those already present for a given deposition process.

HYDROGENATED AMORPHOUS SILICON

From a study of absorption tails in gd a-Si:H materials, Cody et al.[12] concluded that the hydrogen incorporation influences only the extent of the disorder within the material and thereby determines the bandgap. As shown in Fig. 2, the Urbach tail (represented by the exponential region of the absorption coefficient, α, with energy, $h\nu$) extropolated to a common point for their sample when it was annealed at different temperatures, T which in turn is related to the hydrogen content, C_H.[13] Because of the appearance of the common intersect, it was concluded that (i) a-Si:H possessed a "standard" bandgap of 2.2eV and (ii) H induced disorder accounted for the large variations in α which is evident at the

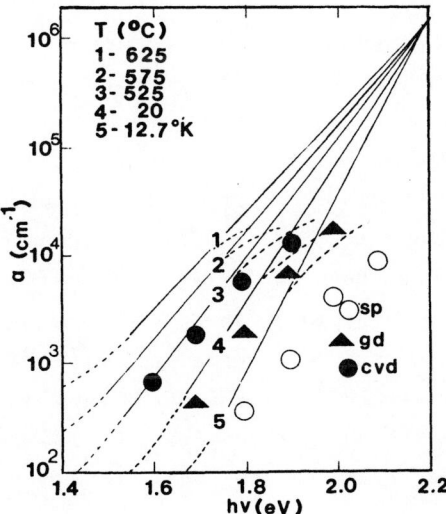

Fig. 2. Absorption coefficient, α, plotted as a function of energy, $h\nu$, for a gd a-Si:H alloy annealed at different temperatures, T and also measured at 12.7K.[12] The figure includes data for sputtered(O), CVD(●),[18] and a SERI device grade gd a-Si:H(▲).

lower energies. However, this interpretation is not unique for a–Si:H type materials since a–Si:H materials produced by other techniques, such as sputtering or CVD, do not extrapolate to a common intersect of $h\nu$ = 2.2eV as indicated in Fig. 2. It should also be mentioned that similar studies performed on a–Ge:H have also proved to be less conclusive.[14] Hence, we conclude that this approach is unable to distinguish between the alloying effect and the ordering effects which can arise from the incorporation of hydrogen. Instead, it is more appropriate that samples of similar E_g and C_H should be compared to examine the extent to which disorder influences the bandgap.

The width of the Raman TO mode (near $500cm^{-1}$ in a–Si:H and $270cm^{-1}$ in a Ge:H) can be taken as direct evidence for the degree of disorder in an amorphous material. Lannin et al.[15] and later Persans et al.[14] have shown a correlation between the width of this Raman mode and the bandgap for a limited number of samples, but unfortunately the data were not normalized to account for variations in the H content. As yet, no systematic Raman studies have been performed on samples prepared by different techniques with comparable C_H. Preliminary unpublished work from Harvard[16] has shown that sputtered samples exhibit wider Raman TO–modes than the glow discharge material; this implies that the sputtered material appears to be more disordered than the gd–or the CVD type samples and is probably due to the violence involved in the sputtering process. The increase in disorder in the sputtered material manifests itself also in the appearance of shallower Urbach edges (as deduced by the Photothermal Deflection Spectroscopy technique which measures the exponential region of α with $h\nu$ from which the parameter E_o, can be extracted). For sputtered material, E_o has been measured to be 70–80meV, whereas for gd material it was found to be smaller and in the 50–60meV range.[17]

Ellis et al.[18] have shown that samples with the same H content show a variation in the bandgap of up to 0.15eV and is dependant on the preparation technique, as shown in Table 1. The variations of the bandgaps (here, the "bandgap" E_{04} is defined as the energy where the absorption coefficient, $\alpha = 10^4 cm^{-1}$) are correlated with the widths of the vibrational modes of the Si–H features. From this, we conclude that disorder can influence the bandgap to within 0.15eV for a–Si:H materials. However, the above considerations are unable to determine whether disorder per se affects the size of the bandgap or whether the extended state distribution is altered by the different growth kinetics involved in the various types of deposition techniques.

Table 1
"Bandgap" E_{04} and Full Width at Half Maximum (FWHM) of the Si-H infrared absorption peaks of a-Si:H samples with similar H contents prepared by different techniques.

Preparation Method	T_s (°C)	C_H (at.%)	E_{04} (eV)	FWHM (630cm^{-1})	FWHM (2000cm^{-1})	FWHM (2090cm^{-1})
CVD	445	14	1.85	∿61	∿59	∿59
gd	300	10	1.92	72–78	70–78	70–72
sp	200	14	2.10	66–80	70–80	80–90

We now consider some theoretical work, based on a study of silanes of the type Si_nH_{2n+2},[19] which could correlate the bandgap with the extent of disorder. In this work, it was shown that the extent of the Si-valence band depends on the number of uninterrupted Si-Si bonds. In addition, the width of the extended states in these molecules was found to be suprisingly wide (unlike in the case of molecular C_nH_{2n+2}) and was related to the frozen-in rotomers which are caused by a low activation energy for SiH_2-SiH_2 rotations between the staggered and the eclipsed orientation. When the last argument is translated to a-Si:H, it appears that disorder could be of negligible effect on the bandgap. The more disordered sputtered a-Si:H should then possess a wider valence band (and thus a smaller E_g) than the more ordered gd or the CVD a-Si:H type materials. Since the opposite is experimentally observed, it is then conjectured that the width of the valence band is determined primarily by the number of unterminated Si-Si bonds. For a given C_H, the number of unterminated (Si-Si) bonds (or average chain-length) in any direction is smallest when the H atoms are randomly distributed rather than clustered. We therefore speculate that in the sputtered case, H is more randomly incorporated, since this is consistent with a larger band gap and wider bandwidths of the infrared features. Hence, it would seem that alloying rather than disorder determines the bandgap of a-Si:H.

Turning now to a discussion of the deep defects, correlations have been found between the bandgap and the density of states at the Fermi level, $g(E_f)$.[10] As shown in Fig. 3, $g(E_f)$ increases as E_{04} (or C_H) is decreased for samples prepared by different techniques such as glow discharge or sputtering; in this figure, we have also included the data for a binary alloy (a-SiGe:H) which will be discussed later. From these curves, it appears that disorder influences $g(E_f)$ and in the case of the more disordered sputtered a-Si:H, a larger amount of H (which leads to a

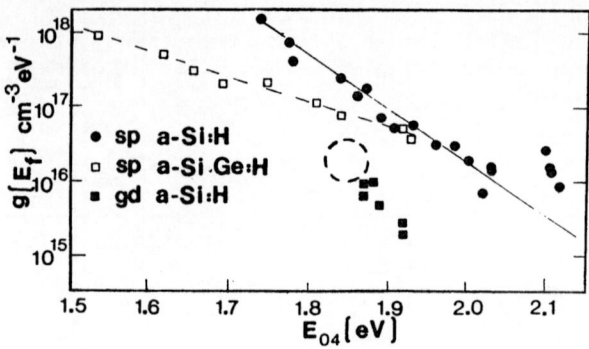

Fig. 3. Density of states at the Fermi level, $g(E_f)$, plotted as a function of "bandgap", E_{04} for sputtered(●) and gd a-Si:H materials(■); also included are the results of sputtered a-SiGe:H material(□).[10] The circle designates the expected $g(E_f)$ values for CVD a-Si:H material.

larger bandgap) is necessary to reach a certain $g(E_f)$ value in comparison with the case of the gd a-Si:H material. No corresponding such data are available for CVD produced material, but judging from other electronic parameters such as the $\mu\tau$ products (where μ is the mobility and τ is the recombination lifetime of the minority carriers) or Schottky barrier performance[18] it is felt that CVD samples produced from Si_2H_6 should possess $g(E_f)$ values in the range 1 to $5 \cdot 10^{16} cm^{-3} eV^{-1}$, thereby falling slightly above the values observed for gd a-Si:H as indicated in Fig. 3. It therefore appears that the number of deep localized states can be minimized when a small (optimum) amount of disorder is introduced into the films during deposition. The above can explain to some extent why, to date, glow discharge a-Si:H has exhibited superior electronic properties when compared with materials prepared by other deposition techniques.

HYDROGENATED AMORPHOUS BINARY ALLOYS

As mentioned previously, there is a quest for even higher conversion efficiencies utilizing the multijunction approach. This requires the synthesis of materials with different bandgaps using the alloying approach. The research has concentrated mainly on a-SiC:H and a-SiGe:H alloys, although some other materials such as a-SiN:H and a-SiSn:H have gained some attention as well. To date, all such hydrogenated binary alloys have turned out to be electronically inferior compared to the a-Si:H material, which is due to an increase in the density of states and lower $\mu\tau$ products for electrons and holes. In all cases, the only parameter which is easily adjustable is the bandgap; it can be varied continuously between that of the two constituent elements by merely adjusting the composition.

It was first proposed by Paul et al[20] that preferential bonding of H to Si rather than to Ge or to C leads to an increased number of dangling bonds on the less hydrogenated constituent of these alloys and hence could explain the deterioration in terms of electronic properties. In these alloys, preferential attachment (PA) factors typically (PA = $C_{Si}/C_Y \cdot C_{Y-H}/C_{Si-H}$ ranged from 2 to 10. However, later studies were not able to confirm a quantitative correlation between the PA values and the electronic properties, such as $g(E_f)$ and the $\mu\tau$ product of the electrons.[21] The variation of $g(E_f)$ with E_{04} of the sputtered a-SiGe:H samples shown in Fig. 3 clearly indicate that alloying actually introduces a defect reducing feature, since these alloys exhibit lowwer $g(E_f)$ values than a-Si:H material with the same bandgap.

For narrow bandgap materials, such as a-SiGe:H or a-SiSn:H, microscopic defect levels have not been identified which are responsible for the observed degradation in the electronic properties. It has been noted that the extent of the band tails in these alloys exhibit significant variations and in most cases are shallower than for a-Si:H material.[22] Only optimized a-SiGe:H materials have yielded bandtails which are almost as sharp as in the a-Si:H case (E_o=52 to 56meV).[17] These findings indicate that for the hydrogenated binary alloys, the bandtails are not limited by the intrinsic effect of mixing atoms, but rather by the disorder and/or microstructural changes that could occur upon alloying. Although the tail state distributions in a-SiGe:H alloy can be relatively sharp, nevertheless these materials possess significantly higher density of states at the Fermi level ($>10^{17}cm^{-3}eV^{-1}$ for a bandgap of 1.5eV), which can pose a considerable limitation to the solar cell performance and will be discussed later.

Some understanding has been achieved about the type of defects in gd a-SiC:H type alloys. Three-fold coordinated (graphitic) Carbon has been identified by the SERI group as the major microscopic configuration which can lead to an increase in $g(E_f)$ and also determines the extent of the tail-state distribution.[23] In brief, deposition conditions can be adjusted for the glow discharge deposition of a-SiC:H such that the graphitic bonding of C can be increased or decreased in the material. In order to force the C into 4-fold coordination, large amounts of H need to be incorporated mainly in the CH_n (n= 2,3) configuration. Satisfactory qualitative and quantitative correlations were found between the graphitic bonding and the electronic properties of the samples. Similar increases of $g(E_f)$ with increasing C content of their films have been reported[24] and presumably, these films were deposited under conditions where C is likely to be incorporated in a graphitic coordination. Thus the conclusion by Schmidt et al[24] that C introduces defects proportional to its concentration is not unique, as deposition conditions can be found which avoid the 3-fold coordinated C configuration and thus can lead to much lower defect densities in the a-SiC:H material.

SOLAR CELL CONSIDERATIONS

The three parameters determining the efficiency η of solar cells are the open circuit voltage (V_{oc}), the short circuit current (J_{sc}) and the fill factor (FF) and all are inextricably linked to basic material parameters such as E_o and $g(E)$. Experimentally, when gd a-Si:H is used in a p-i-n type configuration, the best device parameters reported thus far have been V_{oc}=0.87V, J_{sc}=18.9mAcm^{-2}, FF=0.70, and η=11.5%.[7] Since J_{sc} and FF exceed 80-90% of theoretical maximum[25] it is then appropriate to focus on the major loss mechanism in these devices, namely V_{oc} which is approximately half the band gap. It should be recognized that in crystalline solar cells such as Si, V_{oc} generally exceeds half the band gap. It is well known that the open circuit voltage is recombination limited, i.e. in crystalline semiconductors, the limitation is by band to band recombination. However, since amorphous semiconductors possess a continuous distribution of states within the gap, it is then to be expected that the recombination via these states would limit the values of V_{oc}. Tiedje[26] has estimated an upper limit of V_{oc} to be 1.1V based on the treatment given by Simmons and Taylor,[27] which considers the recombination that occurs between band tail states. Inclusion of deep level defect states in that analysis would reduce the estimate of V_{oc} even further. In general, the value of V_{oc} can be estimated for amorphous semiconductor if we consider that the density of localized states spectrum can be written in the form,

$$g(E) = g_o \cosh \{(E-E_f)/E_{ch}\} + g_{min} \qquad (1)$$

where the first term is composed of acceptor and donor type tail states and g_{min} is a constant density of states in the midgap portion and represents deep defects; E_{ch} is the characteristic energy which defines the slope of the conduction and valence band tails. (In this E_{ch} is to be associated with the extent of disorder and is related to E_o, defined earlier). Under uniform generation, steady state conditions and by considering the capture and emission of electrons, one can deduce the quasi Fermi levels for electrons and holes from which V_{oc} can be extracted. In Fig. 4, we plot V_{oc} as a function of E_g with differing E_{ch} and with $g_{min}=10^{16}cm^{-3}eV^{-1}$.[28] It should be noted that for $E_g=1.7eV$ (corresponding to a-Si:H), the upper limit to V_{oc} is 0.92V, which is lower than the estimate derived by Tiedje;[26] this is primarily due to the presence of deep defect states in the analysis, represented by g_{min} in Equation (1), whose effect will be to increase the recombination traffic for a given generation rate within the mobility gap.

From Fig. 4, it should be noted that the value of V_{oc} falls off sharply when the disorder parameter E_{ch} is increased. This is to be expected for increasingly complex alloys, where positional as well as compositional disorder is present. Therefore, an increase in J_{sc} with bandgap reductions could be more than offset by a decrease in V_{oc} due to this increased recombination traffic. It is interesting to note that this trade-off between voltage loss and increased J_{sc} has been reported for the case of unalloyed sputtered a-Si:H solar cells, where Moustakas and Maruska[29] reported that when E_g was altered from 1.71 to 1.84eV, the

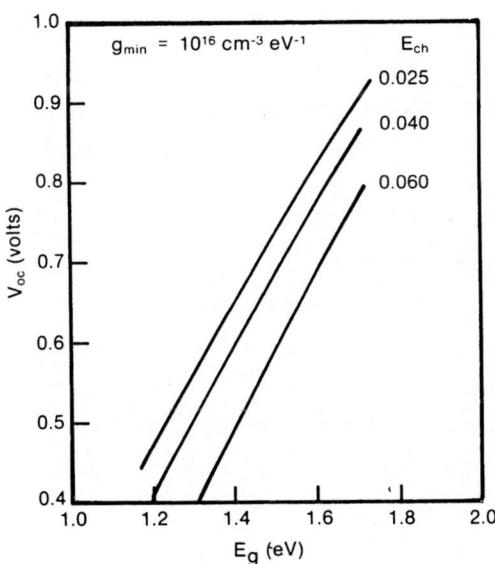

Fig. 4. V_{oc} is plotted as a function of E_g for various values of the disorder parameter, E_{ch} (eV).[28]

larger bandgap material had an overall better performance because of its larger V_{oc}. These considerations also reveal that the quest for low g_{min} values is perhaps insufficient and that attention should also be paid to the extent and width of the band tails, if the alloying techniques are to succeed.

In the context of above, we now briefly review the solar cell performance using different hydrogenated binary alloys. Solar cells have been fabricated in the p-i-n configuration with the use of SiH_4 and $Sn(CH_3)_4$ for the deposition of the a-SiSn:H intrinsic i-layer. This led to a conversion efficiency of only 2.2% with J_{sc}=8mAcm^{-2} and V_{oc}=0.60V.[30] The relatively low short circuit current density indicates a narrow space charge region and a low diffusion length for the minority carriers and is primarily due to an increase in the value of g_{min} and wider band tails. Recent work on material properties of a-SiSn:H type alloys has confirmed that significant amounts of acceptor states are introduced into the material upon alloying. This leads to a drastic decrease in the $\mu\tau$ products of electrons and holes, as well as a very significant broadening of the Urbach tail.[22] Because of these negative results the major emphasis has turned more towards the gd a-SiGe:H alloy. For them, defect densities of the order of $1-2 \cdot 10^{17}$cm^{-3}eV^{-1} at E_f have been obtained for E_g=1.5eV together with the band tails which are sharp (E_o=55meV). p-i(a-SiGe:H)-n junctions have exhibited the following characteristics: J_{sc}=17.5mAcm^{-2}, V_{oc}=0.58V, FF=0.49, η=5.0%.[31] The dark current-voltage characteristics[32] have led to a rectification ratio of 10^4 at V_{oc}=0.6V and a diode quality factor of 1.75,which is representative of recombination limited transport.[33] In this, the major limitation to the device performance also seems to be the V_{oc} since the highest voltage obtained is only about 0.6V, and is a consequence of recombination due to a large $g(E_f)$ value, as discussed above.

Attempts have also been made to use wide band gap materials in a device configuration. Using a structure p(SiC)-i(SiN)-n+, cells of efficiency 5.02% with V_{oc}= 0.8V and J_{sc}=10mAcm^{-2} have been reported.[30] One of the major losses, once again, is in the V_{oc} since this voltage is far below the voltage expected for a material with a bandgap of 2.0eV.

We next consider the potential of tandem type junctions for efficient conversion of sunlight. In Table 2a, we first cite the performance characteristics of single and tandem junctions. It should come as no surprise that the efficiency of tandem cells do not surpass the single cell efficiency with the currently available materials.

Table 2(a): Experimental results of single and tandem junctions.

Type	E_{g1} (eV)	E_{g2} (eV)	V_{oc} (V)	J_{sc} (mAcm^{-2})	FF	η (%)	Ref.
Single	1.7	---	0.85	18.5	0.73	11.5	(7)
Single	1.5	---	0.58	17.5	0.49	5.0	(31)
Tandem	1.7	1.4	1.40	9.7	0.57	7.7	(34)
Tandem	2.2	1.7	1.65	7.0	0.61	7.0	(30)

Table 2(b): Hypothetical conversion efficiency of single and tandem junctions.

Type	E_{g1} (eV)	E_{g2} (eV)	V_{oc} (V)	J_{sc} (mAcm^{-2})	FF	η (%)	Ref.
Single	1.7	---	1.10	21.3	0.80	18.7	(25)
Tandem	1.7	1.0	1.56	21.3	0.80	26.6	(25)
Tandem	2.245	1.7	2.64	10.8	0.80	22.8	(25)

In Table 2b, we reproduce here[25] the ultimate performance of single and two stacked configurations when and if better materials are synthesized. It should be recognized that the results in the table were derived assuming total absorption and based on the recombination arguments given earlier. In principle, various combinations of the bandgaps are possible when the thicknesses of the constituent cells are altered. Also, it should be emphasized that the performance of the cells is critically dependent on the spectral content which in turn is dependent on location, time of day etc.[35] and could affect the tandem junction performance severely. Finally, stacked cells using the same bandgap (such as for the a-Si:H case) but with cells of different thickness could be of interest in circumventing the small instabilities which have been noted in amorphous silicon type alloys.

CONCLUSIONS

We have made an attempt to outline some of the limitations imposed by disorder on the solar cell performance of hydrogenated amorphous semiconductor materials. There are substantial indications that V_{oc} is limited by recombination within the tail states which in turn is determined by the disorder and possibly by structural or electrical inhomogeneity due to variations of the H-content or by the effective electronic compensation. The midgap density of states affects mainly J_{sc} and FF via the decrease of the depletion width and the minority carrier diffusion length and to some extent V_{oc}.

In the future, the challenging frontier of materials research in amorphous semiconductors is the development of materials with sharp tail states distributions. This problem is far from trivial, as we have outlined above. The steepest tails can be expected in material with the least amount of disorder, whereas the deep localized states might be minimized when "some" additional disorder is present. Conceptually, optimized materials will be achieved more easily in unalloyed a-Si:H or a-Ge:H than in hydrogenated binaries. In addition, device engineering may circumvent some of the limitations encountered with the non-optimized materials.

REFERENCES

1. I. Schimizu, T. Komatsu, K. Saito and E. Inoue, J. Non. Cryst. Solids 35 & 36, 773, (1980).
2. Y. Imamura, S. Ataka, Y. Takasaki, C. Kusana, T. Hirai and E. Maruyama, Appl. Phys. Lett. 35, 349 (1979).
3. P.G. LeComber, W. Spear and A. Ghaith, Electron Lett., 15, 179 (1979).
4. W.E. Spear and P.G. LeComber, J. Non Cryst. Solids, 8-10, 727 (1972).
5. A. Madan, P.G. LeComber and W.E. Spear, J. Non Cryst. Solids, 20, 239 (1976)
6. W.E. Spear and P.G. LeComber, Phil, Mag. 33, 935 (1976).
7. S. Nakano, H. Kawada, T. Matsuoka, S. Kiyama, S. Sakai, K. Murata, H. Shibuya, Y. Kishi, I. Nagaoka and Y. Kuwano, Technical Digest of the International PVSEC-1, Kobe, p.584 (1984).
8. B. von Roedern, L. Ley, M. Cardona and F.W. Smith, Phil. Mag. B, 40, 433, (1979).
9. A.J. Lewis, G.A.N. Connell, W. Paul, J.R. Pawlik and R.J. Temkin: Tetrahedrally Bonded Amorphous Semiconeductors, (AIP Conference Proc. No. 20) 27 (1974).
10. See for instance R.L. Weisfield, J. Appl. Phys. 54, 6401 (1983).
11. A. Madan, S.R. Ovshinsky and E. Benn, Phil. Mag. 40, 259 (1979).
12. G.D. Cody, T. Tiedje, B. Abeles, B. Brooks and Y. Goldstein, Phys. Rev. Letters, 47, 1480 (1981).
13. E.C. Freeman and W. Paul, Phys. Rev. B 18, 4288 (1978).
14. P.D. Persans, A.F. Ruppert, S.S. Chan and G.D. Cody, Solid State Commun., 51, 203 (1984).
15. J.S. Lannin, L.J. Pilione, S.T. Kshirsagar, R. Messier, and R.C. Ross, Phys. Rev. B 26, 3506 (1982).
16. E. Mytilineou, Harvard Univ., Private Communication.
17. A.H. Mahan, Private Communication.
18. F.B. Ellis, R.G. Gordon, W. Paul and B.G. Yacobi, J. Appl. Phys. 55, 4309 (1984).
19. H. Bock, W. Ensslin, F. Feher, and R. Freund, J. Am. Chem. Soc., 98, 668 (1976).
20. W. Paul, D.K. Paul, B. von Roedern, J. Blake and S. Oguz, Phys. Rev. Lett., 46, 1016 (1981).
21. W. Paul, B.G. Yacobi and R. Weisfield, Final SERI Subcontractor Report, XW-r-9358-1-6, (1983).

22. B. von Roedern, A.H. Mahan, D.L. Williamson and A. Madan; to be published in the Proceedings of the Fifth Symposium on Materials and New Processing Technologies for Photovoltiacs, New Orleans, (1984).
23. A.H. Mahan, B. von Roedern, D.L. Williamson and A. Madan; to be published in J. Appl. Phys, (1985).
24. M.P. Schmidt, I. Solomon, H. Tran-Quoc, J. Bullot, M. Gauthier and P. Cordier; to be published in Phil. Mag. B (1985).
25. See for instance, A. Madan in Silicon Processing for Photovoltaics; eds. K.V. Ravi and C.P. Khattak (North Holland), in press.
26. T. Tiedje, Appl. Phys. Lett. 40, 627 (1982).
27. J.G. Simmons and G.W. Taylor, Phys. Rev. B4, 502 (1971).
28. A.H. Mahan, A. Sanchez, D.L. Williamson, B. von Roedern and A. Madan, Porc. of the 17th IEEE Photovoltaic Specialist Conference, p.92 (1984)
29. T.D. Moustakas and H.P. Marsuka, Appl. Phys. Lett., 43, 1037 (1983).
30. Y. Kuwano, M. Ohnishi, S. Nishiwaki, T. Tsuda, T. Fukatsu, K. Enomoto, Y. Nakashima and H. Tarui, Proc. 16th IEEE PV Specialist Conf., p.1338 (1982).
31. Mitsubishi Electric Company, Proc. 16th Solar Energy Committee and 10th Amorphous Conference, Tokyo, Keidanren-Kailcan (1984).
32. B. von Roedern, A.H. Mahan, T. McMahon and A. Madan; to be presented at the MRS Symposium on Materials Issues in Application of Amorphous Silicon Technology, (San Francisco, April 1985).
33. T.J. McMahon, B.G. Yacobi, K. Sadlon, J. Dick and A. Madan, J. Non Cryst. Solids, 66, 375 (1984).
34. G. Nakamura, K. Sato and Y. Yukimoto, Proc. 16th IEEE PV Conf., p. 1331 (1982).
35. I. Chambouleyron, Solar Cells, 12, 393 (1984).

GROWTH AND CRYSTALLIZATION MECHANISM OF MICROCRYSTALLINE SILICON

FILMS PRODUCED BY REACTIVE RF SPUTTERING

T. D. Moustakas

Exxon Research and Engineering Company
Corporate Research Science Laboratories
Route 22 East
Annandale, New Jersey 08801

INTRODUCTION

Hydrogenated microcrystalline silicon is a mixed-phase material consisting of submicron size silicon crystallites embedded into an amorphous matrix[1]. This new phase material has received increased attention because of its potential application in thin film devices[2,3]. Such material was first prepared by Veprek and his co-workers[4-6] by hydrogen-plasma assisted silicon transport. In this method, the silicon charge, kept at relatively low temperatures, reacts with atomic hydrogen and forms volatile hydrides, which are transported and decomposed on a substrate, kept at higher temperatures. More recently hydrogenated microcrystalline silicon has been prepared by high power glow discharge decomposition of silane, diluted in hydrogen[7,8]. Material produced by these two methods has been investigated by a number of groups and models regarding the kinetics of growth have been proposed[5,8]. Limited progress has also been made in the deposition of such material by the method of sputtering[3,9,10,11]. However, the mechanism of thin film growth and crystallization has only been briefly addressed[11].

In this paper we report on a systematic study of the structure of microcrystalline and amorphous silicon films produced by reactive RF sputtering. Based on these data, a mechanism of film growth and crystallization is proposed.

EXPERIMENTAL METHODS

The films were deposited in an RF diode sputtering system employing a water-cooled polycrystalline silicon target, 13 cm in diameter and 0.64 cm thick. The anode assembly, 65 cm in diameter, was held at ground potential and a portion of it under the target was heated radiatively. The system was pumped by a combination of a turbomolecular and a mechanical pump to a base pressure of 1×10^{-7} Torr and total leak rate of approximately $1-2 \times 10^{-5}$ Torr $.1.s^{-1}$. The flow rates of the gases (Ar and H_2) were determined with calibrated flowmeters. The total pressure of the two gases was monitored with a capacitance manometer, while the

of the two gases was monitored with a capacitance manometer, while the partial pressures of each was measured with a differentially pumped mass spectrometer.

The film growth and crystallization mechanism was investigated by studying the structure of three series of films which were deposited with the partial pressure of hydrogen, the substrate temperature and the target voltage as the variables. This parameter space is expected to affect the kinetics of thin film growth and the mechanism of crystallization the most. Details of the deposition parameters of the three series of films are given in Table 1. In the first series of films, the only variable was the substrate temperature which was varied from 40°C to 550°C. In the second series the substrate temperature and the target induced-voltage were kept constant and the partial pressures of hydrogen and argon were varied while maintaining a constant total sputtering pressure. In this series of films, the ratio of $PH_2/(PH_2 + P_{Ar})$ was varied from 0.1 to 0.95. Finally, in the third series, the only variable was the target voltage which was varied from −800 volts to −2000 volts. The thicknesses of the investigated films was varied from 200Å to 3μm depending on the type of measurement.

The structure of the films was probed by X-ray diffraction, transmission and scanning electron microscopy and infrared vibrational spectroscopy. These data were also supported by Raman spectroscopy studies, which were published earlier[11]. X-Ray diffraction was used to probe the size and orientation of the crystallites. Films for these measurements were deposited on fused quartz substrates at thicknesses of 1.0 to 3.0μm and studied with Cu-Kα radiation. Scanning electron microscopy (SEM) was used to characterize the surface topography and cross-sectional structure of the films. Cross-sections of the films were exposed by scribing the rear of the substrate (fused quartz or 7059 Corning glass) and cracking the sample. Transmission electron microscopy was used to study the early stages of the film structure by depositing 200Å to 800Å thick films on copper grids, coated first with 50Å of carbon or silica. Infrared vibrational spectroscopy

Table 1. Deposition Parameters of the Investigated Films

Sample	P_{H_2}(mT)	P_{Ar}(mT)	$P_{Ar}+P_{H_2}$(mT)	T_S(°C)	$-V_T$(Volts)	Thickness(μm)
701	28	14	42	40	1200	3.2
689	28	14	42	100	1200	1.21
688	28	14	42	120	1200	1.1
682	28	14	42	150	1200	1.7
681	28	14	42	200	1200	1.7
678	28	14	42	275	1200	1.6
683	28	14	42	330	1200	1.7
684	28	14	42	425	1200	1.4
696	28	14	42	475	1200	1.5
697	28	14	42	550	1200	1.2
680	5	37	42	280	1200	2.6
679	17	25	42	280	1200	2.7
678	28	14	42	280	1200	1.6
677	34	8	42	280	1200	0.8
690	37	5	42	280	1200	1.6
675	40	2	42	280	1200	0.25
687	28	14	42	330	800	0.71
683	28	14	42	330	1200	1.7
686	28	14	42	330	1500	1.9
685	28	14	42	330	2000	1.55

was used to study the nature of the hydrogen in the matrix. The films for this measurement were deposited on high purity crystalline silicon.

EXPERIMENTAL RESULTS

Films Produced as a Function of Substrate Temperature

The dependence of the deposition rate on the substrate temperature is shown in Figure 1. Note that the film growth rate is independent of substrate temperature from 40°C to 550°C.

Figure 2 shows the X-ray diffraction spectra of representative samples described in Figure 1. Besides the broad peak at the shorter

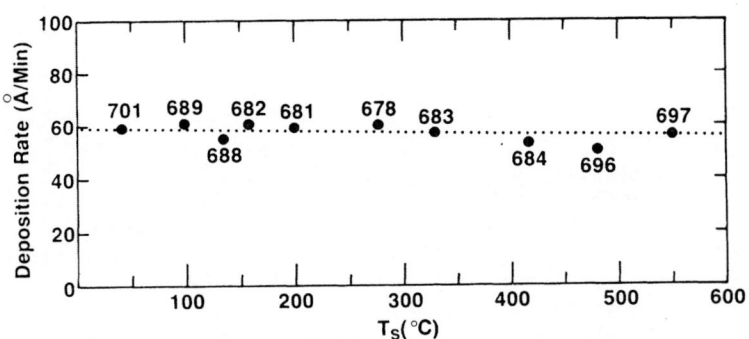

Figure 1. Deposition Rate vs. Substrate Temperature

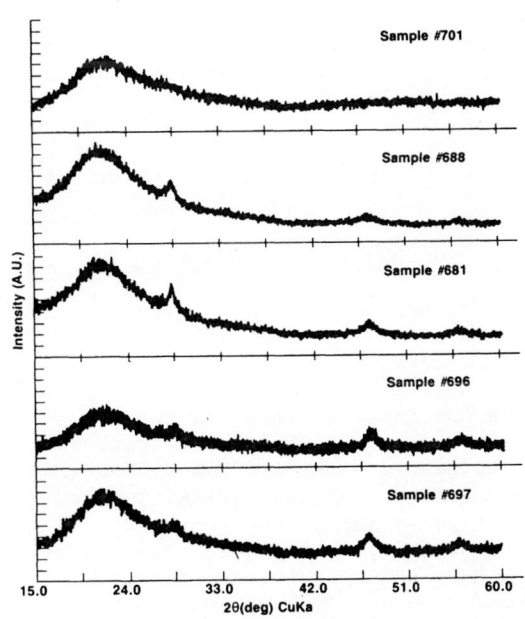

Figure 2. X-Ray Diffraction Spectra of Films Produced at Different Substrate Tempertures

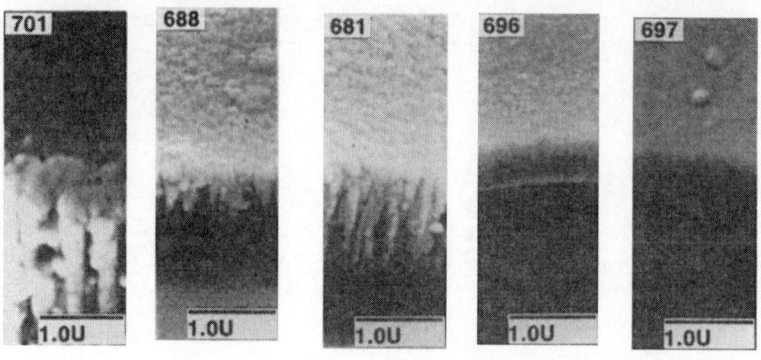

Figure 3.　SEM Surface Topographies and Cross-Sectional
Structures of Films Described in Figure 2

angles, due to diffraction from the fused quartz substrate, some of the
spectra exhibit relatively sharp peaks corresponding to the (111), (220)
and (311) orientations of the crystalline silicon powder. From the
relative intensity of the different peaks, one can deduce the preferred
orientation of the sample grains and from their half-widths the average
grain size. Note that samples produced at 40°C (sample #701) do not
show evidence of crystallinity, while those produced at temperatures
higher than 100°C show evidence of crystallinity. The data also suggest
that the (111) orientation is preferred for samples grown at low
temperatures and the (220) orientation is preferred for samples grown at
high temperatures.

Figure 3 shows the SEM surface topographies and cross-sectional
structures of the same films described in Figure 2. Note that the
growth habit in these films is columnar and that the columnar microstructure becomes less pronounced as the substrate temperature
increases. Based on the size of the columns, these data reveal three
regimes of microstructures. At temperatures below 100°C, the size of
the columns is about 2000 to 3000 Å. At temperatures between 100 to
300°C, the columns become finer (~500Å). Finally, at temperatures above
400°C the columnar morphology is not discernible.

Films Produced as a Function of $PH_2/(PH_2 + P_{Ar})$

The dependence of the deposition rate on the ratio of $PH_2/(PH_2 +
P_{Ar})$ is shown in Figure 4. Note that the deposition rate decreases
monotonically as this ratio increases.

Figure 5 shows the X-ray diffraction spectra of representative
samples described in Figure 4. These data indicate that all films,
including the ones produced with only 10% hydrogen in the discharge
(sample #680), show evidence of crystallinity. In addition, the films
produced at low or high values of $PH_2/(PH_2 + P_{Ar})$ show a pronounced
(111) preferred orientation, while those produced at intermediate values
of $PH_2/(PH_2 + P_{Ar})$ have their grains at random orientations. The
average grain size for sample #679 was determined from the half-width of
the peak (111) to be 140Å.

Figure 6 shows the SEM surface topographies and cross-sectional structures of the same films described in Figure 5. Note that the films produced at low and high values of $PH_2/(PH_2 + P_{Ar})$ show a strong columnar and rough surface morphology, while those produced at intermediate values of $PH_2/(PH_2 + P_{Ar})$ show a less pronounced columnar morphology.

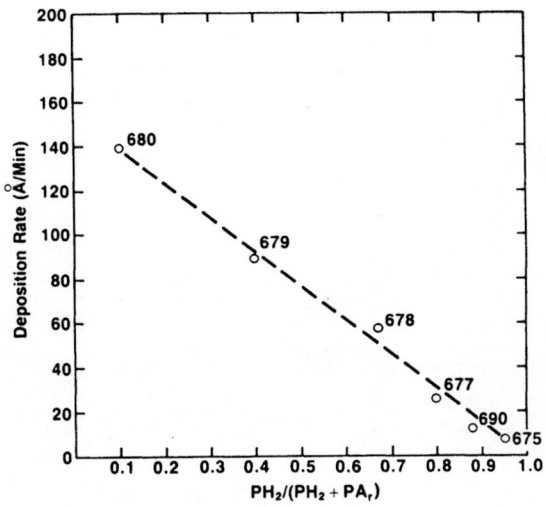

Figure 4.   Deposition Rate vs. $PH_2/(PH_2 + P_{Ar})$

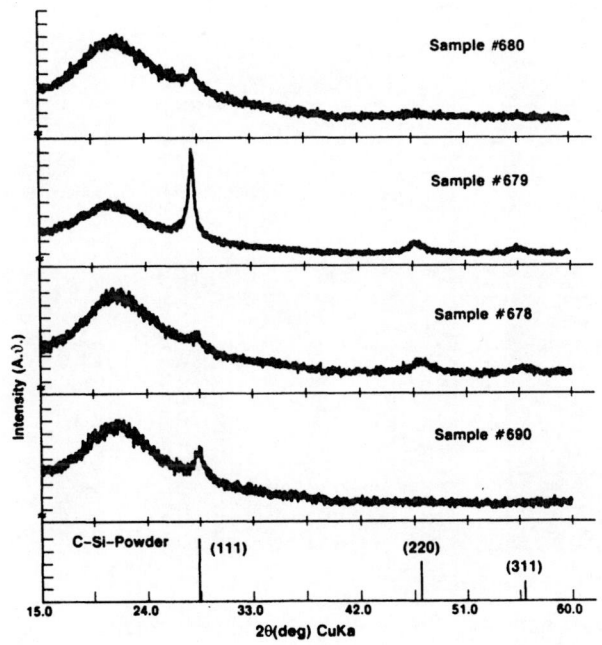

Figure 5.   X-ray Diffraction Spectra of Some of the
Samples Described in Figure 4

Figure 7 shows the bright field TEM image and the electron diffraction pattern for three samples produced under identical conditions as sample #679. The three films differ only in their thickness. Electron diffraction reveals that the black spots in the bright field images are crystallized, while the rest of the film is amorphous. Note that the crystallites are confined to individual islands and that the number of crystallites increases with the thickness of the film.

The infrared spectrum of Si-H stretching vibrations for sample #679 is shown in Figure 8. This spectrum shows sharp absorption bands at 2100 cm^{-1}. These data differ significantly from the spectra of hydrogenated amorphous silicon[12].

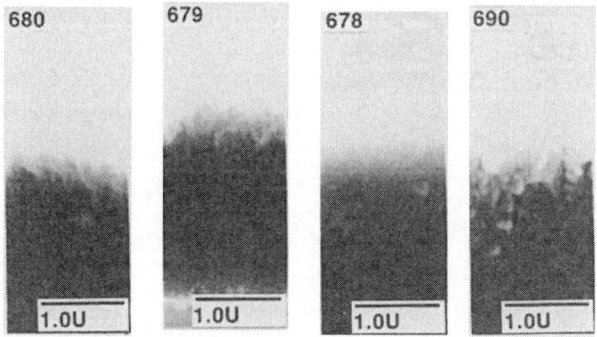

Figure 6. SEM Surface Topographies and Cross-Sectional Structures of the Films Described in Figure 5

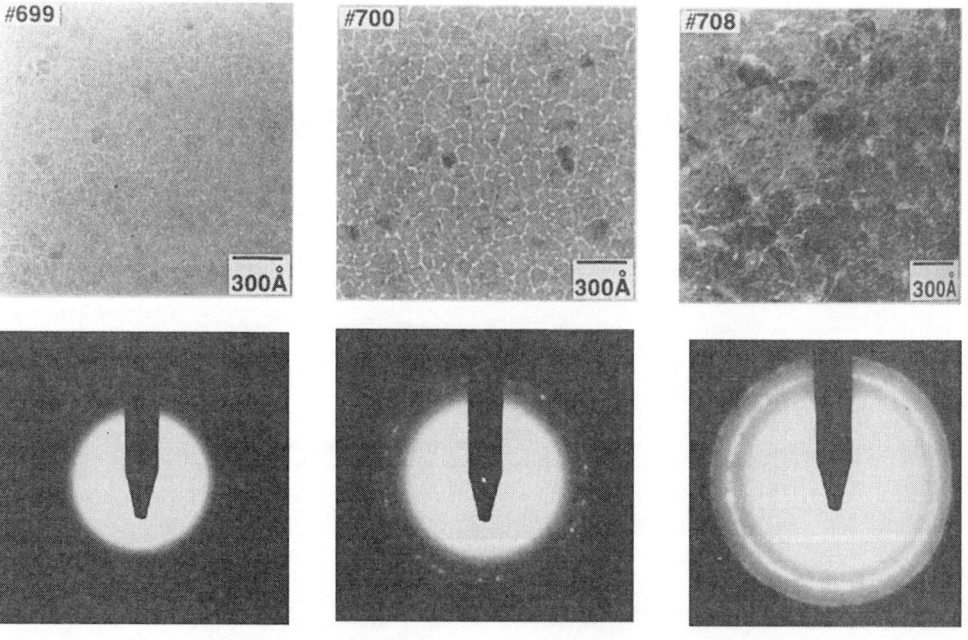

Figure 7. Bright Field TEM Images and Electron Diffraction Patterns of Three Films Produced Under Identical Conditions as Sample #679 with their Thickness as the Only Variable. Sample #699 is 200Å Thick, Sample #700 is 400Å Thick and Sample #708 is 800Å Thick.

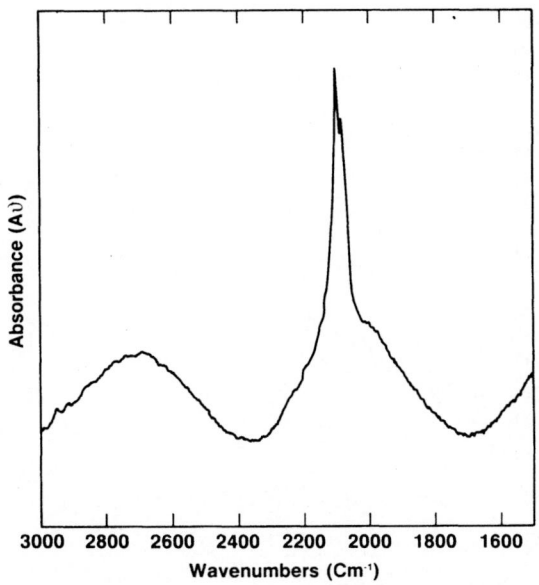

Figure 8. The Si-H Stretching Vibrations for Sample #679

Films Produced as a Function of Target Voltage

The dependence of the deposition rate on the target induced DC voltage is shown in Figure 9. The growth rate increases monotonically with the target voltage.

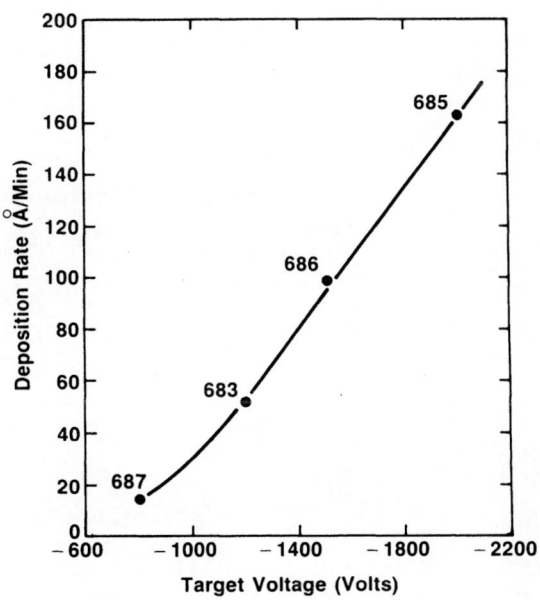

Figure 9. Deposition Rate vs. Target Voltage

Figure 10 shows the X-ray diffraction spectra of the samples described in Figure 9. According to these data, crystallization is more pronounced on films produced at higher target voltages.

Figure 11 shows the SEM surface topography of the films described in Figures 9 and 10. Films produced at low and high target voltage have rough surface morphology, while those produced at intermediate target voltage (sample #683) have much smoother surface morphology.

Totally Amorphous Silicon Films

For comparison with the data presented previously, we show in Figure 12 electron microscopy data of a totally amorphous silicon film. Sample #710 was deposited in an atmosphere of 4.5 mT of argon and 0.5 mT of hydrogen at 325°C and target voltage of −1000 volts. These data show structureless TEM image and no evidence of crystallization.

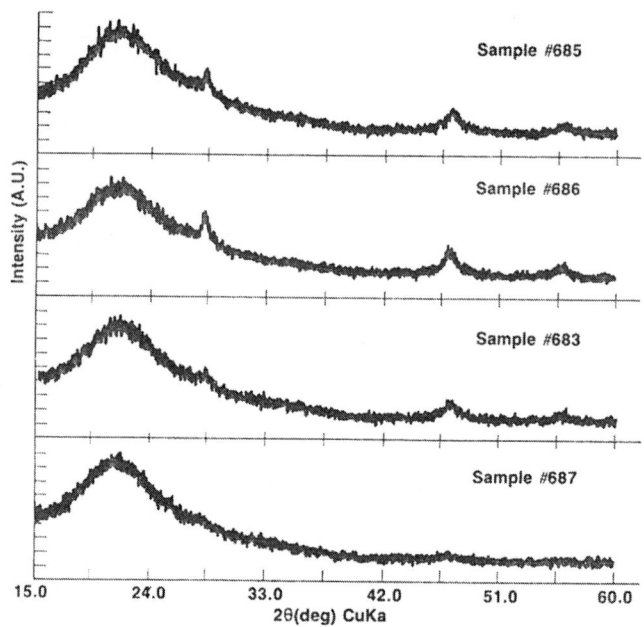

Figure 10. X-Ray Diffraction Spectra of the Samples Described in Figure 9

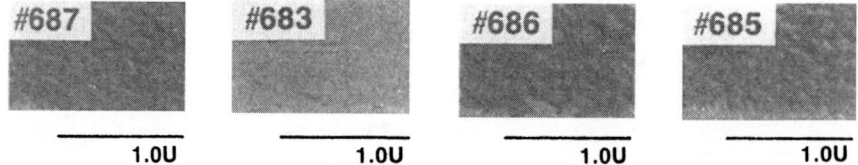

Figure 11. SEM Surface Morphology of the Samples Described in Figure 10

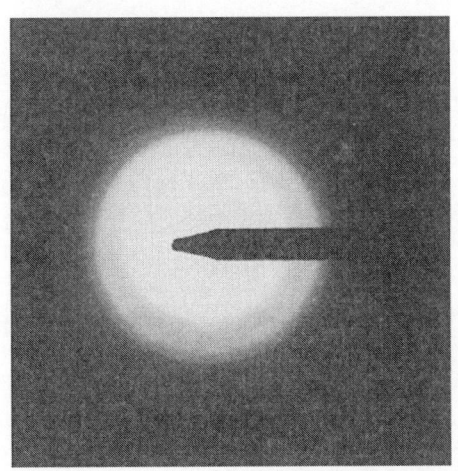

Figure 12. Bright Field TEM Image and Electron Diffration
Pattern for an Amorphous Film Produced as
Described in the Text.

MODEL OF FILM GROWTH AND CRYSTALLIZATION

The growth rate studies reported in Figures 1, 4 and 9 indicate
that the film growth and the mechanism of hydrogen incorporation, under
the investigated deposition parameter space, can be accounted for by
classical reactive sputtering[12,13] rather than by H-plasma transport
phenomema[4]. In reactive sputtering, reactions between the target
material and the reactive gas occurs either at the surface of the substrate or at the surface of the target or both. Reactions in the gas
phase are considered unfavorable because the simultaneous conservation
of energy and momentum requires the presence of a third body. At low
reactive gas partial pressure and high target sputtering rate, virtually
all of the compound formation occurs at the substrate, whereas in the
opposite regime, high reactive gas partial pressure and low target
sputtering rate, the compound formation occurs at the target.

The films of the first series were deposited under P_{H_2}/P_{Ar} = 2 and
relatively high sputtering rate (Figure 1). Therefore, one would expect
that the Si-H compound formation would occur both at the target and at
the substrate. However, the lack of temperature dependence in the
deposition rate suggest that the etching of the film by the hydrogen
plasma is not significant[12].

The films of the second series were deposited under P_{H_2}/P_{Ar} values
varying from 0.1 to 20. This suggests that the Si-H compound formation
in this series of films moves progressively from the substrate to the
target. In the latter case, the target surface is transformed into
disordered silicon hydride. The physical sputtering by Ar-ions from
such a surface would be slower, since, in general, compounds have higher
secondary electron emission yields than metals. The alternative model
in which the film growth is due to H-plasma silicon transport from the

target to the substrate predicts an increase in the deposition rate with $PH_2/(PH_2 + P_{Ar})$, contrary to the experimental results of Figure 4. Thus, although some H-plasma etching of the growing film and target is unavoidable, it appears that physical reactive sputtering is the major mechanism of film growth.

The films of the third series were deposited under $PH_2/P_{Ar} = 2$ and variable deposition rate. At low target voltage, the target sputtering rate is quite low and thus reactions between Si and H at the surface of the target will probably dominate. At high target voltage, the target sputtering rate is fast enough that the most of the H and Si reactions will take place at the substrate. At intermediate target voltages, Si-H compound formation occurs both at the surface of the target and substrate.

The microstructure of the films, reported in Figures 3, 6, and 11, is discussed next. Such microstructures have been observed in many materials produced by physical vapor deposition techniques and led to the development of the structure zone model[14-16]. This model relates the film's microstructure to the ratio of substrate temperature to the melting point of the evaporant (Ts/Tm). Three characteristic regimes (zones) have been recognized. At low values of Ts/Tm (<0.3), there is little adatom mobility and every atom tends to stick where it arrives. In this case if the arriving flux has an oblique component the high points on the growing film receive more coating flux than do the valleys. This leads to the growth of isolated columns (zone 1). Such microstructure is, in general, enhanced by sputtering at high gas pressure, since gas scattering induces an oblique component in the arriving coating flux. At intermediate values of Ts/Tm (0.3-0.5) the thin film growth is dominated by surface diffusion. In this regime the adatoms diffuse over the substrate surface and join, during the initial stages of growth, with other adatoms to form islands. These islands form columns during the subsequent growth (zone 2). A transition regime (zone T) between zone 1 and zone 2 was also reported.[15,16] In this regime the films are subjected during the growth to bombardment by neutral or charged particles, which results in partial coalescence of the islands. These films show a fibrous structure, which can be viewed as the internal structure of the larger columns. Finally, at higher values of Ts/Tm (>0.5) (zone 3) the thin film growth is dominated by bulk diffusion leading to structures with interpenetrating crystal grains.

The microstructures reported in Figure 3 can be qualitatively accounted for in the structure zone model. All films were deposited at relatively high gas pressure (42 mTorr). Sample #701, which was deposited at Ts = 40°C, shows large isolated columns, a characteristic example of zone 1 structure. The size of the columns becomes progressively smaller for sample #688 and #681, which were deposited at Ts = 120 and 200°C respectively. Finally, the samples #696 and #697, which were deposited at Ts/Tm ~0.5 show evidence of pebble-like structure. The microstructures reported in Figure 6 are much more difficult to understand. At the ends of the spectrum of $PH_2/(PH_2 + P_{Ar})$, the films are strongly columnar, while at intermediate values of $PH_2/(PH_2 + P_{Ar})$ the films are less columnar. This could be the result that different mixtures of argon and hydrogen lead to varying amounts of energetic ion bombardment of the growing film. For example, it has been shown that the substrate self-bias voltage depends on the ratio of H/Ar in the discharge[17] and that the amount of atomic hydrogen decreases in the absence of argon[18]. The microstructures of Figure 11 are also consistent with the structure zone model. All films reported in this figure were deposited at an average Ts/Tm which corresponds to the zone 2 regime. Therefore, all these films should grow columnar and have rough surface morphology. The observed morphologies in Fig. 11 are attributed to charged particle bombardment. At low target voltage such bombardment is negligible and the films grow columnar (sample #687). At intermediate target voltages the charged particle bombardment is sufficient to induce partial coalescence of the islands (sample #683). Finally, at very high target voltage the charged particle bombardment is so strong that the growth is dominated by bulk diffusion (zone 3) and such films have rough surface morphology (samples #686 and #687).

Our observations regarding the crystallinity of the investigated films can be summarized as follows: a) All films consist of a mixed phase material of silicon crystallites embedded into an amorphous matrix. b) The film growth proceeds via island formation with the crystallites confined to individual islands. c) The crystalline volume fraction increases with the thickness of the film. Beside the data in Figure 7, which indicate that the number of crystallites increases with the thickness of the film, we have reported previously [11] that Raman Spectra of thick films (> 0.5μm) indicate that the crystalline volume fraction of our films is between 30 to 60%. d) Crystallization depends on the substrate temperature. Films produced at substrate temperatures below 100°C are amorphous, while those produced at temperatures between 100 to 550°C are partially crystallized. e) The lattice orientation appears to correlate with the microstructure of the films. In columnar films the (111) lattice orientation is perferred, while in films with fibrous microstructure the crystallites are randomly oriented with some preference of the (220) orientation. f) The size of the grains is less than 200Å in all investigated films.

These data can now be accounted for as follows: All investigated films are deposited initially as amorphous layers, which are partially crystallized during the subsequent growth of the films to their final thickness. A similar observation has been reported for LPCVD silicon films deposited at 591°C.[19] Such solid phase crystallization is governed by the rates of nucleation and crystal growth during the film deposition. For example, the film will be totally amorphous if the nucleation rate is very low and the crystallization growth rate is lower than the deposition rate. The fact that our films are partially crystallized suggest that either the rate of solid phase crystallization is larger than the rate of film growth or that the nucleation rate is very high.

The nucleation and recrystallization of amorphous silicon films, produced either by e-gun evaporation or by LPCVD, have previously been investigated.[20-22] These studies have shown that there is a characteristic time for the nucleus to reach a critical size (incubation time) before crystallization can take place. The incubation time was estimated to be of the order of days for films produced at temperatures less than 550°C. the activation energy of such solid phase crystallization was found to be about 2.5 eV and the crystallization rate much less than 10Å/min at temperatures below 550°C,

The films reported in the present study were deposited at rates much higher than the crystallization rates mentioned previously. This suggests that the partial crystallization of our films can only be accounted for by high nucleation rates. However, the nucleation mechanism in our films should be diferent than that observed in evaporated films, since even 200Å films show evidence of crystallinity. This suggests that, under the preparation conditions used in our studies, crystalline nuclei with the required critical size exist from the beginning of the growth. We postulate that these nuclei are the islands shown in Figure 7. In these islands, whose lateral dimensions are about 100Å, the internal strain, due to bond angle and dihedral disorder, is by and large relieved at the perimeter of the islands. Thus, the material within the islands may have a better short range order, which could result in the formation of stable crystalline nuclei. This argument is supported by the Raman data,[11] which show that the amorphous component of the scattered intensity is shifted to higher frequencies, an evidence that the amorphous part of the micro-crystalline silicon has better order than the pure amorphous silicon. Therefore, the nucleation starts at the surface of the substrate (heteroeneous) and the crystallization is an epitaxial-like growth along the columns. If the islands have lateral dimensions of a few thousand Angstroms, as for example in Sample #701 of Figure 3, the material within the island is strained and it cannot form stable crystalline nuclei. The same arguments also apply to the totally amorphous silicon film (Figure 12) in which the coalesence of the islands has led to a highly strained material.

The observed preferred orientation of the grains in the (111) direction in columnar films is intuitively expected, since growth in the (111) direction is faster.[23] The sharp peaks of the S-H stretching vibrations suggests that the hydrogen is attached to the crystallites. The termination of the islands by hydrogen could facilitate the low temperature crystallization by prohibiting the coalescence of the islands to larger ones. At temperatures higher than 550°C (hydrogen evolution) the solid phase crystallization will be dominated by crystallization growth rates rather than nucleation rates.

CONCLUSIONS

The kinetics of growth and the mechanism of low temperature crystallization of hydrogenated microcrystalline silicon films produced by reactive RF sputtering in a plasma containing argon and hydrogen have been investigated. Growth rate studies indicate that the film growth and the mechanism of hydrogen incorporation is consistent with classical reactive sputtering rather than H-plasma silicon transport. Structural studies reveal that the film's microstructure is consistent with the structure zone model.[15] The film growth proceeds via amorphous islands formation and subsequent crystallization of some of these islands during the growth of the film. The re-crystallization at low temperatures is dominated by nucleation rather than crystallization growth rates.

ACKNOWLEDGEMENT

I wish to express my gratitude to Sir Nevill Mott for the help he provided me in a number of areas of my work, and also for the inspiration and excellence he has contributed to our field.

REFERENCES

1. S. Veprek, Z. Iqbal and F. A. Sarott, Philos. Mag. B, 45, 137 (1982).
2. Y. Uchida, T. Ichimura, M. Ueno, and H. Haruki, Jap. J. Appl. Phys. 21, L586 (1982).
3. T. D. Moustakas, H. P. Maruska and R. Friedman, J. Appl. Phys., (in press).
4. S. Veprek and V. Marecek, Solid State Electronics, 11, 683 (1968).
5. S. Veprek, Z. Iqbal, H. R. Oswald, F. A. Sarott and J. J. Wagner,

J. de Physique, Tome 42, p. C4-251 (1981).

6. Z. Iqbal and S. Veprek, J. Phys. C., Solid State Phys. 15, 377 (1982).

7. A. Matsuda, T. Yoshida, S. Yamasaki and K. Tanaka, Jap. J. Appl. Phys. 20, L439 (1981).

8. A. Matsuda, J. Non-Cryst. Solids, 59-60, 767 (1983).

9. T. Imara, K. Mogi, A. Hiraki, S. Nakashimer, A. Mitsuishi, Solid State Comm. 40, 161 (1981).

10. H. Ishida, M. Noder, H. Shimizu, Jap. J. of Appl. Phys. 22, L73 (1983).

11. T. D. Moustakas, D. A. Weitz, E. B. Prestridge and R. Friedman, In "Plasma Synthesis and Etching of Electronic Materials," Vol. 38 of Material Research Society (1985).

12. T..D. Moustakas, In "Semiconductors and Semimetals," Vol. 21A, (J. I. Pankove Ed.), Chapter 4, Academic Press, New York, 1984.

13. J. L. Vossen and J. J. Cuomo, In "Thin Film Processes," (J. L. Vossen and W. Kern Eds.) Chapter 2, Academic Press, New York, 1978.

14. B. A. Movchan and A. V. Demchishin, Phys. Met. Metalloved, 28, 83 (1969).

15. J. A. Thornton, J. Vac. Sci. Technol. 11, 666 (1974).

16. J. A. Thornton, Ann. Rev. Mater. Sci., 239 (1977).

17. T. D. Moustakas, Solar Energy Materials 8, 187 (1982).

18. A. Matsuda, K. Nakagawa, K. Tanaka, M. Matsumura, S. Yamasaki, H. Okushi and S. Iizima, J. Non-Crystall. Solids 35-36, 183 (1980).

19. E. Kinsbron, M. Sternheim and R. Knoell, Appl. Phys. Lett. 42, 835 (1983).

20 K. Zellama, P. Germain, S. Squelard, J. C. Bourgoin and P. A. Thomas, J. Appl. Phys. 50, 6995 (1979).

21. N. Proust, R. Bisaro, J. Magarino, D. Kaplan and K. Zellama, Proc. of the Symposium on Material and New Processing Technologies for Photovoltaics, Vol. 83-11 p. 291 (J.Amick, V. K. Kapur and J. Dietl Eds). The Electrochemical Society, Pennington, N.J. (1983).

22. V. Koster, Phys. Status Solidi, A 48, 313 (1978).

23. E. Billig, J. Inst. Metals 83, 53 (1954).

STRUCTURE AND H BONDING IN DEVICE QUALITY a-Si:H

P. John[*] and J.I.B. Wilson[**]
Heriot-Watt University
[*]Chemistry Dept., [**]Physics Dept.
Edinburgh EH14 4AS, U.K.

ABSTRACT

A brief account is given of the applications of hydrogenated amorphous silicon, a-Si:H, to devices including photovoltaic cells. The role of impurities and defects in modifying device quality is given an historical perspective. Whilst the deliberate incorporation of dopants into the amorphous network promises further advances, the simplistic picture of H reducing free spin density is insufficient. To utilise fully the electronic properties of a-Si:H, a deeper understanding of the structure at the molecular level is required, especially those structures including H in close proximity which are unique to amorphous solids.

INTRODUCTION

Interest in amorphous materials at Heriot-Watt University was initially stimulated by their suitability for solid state devices such as photovoltaic cells for which crystalline semiconductors are too expensive. From this beginning, a more fundamental interest in the structure and composition of amorphous silicon alloys has led to an appreciation of the preparation dependence of their properties. It has always been obvious to us that despite the amorphous nature of these thin-film materials, they would still be sensitive to unintentional contaminants, notwithstanding the early success of devices manufactured by simple deposition systems. However, this has not returned us full circle to a device technology having all the unavoidable sophistication of the present-day integrated circuit industry; although plasmas and lasers may be an advantage in low temperature VLSI circuit fabrication, they may be of even more use for materials which have yet to attain industrial production. The basis of all of this must be a thorough study and understanding of the physics and chemistry of amorphous materials in general.

A major goal of our work has been to elucidate the structure of hydrogenated amorphous silicon a-Si:H in order to improve device performance. Recognition that the structure is dependent on the preparation conditions has meant, in practice, that the studies have been undertaken with material prepared with a restricted range of conditions. A limited parameter space is dictated by the need for device quality a-Si:H films. From the structural viewpoint such an artificial division can be misleading unless the entire stoichiometric range for a-Si:H_x is investigated. Within the limits of amorphous silicon (x = 0) to catenated polysilane chains (x = 2), emphasis has been placed on a-Si:H containing 2 - 20 at % H. Even from a pragmatic viewpoint, the need for further knowledge regarding highly hydrogenated regions in a-Si:H is becoming evident. Recent experimental evidence, notably nuclear magnetic resonance (nmr) studies[1], suggest structural inhomogeneities in a-Si:H which exhibit good electronic characteristics. The conditions under which biphasic columnar microstructure growth is produced was established in early investigations[2]. For films prepared under conditions where the fibrils are too small to be evident by electron microscopy, nmr evidence points to residual inhomogeneities in which regions of low and high H content are present. Allen and Joannopoulos[3] have categorised these environments as i) clustered monohydrides as might be present on microvoid surfaces or multiply bonded H configurations e.g. $(SiH_2)_n$ and ii) randomly distributed non-interacting monohydrides. Cognisance of the presence of molecular hydrogen in microvoids[4] is a further strand to the overall model for H in a-Si:H. The influence of the interstitial regions containing $-(SiH_2)_n-$ strands is both chemical and mechanical. The low density[5] and porosity of these regions leads to predominantly oxidation[6] after deposition. Chemical reactivity of $-(SiH_2)_n-$ towards O_2 appears to be specific in that the contiguously bound silicon chains are replaced by $-(SiH_2O)_n-$ groupings[7]. Clearly, polysilane chains influence morphology, chemical reactivity, network flexibility and influence the transport and electronic properties of a-Si:H.

Subsequent sections describe the bonding characteristics of H in a-Si:H especially in those regions of high H content. The inter-relationship between structure and electronic properties is of paramount importance. Ultimately, this knowledge will enable fabrication of devices with optimum performance based on structural information rather than the empirical approach adopted hitherto.

a-Si:H FOR METAL-SEMICONDUCTOR JUNCTIONS

The successes and problems in using hydrogenated amorphous silicon for devices are demonstrated by the "simple" metal-semiconductor diode, the structure used for our first photovoltaic cells[8,9], with undoped a-Si:H over a thin n^+-layer on a stainless steel substrate and a top barrier contact of semitransparent metal. Although all of our early work used material from the Dundee group, it was apparent that we needed an in-house source of undoped and n-type a-Si:H and a contract with the CEC enabled this facility to be established for the preparation of MIS photovoltaic diodes. Following common practice in the semiconductor industry we used mass flow controllers to ensure a stable supply of the gases to the reaction chamber (a quartz tube with external electrodes). Only a rotary pump was available to evacuate the tube and extensive purging and pumping were used to eliminate residual air as far as possible. Good solar energy conversion efficiencies were achieved with small area cells[9,10,11] using our MIS structure, which was later modified at ECD Inc. with similar results. We were also preparing films for other groups who were interested in a range of optical and electrical properties of this novel thin-film semiconductor.

Despite the attractions of such a simple diode construction for opto-electronic applications, it was becoming apparent that the future for photovoltaic cells lay in the more complicated pin device and that MS and MIS junctions would mainly be used for assessment of material quality. Industrial and SERC support for our group meant that we were eventually able to install additional gas lines to allow more film compositions as well as to design a larger system with a more efficiently pumped chamber containing parallel plate electrodes. We also decided to install a quadrupole gas analyser and sampling port to study the actual gas content during plasma deposition, following work by several Ph.D. students who identified a cross-contamination effect between different thin-film layers[12]. This emphasised that material assessment should be performed on structures having the same layers as the devices for which the assessment was required.

a-Si:H AND LASER IRRADIATION

Our work with Schottky type solar cells showed that n-type films doped even with 1% PH_3 were not sufficiently conducting to make the low resistance contacts required in photovoltaic cells, especially when the conductivity through the film was measured instead of that between surface contacts. Since there was a wide range of lasers available at Heriot-Watt we had already decided to use them for semiconductor annealing and in a-Si:H there was the possibility of modifying the hydrogen content as well as changing the structure, all the way to crystalline. Early trials with a Nd:glass laser at 1.06 μm showed from infrared spectroscopy that the hydrogen bonding could be changed in high hydrogen content films, but with little effect on the electrical properties. In contrast, an argon ion laser could produce large changes in the electrical conductivity of even thin films due to the strong absorption of green and blue light in a-Si:H. Heavily doped n^+ films were found to have their dopant content activated more effectively after brief laser irradiation and after further irradiation the films became polycrystalline[13].

When we attempted to improve the conductivity in situ of n^+ films below undoped films by laser irradiation, changes were obviously being produced in the undoped layer as well. These were found to be similar to the Staebler-Wronski reversible conductivity changes which follow prolonged white-light irradiation. By using a Q-switched ruby laser we were able to test films for the presence of absence or Staebler-Wronski instabilities without lengthy illumination periods[14].

IMPURITIES IN a-Si:H

In this way we discovered that undoped a-Si:H films grown in a clean chamber were devoid of these metastable changes whereas both n-type films and "undoped" films grown on top of an n^+ layer using our normal procedure had a pronounced Staebler-Wronski effect. The implication was that the undoped layer was unintentionally doped with phosphorus from the gas lines or chamber, or from the coated substrate itself. Using a series of purging and pumping cycles between layers established that this cross-contamination could be eliminated and so must have arisen from residual gas in the chamber and supply lines[12]. Now that this effect is well known, multiple chambers are commonly used to separate the reaction products and gas feedstock for each layer in a stack, however a single chamber can be used for multiple layers if adequate purging follows each layer. This suggests that chamber wall, electrode and substrate coatings are not much sputtered by the plasma into succeeding plasmas, as is supported by mass spectrometer evidence.

The effect of removing cross-contamination from our Schottky type diodes was an improvement in their Staebler-Wronski stability but a poorer photovoltaic effect. The latter consequence was because the built-in potential at the barrier metal i-type a-Si:H interface was less than that for slightly n-type a-Si:H and so the major band-bending then occurred towards the back of the cell at the i-n^+ interface. It was also apparent why our previous attempts to add a drift field to these cells by a doping gradient from back to front had little effect on cell behaviour: there was already an unintentional doping gradient at levels above the desired values.

With the commissioning of the better reaction chamber we were able to examine these effects in pin structure diodes and. to extend our measurements of minority carrier lifetime, $\mu\tau$ product and gap state density [14,15] to such structures which could also use alloys of Si and N or Si and C. The problems of defining a good quality interface between layers continue to be studied by using the a-Si:H/a-Si$_x$N$_y$:H interface in a MISFET structure and by a-Si$_x$C$_y$:H,B as the p-type layer in pin photovoltaic diodes. Indeed, the range of alloys made possible by using (SiH$_4$ + NH$_3$) or (SiH$_4$ + C$_3$H$_8$) is attractive for optical devices which have to be tuned to various visible wavelength laser lines. This will require a more detailed examination of the visible and infrared optical absorption in these disordered materials.

VIBRATIONAL CHARACTERISTICS OF a-Si:H

The significance of infrared spectroscopy lies in its ability to distinguish local covalently bonded sites of H in a-Si:H and other binary alloys. In retrospect, the unequivocal assignment of the infrared bands to local modes of SiH$_x$ groupings has been more difficult that originally envisaged for a material which simply comprises silicon and hydrogen. The reasons lie with the absence of long range periodicity in the amorphous state, the presence of defect sites and the absence of hydrogen rich model compounds.

The seminal paper of Brodsky et al.[16] proposed one interpretation of the vibrational spectrum of a-Si:H. Normal valence bond attributes require that the monohydride, SiH, and dihydride, SiH$_2$, are possible groupings. Further, for 'dilute' and thus non-interacting SiH$_x$ groups in the amorphous matrix, the vibrational properties could be understood in terms of a 'local mode' model[17]. The ramifications of this are discussed in detail by Lucovsky and Pollard[18]. What was not recognised at the time was the presence of polysilane chains in a-Si:H. As a result, the correct interpretation of the bands in the 800 - 900 cm^{-1} region awaited the work of Knights et al.[19]. Confirmation of the empirical predictions of Lucovsky[20] resulted from the infrared and Raman scattering studies[21] of bulk polysilane.

In these laboratories, a-Si:H films prepared by glow discharge techniques, contain H in the range 2 - 20 at % under normal operating conditions. In the limit that the H content is $\sim$ 67 at %, the structure comprises infinite chains -(SiH$_2$)$_n$- and the 3-dimensional character is destroyed. The extent to which -(SiH$_2$)$_n$- is incorporated depends on the critical deposition parameters, especially substrate temperature. Process control has developed to the point that total H and the distribution of SiH$_x$ can be predetermined. Some progress has been made in understanding the interplay of gas phase kinetics and deposition mechanism[22]. We foresee the role of infrared spectroscopy in terms of material characterisation, possibly spatially, and quantifying the H content[23]. For the former, one relies on spectral assignments for identifying particular local groupings and their immediate environment. A consensus has been reached for the majority of the infrared active bands in a-Si:H. The present status is summarised in the Table. At least two problems remain unresolved.

First, the reasons for the observed shifts in the doublet at 800 - 900 cm^{-1} is unclear. Are they associated with the stereoregularity of the polymer chain or is microcrystallinity responsible? Secondly, not all the groupings contributing to the 2100 cm^{-1} band have been unequivocally identified. Suggested species include SiH groups located on the internal surfaces of voids[24]. This has gained support from the direct observation of occluded molecular H$_2$ in such microvoids[25]. Alternatively, SiH groups in close proximity might be the cause of this particularly controversial band.

110

Table: Vibrational Assignments of Infrared Active Local Modes in a-Si:H

Structural Unit	Description	Wavenumber/cm^{-1} (symmetry)
SiH (C_{3v})	stretching	2000 (A_1)
SiH$_2$ (C_{2v})	symmetric stretch	2100 (A_1)
	assymmetric stretch	2100 (B_1)
	bending	840 (A_1)
	rocking	630 (B_1)
	wagging	(a)
(SiH$_2$)$_n$ (D_{2h})	symmetric stretch	2100 (B_{3U})
	asymmetric stretch	2100 (B_{2U})
	bending	890–910 (B_{3U})
	wagging	840–865 (B_{1U})
	rocking	650 (B_{2U})

(a)   see W B Pollard and G Lucovsky, Phys. Rev. B, 26: 3172 (1982) for a discussion of this mode.

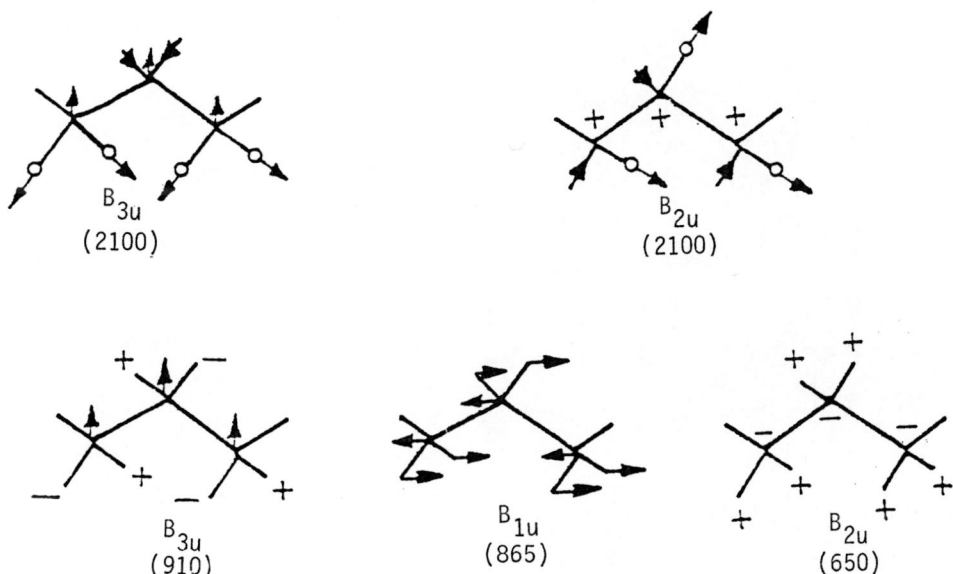

Figure 1.   Infrared active normal modes of –(SiH$_2$)$_n$– indicating the atomic displacements.

The infrared spectrum and Raman scattering of polysilane $-(SiH_2)_x-$ chains in the bulk, has exhibited characteristic spectral features that mimic those in a-Si:H films. In particular, the Si-H stretching mode is at 2100 cm^{-1}, the bending and wagging doublet at 910 cm^{-1} and 865 cm^{-1} respectively and the rocking mode at 650 cm^{-1}. For clarity, the atomic displacements for the infrared active normal modes are displayed in Fig. 1. A comprehensive tabulation is given elsewhere[26]. The conformation of the polymer is not known and whilst it is unlikely to be 100% stereoregular the higher frequency doublet could conceivably be due to the predominant trans conformer. It has been argued that a random distribution of dihedral angle in short chains embedded in an amorphous Si matrix gives rise to the doublet at 907 cm^{-1} and 862 cm^{-1}. The thermal dehydrogenation of $-(SiH_2)_x-$ takes place[27] without a significant shift of the ω_b and ω_w band positions. Observations of a low temperature ($\sim$ 300°C) rupture of $-(SiH_2)_n-$ chains in a-Si:H and polysilane[28] concur.

Random network models built by Weaire and students at Heriot-Watt University[29] produced a striking feature, namely, the clustered arrangement of H sites. The challenge to predict the physical and spectroscopic consequences of a clustered hydrogen stimulated our efforts. At the time, we were completing an infrared study of the thermal dehydrogenation of a-Si:H films. It dawned on us that the observation of a lower activation process[30], non specificity in the loss of SiH and SiH_2 groups, a thermo-stationary state on isothermal heating, and the vexed question of the 2100 cm^{-1} component might be rationalised by a simple structural unit. In attempting to describe the effect on the vibrational modes we adopted a

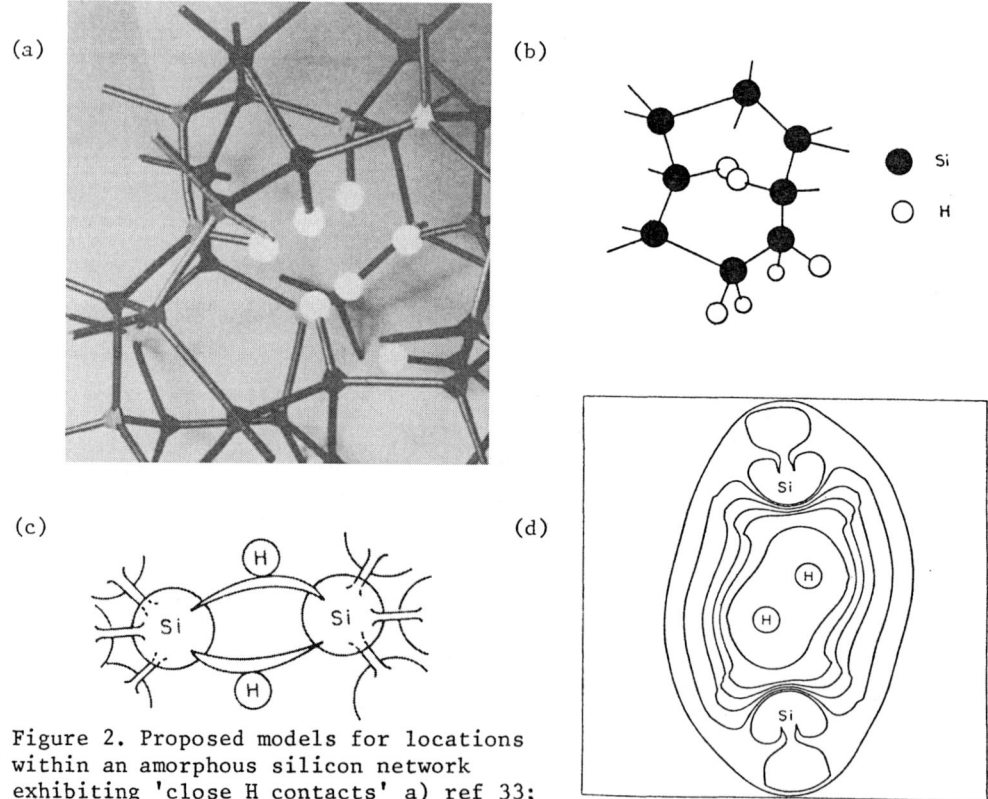

(a)

(b)

(c)

(d)

Figure 2. Proposed models for locations within an amorphous silicon network exhibiting 'close H contacts' a) ref 33; b) ref 33; c) ref 3; d) ref 35.

112

semiempirical FG matrix approach[31]. The type of structure originally
envisaged is shown in Fig. 2. A structure with r_0 (Si-H) = 1.35 Å was
predicted to exhibit a stretching frequency in the region of 2100 cm^{-1}.
As expected changing the bond angle by 10° had a negligible effect on the
Si-H stretching frequency. Alternatively, arguments associating monohydride groups at the surfaces of microvoids with the band at 2100 cm^{-1}
are compelling. Indeed observation of such voids through the pressure
induced infrared absorption of molecular hydrogen adds greater weight to
the argument. The origin of the 2100 cm^{-1} band lies in a number of
contributory groupings including defect structures of the type described
here.

Related studies of the thermal dehydrogenation of a-Si:H were directed
at providing a molecular description of the two stage hydrogen loss process.
Particularly intriguing was the lower temperature process occurring at ca.
300 - 350° with an activation energy around 1.5 eV[32]. An early postulate
that the lower and upper temperature dehydrogenation processes were
attributable to loss of SiH$_2$ and SiH groupings respectively was not
substantiated by the experimental evidence[33]. Isothermal infrared
investigations showed no preferential loss of either group as witnessed
by band intensity changes. We originally proposed the presence of
'incipient hydrogen molecules' within the matrix schematically depicted in
Fig. 2. The local strain induced by bond angle and bond length distortions
would imply an activation energy less than the thermochemical Si-H bond
strength. Associated esr evidence further points to relaxation of the
network to produce silicon-silicon bond formation[34] after or during H$_2$
release. Similar conceptual ideas have emerged on different grounds[35].
In Fig. 2 (a-d), the montage of diagrams illustrate the common theme of
SiH groupings constrained by the local amorphous structure to produce
groupings unique to the amorphous state of matter. The structures, in
terms of atomic coordinates, are similar. However, recent SCF X_α m.o.
treatments suggest that the valence band approach with strong van der Waals
non-bonding interactions might be inadequate. The electronic description
of the bonding favours a greater degree of H-H covalent bonding affecting,
in turn, the Si-H stretching force constant.

CONCLUSIONS

Collaboration between chemists and physicists has made each examine
this novel semiconducting material in new ways. An appreciation of the
complexity of chemical bonding in even a supposedly simple, two element
material as revealed by infrared spectroscopy has yet to explain the origin
of such phenomena as photoinduced conductivity change, despite a belief
that it is related to the hydrogen content. With the use of other amorphous
silicon-based alloys there is even greater need to examine the interplay
between chemical structure and optical or electrical behaviour. This is
particularly important if the operating lifetime and possible degradation
mechanisms of devices are to be predicted. The subject continues to be an
intriguing interdisciplinary one, having both academic and applied interests.
It does not promise to become any simpler with the current moves towards
combinations of plasma and laser deposition techniques.

ACKNOWLEDGEMENTS

We are pleased to acknowledge the interest and assistance of many members
of our departments including our graduate students who have produced Ph.D.
theses in this subject. It is also a pleasure to thank the various
companies and funding agencies who have supported our work, and provided
fruitful collaborations at home and overseas.

REFERENCES

1. P.C. Taylor, in "Semiconductors and Semimetals, vol. 21, part C", J.I. Pankove, ed., Academic Press, 1984.
2. J.C. Knights and R.A. Lujan, Appl. Phys. Lett., 35:244 (1979).
3. D. Allan and J.D. Joannopoulos, in "The Physics of Hydrogenated Amorphous Silicon II", J.D. Joannopoulos and G. Lucovsky, ed., Springer-Verlag, 1984.
4. Y.J. Chabal and C.K.N. Patel, Phys. Rev. Lett., 53:210 (1984).
5. P. John, I.M. Odeh, M.J.K. Thomas, M.J. Tricker, J.I.B. Wilson, J.B.A. England and D. Newton, J. Phys. C., 14:309 (1981).
6. P. John, I.M. Odeh, K.J.K. Thomas, M.J. Tricker and J.I.B. Wilson Phys. Stat. Sol.(b), 105:499 (1981).
7. P. John, B.C. Cowie and I.M. Odeh, Phil. Mag. B, 49:559 (1984).
8. J.I.B. Wilson and P. Robinson, 1st European Photovoltaic Solar Energy Conference, 223-230 (1977).
9. J.I.B. Wilson, J.C. McGill and S. Kinmond, Nature, 272:152 (1978).
10. J.I.B. Wilson and J.C. McGill, Sol. Stat. & Elect. Dev., 2:S7 (1978).
11. J.I.B. Wilson, J.C. McGill and P. Robinson, 13th IEEE Photovoltaic Specialists Conference, 751-754 (1978).
12. R.W. Thompson, Y.M. Hassan and J.I.B. Wilson, Solar Cells, 10:189 (1983).
13. J.I.B. Wilson, NATO Advanced Study Institute on Physical Processes in Laser-Materials Interaction (July 1980), Plenum Press, 413-442 (1983).
14. Y.M. Hassan, I.W. Boyd, F. Riddoch and J.I.B. Wilson, IEE Proc. Part I, 129:180 (1982).
15. Y.M. Hassan and J.I.B. Wilson, Sol. Stat. Commun., 49:771 (1984).
16. M.H. Brodsky, M. Cardona and J.J. Cuomo, Phys. Rev. B, 16:3556 (1977).
17. B.R. Henry and W. Siebrand, J. Chem. Phys., 49:5369 (1968).
18. G. Lucovsky and W.B. Pollard, in "Topics in Applied Physics, vol. 56", J.D. Joannopoulos and G. Lucovsky, ed., Springer-Verlag, 1984.
19. G. Lucovsky, R.J. Nemanich and J.C. Knights, Phys. Rev. B, 19:2064 (1979).
20. G. Lucovsky, in "Fundamental Physics of Amorphous Semiconductors", F. Yonezawa, ed., Springer-Verlag, 1981.
21. P. Vora, S.A. Solin and P. John, Phys. Rev. B., 29:3423 (1984).
22. F.J. Kampas, in "Semiconductors and Semimetals, vol. 21", J.I. Pankove, ed., Academic Press, 1984.
23. P. John, I.M. Odeh, M.J.K. Thomas, M.J. Tricker and J.I.B. Wilson, Phys. Stat. Sol., (b), 104:607 (1981).
24. H. Wagner and W. Beyer, Sol. State Commun., 48:585 (1983).
25. Y.J. Chabal and C.K.N. Patel, Phys. Rev. Lett., 53:1771 (1984).
26. A. Elliott, in "Infrared Spectra and Structure of Organic Long Chain Polymer, Edward Arnold Ltd., 1969.
27. P. John, B.C. Cowie and I.M. Odeh, Phil. Mag. B, 49:559 (1984).
28. The polysilane material was produced chemically. The spectroscopic features of $(SiH_2)_n$ in bulk and in a-Si:H are identical.
29. D. Weaire, N. Higgins, P. Moore and I. Marshall, Phil. Mag. B, 40:243 (1979).
30. J.A. McMillan and E.M. Peterson, J. Appl. Phys., 50:5238 (1979).
31. P. John, I.M. Odeh, M.J.K. Thomas, M.J. Tricker and J.I.B. Wilson, Phys. Stat. Sol., 103:K141 (1981).
32. This value should be regarded as approximate since the dehydrogenation kinetics is film dependent, see ref. 33.
33. P. John, I.M. Odeh, M.J. K. Thomas, M.J. Tricker, F. Riddoch and J.I.B. Wilson, Phil. Mag, B, 42:671 (1980).

34. D.K. Biegelsen, Solar Cells, 2:421 (1980).
35. D. Adler, in "Semiconductors and Semimetals, vol. 21", J.I. Pankove, ed., Academic Press, 1984.

HYDROGEN NEUTRALIZATION OF DEFECTS

IN SILICON

J. I. Pankove

RCA Laboratories
Princeton, NJ 08540

I. INTRODUCTION

Considerable attention has been given to hydrogenated amorphous silicon, a-Si:H. The insight derived from the studies of amorphous materials can be applied to the better known crystalline form of Si with interesting consequences.

The role of hydrogen in a-Si:H is now reasonably well understood.[1-4] Hydrogen binds to Si more strongly than Si to Si. Since the energy gap of a semiconductor reflects its average binding energy, it is no surprise that the energy gap of a-Si:H is larger than that of crystalline Si (c-Si)-i.e. $1.7 \pm .2$ vs. 1.1 eV.

Although in amorphous silicon almost every Si atom is surrounded by four other Si atoms - just as in a single crystal - there is no long-range order or periodicity because bonds are stretched and twisted. In fact, many bonds in a-Si are broken and left dangling. It is also known that dangling bonds create states in the gap that are responsible for recombination and generation of carriers. In fact, the density of these states is so large ($\sim 10^{20}$/eV cm^3) that the material is insulating and non-photoconducting--any optically excited electron-hole pair would immediately recombine. In a-Si:H, on the other hand, all the hydrogenated bonds are removed from the energy gap, thus reducing the concentration of gap states to as low as 10^{15}/eV cm^3, a concentration that allows one to move the Fermi level by doping. It is this possibility to control the conductivity of the material and the fact that, with a reduced density of gap states, usably long carrier lifetimes and diffusion lengths are possible, that a-Si:H has commanded the attention of device technologists.

Can hydrogen neutralize dangling bonds in c-Si? Can various properties of c-Si be dramatically altered by the presence of hydrogen? These are the questions we shall address in what follows.

II. HYDROGENATION TECHNIQUE

To neutralize dangling bonds one needs atomic hydrogen. This was shown in the case of a-Si.[5,6] We have dissociated H_2 by a glow discharge in

117

$\sim$0.2 Torr of H_2 at $\sim$100 MHz. The sample was keep out of the plasma to avoid ion bombardment-induced damage. The sample must be heated to promote the diffusion of H_1 into the bulk of the sample. However, one must not exceed the dissociation-temperature of the H-Si bond, as will be pointed out in the text.

III. PASSIVATION OF SURFACE STATES

The surface of a crystal is always covered with dangling bonds that provide the surface states responsible for surface recombination of minority carriers. These same surface states, in the presence of an electric field, act as generators of unwanted carriers that cause leakage current and noise in electronic devices. A variety of passivation techniques such as oxidation are used to eliminate surface states, although success has never been complete.

To test the surface passivating property of atomic hydrogen, an array[6] of p-n junctions was tested before and after the hydrogenation treatment. Fig. 1 shows the reverse bias I(V) characteristic of such a p-n junction. Exposure to atomic hydrogen for 1/2 hour at 300°C resulted in a substantial decrease in leakage current. Heating the silicon at 550°C drives off most of the hydrogen, thus restoring most of the surface states and returning the I(V) characteristics to a leaky condition.

Note that even an oxide-passivated diode can benefit from atomic hydrogen treatment because oxygen, being a large atom, cannot passivate every surface bond. Most of the remaining surface dangling bonds can be passivated by atomic hydrogen as evidenced by a $\sim$10-fold drop in leakage current when an oxide-passivated diode was hydrogenated.[7] Actually, encapsulating a p-n junction under a-Si:H was found to be beneficial by more than one order of magnitude compared to the standard oxidation treatment, provided the device is not heated above 500°C where hydrogen can escape from the a-Si:H layer.[8]

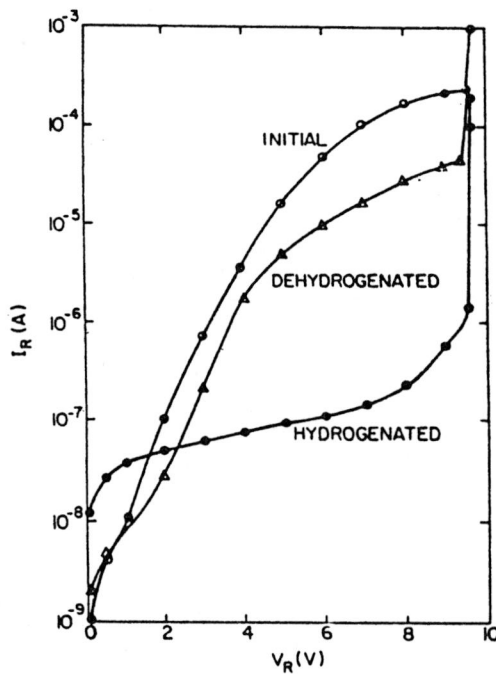

Figure 1.

Reverse I(V) characteristics for a p-n junction in crystalline Si before and after hydrogenation, and also after dehydrogenation.

IV. PASSIVATION OF BULK DANGLING BONDS

A. Grain Boundaries.

A grain boundary, the interface between two crystals, is the site of many dangling bonds. These dangling bonds pin the Fermi level near midgap at the interface. Consequently, the grain boundary comprises two back-to-back Schottky barriers. If a voltage is applied across the grain boundary, one of these Schottky barriers will be reverse biased. A scanning light spot will generate the largest photocurrent at the reverse biased Schottky barrier. In this way Faughnan[9] generated a map of photocurrents in a wafer of polycrystalline silicon (see Fig. 2a). The peaks in photocurrent clearly depict the contours of each crystalline grain. After hydrogenation, as shown in Fig. 2b, the intensity of the photocurrent is substantially reduced, indicating that hydrogenating some of the dangling bonds in the grain boundaries has unpinned the Fermi level, allowing the bands to flatten at those boundaries.

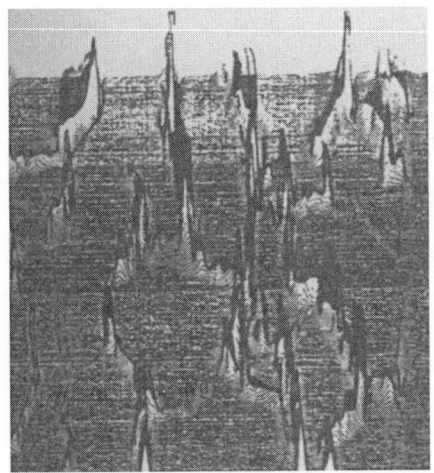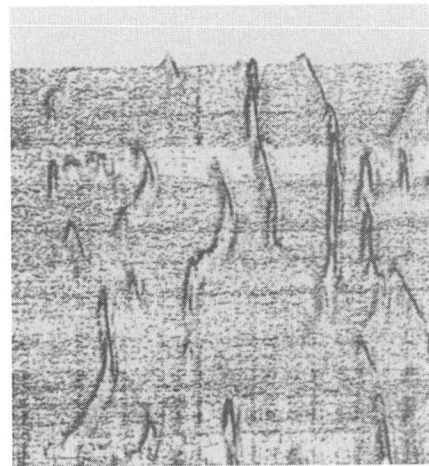

Figure 2. Photocurrent map of polycrystalline Si with voltage applied between right and left. The signal was generated by scanning a focussed light spot. a) before and b) after hydrogenation at 300°C for 1/2 hour. Scan length is 1 cm. Courtesy of B. W. Faughnan.

Solar cells can be made on inexpensive coarse grain polycrystalline silicon by diffusing phosphorus into one surface of a p-type wafer. Grain boundaries that intersect the p-n junction are sites of recombination that removes photogenerated carriers from participating in the photovoltaic effect. This results in lowered short-circuit current and open-circuit voltage. Atomic hydrogen, by passivating the dangling bonds, reduces the loss of carriers in the grain boundaries. As shown in Fig. 3, both short-circuit current and open-circuit voltage increase. Typically the efficiency is increased from ∿8% to ∿11%.[9]

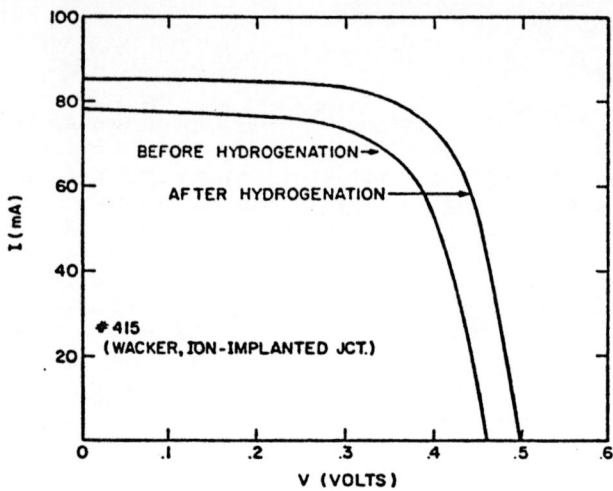

Figure 3. I(V) characteristics at AMI illumination of solar cell
consisting of a p-n junction in polycrystalline Si.
Courtesy of B. W. Faughnan.

B. Dislocations

Dislocations are localized interruptions in a crystal's periodic
network. These interruptions result in dangling bonds. Dislocations can be
localized along a line, or collect in an area. In the latter case, with the
Fermi level pinned near midgap, an areal dislocation forms two Schottky

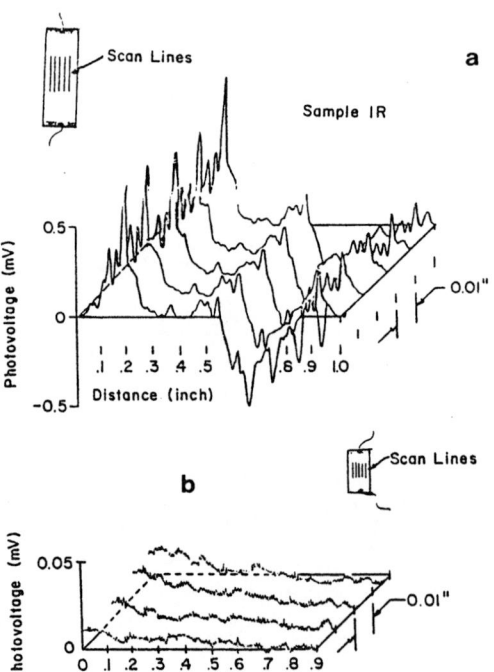

Figure 4.

Photovoltaic map of silicon on
sapphire scanned with a focussed
light spot a) before and b) after
hydrogenation.

120

barriers back-to-back (like the grain boundary discussed above). If such a
region is scanned with a fine spot of light, the photogenerated carriers
will move in opposite directions when the light spot crosses the dislocation. Hence the photovoltage will swing through two peaks of opposite
polarities. However, dislocations usually come in clusters, so that the
superposition of the photovoltages contributed by a cluster of dislocations
can have a complex shape. When silicon is grown on sapphire, the difference
in lattice constants of two materials results in severe stresses that are
partly relieved by the formation of dislocations. Fig. 4a shows the
photovoltaic pattern generated by scanning a small light spot along several
lines along the layer of silicon-on-sapphire between two electrodes as shown
in the inset.[10] After hydrogenation for 1/2 hour at 300°C, the photovoltaic signal nearly vanished as shown in Fig. 4b (note the tenfold
increase in sensitivity). Evidently the hydrogenation of dangling bonds at
the dislocations flattened the bands, eliminating the local field that
produced a photovoltage.

C. Implantation-Induced Defects

Ion implantation breaks many bonds.[11] Usually they are reconstructed by
a thermal anneal. In our experiment, the purpose of generating so much
damage was to explore the possibility of making a-Si:H by first amorphizing
the surface of the crystal and then hydrogenating the amorphized surface
layer. The test as to whether or not we succeeded in forming a-Si:H
consisted in measuring the photoluminescence spectrum. Indeed, we found
that when the implantation dose exceeded $\sim 2 \times 10^{14}$ cm^{-2}, the hydrogenated
amorphous layer emitted the characteristic spectrum of a-Si:H, broad,
efficient and peaking at about 1.3 eV. Fig. 5 shows such an emission. For
comparison, the photoluminescence peak of c-Si at 1.1 eV is also shown; its
efficiency is very low. When the implantation dose was reduced below $\sim 2 \times 10^{14}$ cm^{-2}, a new luminescence peak appeared at 1.0 eV. This emission did not

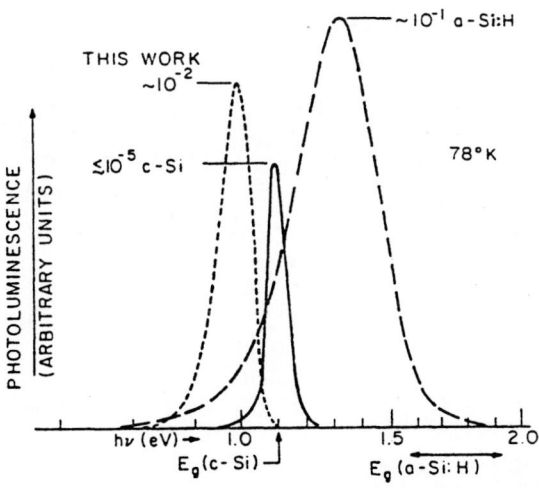

Figure 5. Emission spectra at 78°K from hydrogenated ion-implanted Si
from crystalline Si (c-Si), and from hydrogenated amorphous
Si (a-Si:H). E_g is the energy gap. The numbers labeling
the peaks are the external emission efficiencies.

depend on the ion implanted, hence it is related to the damage. The role of hydrogen is merely to passivate the non-radiative recombination centers that compete with the radiative transition.

An important conclusion from this experiment is that whenever a new luminescence peak is obtained in silicon, it is advisable to first hydrogenate the non-radiative recombination centers before studying the luminescence, thus enhancing the signal by eliminating a competing process.

V. PASSIVATION OF ACCEPTORS

The common acceptors in silicon are trivalent atoms in substitutional sites. Their three electrons can bond to only three of the four surrounding Si atoms. One of these neighbors is then left with a dangling bond which, by capturing a valence band electron produces a hole. If instead, a hydrogen could be attached to this dangling bond, the complex would be neutral and would not take an electron from the valence band. One consequence would be an increase in resistivity. This possibility was verified as described below.[12]

A. Resistivity Change

Samples of B-doped silicon were hydrogenated at various temperatures, then angle lapped and the spreading resistance profile measured using an automatic probe. It was soon found that the optimum temperature for hydrogenating acceptors is in the range 100-130°C. Below 100°C, the diffusivity of hydrogen is very low. Above 160°C, the BH complex dissociates. The spreading resistance data of Fig. 6 was obtained on four differently B-doped samples of silicon that were hydrogenated at 122°C: the upper three samples for one hour, the most heavily doped sample was hydrogenated for 4 hours. Note that the spreading resistance increases toward the surface. The value of resistivity deep in the material is the original bulk resistivity of the sample. If the sample is heated in vacuum or in molecular hydrogen (H_2), the spreading resistance profile remains flat; if the hydrogenated sample is heated at 300°C to drive off the hydrogen, the resistivity profile becomes flat again at the original bulk value. Subsequent hydrogenation restores the high resistivity at the surface. Hence the process is reversible and reproducible, and requires the presence of atomic hydrogen.

The possibility of neutralizing acceptors other than B, i.e., Al, Ga and In was verified as demonstrated in Fig. 7.[13] Thalium could also be passivated. However, donors such as P, Sb and As appeared unaffected by hydrogenation.

Another demonstration that hydrogenation neutralizes acceptors is the decreases in free-carrier absorption observed during infra-red transmission experiments.[14]

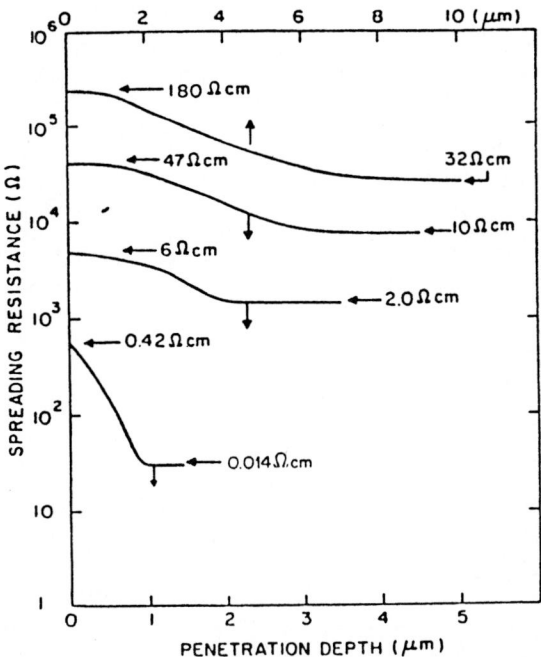

Figure 6. Spreading resistance profile of four B-doped samples of
(100) Si hydrogenated at 122°C. The three higher resistivity samples were hydrogenated for 1 h; the lowest resistivity sample was treated for 4 hrs. The resistivities were
obtained from a calibration curve. Note the greater penetration depth of atomic hydrogen as the boron concentration
decreases.

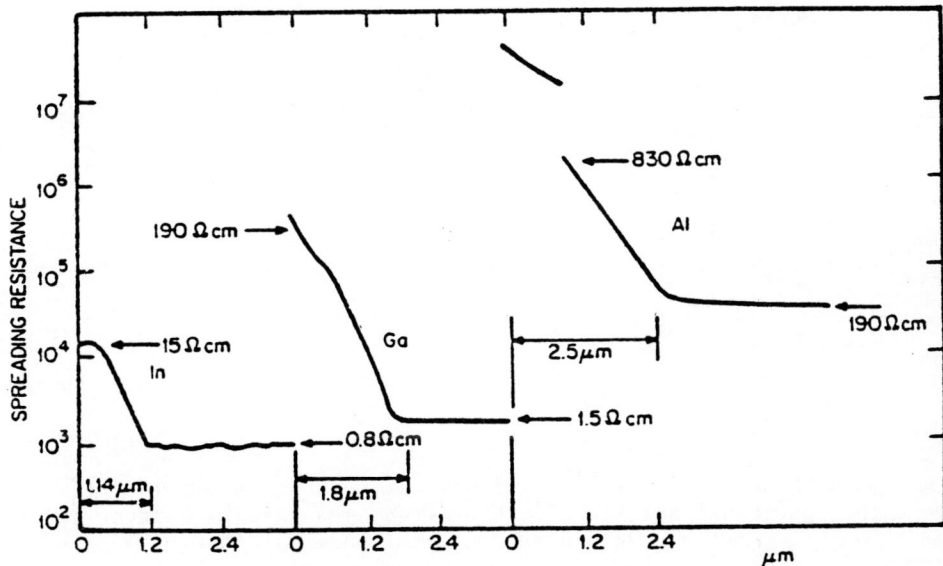

Figure 7. Spreading resistance profiles for silicon doped with indium,
gallium, and aluminum.

123

B. Effect on Bound Excitons

At low temperatures, donors and acceptors remain neutral when they trap an electron hole pair, forming a bound exciton. Bound exciton recombination emits a characteristic luminescence peak, the energy of which is so specific that it can be used to identify the impurities present. Thewalt et al.[15] measured the luminescence spectrum of Si samples doped with B, P, In and Tl before and after hydrogenation. As shown in Fig. 8, the excitons bound to acceptors vanish after hydrogenation. This indicates that the acceptor-hydrogen complex is not prone to attract either a hole or an electron. Furthermore, the binding energy of the hydrogenated complex, an isoelectronic center, appears larger than the bandgap energy since no luminescence peak attributable to luminescence from isoelectronic centers could be found.

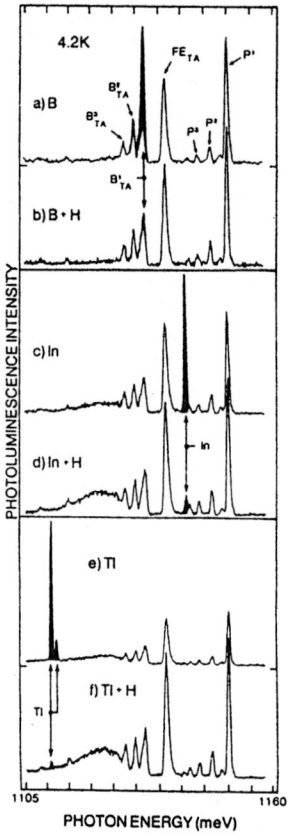

Figure 8. Photoluminescence spectra covering the no-phonon and TA phonon-replica energy regions taken at 4.2 K. The spectra show the bound exciton luminescence of samples implanted with B, In and Tl before (a,c,e,) and after (b,d,f,) treatment in atomic H. Bound exciton luminescence due to the implanted impurities has been shaded in to distinguish it from the substrate luminescence.

C. Capacitive Studies

In the work of Sah et al.[16] and Gale et al.[17], a substantial decrease in capacitance was found in MOS structures after a current had been passed through the oxide. This drop in capacitance was attributed to electrolytic transport of protons from moisture either on or in the oxide. The hydrogen migration through the oxide and its pile up at the oxide-Si interface is well documented.[17] The decrease in capacitance results from the hydrogenation of the B-doped silicon: the reduced free carrier concentration near the interface causes a widening of the depletion layer and therefore a drop in capacitance.

124

D. Effect on Infrared Absorption

Other models where the hydrogen is bonded to B rather than to Si,[14] led us to look for a vibrational signature in the IR absorption spectrum of the hydrogenated material. Fortunately, the earlier work of Tsai[18] on the a-Si$_{1-x}$B$_x$:H system provided some guidance as to what spectral region to study. In Tsai's data the stretch mode of SiH is at $\sim$2100cm^{-1} while the stretch mode of BH is at $\sim$2600cm^{-1}. This is the spectral range we observed and we found that our hydrogenated sample had neither a peak at $\sim$2100cm^{-1} nor at $\sim$2600cm^{-1} but rather a narrow one at 1875cm^{-1} (Fig. 9).[14] There is no other report of an absorption peak in Si at room temperature at this spectral value. The explanation for this peak at 1875cm^{-1} is credited to Lucovsky.[14] The model is shown in Fig. 10. Inserting hydrogen along the <111> direction causes the B atom to collapse inside the plane formed by the three nearest Si neighbors. Since the three electrons of B are involved in bonding to these three Si atoms, the hydrogen sees the slightly positive charge of B which weakens the force constant between Si and H, thus causing a "mode softening" or shift from 2000 to 1875cm^{-1}. As to the bridging bond

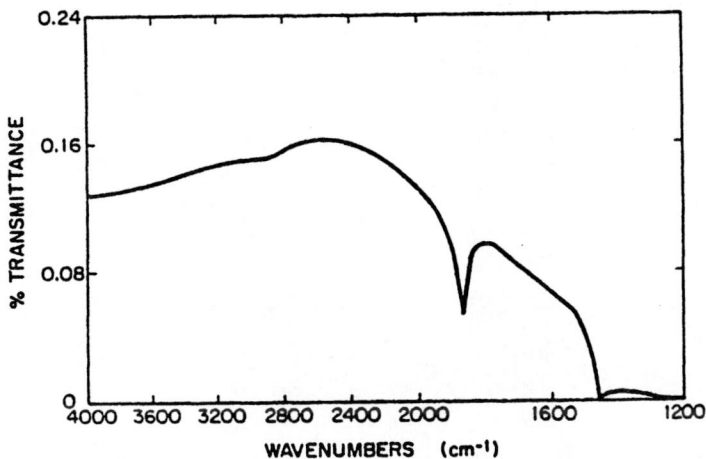

Figure 9. IR transmission spectrum of H-neutralized boron in silicon.

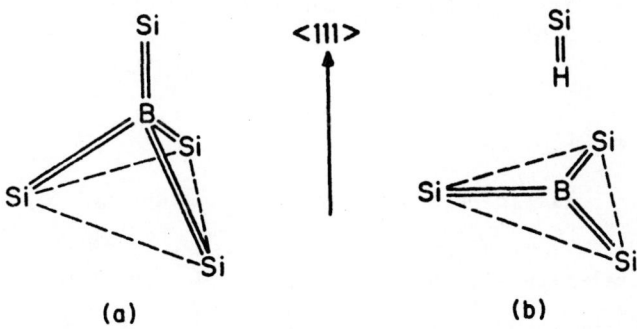

Figure 10. Model of substitutional boron in silicon a) before and b) after H-neutralization.

model, it can be dismissed on the basis that the absorption band is much narrower ($<50cm^{-1}$) than would be expected from a bridging bond ($> 100cm^{-1}$).[14]

E. N–P–N Structure

If one starts with an n-type crystal and forms a p-type region by adding an excess of acceptors, such as by B-implantation followed by annealing, then the acceptors nearest the surface can be neutralized by hydrogenation. This process allows the donors to dominate near the surface, making the layer nearest the surface n-type. This is a low-temperature method for making n-p-n structures. Fig. 11 shows that we have achieved such a structure starting with n-type 0.15Ω cm Si into which 4×10^{13} B/cm^2 were implanted at 100 KeV. This sample was then annealed at $1200°C$ for 16 hours and hydrogenated at $100°C$ for 94 hours. Finally the sample was angle-lapped and stained with a $CuSO_4$ solution that decorated the n-type regions.

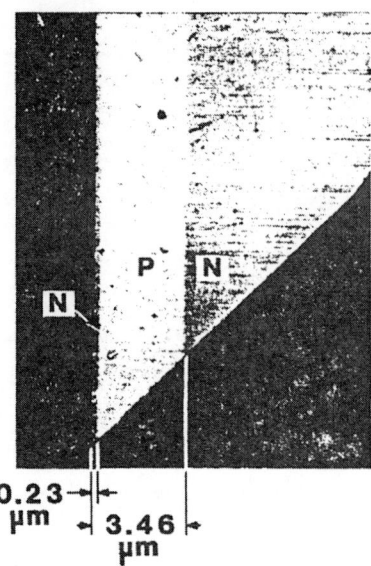

Figure 11. Stained angle-lapped section of B-doped Si that has been partly hydrogenated to form a 0.23 μm-thick layer of n-type B-neutralized material. The n-type regions are preferentially stained.

VI. Acceptor Activation in a–Si:H,B

a-Si:H cannot readily be made p-type and conducting. The doping efficiency is very low. It is frequently assumed that boron is mostly in a trigonal bonding configuration and therefore neutral. However B could also be in a tetrahedral site with a neutralizing hydrogen atom.

B-doped a-Si:H was deposited by glow discharge in silane + 1% B_2H_6 onto a glass substrate at $100°C$. The resulting film is rich in H that passivates

all the Si dangling bonds including those in the vicinity of B-atoms. This
sample was broken into two pieces. One of these pieces was annealed in
vacuum at 200°C for 1 hour. This annealing treatment caused the
dissociation of those H atoms that are near the acceptors, but not those H
atoms that passivate silicon dangling bonds. (Removing H from dangling
bonds at external and internal surfaces (microvoids) requires heating at >
350°C). After the 200°C anneal, all the tetrahedrally bonded B-atoms become
active acceptors, and the a-Si:H,B becomes more conducting by a factor of
600.

VII. CONCLUSION

We have demonstrated that atomic hydrogen can attach itself to dangling
bonds in crystalline silicon. These dangling bonds can be those on the
surface of a crystal, or those lining grain boundaries in polycrystalline
material, and also those present in dislocations. Atomic hydrogen can also
neutralize dangling bonds generated by ion implantation. The major useful
consequence of dangling bond passivation is the reduction of recombination-
generation centers responsible for leakage current and noise in devices and
also responsible for loss of carriers by non-radiative recombination. This
last effect, neutralization of non-radiative recombination centers, permits
the observation of luminescent transitions with a much greater efficiency
than obtainable without the hydrogenation treatment.

Finally, recognizing that a silicon dangling bond is associated with
every acceptor, it is possible to neutralize the acceptors with the
consequent increase in resistivity, decrease in free carrier absorption,
disappearance of bound excitons, and discovery of a softened Si-H
vibrational mode at 1875cm^{-1} along a <111> direction near the acceptor.

It is conceivable that new properties may be found for other impurities
in Si after hydrogenation, just as those found by McMurray et al.[19] in Ge:
double acceptors becoming single acceptors, double donors becoming single
donors and an isoelectronic center (C or Ge) becoming a single donor.

References

1. M. H. Brodsky, ed. Amorphous Semiconductors, Topics in Applied
      Phys., Vol. 36, Springer, Berlin, 1979.
2. Y. Hamakawa, Amorphous Semiconductor Technologies and Devices,
      North Holland, Amsterdam, 1982.
3. J. D. Joamopoulos and G. Lucovsky, eds. The Physics of
      Hydrogenated Amorphous Silicon, Topics in Applied Phys., Vol.
      55-56, Springer, Berlin, 1984.
4. J. I. Pankove, Hydrogenated Amorphous Silicon, Semiconductors and
      Semimetals, Vol. 21, Willardson and Beer, eds., Academic Press,
      1984.
5. D. Kaplan, N. Sol, G. Velasco and P. A. Thomas, Appl. Phys. Lett.,
      33, 440 (1978).
6. J. I. Pankove, M. A. Lampert and M. L. Tarng, Appl. Phys. Lett.,
      32, 439 (1978).
7. J. I. Pankove and M. L. Tarng, unpublished results.
8. J. I. Pankove and M. L. Tarng, Appl. Phys. Lett., 34, 156 (1979).
9. B. W. Faughnan, personal communication.
10. L. Jastrzebski, J. Lagowski, G. W. Cullen and J. I. Pankove, Appl.
      Phys. Lett., 40, 713 (1982).
11. J. I. Pankove and C. P. Wu, Appl. Phys. Lett., 35, 937 (1979).

12. J. I. Pankove, D. E. Carlson, J. E. Berkeyheiser and R. O. Wance, Phys. Rev. Lett., $\underline{51}$, 2224 (1983).
13. J. I. Pankove, R. O. Wance and J. E. Berkeyheiser, Appl. Phys. Lett., $\underline{45}$, 1100 (1984).
14. J. I. Pankove, P. J. Zanzucchi, C. W. Magee and G. Lucovsky, Appl. Phys. Lett., $\underline{46}$,   (1985).
15. M. L. W. Thewalt, E. C. Lightowlers and J. I. Pankove, to·be published.
16. C. T. Sah, J. Y. C. Sun and J. J. Tzou, Appl. Phys. Lett. $\underline{43}$, 204 (1983) and J. Appl. Phys., $\underline{55}$, 1525 (1984).
17. R. Gale, F. J. Feigl, C. W. Magee and D. R. Young, J. Appl. Phys., $\underline{54}$, 6938 (1983).
18. C. C. Tsai, Phys. Rev., $\underline{B19}$, 2041 (1979).
19. R. McMurray et al., to be published.

HYDROGEN INCORPORATION IN AMORPHOUS SILICON

AND PROCESSES OF ITS RELEASE

W. Beyer

Institut für Grenzflächenforschung
und Vakuumphysik
Kernforschungsanlage Jülich GmbH
D-5170 Jülich, F.R. Germany

INTRODUCTION

It is well known that the unusual properties of hydrogenated amorphous silicon (a-Si:H) like its high photo-to-dark conductivity ratio[1] and its for amorphous semiconductors surprisingly[2] wide controlability of the electronic properties by doping[3] are related to the hydrogen content. Unhydrogenated, pure amorphous silicon films exhibit rather poor electronic properties due to a high concentration of unsaturated (dangling) bonds[4]. The understanding of the role of hydrogen in the amorphous material, therefore, is highly desirable. Equally of interest are the mechanisms governing the release of hydrogen from the amorphous network at elevated temperatures since they may limit the stability of any electronic device made from this material.

Since the first direct evidence for the presence of hydrogen in glow-discharge[1] fabricated amorphous silicon by the Chicago group[5] and the demonstration that the presence of hydrogen affects the electronic properties of evaporated and sputtered amorphous silicon and germanium strongly[6,7,8], a huge amount of data has accumulated dealing with hydrogen in a-Si:H. A multitude of techniques for the hydrogen detection have been applied, like hydrogen evolution[5,9,10,11], infrared absorption[9,10], nuclear reaction[9], nuclear magnetic resonance (NMR)[12], Compton scattering[13], α particle elastic scattering[14], secondary ion mass spectroscopy (SIMS)[15], calorimetric measurements[16,17], yet several aspects of hydrogen incorporation are far from being understood. It is the aim of this paper to discuss the role of hydrogen as it is revealed from an extensive study of a-Si:H films employing mainly hydrogen evolution experiments. Infrared absorption, SIMS, calorimetric measurements have also been applied as well as measurements of the photo- and dark conductivity, the thermoelectric power and the optical absorption for studying the hydrogen influence on the electronic properties. We shall confine ourselves mainly on a-Si:H films deposited by the glow discharge (GD) technique since at present such films are of highest interest for applicative uses. For comparison, however, we shall also discuss data from films prepared by reactive sputtering (SP) and by chemical vapor deposition (CVD).

EXPERIMENTAL METHODS

The glow discharge amorphous silicon films discussed in this article were prepared at Marburg and Jülich mainly in capacitively coupled systems from undiluted silane gases (SiH_4, SiD_4, Si_2H_6). As a substrate we used sapphire or crystalline silicon platelets mounted on the grounded electrode of the reactor. Typical deposition conditions were: pressure 0.4-0.5 mbar, gas flow 6-7 sccm, rf power ~ 5 W yielding deposition rates between 1 and 5 Å/s. The measuring techniques for SIMS[18], the calorimetric detection of H_2[16], and the electronic transport measurements[19] have been described elsewhere. The infrared transmission was determined using a Digilab-20 Fourier-Transform-Spectrometer. For the hydrogen evolution measurements the samples were inserted into a (previously outgassed) quartz tube evacuated by a turbomolecular pump. While heating to 1000°C at a constant heating rate of typically 20°C/min the partial pressures of H, H_2 and other species in the recipient were monitored by a quadrupole mass analyzer. As the pumping speed of the turbomolecular pump is constant over a wide range of pressure, the partial pressure is a measure of the evolution rate. Absolute calibration is achieved either by measuring hydrogen implanted samples or by inserting a known flow of hydrogen into the apparatus through a calibrated leak.

RESULTS

The Absolute Hydrogen Content

Under the preparation conditions described in the preceding paragraph, the free parameter determining the hydrogen density N_H in our a-Si:H films is the substrate temperature T_s. Fig. 1 shows the results for N_H as a function of T_s for undoped films prepared from silane and disilane. N_H has been obtained from hydrogen evolution measurements integrating over the amount of hydrogen released when the films are heated to 1000°C. Also indicated in Fig. 1 is the hydrogen concentration determined from N_H accounting for the

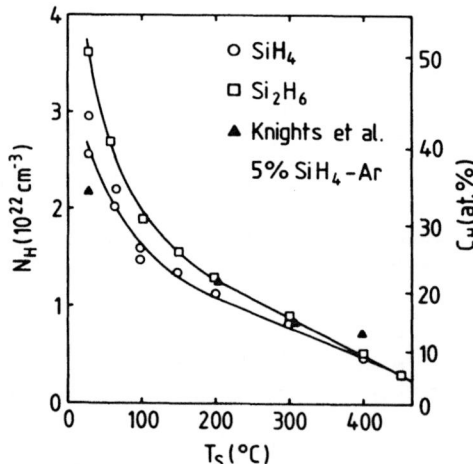

Fig. 1. Hydrogen density N_H and hydrogen concentration C_H for glow discharge a-Si:H films as a function of substrate temperature T_s.

density variation from $\rho \sim 1.7$ g cm^{-3} for highly hydrogenated films to $\rho \sim 2.2$ g cm^{-3} for films with $N_H \sim 7 \times 10^{22}$ cm^{-3}. The results show a variation of C_H from values > 40 at. % for $T_s = 25°C$ to values < 10 at.% for $T_s > 400°C$. The values for films deposited from Si_2H_6 tend to lie somewhat higher than those from SiH_4. The rather smooth variation of C_H as well as the close agreement with data of Knights et al.[20] suggest that essentially one process determines the hydrogen content in the amorphous material. According to the growth model by Kampas and Griffith[21] this process is the H_2 elimination during the film growth, in the model of Longeway[22] the abstraction of hydrogen due to SiH_3 radicals. Only at substrate temperatures $T_s > 400°C$ the process of bulk hydrogen evolution may lead to an additional decrease of the hydrogen content.

Molecular Hydrogen in a-Si:H

The possibility that some fraction of the hydrogen content in a-Si:H may be incorporated in molecular form has been discussed as early as 1977[23]. Yet, the null result for the 4156 cm^{-1} Raman stretching vibration of H_2 placed an upper limit in the range of 1 %. More recently, the observation of a minimum in the proton NMR spin-lattice relaxation time T_1, (when measured as a function of temperature) has been associated with the presence of molecular hydrogen[24]. We were able to detect H_2 by calorimetric measurements of our films[16]. Used is the ortho-to-para conversion of molecular hydrogen at low temperatures which inserts a heat leak in a specific heat experiment. The results for a-Si:H films of ~ 5 μm thickness are shown in Fig. 2. Plotted is the density N of molecular hydrogen and the total hydrogen density for films deposited at 25 and 250°C. The rather paradox behavior of a considerably lower H_2 content of the sample with higher total hydrogen concentration is observed. Annealing experiments, moreover, are found to change the H_2 density in the highly hydrogenated film (up to an annealing temperature of $T_a = 210°C$) only slightly whereas for the film of lower total hydrogen content the H_2 density (for $T_a = 400°C$) increases considerably.

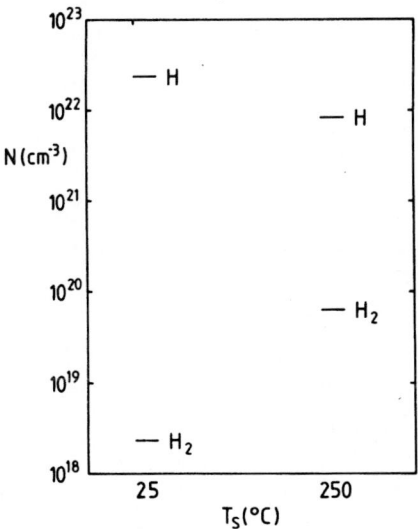

Fig. 2. Total hydrogen density and density of molecular hydrogen in a-Si:H for two substrate temperatures T_s.

After annealing at 500°C for 15 min. Graebner et al.[17] determined the H_2 concentration to about 0.5 %, or approximately 5 % of the total hydrogen concentration for a T_s = 230°C sample.

The Hydrogen Bonding

Taking from the previous paragraph that most of the hydrogen incorporated in a-Si:H is bound to silicon, this paragraph focusses on our knowledge on the bonding of hydrogen to the amorphous network. Most information on this subject is gained from infrared (IR) absorption experiments, where the energetic position and the strength of the vibrational modes give information on bonding states of hydrogen to the amorphous network and the amount of hydrogen incorporated in each state. Fig. 3 shows IR transmission results for two a-Si:H films prepared at T_s = 25°C ($C_H \sim$ 38 at.%) and T_s = 300°C ($C_H \sim$ 13 at.%). Major differences between these films occur in the range of the bond bending modes: The scissor bending mode at 890 cm^{-1} in highly hydrogenated a-Si:H (Fig. 3a) proves the presence of SiH_2 species in this material; the observation of the 845 cm^{-1} mode, moreover, suggests the presence of polymeric SiH_2[25,26]. Yet, there has been some controversy about the assignment of the silicon-hydrogen stretching modes observed in Figs. 3a and b at 2100 and 2000 cm^{-1}, respectively. The original assignment[10,23,27] attributed the former mode to stretching vibrations of SiH_2, the latter to SiH. The poor correlation between the integrated absorptions at 2100 and 890 cm^{-1} and the observation of the 2100 cm^{-1} vibration even without the 845-890 cm^{-1} bond bending modes for certain a-Si:H samples, however, has cast doubts if this assignment is always correct[28,29]. Recent high resolution electron loss measurements of hydrogen on the crystalline silicon surface show a silicon-hydrogen stretching mode near 2100 cm^{-1} independent if SiH_2 (as detected by the 900 cm^{-1} bending mode) or SiH species are present.[30] Accordingly, we have attributed the 2100 cm^{-1} mode to stretching modes of hydrogen on internal (void) surfaces independent if the hydrogen is bonded as SiH or SiH_2[31]. The 2000 cm^{-1} absorption in low hydrogen (high T_s) material is then considered to arise from Si-H oscillators embedded in a dielectric medium (i.e. in the bulk). Basically the same explanation has been proposed by Cardona[26]. The results in Fig. 3 then suggest that highly hydrogenated (low T_s) a-Si:H grows as a void-rich material whereas low hydrogen (T_s = 200-400°C) material is rather compact.

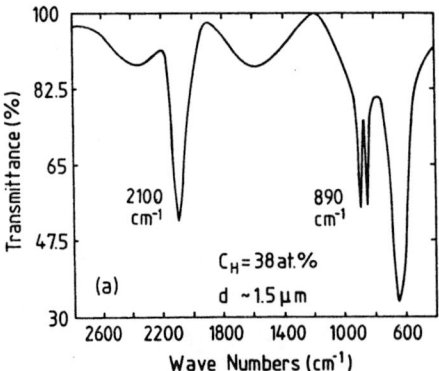 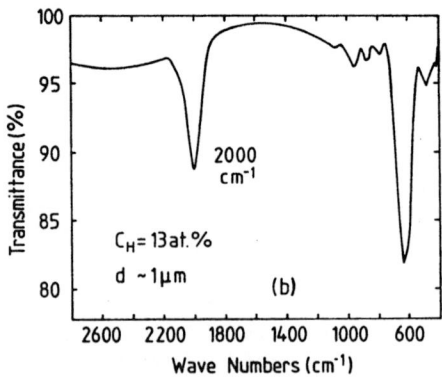

Fig. 3.   Infrared transmission spectra of a-Si:H.
Substrate temperatures: T_s = 25°C (a), T_s = 300°C (b).

Analysis of the Evolution Curves. Hydrogen evolution measurements can yield information not only on the absolute hydrogen concentration (see Fig. 1) but also on the hydrogen stability and on processes of hydrogen release, if the hydrogen evolution rate is measured as a function of temperature. Essentially three processes may influence the evolution spectra:(1) the Si-H bond rupture, (2) the diffusion of atomic or molecular hydrogen to the film surface and (3) in case of diffusion of atomic hydrogen a surface recombination and H_2-desorption step. (Note that the major effusing species is H_2).

Diffusion as the rate-limiting step for hydrogen evolution can easily be recognized by measuring a series of identical samples prepared with a different film thickness[32]. For a diffusion-limited process the evolution peaks should shift to higher temperatures with increasing film thickness since the diffusion length for out-diffusion increases. The diffusion coefficient $D = D_0 \exp(-E_D/kT)$ (with D_0 the diffusion prefactor, E_D the diffusion energy and the Boltzmann constant k) can then be determined from an evolution experiment. Solving the diffusion equation of out-diffusion from a thin film at a constant heating rate ß one obtains the following relation:[33]

$$\ln(D/E_D) = \ln(d^2 ß/\pi^2 kT_M^2) = \ln(D_0/E_D) - E_D/kT_M \qquad (1)$$

with the film thickness d and a temperature T_M of maximum effusion rate. Thus, measurements of T_M as a function of heating rate ß or thickness d yield the diffusion parameters D_0 and E_D.

If diffusion is not rate-limiting the evolution, the effusion rate for a single process can be described by[33-35]

$$d(N/N_0)/dt = \bar{r}(kT/h)(1-N/N_0)^n \exp(-\Delta G/kT) \qquad (2)$$

with a free energy of activation ΔG, an order of reaction n, a transmission factor $\bar{r}$, an initial and evolved hydrogen concentration N_0 and N, respectively, and Planck's constant h. The actual activation energy measured will be smaller than ΔG if there is a change in entropy during the reaction. With

$$\Delta G = \Delta H - T \Delta S \qquad (3)$$

(ΔH the enthalpy and ΔS the entropy of reaction and $\Delta S^* = \Delta S + k \ln \bar{r}$ the experimental entropy of reaction) one obtains

$$d(N/N_0)/dt = (kT/h)(1-N/N_0)^n \exp(\Delta S^*/k) \exp(-\Delta H/kT). \qquad (4)$$

The order of reaction can be determined from an analysis of the peak shape.[35,36] For immobile hydrogen species prior to desorption we expect n = 1 whereas for a surface recombination process of atomic hydrogen to form H_2 the order of reaction should be n = 2. When the order of reaction is known, ΔS^* and ΔH follow from a plot of $\log(d(N/N_0)/dt (1-N/N_0)^{-n})$ versus 1/T. Alternatively, ΔH and ΔS^* can be determined by measuring the temperature T_M of maximum effusion rate as a function of the heating rate. For a first order reaction and a linear heating rate ß one obtains:

$$ß(\Delta H + kT_M)/kT_M^2 = (kT_M/h)\exp(\Delta S^*/k) \exp(-\Delta H/kT_M). \qquad (5)$$

The analysis of the evolution curves in terms of kinetic parameters becomes difficult or impossible when we deal with several overlapping processes instead of a single one. In particular, a peak shape analysis in terms of eq. (4) will then become rather meaningless. In contrast, eq. (5) may still be applicable describing then the dominant effusion process only.

For a further understanding of the diffusion or effusion processes, the knowledge of the diffusing species in the amorphous film , whether it is atomic or molecular hydrogen, is required. This information can be obtained by measuring the H_2, D_2 and HD effusion from a-Si:H/a-Si:D/a-Si:H sandwich samples[37,38]. If diffusion proceeds by molecular hydrogen (deuterium) we expect little HD which may arise from some intermixing or interdiffusion in the boundary zone between Si:H and Si:D layers as well as from interaction of diffusing $H_2(D_2)$ with bound deuterium (hydrogen). In the other case of diffusion of atomic hydrogen, a large fraction of HD is expected which is formed in a surface recombination process. If dN_H/dt and dN_D/dt are the effusion rates for H_2 and D_2, respectively, the effusion rate for HD follows to

$$dN_{HD}/dt = 2 \sqrt{(dN_H/dt)(dN_D/dt)}. \qquad (6)$$

A comparison between calculated and measured HD evolution rates therefore allows a discrimination between diffusion of atomic or molecular hydrogen.

Undoped a-Si:H Films. The evolution spectra for a series of films deposited at different substrate temperatures and, accordingly, with different hydrogen content (see Fig. 1) are shown in Fig. 4[39]. Plotted is the hydrogen evolution rate dN/dt as a function of temperature for a heating rate of ß = 20°C/min. Characteristic for a-Si:H films deposited at low substrate temperatures is a double peak structure. When the substrate temperature is raised, the low temperature (LT) peak disappears near T_s = 150°C, and only the high temperature (HT) process remains. Fig. 5 shows the temperature T_M of maximum evolution rate as a function of film thickness d for a series of films deposited at the substrate temperatures T_s = 25 and 300°C. The high temperature evolution peaks (curves 1b and 2) are found to be limited by hydrogen diffusion whereas for the low temperature peak (curve 1a) no influence of diffusion is observed.

In Fig. 6 we plot for the HT evolution data of Fig. 5 the quantity $\ln(d/T_M)^2$ as a function of $1/T_M$. (We neglect that for T_s = 25°C material the film thickness shrinks during the LT evolution by roughly a factor of

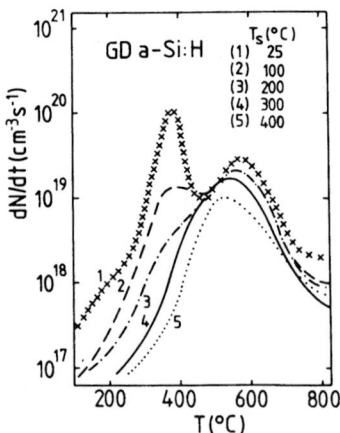

Fig. 4.   Hydrogen evolution rate dN/dt versus temperature for a-Si:H films. Film thickness ~ 0.6 μm; heating rate ß = 20°C/min.

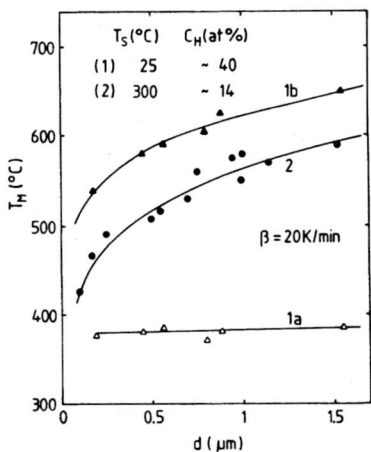

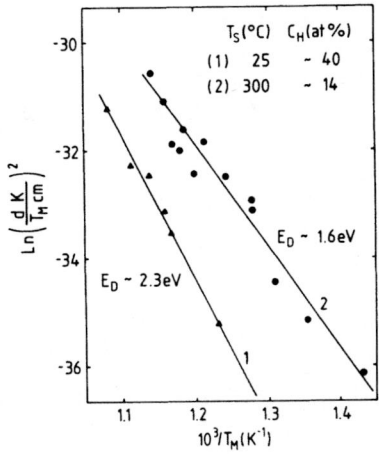

Fig. 5. Temperature T_M of maximum evolution rate versus film thickness d.

Fig. 6. Plot of $\ln(d/T_M)^2$ versus $1/T_M$ for data of Fig. 5.

0.9[39]). Straight lines are obtained yielding for T_s = 25°C material $E_D \sim 2.3$ eV and $D_o \sim 100$ cm^2/s and for T_s = 300°C: $E_D \sim 1.6$ eV and $D_o \sim 3 \times 10^{-2}$ cm^2/s. The latter data agree closely with results by Carlson and Magee[15] for diffusion of deuterium in a-Si:H ($T_s \sim 300$°C). Hydrogen diffusion data for (annealed) low T_s a-Si:H are, to our knowledge, so far not available in literature. We note, however, that we obtained a similar hydrogen diffusion energy for sputtered a-Si:H:F films (T_s = 300°C).[40]

In order to discriminate between diffusion by atomic and molecular hydrogen we measured the evolution from a-Si:H/a-Si:D/a-Si:H sandwich samples. Fig. 7 shows the results[38] for a sandwich sample deposited at 300°C. Plotted are the H_2, D_2 and HD evolution rates as a function of temperature. Also shown is the HD evolution rate (HD calc.) calculated according to eq. (6) from the H_2 and D_2 rates assuming that diffusion proceeds by the atomic species. A close agreement of the calculated and experimental HD evolution rates is obtained for the evolution maximum indicating that most of the hydrogen in T_s = 300°C a-Si:H diffuses in the effusion process as atomic hydrogen.

Since molecular hydrogen according to these results apparently cannot diffuse in T_s = 300°C a-Si:H up to 700°C we also explored the diffusion of rare gas atoms like He, Ne and Ar which were brought into the films by ion implantation. The effusion measurements (Fig. 8) practically show no diffusion of Ne and Ar up to 900°C whereas He is released already near 500°C. These results demonstrate that the diffusion of atoms or molecules in a-Si:H depends rather critically on the size of the atomic species. On the other hand, one may expect drastic changes for the diffusion of a given species if the density of the amorphous material is changed.

Models for the diffusion process in T_s = 300°C a-Si:H must account for the fact that the diffusing hydrogen atoms were originally bound to silicon. The literature data for the binding energy E_B of hydrogen to silicon range between E_B(Si-H) $\sim$ 3.1 eV[41,42] and 3.6 eV[43]. In any event, if the probability per unit time for a Si-H bond rupture is estimated to be (see eq. (2))

$$P \sim \nu_o \exp(-E_B/kT); \quad \nu_o \sim 10^{13} s^{-1}, \quad (7)$$

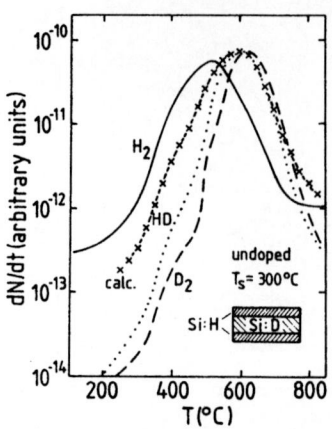

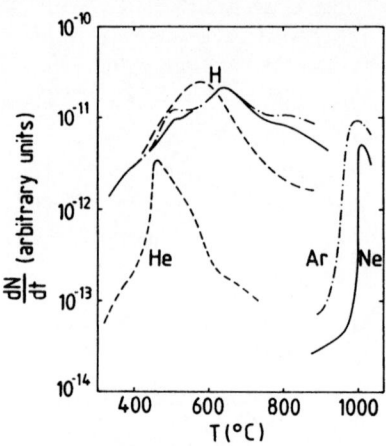

Fig. 7. H_2, D_2 and HD evolution rate versus temperature for sandwich sample of a-Si:H and a-Si:D. (T_s = 300°C; d ~ 1 μm).

Fig. 8. Evolution rate versus temperature for He, Ne and Ar implanted a-Si:H (T_s = 300°C; d ~ 0.7μm. Dose: ~ 10^{17} cm^{-2}; implantation energy 120 – 300 keV.

one obtains for a temperature T_M = 500–600°C (typical for a film of 1 μm thickness) and a total hydrogen concentration of N ~ 10^{22} cm^{-3}: dN/dt = 10^{11}– 10^{17} cm^{-3} s^{-1}, i.e. a much too low bond rupture rate to explain an experimentally observed maximum evolution rate of dN/dt ~ 10^{19} cm^{-3} s^{-1} (see Fig. 4). Accordingly, energy gain mechanisms must be involved reducing the total energy amount necessary for a Si–H bond rupture. Conceivable are the simultaneous reconstruction of Si–Si bonds and the formation of H_2 concomitant with the breaking of the Si–H bonds. Since the diffusing species in T_s = 300°C a-Si:H is atomic hydrogen, the energy gain mechanism apparently proceeds via the Si–Si bond reconstruction. With Si–H and Si–Si binding energies of both E_B ~ 3.1 eV[41] (note, however, that there is a wide range of literature data not only for the Si–H but also for the Si–Si binding energy[42,43]), one obtains for the energy to free a hydrogen atom E ~ 1.6 eV. It is tempting to associate this bond rupture energy also with the diffusion energy.[39] Necessary for such a diffusion process is that most of the hydrogen atoms are incorporated into the silicon lattice at positions breaking Si–Si bonds. After the Si–H bond rupture, the free hydrogen atom would break and hydrogenate another (not necessarily neighboring) Si–Si bond. Such a process would not lead to the formation of dangling bonds. In agreement, it is well known that in the initial phase of hydrogen evolution the ESR signal rises rather slowly at a rate of about one dangling bond per > 100 effused hydrogen atoms[32]. Also in agreement, more than half of the hydrogen present in a T_s = 300°C a-Si:H film can be evolved without great decay of the (dangling bond sensitive) photoconductivity[39].

In this model, the evolution process will set in when the temperature is high enough for Si–H bond rupture. With eq. (5) we estimate this temperature (assuming ΔS* = 0, ΔH = ΔG* ~ 1.6 eV) to T ~ 280°C in agreement with the experimental data. The diffusion-limited process will proceed until all bond-breaking hydrogen has effused. Hydrogen in isolated Si–H bonds, however, will remain in the film and will effuse only if the ~ 3 eV binding energy is brought up or if there is enough lattice reorganization that energy gain mechanisms like Si–Si reconstruction or H formation can become active. In order to estimate the fraction of isolated bonded hydrogen in our a-Si:H films we calculated the hydrogen effusion rate according to eq. (8)

136

of Ref. 33, using the hydrogen diffusion coefficient of $D = 1.1 \times 10^{-2} \text{cm}^2\text{s}^{-1}$ x exp(-1.49 eV/kT) determined for the particular sample[33]. In Fig. 9 this calculated rate is plotted as a dashed line as a function of temperature along with the experimental curve (full line). The calculated curve has been normalized to fit the experimental data in the effusion maximum. A major deviation between experimental and calculated curve occurs above 600°C. Attributing this to the rupture of isolated Si-H bonds we obtain a concentration of ~ 4 at.% of such bonds compared to a total hydrogen concentration of ~ 14 at.%. The evolution maxima for this isolated bonded hydrogen lies according to Fig. 9 near 650°C. If we assume that hydrogen atoms, after bond rupture, will diffuse out rather rapidly, we can use eq. (5) to estimate the reaction energy. With $\Delta S^* = 0$, a value of $\Delta H = \Delta G^* \sim 2.8$ eV is obtained close to literature data of the Si-H binding energy of ~ 3.1 eV [41,42]. It is tempting to associate this tightly bound hydrogen with the dilute hydrogen phase (narrow line) observed in NMR[12]. The broad NMR line (clustered phase) would then be related to the bond-breaking hydrogen.

Turning to samples prepared at low substrate temperatures, the evolution results for an a-Si:H/Si:D/Si:H sandwich sample prepared at $T_s = 25°C$ are shown in Fig. 10. In the low temperature evolution process the calculated HD effusion rate is found to exceed considerably the experimental one supporting that molecular hydrogen is here the main diffusing species. In the HT peak, on the other hand, both calculated and experimental HD data agree as observed for $T_s = 300°C$ material (see Fig. 7). These results demonstrate that low T_s a-Si:H consists of an open void-rich structure allowing for rapid diffusion of molecular hydrogen. Subsequently to the low temperature effusion peak, the material apparently contracts so that diffusion of H_2 becomes impossible. We verified this shrinkage effect which is well known from literature[11] for our samples by measuring the film thickness as a function of the annealing temperature[39]. As the LT evolution peak is not diffusion-limited, it may be analyzed in terms of eq. (4). Fig. 11 shows the results for a $T_s = 25°C$ film suggesting that we deal with a first order reaction governed by a free energy of activation of $\Delta G^* = 1.95$ eV. Literature data by Brodsky et al.[9], McMillan and Peterson[35] and Oguz and Paesler[44] agree with this result closely.

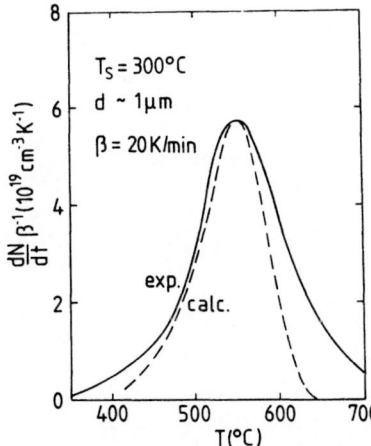

Fig. 9. Calculated and experimental hydrogen evolution rate versus temperature.

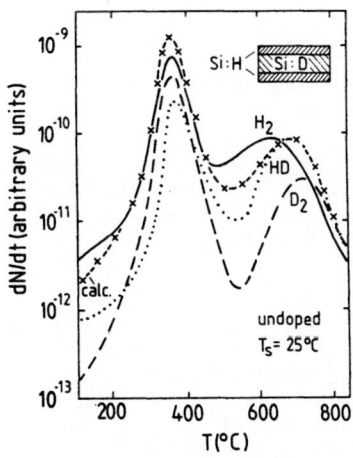

Fig. 10. H_2, D_2 and HD evolution rate versus temperature for sandwich sample of a-Si:H and a-Si:D (T_s = 25°C; d ~ 1.4 μm).

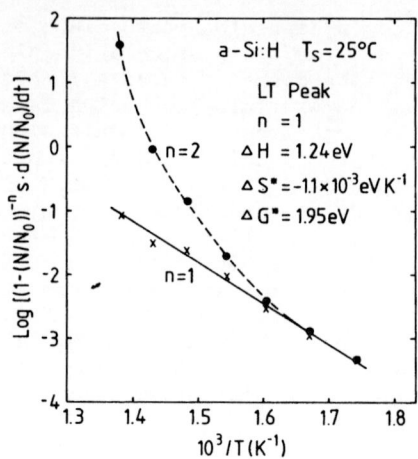

Fig. 11. Analysis of LT evolution peak of a–Si:H (T_s = 25°C)
in terms of eq. (4).

The most likely reaction accounting for the LT evolution peak is the
rupture of two neighboring Si–H bonds simultaneous with the formation of
H_2[45]. With literature data for Si–H and H_2 binding energies[41,42] of ~ 3.1
and 4.5 eV, respectively, one gets an energy step of ~1.7 eV for the reaction. If this assignment is correct, the LT evolution should lead to the
formation of two silicon dangling bonds for each desorbed hydrogen molecule.
Indeed, for low T_s a–Si:H showing a low temperature evolution process only
(i.e. for films with a microstructure) a steep increase of the ESR signal
has been reported for annealing temperatures above 250°C[32]. Films without a
microstructure (with hydrogen evolution spectra similar to our results) show
only a slight increase of the ESR signal when annealed at 200–400°C[32]. We
must conclude accordingly that the dangling bonds formed during the H_2 desorption reconstruct rapidly, and it is likely this reconstruction which
leads to the formation of a rather compact material subsequent to the LT
evolution. This reconstruction effect along with the result that our films
do not oxidize after long storage in air suggest that the voids are not very
large, possibly after hydrogen removal just large enough to allow for diffusion of H_2 (i.e. of the order ~ 4 Å). For films exhibiting a microstructure, the voids are likely to be much larger: These films are known to oxidize upon storage in air[46].

Also the origin of the voids in our films is likely to be different
from that of the microstructure. While the latter has been attributed[47] to a
particular nucleation and growth pattern also observed in unhydrogenated a-
Si films[48] the voids in our films appear to be related to the hydrogen content. This statement is based on evolution results on a series of sputtered
a–Si:H films[39] prepared at T_s = 25°C showing for low hydrogen concentration
only the HT evolution peak and for high hydrogen content the double peak
structure. Plotting the amount of hydrogen evolving under the LT and HT
evolution peaks as a function of the total hydrogen density (Fig. 12) these
sputtered films show a dependence which agrees closely with data of glow-
discharge a–Si:H films (here N_H is varied by changing the substrate temperature): Up to N_H ~ 10^{22} cm^{-3} we have only the HT evolution, i.e. no (interconnected) void structure; above 10^{22} cm^{-3} the LT evolution, i.e. the void
structure appears. According to the model of hydrogen incorporation at
positions breaking Si–Si bonds, this result is easily understood: From a
certain density of broken Si–Si bonds the lattice will lose its connectiveness and voids will be formed. On the other hand, if hydrogen is removed

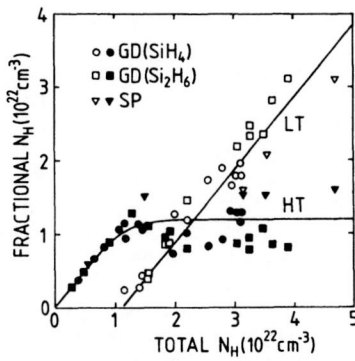

Fig. 12. Hydrogen density released under LT and HT peaks versus total hydrogen density.

from the lattice by evolution, Si-Si bonds will reconstruct so that a compact material (allowing only for diffusion of atomic hydrogen) is obtained. This reconstructed a-Si:H, however, is likely to be quite defective. The rather high hydrogen diffusion energy of ~ 2.3 eV (see Fig. 6) suggests that the diffusion process involves deep trapping sites for hydrogen, presumably isolated silicon dangling bonds or highly strained Si-Si bonds.

Doped-a-Si:H Films. While phosphorus-doping leads only to slight changes of the evolution spectra, the influence of doping by boron is quite dramatic, as is illustrated in Fig. 13[49]. The addition of only 10 ppm of B_2H_6 (diborane) changes the shape of the evolution curve; a further increase of the boron concentration leads to a shift of the evolution peak down to T_M ~ 370°C, i.e. close to the deposition temperature: films doped with 0.1-1% of boron lose part of the incorporated hydrogen content already during deposition[50].

Measurements of the evolution peak as a function of film thickness (Fig. 14) show a considerable scatter of the data but no indication of a

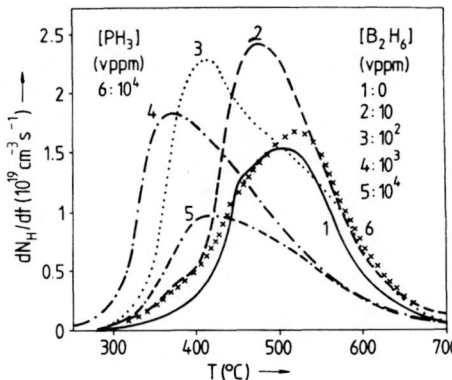

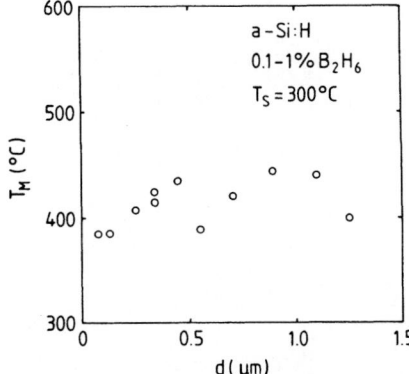

Fig. 13. Evolution rate versus temperature for doped a-Si:H films (T_s = 300°C; d ~ 0.6 μm).

Fig. 14. Temperature T_M of maximum evolution rate versus film thickness for boron-doped a-Si:H films.

diffusion-limited process. Accordingly, eq. (4) should be applicable; however, a plot of $\log((1-N/N_o)^{-n}\, d(N/N_o)/dt)$ versus $1/T$ (Fig. 15) does not lead to straight lines for either $n = 1$ or 2, suggesting that two or more evolution processes are involved. An analysis of the peak temperature T_M as a function of the heating rate ß indicates a surface desorption process governed by a free energy of activation $\Delta G^* = 1.7$ eV [33] similar to the LT evolution process in undoped material (see preceding paragraph). Measurements of the H_2, D_2 and HD evolution from boron-doped a-Si:H/a-Si:D/a-Si:H sandwich samples show that diffusion proceeds by atomic hydrogen. The diffusion process is apparently rather fast already at $T = 200-300°C$. SIMS profiles of the a-Si:H/a-Si:D/a-Si:H sandwich films show an almost complete mixture of hydrogen and deuterium after deposition at $300°C$[39]. Applying a deuterium plasma to a boron-doped a-Si:H film almost half of the hydrogen content can be replaced by deuterium within one hour at a temperature of $300°C$.

All these results can be understood if we assume that boron-doped a-Si:H grows as a granular material and that there exist fast hydrogen diffusion paths. In this case, the hydrogen evolution would be limited either by diffusion from the interior of the grains or by recombination/desorption at the film surface, i.e. in neither case a dependence on the film thickness is expected. The model can explain why infrared measurements show little difference between undoped and boron-doped material[49,51] and why the evolution curves of boron doped material are strongly dependent on the preparation conditions[52].

The nature of the grain structure as well as of the fast diffusion paths of hydrogen in boron-doped a-Si:H remains obscure. One can speculate that boron acts as a nucleation center for silicon during the film growth. Possibly the fast diffusion paths are located at the grain boundaries. Post-deuteration experiments[39] suggest that these paths are related to the presence of hydrogen, i.e. one could think of a hydrogen exchange along hydrogenated grain boundaries.

Under certain conditions, such highly diffusive paths also seem to exist in undoped a-Si:H films. SIMS experiments on films deposited below $150°C$ (i.e. with a void structure) show a fast penetration by deuterium when treated in a deuterium plasma at $200-300°C$[53].

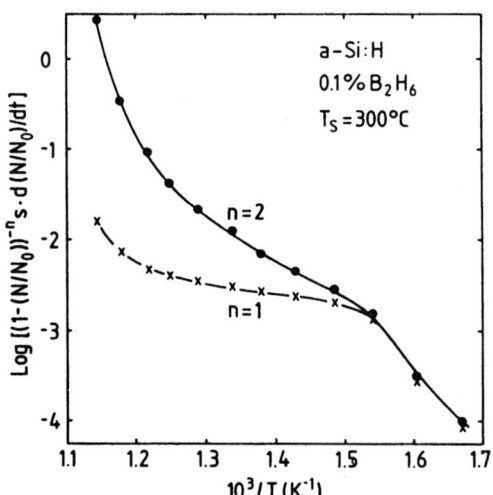

Fig. 15.   Analysis of evolution peak of boron-doped a-Si:H film in terms of eq. (4).

A most pronounced effect of hydrogen incorporation is the increase of the optical gap of a-Si:H. This effect has been widely observed and essentially two models have been proposed for its explanation. Cody[54] suggests that up to about 20% of hydrogen the gap widening is due to a removal of disorder, above 20 % it is tentatively assigned to a loss of connectiveness of the amorphous network. Ley[55], on the other hand, relates the gap widening mainly to the recession of the valence band edge from ~ 0.75 eV below the conduction band edge E_{CB} in unhydrogenated amorphous silicon to about 1.65 eV below E_{CB} for highly hydrogenated material, inferred from photoemission data. This recession has been ascribed to the replacement of the Si-Si bonds by stronger Si-H bonds moving states from the top of the valence bands to a position deep inside the valence bands. Fig. 16 shows the effect for our samples. Plotted is the optical gap E_G as determined by a Tauc plot[56] as a function of C_H. As a reference for unhydrogenated a-Si we use a film prepared by low-pressure CVD[57] at T_s = 560°C with a hydrogen content of less than 0.3% estimated by SIMS. Our data agree closely with those of Cody et al.[58].

In contrast to the widening of the optical gap, the Fermi level position with reference to the conduction band edge in undoped films is only slightly changed by hydrogen incorporation. Fig. 17 shows the conductivity and thermopower as a function of 1/T for the CVD film (C_H < 0.3 at.%) as well as for two undoped glow-discharge a-Si:H films with C_H = 14 and 40 at.%. While the CVD film shows hopping conduction at low temperatures due to a large concentration of unsaturated silicon bonds, at high temperatures all three curves fall rather closely together suggesting that the energetic distance between Fermi level and conduction band edge is always in the range of 0.8-0.9 eV. As we have n-type conduction according to the thermopower, the transport gap must be at least twice as large, i.e. E_G > 1.6 - 1.8 eV.

The influence of hydrogen incorporation on the actual transport process is also rather small. Previously we have shown that the quantity Q derived from thermopower S and conductivity σ measurements by

$$Q = \ln\sigma + |(e/k) \, S| \qquad (8)$$

contains the major information on the transport process but is independent

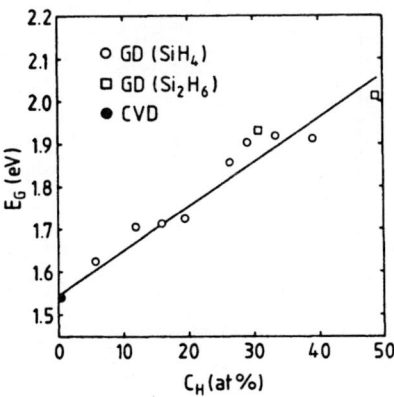

Fig. 16. Optical gap E_G versus hydrogen concentration C_H.

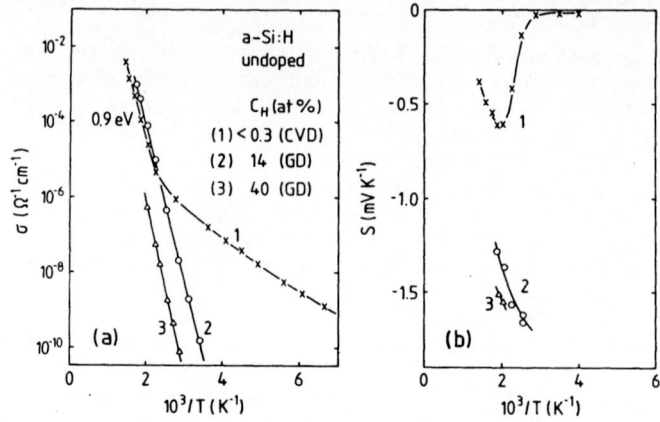

Fig. 17.   Conductivity σ (a) and thermopower S (b) versus 1/T for
           undoped a-Si:H films with different hydrogen concentrations.

of the Fermi level position[19]. In Fig. 18 Q is plotted as a function of 1/T
for a series of phosphorus-doped films with different hydrogen concentrations. Samples with T_s < 300°C have been annealed at 250°C (10 min) to remove dangling bonds created due to deposition at low substrate temperature.
The results show that up to C_H ~ 20 at.% the Q-function is changed only
slightly. Above 20 at.% the slope of Q(1/T) rises in agreement with increased potential fluctuations[19].

A most dramatic effect of hydrogen incorporation is observed for the
photoconductivity as well as for the doping effect. Fig. 19 shows the photoconductivity $\Delta\sigma_{ph}$ of undoped films (measured under illumination with 50
mW/cm² of white light) as a function of the hydrogen density N_H (full line).
Again, films prepared at substrate temperatures below 300°C were annealed at
250°C. For comparison we also show data obtained for (T_s = 250°C) a-Si:C:H
films where the hydrogen concentration was varied by changing the carbon

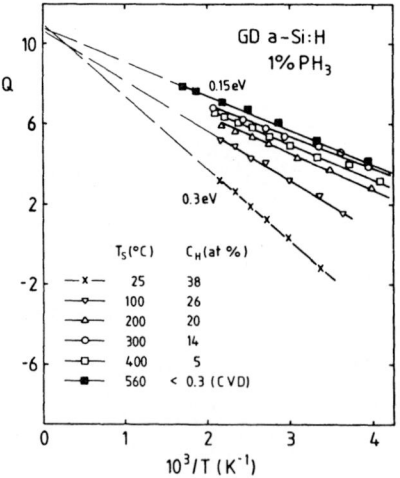

Fig. 18.   Q versus 1/T for phosphorus-doped glow-discharge (1 % PH$_3$)
            and CVD (0.08 % PH$_3$) amorphous silicon films.

142

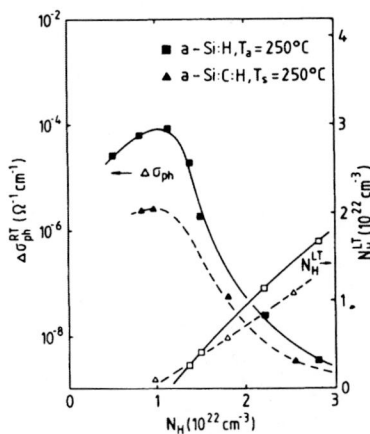

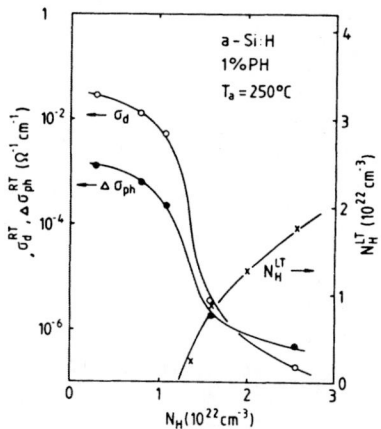

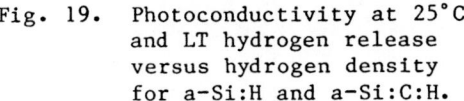

Fig. 19. Photoconductivity at 25°C
          and LT hydrogen release
          versus hydrogen density
          for a-Si:H and a-Si:C:H.

Fig. 20. Dark conductivity and
          photoconductivity at 25°C
          and LT hydrogen release
          versus hydrogen density
          for P-doped a-Si:H.

content[59] (dashed line). The close agreement between the two curves suggests a common reason for the strong decay of both photoconductivity data above N_H = 10^{22} cm^{-3}: the presence of a void structure as indicated by the onset of the low temperature hydrogen evolution. Rather similar, the doping effect of phosphorus in films prepared from mixtures of SiH_4 with 1% of phosphine drops rapidly when the LT hydrogen evolution appears (Fig. 20).

DISCUSSION

     Our evolution results suggest for undoped glow discharge a-Si:H films hydrogen incorporation mainly at positions breaking Si-Si bonds. Thus, one can understand why a large fraction of hydrogen can be removed from the silicon lattice in the evolution process without simultaneous formation of dangling bonds. In the evolution process dominant in C_H < 20 at.% films, atomic hydrogen is considered to diffuse away from its original site leaving behind a reconstructed Si-Si bond. For C_H > 20 at.% hydrogen mainly desorbs as H_2 into voids present in such material due to the hydrogen-related reduced connectiveness of the amorphous network. In this process, silicon dangling bonds are created which, however, will subsequently reconstruct to the major part. The calorimetric measurements of H_2 demonstrate that this latter evolution process occurs to some degree also in films with C_H < 20 at.%. The presence of a lower H_2 concentration in C_H > 20 at.% material than in C_H < 20 at.% films (see Fig. 2) can easily be understood by the different permeability for H_2.

     Hydrogen incorporation at bond breaking position in amorphous silicon has first been proposed by Ching et al[60]. As the distance of the hydrogen atoms in this bonding configuration is likely to be smaller than 1.9 Å, it will contribute to the broad line (clustered phase) in NMR[61]. The assignment to infrared absorption bands, however, is not so clear. Judging from the relative strength of the infrared stretching modes at 2000 and 2100 cm^{-1} as a function of substrate temperature (Fig. 21)[39] the disappearance of the low temperature evolution process (i.e. of a void structure) is clearly correlated with the transition from the 2100 to the 2000 cm^{-1} mode. This would suggest that bond-breaking hydrogen, when no voids are formed, contributes

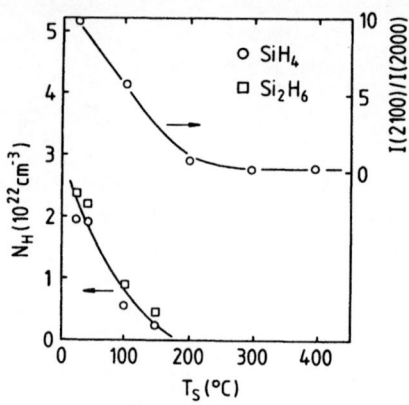

Fig. 21. LT hydrogen release and ratio of IR absorption
strengths versus substrate temperature.

to the 2000 cm^{-1} absorption band. If voids are formed, bond-breaking hydrogen would contribute to the 2100 cm^{-1} absorption. However, if the strengths
of the infrared absorption bands are converted to hydrogen densities in
bonding configurations using the oscillator strengths published by Cardona[26]
a completely different picture emerges. As the oscillator strength of the
2100 cm^{-1} vibration is almost a factor of 10 larger than that of the 2000
cm^{-1} band and since the 2000 cm^{-1} absorption has always a certain tailing
towards 2100 cm^{-1} (see Fig. 3b), the hydrogen concentration accounting for
the 2000 cm^{-1} band then turns out not to exceed ~ 3×10^{21} cm^{-3} (6 %)[26]. This
would suggest the attribution of isolated Si-H bonds to the 2000 cm^{-1} band[26]
and of any bond-breaking hydrogen to the 2100 cm^{-1} absorption.

The predominance of bond-breaking hydrogen in glow-discharge a-Si:H is
in agreement with Cody's model[54] for a widening of the optical gap due to
hydrogen-induced strain relief. The latter model implying the preferential
removal of band tail states by hydrogen incorporation, a large influence of
hydrogen on the electronic transport at the mobility edges is not expected.
The effect of hydrogen incorporation on the electronic transport (apart from
Si dangling bond saturation by a small initial amount of hydrogen) rather
seems to be an indirect one, mainly occurring via the formation of voids.
The presence of voids then leads to dangling bonds due to H_2 desorption from
void surfaces, to a decrease of the doping effect of substitutional dopants
due to a reduced dimensionality of the silicon lattice and to increased
potential fluctuations. Still, the experimentally observed increase of the
slope of $Q(1/T)$ at high hydrogen content (Fig. 18) is surprisingly small
considering the strong dependence of the doping effect on the total hydrogen
concentration (Fig. 20). This result suggests that we have to regard even
the highly hydrogenated void-rich a-Si:H as a rather homogeneous material
since inhomogeneities in the hydrogen concentration should result in a
strong spacial variation of the Fermi level, i.e. in a steep slope of
$Q(1/T)$. The explanation is given by the size of the voids which has to be
rather small also from the consideration that there is lattice reconstruction upon hydrogen removal. Of atomic dimensions, these voids should be regarded just as spacings in the amorphous silicon network large enough to
allow diffusion of H_2.

Thus, the improvement of the electronic quality of a-Si:H films when
the hydrogen concentration is decreased below 20 at.% appears to be mainly
connected to the increased density of the material making the low-temperature evolution process less likely. Still, from the calorimetric measurements of H_2 it is quite clear that also films with $C_H < 20$ at.% contain

voids. This may be due to local fluctuations of the silicon or hydrogen density or to an incomplete reconstruction of Si-Si bonds with or without involvement of impurity atoms. It is tempting to associate the occurrence of dangling bonds even in best a-Si:H films with the presence of voids. The probability p of surface hydrogen desorption can be inferred from Fig. 11 yielding $p \sim 10^{-4}-10^{-3}$ s^{-1} at 300°C, i.e. for a typical film deposition run of one hour or more practically all hydrogen bound to internal surfaces will be desorbed and, accordingly, dangling bonds are likely to be produced. Therefore, a better understanding of processes stabilizing voids or inhibiting a Si-Si reconstruction at internal surfaces could help for a further reduction of the density of dangling bonds and for an improvement of the thermal stability of a-Si:H.

ACKNOWLEDGEMENTS

The author wishes to thank G. Wichner for carrying out the IR transmission measurements, M. Gebauer for performing the ion implantation and W. Hilgers and W. Knörchen for technical assistance. The CVD a-Si films were kindly supplied by G. Harbeke and A.E. Widmer, RCA Zürich. Valuable discussions with H. Wagner are gratefully acknowledged.

REFERENCES

1. R.C. Chittick, J.H. Alexander and H.F. Sterling, J. Electrochem. Soc. 116:77 (1969).
2. N.F. Mott, Adv. Phys. 16:49 (1967).
    N.F. Mott, Philos. Mag. 19:835 (1969).
3. W.E. Spear and P.G. LeComber, Solid State Commun. 17:1193 (1975).
    W.E. Spear and P.G. LeComber, Philos. Mag. 33:935 (1976).
4. M.H. Brodsky and R.S. Title, Phys. Rev. Lett. 23:581 (1969).
5. A. Triska, D. Dennison, H. Fritzsche, Bull. Am. Phys. Soc. 20:392 (1975).
6. W. Beyer, J. Stuke and H. Wagner, Phys. Stat. Solidi (a) 30:231 (1975).
7. A.K. Malhotra and G.W. Neudeck, Appl. Phys. Lett. 28:47 (1976).
8. W. Paul, A.J. Lewis, G.A.N. Connell and D. Moustakas, Solid State Commun. 20:969 (1976).
9. M.H. Brodsky, M.A. Frisch, J.F. Ziegler and W.A. Lanford, Appl. Phys. Lett. 30:561 (1977).
10. C.C. Tsai, H. Fritzsche, M.H. Tanielian, P.J. Gaczi, P.D. Persans and M.A. Vesaghi, in: "Amorphous and Liquid Semiconductors", W.E. Spear, ed., CICL, Edinburgh (1977), p. 339.
11. H. Fritzsche, M. Tanielian, C.C. Tsai and P.J. Gaczi, J. Appl. Phys. 50:3366 (1979).
12. J.A. Reimer, R.W. Vaughan, J.C. Knights, Phys. Rev. Lett. 44:193 (1980).
13. U. Bonse, W. Schülke, G. Wolf, Philos. Mag. B 42:499 (1980).
14. P. John, I.M. Odeh, M.J.K. Thomas, M.J. Tricker, J.I.B. Wilson, J.B.A. England and D. Newton, J. Phys. C 14:309 (1981).
15. D.E. Carlson and C.W. Magee, Appl. Phys. Lett. 33:81 (1978).
16. H. v. Löhneysen, H.J. Schink and W. Beyer, Phys. Rev. Lett. 52:549 (1984).
17. J.E. Graebner, B. Golding, L.C. Allen, D.K. Biegelsen and M. Stutzmann, Phys. Rev. Lett. 52:553 (1984).
18. J. Herion, W. Beyer and H. Wagner, Fresenius Z. Anal. Chem. 319:714 (1984).
19. W. Beyer and H. Overhof, in: "Semiconductors and Semimetals 21C", R.K. Willardson and A.C. Beer, eds., Academic Press, Orlando (1984). p. 257.
20. J.C. Knights, G. Lucovsky and R.J. Nemanich, J. Non-Cryst. Solids 32:393 (1979).

21. F.J. Kampas and R.W. Griffith, Appl. Phys. Lett. 39:407 (1981).
22. P.A. Longeway, in: "Semiconductors and Semimetals 21A", R.K. Willardson and A.C. Beer, eds., Academic Press, Orlando (1984) p. 179.
23. M.H. Brodsky, M. Cardona and J.J. Cuomo, Phys. Rev. B 16:355 (1977).
24. M.S. Conradi and R.E. Norberg, Phys. Rev. B 24:2285 (1981).
25. J.C. Knights, G. Lucovsky, and R.J. Nemanich, Philos. Mag. B 37:467 (1978).
26. M. Cardona, Phys. Stat. Solidi(b) 118, 463 (1983).
27. G. Lucovsky, R.J. Nemanich, and J.C. Knights, Phys. Rev B 19:2064 (1979).
28. W. Paul, Solid State Commun. 34:283 (1980).
29. H. Shanks, C.J. Fang, L. Ley, M. Cardona, F.J. Demond and S. Kalbitzer, Phys. Stat. Solidi (b) 100:43 (1980).
30. H. Wagner and H. Ibach, in: "Festkörperprobleme XXIII", Vieweg Verlag, Wiesbaden (1983).
31. H. Wagner and W. Beyer, Solid State Commun. 48:585 (1983).
32. D.K. Biegelsen, R.A. Street, C.C. Tsai and J.C. Knights, Phys. Rev. B 20:4839 (1979).
33. W. Beyer and H. Wagner, J. Appl. Phys. 53:8745 (1982).
34. L.A. Pétermann, Prog. Surf. Sci. 3:1 (1972)
35. J.A. McMillan and E.M. Peterson, J. Appl. Phys. 50:5238 (1979).
36. R.A. Redhead, Vacuum 12:203 (1962).
37. W. Beyer, H. Wagner, J. Chevallier and K. Reichelt, Thin Solid Films 90:145 (1982).
38. W. Beyer and E. Holzenkämpfer, to be published.
39. W. Beyer and H. Wagner, J. Non-Cryst. Solids 59-60:161 (1983).
40. W. Beyer, J. Chevallier and K. Reichelt, Solar Energy Mater. 9:229 (1983).
41. K.P. Huber, in: "AIP Handbook of Physics", D.E. Gray, ed., McGraw-Hill, New York (1972) p. 7-168.
42. J.A. Kerr and A.F. Trotman-Dickenson, in: "CRC Handbook of Chemistry & Physics", R.C. Weast, ed., CRC, West Palm Beach, Fl (1977) p. F-219.
43. D.C. Allan, J.D. Joannopoulos and W.B. Pollard, Phys. Rev. B 25:1065 (1982).
44. S. Oguz and M.A. Paesler, Phys. Rev. B 22:6213 (1980).
45. P. John, I.M. Odeh, M.J.K. Thomas, M.J. Tricker, F. Riddoch and J.I.B. Wilson, Philos. Mag. B 42:671 (1980).
46. R.A. Street and J.C. Knights, Philos. Mag. B 43:1091 (1981).
47. J.C. Knights, J. Non-Cryst. Solids 35-36:159 (1980).
48. A. Barna, P.B. Barna, G. Radnoczi, L. Toth and P. Thomas, Phys. Stat. Solidi (a) 41:81 (1979).
49. W. Beyer, H. Wagner and H. Mell, Solid State Commun. 39:375 (1981).
50. W. Beyer and H. Wagner, J. Phys. Colloq. Orsay, France, 42:C4-783 (1981).
51. S.C. Shen and Cardona, Phys. Rev. B 23:5322 (1981).
52. K. Chen and H. Fritzsche, Solar Energy Mater. 8:205 (1982).
53. W. Beyer and J. Herion, to be published.
54. C.D. Cody, in: "Semiconductors and Semimetals 21B", R.K. Willardson and A.C. Beer, eds., Academic Press, Orlando (1984) p. 11.
55. L. Ley, in: "The Physics of Hydrogenated Amorphous Silicon II", J.D. Joannopoulos and G. Lucovsky, eds., Springer, Berlin (1983) p. 61.
56. J. Tauc, R. Grigorovici and A. Vancu, Phys. Stat. Solidi 15:627 (1966).
57. G. Harbeke, L. Krausbauer, E.F. Steigmeier, A.E. Widmer, H.F. Kappert and G. Neugebauer, Appl. Phys. Lett. 42:249 (1983).
58. G.D. Cody, C.R. Wronski, B. Abeles, R.B. Stephens and B. Brooks, Solar Cells 2:227 (1980).
59. W. Beyer, H. Wagner and H. Mell, MRS Proceed. (1985), to be published.
60. M.Y. Ching, D.J. Lam and C.C. Lin, Phys. Rev. B 21:2378 (1980).
61. M. Janai, R. Weil and B. Pratt, Phys. Rev. B (1985), to be published.

DEFECT PASSIVATION AND PHOTOCONDUCTION IN SPUTTERED a-Ge:H

P. D. Persans, A. F. Ruppert and C. B. Roxlo
Corporate Research-Science Laboratories
Exxon Research and Engineering Co.
Annandale, New Jersey 08801

        Studies of reactively sputtered a-Ge:H show that the dangling bond density measured by optical absorption and electron spin resonance decreases exponentially with increasing hydrogen partial pressure. The magnitude of steady-state photoconductivity is inversely related to defect density in materials deposited or annealed above 350K, indicating that photoconduction is limited by defect mediated recombination. Materials prepared at lower temperatures are poor photoconductors despite low defect density. A residual mid-gap optical absorption of ~ 10 cm^{-1} is present which cannot be removed by H_2 during growth.

INTRODUCTION

        Dangling bond defects near the center of the energy gap in amorphous tetrahedral semiconductors such as a-Si and a-Ge play a major role in determining their transport properties. For example they can fix the Fermi level position and thus control dark conductivity or they can mediate recombination to limit photocarrier lifetimes. Hydrogenation of amorphous Si (a-Si:H) leads to a decrease in dangling bond defect density by up to five orders of magnitude. The resulting scientific and technological understanding has led to the development of devices such as thin film transistors, photovoltaics and photocopiers.[1-3] Our understanding of defects and transport in a-Ge:H is less well developed despite early recognition of the ability to passivate defects with hydrogen.[4,5] In the present paper we discuss the application of optical absorption techniques to the determination of the deep defect density and its relationship to materials preparation conditions. Additionally, we demonstrate that defect density correlates inversely with the magnitude of the steady-state photoconductivity in all diode sputtered films prepared above 350K.

EXPERIMENTAL DETAILS

Samples 1-2 μm thick were prepared by rf diode sputtering of pure polycrystalline Ge in an Ar:H_2 gas mixture. The deposition system was baked prior to each deposition and had a base pressure of ~5 x 10^{-8} Torr. The Ar and H_2 gases had nominal purity of 99.9995%. For the present study Ar partial pressure and flow rate were fixed at 5 mTorr and 20 sccm respectively and the H_2 partial pressure p[H_2] was varied between 0 and 1 mTorr. Fused silica and crystalline Si substrates were clamped to a stainless steel anode which was heated radiatively. Deposition rate was ~ 1.5Å sec^{-1} for an input power of 75W and a target self-bias voltage of 1000V. The target diameter was 15 cm and the target-substrate distance was 7 cm.

Coplanar electrical contact was made with carbon paint on the top surface 5mm apart. In some cases conductivity was also measured using pre-deposited Cr strips ~ 0.5mm apart. Both photo and dark conductance were independent of applied field up to 10^3V/cm. Above this value, joule heating leads to supra-linear I-V relationships in samples with R < 10^8 Ω. Most conductivity and photoconductivity measurements were performed using a two-probe configuration and a Keithley 616 electrometer to measure current. Four probe measurements were also performed to ensure that contact resistance was negligible.

Optical absorption for 2 x 10^2 cm^{-1} < α < 10^5 cm^{-1} was measured by transmission on fused silica substrates using a Varian 2390 spectrophotometer. Absorption measurements were extended to lower values of α using photo-thermal deflection spectroscopy[6] and also by measuring photo-thermal conductance changes.[7] Photothermal measurements were matched to transmission measurements in an overlap region of at least one decade in α in the range 5 x 10^2 < α < 10^4 cm^{-1}.

Photoconductivity was excited with the monochromatic output of an interchangeable grating IR monchromator with a tungsten-halide source. The light was chopped at several frequencies between 5 Hz and 1000 Hz and modulation in the current was detected using a resistor in series with the sample placed across the input of a lock-in in order to separate photo-thermal and photo-conductive signals. Photon flux was varied between 10^{13} and 10^{15} photons cm^{-2} sec^{-1}. Photoconductance was linear in flux to within 5% over this range for all samples.

EXPERIMENTAL RESULTS AND DISCUSSION

Defect Density

In Fig. 1 we show the optical absorption coefficient α plotted

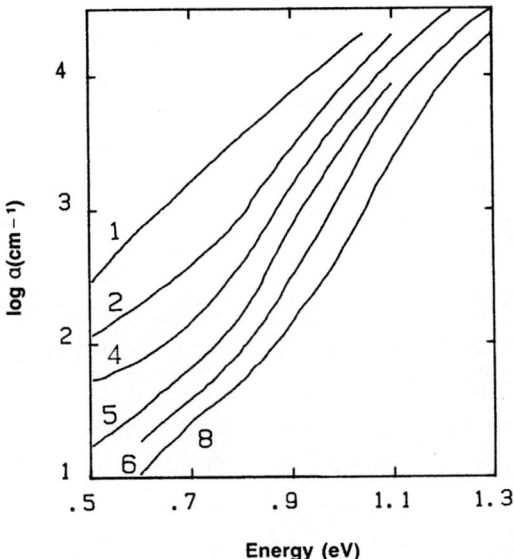

Fig. 1.   Optical absorption coefficient
α for films prepared at T_D =
410K. Solid line sections were
determined by transmission
spectroscopy and dashed
sections by photothermal
deflection and photothermal
conductance changes.

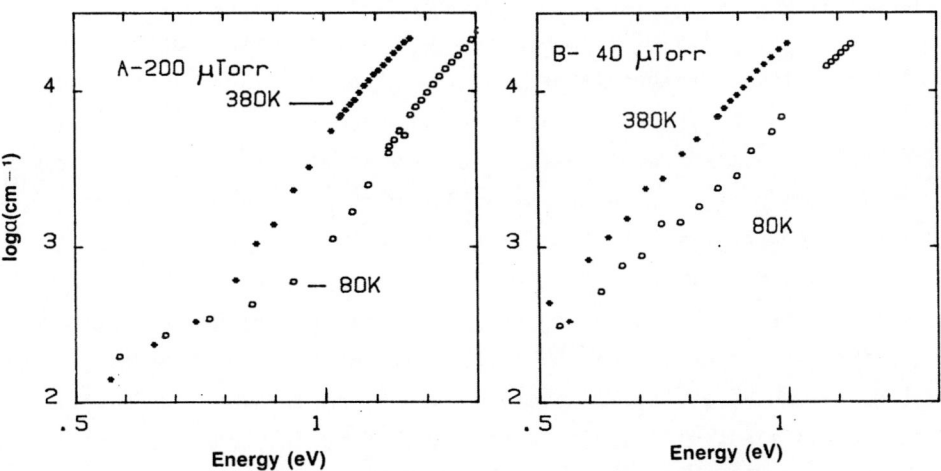

Fig. 2.   Optical absorption coefficient
α measured by transmission at
high and low temperature.

against photon energy for films prepared at T_D = 410K. Hydrogenation shifts the energy gap E_G to higher energy, sharpens the absorption edge, and causes a decrease in mid-gap absorption α_{MG}.[8] The increase in E_G arises from an increase in network order which also sharpens the absorption tail.[9] An increase in E_G also arises from H alloying.[10] The low energy absorption is decreased both by sharpening and shift of the edge and by direct defect passivation. Additional information is given in Table 1.

The separation of band tail absorption from deep defect absorption is difficult, especially for low H content samples with small energy gaps and broad band tails. We have found however that there is a distinct difference between the thermal behavior of deep defect levels and states associated with the band edge. In Fig. 2 we plot the absorption spectrum for two samples measured at 80K and 380K. The band edge absorption ($\alpha > 3 \times 10^3$ cm^{-1}) shifts upward by $\sim$ 150 meV upon cooling but the low energy absorption remains nearly constant. This temperature variation technique enables us to separate absorption associated with defects from band edges and tails even when the distinction is not obvious in room temperature measurements. In samples with the highest defect densities (i.e. - room temperature deposited unhydrogenated a-Ge), band edge and defect absorption cannot be easily separated even by temperature variation.[11]

We estimate the number of deep defects from the magnitude of the low energy absorption. In Fig. 3 we show the mid-gap absorption α_{MG} measured at one-half of the extrapolated gap E_G plotted against H_2 partial pressure for a series of films prepared at 410K. Defect absorption measured in this way correlates linearly with dangling bond spin density Ns for films with $p[H_2] < 200\mu$Torr, as we show in Fig. 4.[12] Similar correlation has been reported for a-Si:H materials.[13]

For $p[H_2] > 200\mu$Torr Ns is lower than the present sensitivity of 10^{18} cm^{-3} whereas extrapolation of α_{MG} in Fig. 4 suggests that the dangling bond density is $\sim 2 \times 10^{18}$ cm^{-3}. The nature of the dominant defect seen by absorption changes as $p[H_2]$ is increased above 200μTorr. The defect energy determined by plotting (α hν)^2 against hν shifts from $\sim$ 0.45 eV to $\sim$ 0.65 eV as $p[H_2]$ is increased from 200μTorr to 350μTorr. The data is consistent with a singly occcupied dangling bond state $\sim$ 0.45 eV from the conduction band edge which can be removed by H plus a deeper defect whose density does not depend on hydrogenation. This picture agrees with the interpretation of the doping dependence of Ns in PACVD material.[14]

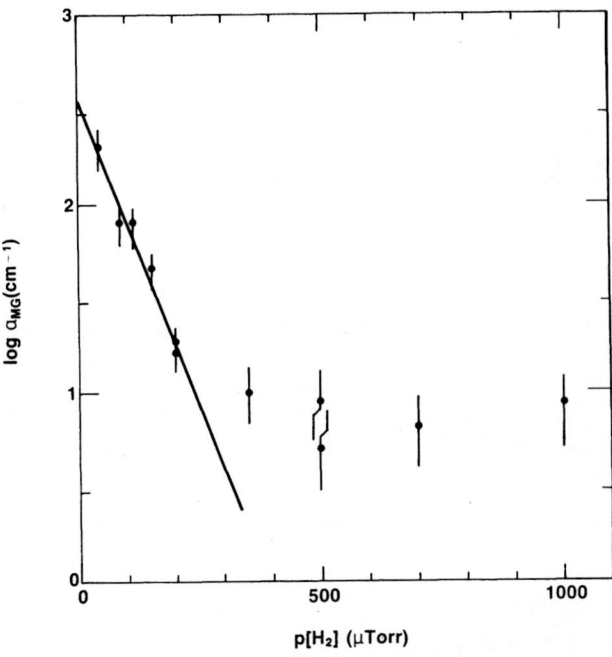

Fig. 3. Absorption coefficient α measured at one-half the extrapolated gap plotted against hydrogen partial pressure $p[H_2]$ during deposition.

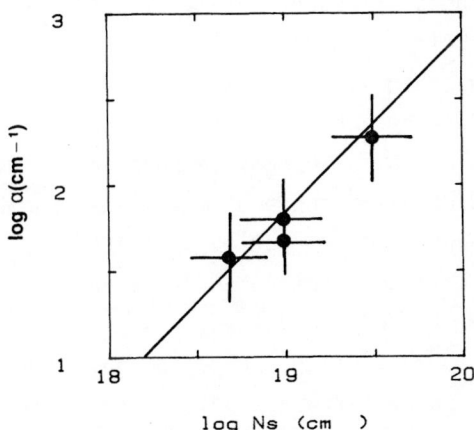

Fig. 4. Dangling bond spin density Ns plotted against mid-gap absorption α_{MG} for low $p[H_2]$ films deposited at 410K.

The dangling bond defect density decreases exponentially with increasing $p[H_2]$ up to 200μTorr for the films in Fig. 3. This suggests that dangling bond passivation is limited by the flux of active H species to the surface as has been proposed for sputtered a-Si:H.[15] We expect that the flux of active H species is proportional to $p[H_2]$ because $p[H_2]$ is a small fraction of the total pressure and all other parameters were kept fixed. In the a-Si:H model the dangling bond density is related exponentially to H_2 pressure $Ns = N_{max} \exp(-p[H_2]/p_0)$. For $T_D = 410K$ we find $p_0 = 60\mu$Torr. Connell and Pawlik[4] reported similar behavior in $T_D = 300K$ material with $p_0 = 260\mu$Torr, indicating that H more effectively passivates dangling bonds for T_D higher than room temperature.

As noted, when $p[H_2]$ is increased above 200μTorr α_{MG} levels off at ~ 5-10 cm^{-1} and is independent of $p[H_2]$. The minimum level of 5 cm^{-1} appears to hold for all materials made with $300K < T_D < 570K$. These remaining defects do not appear to have spin and suggest that these remaining defects are extrinsic or at the film surface. Hydrocarbons and oxygen are likely low level impurities and we cannot exclude the possibility of ppm levels of metallic impurities.

Photoconductivity

Measurement of the mobility-lifetime ($\mu\tau$) product in a-Ge:H necessitates subtraction of photo-thermal (bolometric) contributions from the total change in sample conductance Δg_{tot} upon illumination. Bolometric effects can be separated from photoconductive effects by using the relatively slow response times of bolometric effects of a sample on a thick substrate.[7] The bolometric contribution to conductance Δg_{bol} is given by $g E_\sigma \Delta T/kT^2$ where g is the dark conductance, E_σ is the conductance activation energy and ΔT is the frequency dependent magnitude of the temperature modulation. The room temperature conductance of intrinsic a-Ge:H decreases by 4 orders of magnitude upon hydrogenation as seen in Table 1; E_σ increases from 0.2 eV to 0.45eV as σ decreases. We assume that the entire in-phase photoresponse signal at 35 Hz is thermal for unhydrogenated material and find $\Delta T = 7 \times 10^{-5}K$ for a fully absorbed flux of ~10^{14} photons $cm^{-2}s^{-1}$. Thus, Δg_{bol} can easily be subtracted from Δg_{tot} to yield Δg_{pc} for all other samples. For $p[H_2] > 200\mu$Torr Δg_{bol} is a small fraction of Δg_{tot}.

We now address the question whether $\mu\tau$ is related to defect density as measured in the previous section. In Fig. 5 we plot $\mu\tau$ against α_{MG} for all films prepared above 370K. We find linear correlation between defect density and $\mu\tau$ over 1 1/2 orders of magnitude. From $Ns = 10^{18}$ cm^{-3} and $\tau = 3 \times 10^{-8}$ cm^2 V^{-1} we calculate a recombination center coefficient

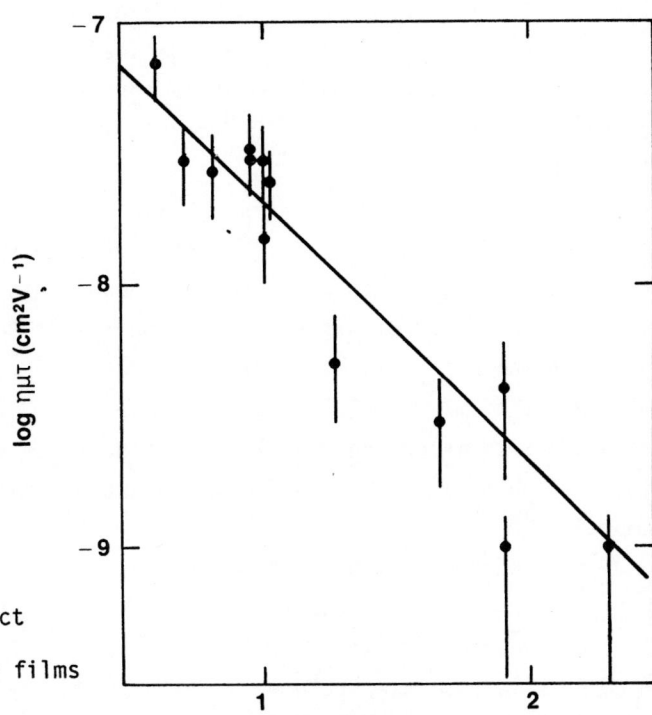

Fig. 5

Photoconductive $\mu\tau$ product plotted against defect absorption level α_{MG} for films prepared at T_D = 410K.

log α_{MG}

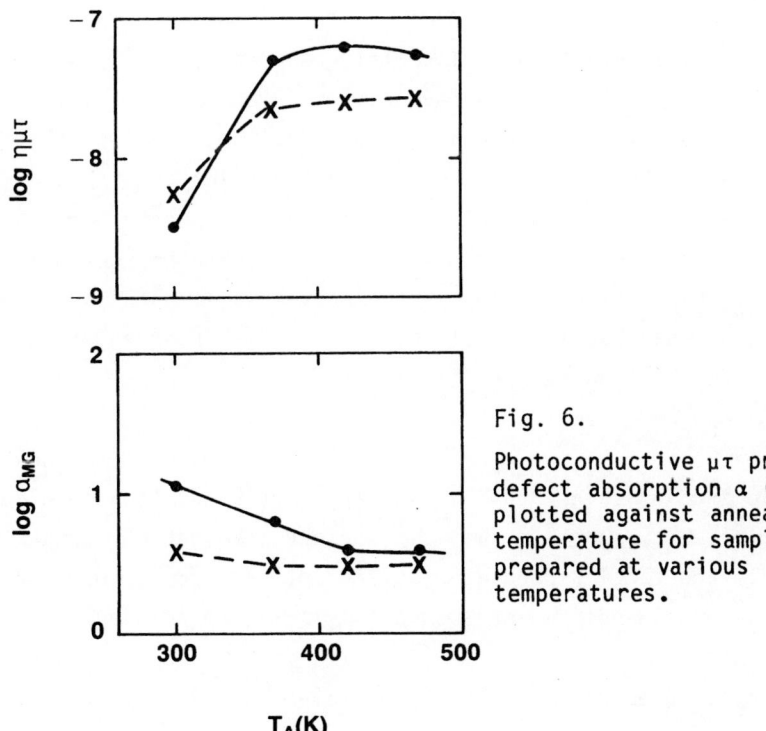

Fig. 6.

Photoconductive $\mu\tau$ product and defect absorption α ($E_G/2$) plotted against annealing temperature for samples prepared at various temperatures.

T_A(K)

$b_r \sim 3 \times 10^{-10}$ cm^3 s^{-1}. From this data alone we would conclude that hydrogen incorporation and passivation leads to improved photoconductive properties. The situation however is more complicated. Samples prepared at room temperature have $\mu\tau \sim 10^{-9}$ cm^{2}V^{-1} in agreement with Moustakas et al.[7,16] but the defect level as measured by mid-gap absorption is also low ($\sim$ 5-10 cm^{-1}). This is inconsistent with the idea that passivation always leads to better photoconductive properties. We suggest that either the microscopic structure of room temperature diode-sputtered films is different from that of higher deposition temperature films or that the nature of the defect is different. We note that annealing of the room temperature films at relatively low temperature leads to dramatic improvement in $\mu\tau$ but relatively little decrease in α_{MG} as is shown in Fig. 6 in which we plot $\mu\tau$ and α_{MG} against annealing temperature for two samples.

The present photoconduction results link early work by Moustakas et al.[7,16] with recent reports of high $\mu\tau$ products in a-Ge:H by Jones et al.[17] and Rudder et al.[18] Moustakas found that hydrogenation did not improve $\mu\tau$ in room temperature diode-sputtered a-Ge:H even though Ns measurements[4] indicated good defect passivation. Jones et al.[17] reported improved $\mu\tau$ of $\sim 10^{-8}$ cm^2 V^{-1} in undoped material prepared by PACVD and attributed the difference to better defect passivation due to a gentler deposition process. More recently, Rudder et al.[18] reported $\mu\tau$ products as large as 10^{-6} cm^2 V^{-1} in intrinsic material prepared by magnetron sputtering and also attributed the improvement to lower defect levels although in neither case was defect density reported. We have established that $\mu\tau$ of diode-sputtered a-Ge:H can be improved drastically by increasing T_D to 400-600K or by annealing in the same temperature range. Both PACVD[17] and magnetron sputtered[18] materials were prepared in this T_D range. We suggest therefore that the major improvement is not intrinsic to the deposition process but is associated with thermal rearrangement during and after growth. The nature of this effect is unclear.

SUMMARY

We have studied defect passivation and photoconduction in diode-sputtered hydrogenated amorphous germanium. Defect passivation appears to be kinetically limited for low partial pressures of hydrogen. For higher partial pressures the mid-gap defect absorption is $\sim$10 cm^{-1} independent of $p[H_2]$.

For samples prepared above 370K, steady-state majority carrier photoconductivity is inversely proportional to the defect absorption

Table 1

Deposition and transport parameters for several a-Ge:H films

Sample	$p[H_2]$ (μTorr)	T_D (K)	E_G (eV)	σ ($\Omega^{-1}cm^{-1}$)	$\mu\tau$ (cm^2V^{-1})	α_{MG} (cm^{-1})
1	40	410	0.82	1×10^{-3}	$<10^{-9}$	200
2	80	410	0.86	4.5×10^{-4}	$<10^{-9}$	80
3	110	410	0.86	4×10^{-4}	4×10^{-9}	80
4	150	410	0.92	3.3×10^{-4}	3×10^{-9}	45
5	200	410	0.95	1.3×10^{-4}	3.5×10^{-9}	17
6	350	410	1.05	6×10^{-5}	3×10^{-8}	10
7	500	410	1.07	3×10^{-5}	3×10^{-8}	9
8	700	410	1.10	3×10^{-5}	2.7×10^{-8}	6.5
9	1000	410	1.14	1.5×19^{-5}	3.3×10^{-8}	9
10	0	300	0.75	3.5×10^{-2}		500
11	500	300	1.09	8.5×10^{-6}	3×10^{-9}	11
12	1000	300	1.18	1×10^{-6}	6×10^{-9}	4
13	500	570	0.99	9×10^{-5}	1.5×10^{-8}	10
14	500	470	1.05	7×10^{-4}	3×10^{-8}	5

level. We suggest that the photoconduction mechanism is simply excitation of carriers to the band edge and transport in extended or band-tail states limited by recombination through deep defects. Photocarrier $\mu\tau$ products of $\sim 3 \times 10^{-8}$ cm^2V^{-1} approach values reported for materials produced by glow discharge or magnetron sputtering. We believe that systematic variation of deposition parameters could lead to further improvements in diode sputtered material. On the other hand, photoconductive behavior in room temperature prepared material cannot be simply interpreted. Good defect passivation is inferred from low mid-gap absorption values but we find that $\mu\tau$ remains low until the samples are annealed. We suggest that photocurrent in room temperature material is limited by a mechanism other than recombination through defects. The fundamental properties and limitations of a-Ge:H have not yet been delineated. More work is clearly warranted.

We thank G. Cody and B. Brooks for assistance with optical transmission measurements and Prof. E. A. Schiff for electron paramagnetic resonance measurements.

REFERENCES

1. The Physics of Hydrogenated Amorphous Silicon, ed. J. Joanopoulos and G. Lucovsky, (Springer-Verlag, 1984) and references therein.
2. Hydrogenated Amorphous Silicon I and II, ed. J. Pankove, (Academic Press, 1984) and references therein.
3. Proceedings of the 10th International Conference on Amorphous and Liquid Semiconductors, ed.
4. G. A. N. Connell and J. Pawlik, Phys. Rev. B, 13, 787, (1976).
5. A. J. Lewis, Phys. Rev. B, 14, 658 (1976).
6. W. B. Jackson, N. M. Amer, A. C. Boccara and D. Fourner, Appl. Opt. 20, 1333 (1981).
7. T. D. Moustakas, G. A. N. Connell and W. Paul, in Electronic Phenomena in Non-Crystalline Solids, ed. B. T. Kolomiets (Acad. of Sciences of the USSR, 1976), p. 310.
8. We define the energy gap E_G by fitting the absorption to $\alpha\, h\nu = (h\nu - E_G)^2$. The mid-gap absorption α_{MG} is the value of α at $h\nu = E_G/2$.
9. P. D. Persans, A. F. Ruppert, G. D. Cody, B. G. Brooks and W. Lanford, AIP Conf. Proc. No. 120, (American Inst. of Physics, 1984), p. 349.
10. P. D. Persans, A. F. Ruppert, S. S. Chan, G. D. Cody, Sol. St. Commun., 54, 203, (1984).
11. M. Zavetova and V. Vorlicek, phys. stat. sol. (b) 48, 113 (1971).
12. We find essentially the same results if we characterize defect density by choosing to measure α at single low energy (e.g. - α (0.6 eV)). A third method, plotting $(\alpha\, h\nu)^2$ against $h\nu$ and characterizing defect density by the slope, also yields quantitatively similar conclusions.
13. W. B. Jackson and N. Amer, Phys. Rev. B, 25, 5321, (1981).
14. M. Stutzmann, J. Stuke and H. Dersch, phys. stat. sol. 115, 141 (1983).
15. T. Tiedje, T. Moustakas, J. Cebulka, Phys. Rev. B 23, 5634, (1981).
16. T. D. Moustakas and W. Paul, Phys. Rev. B, 16, 1564, (1977).
17. D. I. Jones, W. E. Spear, P. G. LeComber, S. Li, R. Martins, Phil. Mag. B, 39, 147 (1979).
18. R. A. Rudder, J. W. Cook, Jr., and G. Lucovsky, Appl. Phys. Lett. 43, 871, (1983).

ION IMPLANTED HYDROGENATED AMORPHOUS SILICON

Milena Závětová and Irena P. Akimchenko[*]

Institute of Physics, Czechoslovak Academy of
Sciences, 180 40 Prague 8, Czechoslovakia

An attempt is made to review our results concerning the influence of some elements (3d transition metals Fe, Co, Ni, Cr, representatives of III and V groups B, Ga, P, As and further Mg, Cl and Ar as representatives of the II, VII and VIII groups), introduced into the a-Si:H by means of the ion implantation, on its physical properties as IR absorption, electrical conductivity, photoconductivity and photoluminescence.

INTRODUCTION

The method of ion implantation has been widely used in the semiconductor technology. It is, however, employed also for the study of some fundamental problems of solid state physics, in particular enabling an exact doping by impurities even in concentrations exceeding the equilibrium ones or introduction of such impurities that are usually not built-in during the crystal growth.

The use of this method for the case of amorphous semiconductors has also been successful. It was almost unambiguosly proved that oxygen, by saturating the dangling bonds in a-Ge, increases the optical gap (E_g) and resistivity (ρ). In this way it was possible to explain the diverse results obtained on a-Ge prepared in various laboratories[1]. It was further shown[2] that even in a-Si the impurities that can form chemical bond with Si, do increase E_g and ρ. Hydrogenated amorphous silicon (a-Si:H) has been studied and used for the last 10 years as a material useful for solar energy conversion. In spite of the fact that in the fundamental paper[3] the ion implantation was shown to be suitable doping method, it is not used yet for this purpose apparently for its stringent demands on technology. On the other hand it is employed for the study of the influence of impurities and defects on the properties of a-Si:H prepared by various methods. The exploitation of a-Si:H for fabrication of semiconductor elements, in particular transistors[4], may make ion implantation useful for this material too.

----

[*] Institute of Physics, Academy of Sciences of the USSR, Moscow

The aim of the present work is to review some of our papers dealing with the influence of various atoms introduced into a-Si:H by ion implantation on its optical, photoelectrical and photoluminescent properties, because they were mostly published in hardly accessible journals and proceedings.

EXPERIMENTAL

Our measurements were mostly performed on a-Si:H layers (d=0.42-1 μm) deposited in glow discharge of the mixture SiH_4+Ar (4%+96%) onto the crystalline Si or glass substrates at temperatures 190-290°C. Some of the measurements were also done on sputtered layers. The implantation was carried through at room temperature using ion accelerator with energies up to 350 keV. We used two different energies in order to divide the implanted ions in the given layer in a sufficiently homogeneous way. According to the type of the ions and their energies the depths of the implanted layers were between 0.23-0.41 μm, the impurity concentration being 10^{16} to 10^{21} cm^{-3}. After the ion implantation the layers were isochronously annealed for 30 minutes in vacuum at various temperatures from 300 to 600°C.

Table I.

X	N 10^{19}cm^{-3}	d μm	d_{impl} μm	C_H at%	E_1 eV	E_2 eV	E_3 eV	ρ Ω cm	$\bar{\rho}$ Ω cm
P	1	0.42	0.41	18	1.83	1.42	1.75	1.5×10^8	1.5×10^5
P	1	1.9	0.41	9	1.60	1.40	1.60		
B	1	0.42	0.41	18	1.83	1.42	1.65	1.5×10^8	2.0×10^6
B	1	1.9	0.41	9	1.60	1.40	1.60		
Cr	1	1.04	0.23	12	1.70	1.55	1.68	2.0×10^8	8.0×10^9
Co	1	1.04	0.23	12	1.70	1.56	1.72	2.0×10^8	5.0×10^7
Ni	1	1.04	0.23	12	1.70	1.40	1.50	2.0×10^8	9.0×10^7
Fe	1	1.04	0.25	10	1.62	1.42	1.60		
Mg	10	0.42	0.27	19	1.85	1.50	1.82	2.0×10^{11}	3.0×10^8
Ga	8	1.04	0.40	15	1.75	1.61	1.76		
As	8	1.04	0.40	15	1.75	1.61	1.76		

E_1, E_2, E_3 is the optical gap of the as-deposited, implanted, and implanted and annealed samples respectively, ρ is the resistivity of the as-deposited and $\bar{\rho}$ implanted and annealed samples.

The following ions were used for implantation: transition metals Fe, Co, Ni, Cr; dopants of III and V groups: B, Ga, P, As, and further Mg, Cl and Ar from II, VII and VIII groups. The rich variety of the implanted elements was motivated by the endeavour to find the influence of the chemical nature of the impurity on the properties of a-Si:H and to estimate the rules of mutual interaction of the impurity with the a-Si:H network.

The hydrogen content was determined either according to the empirical relation[5] E_g = 1.37 + 2.5 X, where E_g is given by the extrapolation of the dependence $(\alpha h \nu)^{1/2} = f(n \nu)$ in the region of the absorption edge, or better from the absorption in the 650 cm^{-1} band with the help of the relation[6] $C_H = A \int \alpha \, d\omega / \omega_L \alpha_{L2}$ is the absorption coefficient in the maximum and $A = 1.6 \times 10^{19}$ cm^{-2}. The agreement of the concentration determined by the two methods was not always satisfactory. The layers deposited at 290°C gave a fair agreement within the limits $\pm$ 2%.

The Tab. 1. summarizes some of the parameters of the a-Si:H layers as-deposited, implanted and subsequently annealed at 350°C for 30 minutes.

IR SPECTRA

The IR spectra were studied on double beam spectrophotometers Perkin Elmer 325 and Zeiss UR 10. In Fig. 1 as-deposited sample (curve 1) contains both Si-H and Si-H$_2$ bonds, the bands in the 2100 cm^{-1} region are broad and mutually overlapping. After implanting Cl (curve 2) and As (curve 3) in concentrations 6×10^{20} cm^{-3} and 5×10^{20} cm^{-3} respectively and subsequent annealing at 400°C (curves 1', 2', 3') the following changes are visible: for Cl both bands are clearly distinguished, which indicates some ordering of the structure, while in the case of As the broad band is not split but a new band at $\sim$ 2200 cm^{-1} is formed on its short wavelength side, probably due to As-H bond. The difference in the influence of As and Cl on the spectra is explained by different electronegativity of As and Cl ions with respect to Si[7]. The samples that were implanted by the P ions within the whole sample volume displayed a lower absorption in the 650 cm^{-1} band than the as-deposited ones. This fact corresponds to the decrease of the number of Si-H bonds[8]. No changes of the spectra (200-4000 cm^{-1}) were caused by the implantation of Mg[9]. The IR spectra indicate interactions of the implanted ions and the a-Si:H network, the changes are, however, visible at minimum concentration of 2 at%, when already a new alloy a-Si:H:M is formed.

ABSORPTION EDGE

The absorption spectra between 1.2 and 2.3 eV were measured at 300 and 78 K. They are illustrated in Fig. 2, where the edges are plotted in the representation $\sqrt{\alpha h \nu}$ vs $h \nu$ for: as-deposited a-Si:H (1), Mg implanted sample ($N = 9.6 \times 10^{20}$ cm^{-3} - curve 2) and the implanted sample subsequently annealed at 350°C (3). It is seen that annealing which follows the implantation of Mg shifts the edge back to higher energies but the initial position of the as-deposited sample is not reached. The role of the anneal is twofold - on the one hand it heals the radiation damage and on the other hand it probably increases the doping effect of the implanted impurity[2].

The dependence of E_g on the concentration of implanted Ni before and after annealing at 350°C is shown in Fig. 3. At low concentrations (10^{16} - 10^{18} cm^{-3}) the restoration of E_g is practically complete. Beginning at appr. 10^{19} cm^{-3}, which we call the "critical" concentration, E_g does not reach its initial value and when further increasing the dose E_g is sharply decreased-down to 1.2 eV for $N = 2 \times 10^{21}$ cm^{-3} - and other physical properties are also changed, in particular the photosensitivity

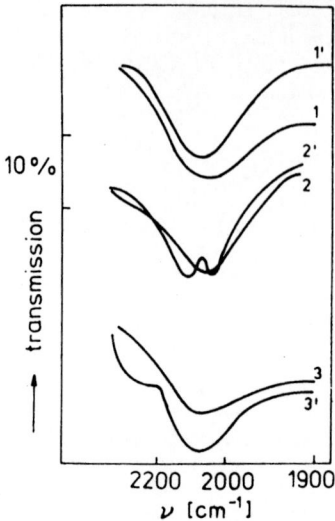

Fig.1. IR spectra of as deposited a-Si:H (1) Cl (2) and As (3) implanted,(1', 2', 3') after 400°C anneal.

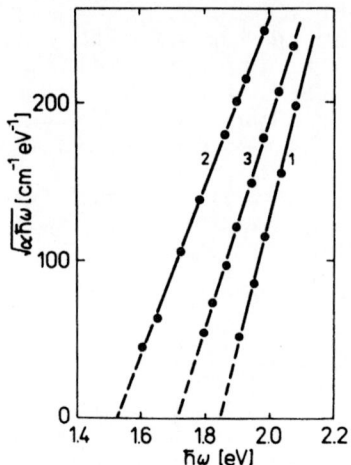

Fig.2. Absorption edges of a-Si:H as-deposited (1), Mg $(9.6 \times 10^{20}$ $cm^{-3})$ implanted before (2) and after (3) 350°C anneal.

is lost and the resistivity drops. For the case of Mg and P the changes are not so large for the same doses. It is possible that the high activity of Ni with respect to H and a high probability of the formation of Ni silicides[10] leads to these catastrophic changes in a-Si:H layers. At high doses of implanted ions, independent of their kind, E_g attains the limiting value of about 1.4 ± 0.2 eV, corresponding to a-Si. This indicates a large amount of broken Si-Si bonds. The degree, to which E_g is restored to its initial value after implantation and subsequent annealing, is given by the chemical nature of the impurity. It is worth noting that e.g. implanted Fe or Ni passivate the Si-Si bonds in a-Si layers and after annealing E_g is increased[10,11]. The presence of H substantially changes the interaction of the impurity with the a-Si net.

The study of the dependence of E_g on the dose and chemical nature of the implanted ions leads to these conclusions: for chemically active (with respect to H) impurities that are able to dope a-Si:H the optical gap is well restored by annealing at ~400°C, when the exodiffusion of hydrogen practically begins and E_g has to decrease. In the case of impurities with weak or no doping effect and impurities that do not form any chemical bond with Si, the changes of E_g with the annealing temperature practically correspond to those in unimplanted layers. Implantation of transition metals diminishes the changes of E_g produced by annealing at high temperatures[10]. All these facts point to the inference that the presence of some impurities stabilizes H in the a-Si:H layers and leads to decrease of Staebler-Wronski effect, which was observed in layers with implanted As and Ga ions[12].

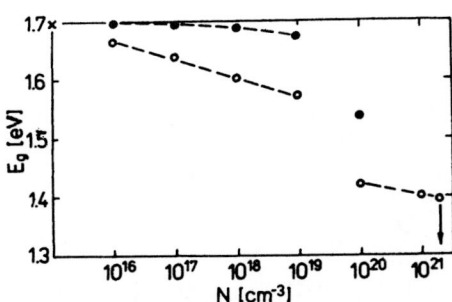

Fig.3. Dependence of the optical gap of a-Si:H on the concentration of implanted Ni ($\times$ as-deposited, $\circ$ implanted, $\bullet$ implanted and annealed).

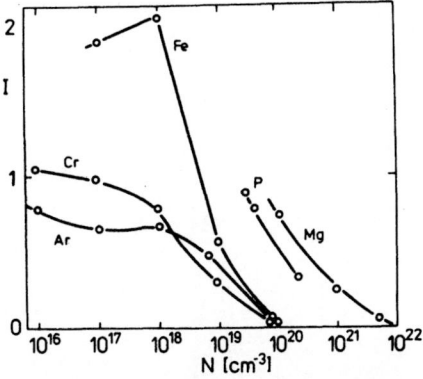

Fig.4. Dependence of the ratio I of the edge luminescence intensity of the implanted + annealed to as-deposited a-Si:H on the concentration of the implanted ions.

PHOTOLUMINESCENCE

The photoluminescence of a-Si:H layers was studied in the range from 0.5 to 1.4 eV at 78 K. It was excited by high-pressure Xe lamp equipped with a set of filters. The luminescence spectrum of as-deposited a-Si:H layers has a single band with a maximum at 1.17 ± 0.02 eV (and a weakly pronounced longwave shoulder), which corresponds to the s.c. edge luminescence (EL) connected with the recombination of carriers between the tails of the valence and conductivity bands[13]. After implantation this EL becomes progressively quenched with increasing dose. The luminescence disappears with concentrations of implanted impurities $\sim 10^{19}$ cm^{-3}. In order to compare the influence of various impurities on EL, all the implanted layers were annealed at identical conditions at 350°C for 30 minutes. The anneal results in a restoration of the EL band, the degree of the reconstruction depending on the type of the ion and the dose employed. Fig.4 presents the dependences of the ratio I (of the EL intensity of the implanted and annealed samples to the EL intensity of the as-deposited ones) on the dose of Cr, Fe, Ar, Mg and P ions. These dependences show the influence of the chemical nature of the implanted ions on EL. While e.g. Cr slightly decreases I at $N = 10^{18}$ cm^{-3}, Fe increases the EL intensity, which points to the decrease of the recombination centre concentration[11]. With Cr, Fe and Ar at the concentration 10^{20} cm^{-3} EL is completely quenched even in annealed samples, while with Mg and P the ratio I stays at ~ 0.8 for this concentration and even for 10^{21} cm^{-3} the samples possess some photoluminescence[14,15] as seen from Fig.4. The decrease of the EL intensity is accompanied by the appearence of a weak maximum on the longwave shoulder at 0.9 or 0.8 eV due to the recombination of the carriers between the tails of the valence band and the defect levels (dangling bonds). The

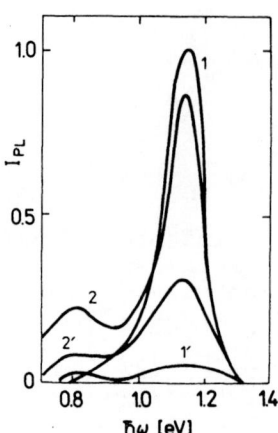

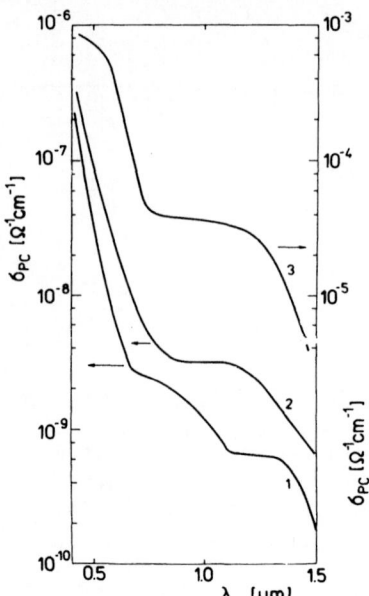

Fig. 5. Photoluminescence
spectra of a-Si:H; (1) as-
deposited, (2) Ni implan-
ted and 300°C anneal,
(1', 2') annealed at
500°C.

Fig. 6. Photoconductivity
of a-Si:H as-deposited (1),
B (2) and P (3) implanted,
both the same concentration
$N = 2\times10^{20}$ cm^{-3}.

situation is illustrated by Fig. 5, where we present the photoluminescent spectra of the as-deposited a-Si:H layer (curve 1),
of the Ni implanted layer annealed at 300°C (2) and of both
layers after annealing at 500°C (1' and 2'). The intensity of
the photoluminescence in the EL band and the defect one after
annealing at 500°C is higher in the implanted samples than in
the as-deposited which indicates a certain stabilization of the
properties at higher temperatures by implantation.

CONDUCTIVITY AND PHOTOCONDUCTIVITY

Conductivity and photoconductivity was studied in samples
implanted by Ga, As, P, B and N[14,16-18]. Impurities of the III
and V groups are effective in doping, the best results were
achieved with layers implanted by P (2×10^{20} cm^{-3}), where the
resistivity was decreased from 1.5×10^{8} Ω cm to 2.3×10^{3} Ω cm by
implantation[14]. The implantation of doping elements into the
a-Si:H layers usually increases the photoconductivity both in
the longwave and shortwave range. Fig. 6 shows the photoconductivity spectrum of the as-deposited layer (curve 1) and the
spectra of layers implanted by the same concentration (2×10^{20}
cm^{-3}) of B (curve 2) and P (curve 3). The curve for the as-deposited layer displays 3 regions of the photocurrent growth connected to the transitions at energies of ~ 0.8 eV, ~ 1.2 eV
and ~ 1.7 eV (in the high-resistivity layers $E_F \simeq 0.8$ eV). In the
a-Si:H:B layer the photoconductivity in the impurity region is
increased more than in the intrinsic region ($E_F \simeq 0.4$ eV). The

transitions in the middle part of the spectrum are smeared and the longwave edge is shifted to 0.9 eV. In the layer implanted by P the photoconductivity in the intrinsic region is increased by almost 5 orders of magnitude. The Fermi level lies by E_C − 0.22 eV and the photoconductivity in the longwave region is due to the transitions from the filled level E_C − 0.8 eV.

The a-Si:H layers prepared by cathode sputtering[2] are photosensitive at longer wavelength both in the as-prepared state and after implantation by Ga and N. At the same time it must be stressed that the ratio of the dark to the conductivity under illumination is lower with implanted than with the as-prepared layers; the same is true also for the layers that were doped during the growth.

REMARK

It is worth noting that during implantation C and O are simultaneously incorporated into the layers[19]. Moreover, at high doses of implanted ions (e.g. $D = 6 \times 10^{16}$ cm^{-2}, $j \sim 4$ μAcm^{-2} of As or Cl) a columnar structure of β-SiC was found by the reflection electron diffractography[20] (the columns of diameter 6-10 nm, distant 20 nm and sorrounded by SiO_2). It was supposed that the synthesis of β-SiC is due to the radiation-accelerated diffusion of C and O atoms which are present at the pressure $\sim 10^{-4}$ Pa. The columnar structure is not found in crystalline Si implanted under analogous conditions, but it was discovered in a-Si:H layers not subjected to implantation and it was found that the column size depends on the growth conditions[21] and that the density of the columns is higher than that of the sorrounding material. It seems therefore probable that the columnar structure, which we found in implanted specimens, is created by the accelerated synthesis on nuclei formed by the columns already present in the a-Si:H layer.

CONCLUSIONS

From the complex study of the a-Si:H layers implanted by various ions the following conclusions may be drawn:

(i) The influence of the chemical nature of the implanted ions (Fe, Co, Ni, Cr, B, Ga, P, As, Mg, Cl and Ar) was expressed in various efficiency of doping and diverse effect on the intensity of edge photoluminescence as well as restoration of the energy gap due to subsequent annealing. A fair restoration of E_g and of the intensity of EL is apparently conditioned by the additional passivation of the dangling bonds. The largest effect of the chemical nature of the implanted ions was found when implanting transition metals.

(ii) The implantation of some impurities (Ni, As, Ga, Cl) orders the structure of a-Si:H network, stabilizes H in the layers which is expressed in smaller changes of E_g and photoluminescence intensity after anneal at high temperatures. Reasonably fair restoration of E_g and of EL in the layers with the concentration of doping impurities $\sim 10^{20}$ cm^{-3} gives evidence about negligible increase of the localized state density, connected to the dangling bonds.

(iii) The ion implantation enables the production of photo-sensitive "low-ohmic" n-type layers (doping by P and As) and p-type layers of higher resistivity (doping by B and Ga).

ACKNOWLEDGEMENTS

We should like to thank V.V. Krasnopevtsev, J. Zemek and many other coworkers, too numerous to list, for their valuable help.

REFERENCES

1. M. Závětová, J. Zemek, I.P. Akimchenko, and J. Miljutin, Czech. J. Phys. B 26:1409 (1976).
2. M. Závětová, J. Zemek, and I.P. Akimchenko, Czech. J. Phys. B 31:744 (1981).
3. S. Kalbitzer, G. Müller, P.C. Le Comber, and W.E. Spear, Phil. Mag. 41:439 (1980).
4. P. C. Le Comber, A.J. Snell, K.D. Mackenzie, and W.E. Spear, J. Phys. Coll. C-4 Suppl. 42:423 (1981).
5. J. P. De Neufville and A.F. Ruppert, Bull. Am. Phys. Soc. 24:400 (1979).
6. M. Cardona, Phys. Stat. Sol. b 118:463 (1983).
7. I. P. Akimchenko, M. Závětová, and D.P. Utkin-Edin, Czech. J. Phys. B 32:827 (1982).
8. I. P. Akimchenko, V.V. Krasnopevtsev, and D.P. Utkin-Edin, Materials 7th Int.Conf.on Ion Impl.,p.176,Vilnius (1983).
9. M. Závětová and I.P. Akimchenko, Materials 7th Czech. Spectr. Conf., vol.2,p.126, České Budějovice (1984).
10. I. P. Akimchenko, A.N. Karryev, V.S. Vavilov, and A.A. Gippius, Kratkie soobchenia po fizike, No.8:16 (1984) in Russian.
11. A. N. Karryev, I.P. Akimchenko, and A.A. Grippius, Materials 7th Int. Conf. on Ion Impl., III-112, Vilnius (1983).
12. I. P. Akimchenko et al., Pisma v ZhETF 9:448 (1981) in Russ.
13. R. A.Street, Adv. in Phys. 30:593 (1981).
14. A. N. Karryev, I.P. Akimchenko, V.S. Vavilov, and A.A. Gippius, Kratkie soobchenia po fizike (1985) in print, in Russian.
15. I. P. Akimchenko, M. Závětová, A.N. Karryev, and V.V. Krasnopevtsev, Poverchnost 1985 in print, in Russian.
16. I. P. Akimchenko et al., Proc. Conf. Amorfnye poluprovodniky '80, ed. A.M. Andriesh, p.268 Kishinew (1980), in Russian.
17. I. P. Akimchenko, V.V. Krasnopevtsev, and D.P. Utkin-Edin, Mater. 7th Int. Conf. on Ion Impl., p.14,176, Vilnius (1983).
18. I. P. Akimchenko et al., Fiz. Tekhn. Polupr. 16:656 (1982).
19. J. Zemek, I.P. Akimchenko, and V.V. Krasnopevtsev, Materials Int. Meeting on Ion Impl., ed. M. Setvák, p.139, Prague (1981).
20. M. Závětová, M. Matyáš, I.P. Akimchenko, V.V. Krasnopevtsev, Proc. Conf. Amorphous Semiconductors '82, ed. R. Grigorovici, p.144, Bucharest 1982.
21. H. Fritzsche, Thin Solid Films 90:119 (1982).

THE ROLE OF IMPURITIES IN HYDROGENATED

AMORPHOUS SILICON

David E. Carlson
Solarex Thin Film Division
826 Newtown-Yardley Road
Newtown, PA 18940

INTRODUCTION

It is well known that impurities can strongly affect the properties of crystalline semiconductors. However, early work with amorphous semiconductors showed that these materials were relatively insensitive to impurities. This result was not totally unexpected since amorphous semiconductors were known to possess a large density of localized states. Moreover, most scientists believed that many of the effects of impurities in crystalline semiconductors would be supressed by the amorphous matrix adjusting so as to satisfy the normal valence bonding requirements of the impurity.

In 1969 Chittick et al. (1) published a landmark paper that disclosed some of the properties of a new type of amorphous semiconductor, hydrogenated amorphous silicon (a-Si:H). This material typically contains about 5-20 at.% of hydrogen. The hydrogen acts as an efficient passivator of dangling bonds so that it is possible to grow a-Si:H films with low densities of gap states ($\sim 10^{+15} cm^{-3}$). As a result of the low density of gap states, a-Si:H is much more sensitive to impurities than other amorphous semiconductors.

In this paper, we will first review the types of impurities found in a-Si:H films and the sources of these impurities. We will then consider the types of defect states produced in crystalline silicon by these impurities as well as the interaction of the impurities with hydrogen. We will also discuss the effects of impurities on the density of gap states and transport in a-Si:H films. We will conclude with a discussion of the role of impurities with respect to metastable states in a-Si:H.

165

IMPURITIES FOUND IN a-Si:H FILMS

Quantitative measurements of impurities in a-Si:H films are usually made by means of SIMS (secondary ion mass spectrometry) (2), but sometimes other techniques are used such as spark mass spectroscopy, Auger electron spectroscopy, infrared absorption and GDOS (glow discharge optical spectroscopy) (3). The most common impurities found in a-Si:H films are oxygen, carbon, and nitrogen (4). Oxygen and carbon concentrations are typically in the range of 10^{+19} to $10^{+20} cm^{-3}$ while nitrogen concentrations are usually one order of magnitude less. Chlorine is often detected in a-Si:H films at a level in the range of 10^{+16} to $10^{+17} cm^{-3}$ (5).

The cleanest a-Si:H films that have been produced to date were deposited by rf glow discharge in a bakeable, ultra-high vacuum system (6). The oxygen content as determined by SIMS was $2 \times 10^{+18} cm^{-3}$; carbon, $4 \times 10^{+17} cm^{-3}$; nitrogen, $5 \times 10^{+16} cm^{-3}$; chlorine, $< 10^{+17} cm^{-3}$; and fluorine, $< 5 \times 10^{+15} cm^{-3}$.

Other impurities that are often found in a-Si:H films grown from discharges in pure silane are dopants such as boron and phosphorus at levels ranging from 10^{+16} to $10^{+17} cm^{-3}$ (7). Occasionally, contaminants such as iron, chromium and aluminum are detected in a-Si:H films at levels on the order of parts per million (ppm) (7,8,9).

SOURCES OF IMPURITIES

A common source of oxygen and nitrogen in a-Si:H films is a leak in either the gas manifold lines or the glow discharge vacuum chamber. The incorporation of oxygen is much more efficient than that of nitrogen so that the oxygen concentration is generally about an order of magnitude greater than that of nitrogen when the source is an air leak. Oxygen and nitrogen impurities can also originate from the outgassing of vacuum chamber walls or manifold lines. Water vapor, carbon monoxide and carbon dioxide are oxygen-containing species that are commonly observed outgassing from the surfaces of vacuum chambers. Tsai et al. (10) observed a decrease in the oxygen content of their a-Si:H films from about 10^{+19} to about $10^{+18} cm^{-3}$ when they baked their gas lines above 110°C.

Nitrogen (as N_2) is occasionally detected in silane gas cylinders

(7). Disiloxane ($(SiH_3)_2O$) is a much more common contaminant in both silane (7,11,12) and disilane (13) gas cylinders; disiloxane levels as high as 3600 ppm have been observed in silane cylinders (11). A level of about 2000 ppm of disiloxane in silane resulted in a concentration of 1.5 x 10^{+20} cm^{-3} of oxygen in an a-Si:H film grown in a dc glow discharge (12). Higher siloxanes such as trisiloxane ($(SiH_3O)_2SiH_2$) have also been detected in silane gas cylinders (7,11).

Carbon contamination can result from the outgassing of carbon monoxide or carbon dioxide from the vacuum chamber walls and/or from hydrocarbons that originate primarily from vacuum pump oils. Occasionally, carbon-containing species such as tetrahydrofuran (C_4H_8O) have been detected in silane cylinders (14), and species such as n-butane and ethyl silane have been found in disilane cylinders (7).

Chlorine contamination often arises from the presence of species such as Si_2H_5Cl and HCl in disilane cylinders (13) and SiH_3Cl, SiH_2Cl_2, $SiCl_4$ and HCl in silane cylinders (11); levels of SiH_3Cl as high as 3000 ppm have been observed in some silane cylinders. A concentration of 1000 ppm of SiH_3Cl in silane led to a chlorine content of 400 to 600 ppm in an a-Si:H film grown in a rf glow discharge (15) while a concentration of 700 ppm of SiH_3Cl in silane led to a chlorine content of about 40 ppm in an a-Si:H film grown in a dc glow discharge (12). The incorporation efficiency depends not only on the molecular species involved, but also on deposition parameters such as discharge power.

Impurities such as boron and phosphorus are often found in a-Si:H films grown in silane discharges as a result of outgassing from chamber walls that were coated with doped a-Si:H films in earlier depositions (16). Similarly, fluorine outgassing will contaminate a-Si:H films long after a-Si:H:F films have been deposited in a reaction chamber. Fluorine may also evolve from materials such as Teflon or fluorine-doped tin oxide films if present in the discharge chamber.

Metal impurities such as tin or indium have been observed to diffuse into the p layer of p-i-n solar cells when substrates such as tin oxide or indium oxide have been used (17). Aluminum has been detected in a-Si:H films grown in discharge chambers containing Al electrodes (8), and both iron and chromium have been observed in films where the electrodes were stainless steel (9).

IMPURITY LEVELS IN CRYSTALLINE SILICON

Since the short-range order of tetrahedral bonding is preserved in a-Si:H films, some impurities may find themselves in an environment similar to that in crystalline silicon. In Figure 1, we show the energy levels in crystalline silicon of some of the impurities mentioned above. We have also shown the levels for vacancy-impurity complexes and for divacancies. Impurity states that are known to act as donors are positively charged when positioned above the Fermi level and neutral when located below the Fermi level. Acceptor-like states are neutral when above and negatively charged when below the Fermi level.

As mentioned in the introduction, the amorphous matrix may be able to adjust so as to satisfy the normal valence requirements of some impurity atoms so that the density of localized states generated by impurities may be very low in a-Si:H. Moreover, any extrapolation from Figure 1 to estimate impurity levels in a-Si:H must take into account the wider band gap in a-Si:H (about 1.7 eV as compared to 1.1 eV in crystalline silicon).

THE INFLUENCE OF HYDROGEN ON IMPURITY LEVELS

Theoretical studies indicate that hydrogen atoms can totally passivate defects such as monovacancies and divacancies (30). Recent DLTS work has shown that a hydrogen plasma cannot only reduce the density of divacancies in crystalline silicon, but also that of oxygen-vacancy complexes and phosphorus-vacancy complexes (31). Moreover, there is also evidence that atomic hydrogen can reduce the density of shallow acceptors such as those produced by boron, aluminum, gallium and indium in crystalline silicon (32, 33). In addition, DLTS work on tellurium impurities in crystalline germanium indicates that atomic hydrogen can reduce the density of certain states, has no effect on others, and can also introduce new gap states (34). Thus, we can expect that the presence of hydrogen in a-Si:H will cause a further distortion in any extrapolations from Figure 1.

Since a-Si:H typically contains about 10 at.% of hydrogen, there is a high probability that many impurities are complexed with hydrogen. Infrared absorption data show that boron-hydrogen complexes are relatively common in p-type films, especially those grown at low substrate temperatures (< 250°C) (35). Infrared studies of a-Si:C:H films show that CH_3 or C_2H_5 groups are common in films grown from

168

discharges containing CH_4 or C_2H_4, respectively (36). Thus, many
impurity complexes in a-Si:H appear to retain the basic molecular
structure present in the impurity gas molecules. It is highly likely
that gap states will occur if dangling bonds are associated with any of
the impurities or impurity-hydrogen complexes.

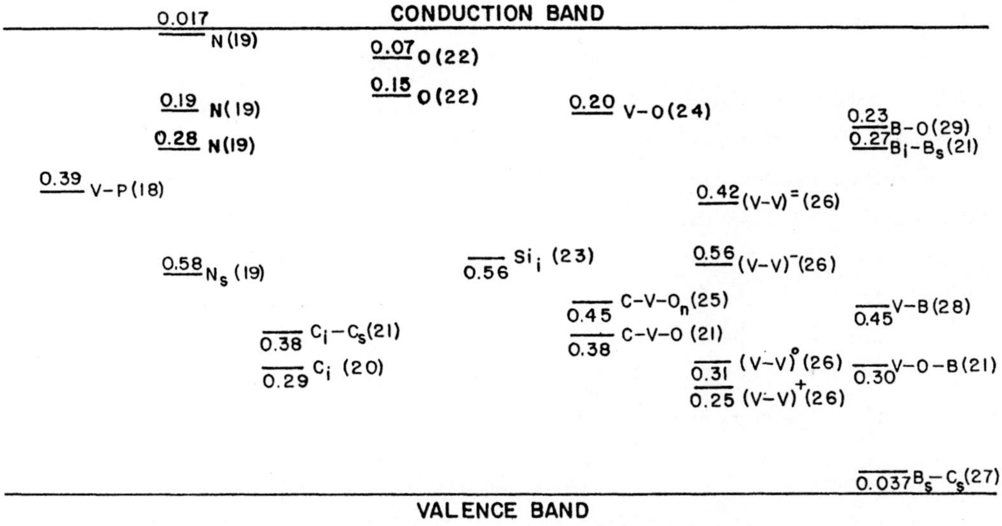

CONDUCTION BAND

```
0.017 N(19)
          0.07 O(22)
          0.15 O(22)
0.19 N(19)                    0.20 V-O(24)
0.28 N(19)                                              0.23 B-O(29)
                                                        0.27 Bi-Bs(21)
0.39 V-P(18)                              0.42 (V-V)=(26)

0.58 Ns(19)          0.56 Sii(23)    0.56 (V-V)-(26)
                          0.45 C-V-On(25)        0.45 V-B(28)
      0.38 Ci-Cs(21)      0.38 C-V-O(21)
      0.29 Ci(20)              0.31 (V-V)°(26)   0.30 V-O-B(21)
                              0.25 (V-V)+(26)

                                        0.037 Bs-Cs(27)
```

VALENCE BAND

Figure 1. Impurity energy levels in crystalline silicon. B_i-B_s is an
interstitial boron-substitutional boron complex, C-V-O_n is a carbon-
vacancy-multiple oxygen complex, $(V$-$V)^-$ is a negatively charged
divacancy, etc. The energy levels are measured either from the
conduction band edge (shown in eV above the lines) or from the valence
band edge (shown in eV below the lines). The references are listed in
parenthesis after the symbols for the impurity or impurity complex.

EFFECTS OF IMPURITIES ON TRANSPORT AND GAP STATES IN a-Si:H

Early work with a-Si:H films indicated that additions of either
oxygen and/or nitrogen created donor-like states leading to increase in
both the dark conductivity and the photoconductivity (37). There is
evidence for the presence of shallow donor levels associated with
nitrogen, especially for sputtered a-Si:H films (38). In the case of
a-Si:H films produced by glow discharge, only a very small fraction of
the nitrogen atoms (< 0.02) appears to be in active donor sites (39), and
in some cases, there is no indication of donor activity (40). On the
other hand, several studies have shown that nitrogen can introduce deep
gap states in a-Si:H (39-42). Thus, there is evidence for states in

a-Si:H similar to those seen in crystalline silicon for nitrogen impurities (see Figure 1).

Examination of a-Si:H films by ESR after X-irradiation at 77 K indicated the presence of NO_2 in films containing 1.80 at.% oxygen and 0.1 at.% nitrogen (43). This same study also found evidence for the presence of a dangling bond on an oxygen atom. Other work has shown that additions of either NO or air to a silane discharge leads to an increase in the positive space charge density, but no change in the density of dangling bonds on silicon atoms (15). Lucovsky and Lin (44) have determined from infrared absorption studies that oxygen atoms are often bonded to the same silicon atom as a hydrogen atom; thus, vibrational modes similar to disiloxane are commonly observed in a-Si:H films contaminated with oxygen. Lucovsky et al. (45) have also shown that nitrogen is incorporated in a 3-fold coordinated planar site with one hydrogen atom as a second nearest neighbor when the substrate temperature (Ts) is > 300°C; for Ts < 200°C, the nitrogen impurity is in a similar site but the second nearest neighbors are all silicon atoms.

Lucovsky and Lin (44) have also shown that the energy level of a silicon dangling bond is shifted by the presence of an impurity atom bonded to that silicon atom. As the electronegativity of the impurity atom increases, the energy level of the dangling bond shifts up in energy. Thus, while the energy level of a dangling bond in pure a-Si:H is estimated to be ~ 0.8 eV above the valence band, the level shifts upward by ~ 0.3 eV if a nitrogen atom is bonded to the silicon atom with the dangling bond (44).

Paul et at. (46) found that adding 1.6 at.% of oxygen caused $(\mu\tau)_p$ to decrease from 2×10^{-10} to $10^{-11} cm^2 V^{-1}$ and caused $(\mu\tau)_n$ to increase from 10^{-8} to $7 \times 10^{-8} cm^2 V^{-1}$. These results imply an increase in the photoconductivity and a decrease in the hole diffusion length. Another study has shown that both oxygen and nitrogen impurities reduce the diffusion length in a-Si:H films (see Table 1) indicating that recombination centers (deep gap states) are formed (5). The same work also showed that both impurities also cause an increase in the net positive space-charge density indicating the formation of donor-like states. As shown in Figure 1, donor-like states have been observed for both oxygen and nitrogen impurities in crystalline silicon.

Adding carbon-containing species such as CH_4 or C_2H_4 to silane discharges also caused the diffusion length to decrease and the space-charge density to increase (5). As shown in Table 1, the film properties are more severely affected by the larger hydrocarbon impurity. This is in agreement with the observation of Tawada et al. (36), who found that C_2H_4 produced inferior p-type a-Si:C:H films when compared to CH_4 as a feedstock material.

Chlorine-containing species such as SiH_3Cl reduce both the dark conductivity and photoconductivity of undoped a-Si:H (15); Cl apparently acts as a p-type dopant. The data in Table 1 show that adding SiH_2Cl_2 to a silane discharge also reduces the diffusion length, but has little effect on the space-charge density so few donor-like states are added.

Fluorine may create gap states in a-Si:H if SiF_3 complexes are present (47). As shown in Table 1, the addition of 0.1 vol.% of SiF_4 to a silane discharge caused some reduction in the diffusion length.

The presence of small amounts (ppm) of phosphorus in a-Si:H causes a significant increase in the dark conductivity and photoconductivity (48) and a decrease in the hole diffusion length (49). The presence of less than a few ppm of boron causes both the dark conductivity and photoconductivity to decrease (48) and the hole diffusion length to increase (50).

With the exception of small amounts of boron, most impurities adversely affect the performance of a-Si:H solar cells because of a reduction in the diffusion length (see Table 1). A simple model based on a uniform internal electric field predicts that a decrease in the diffusion length from 0.5μm to 0.2μm would cause a 40% decrease in conversion efficiency (5). The performance of a-Si:H solar cells can be degraded even more if the impurities increase the space-charge density to a level where a quasi-neutral region is created (51). Impurities often accumulate at interfaces and may also reduce the efficiency of solar cells by increasing surface recombination.

IMPURITIES AND METASTABLE CENTERS

There is considerable evidence that impurities such as oxygen (41,42,52,53), nitrogen (41,42) and carbon (5,54) create additional metastable centers when present in concentrations $>10^{+19}cm^{-3}$. In all cases, the density of defect centers increases after the a-Si:H films are exposed to prolonged illumination, and these centers can be removed by annealing at temperatures of about 200°C for 30 minutes.

The activation energy for annealing out centers associated with oxygen is about 1.0 eV (52). Nakamura et al. (42) used Isothermal Capacitance Transient Spectroscopy (ICTS) to show that both oxygen and nitrogen cause an increase in the density of states approximately 0.5 eV below the conduction band edge. In both cases, prolonged illumination caused the peak to increase. They also found that adding oxygen or nitrogen caused an increase in the space charge induced by light soaking.

Taylor and Ohlsen (55) observed an increase in an ESR line associated with oxygen after light soaking a-Si:H films containing about 1 at.% oxygen at 77 K. This ESR line was attributed to a hole trapped on a singly coordinated oxygen atom in a local SiO_x matrix. They also observed an increase in the ESR line associated with the Si dangling bond after light soaking. On heating to 300 K, they observed some annealing in the oxygen-related line but not in the Si dangling bond line.

Crandall (52) used Deep Level Transient Spectroscopy (DLTS) to show that the density of metastable centers increased as the oxygen and nitrogen concentrations increased, but only about 1 in 10^{+4} impurities appeared to be associated with a metastable center. Tsai et al. (41) showed that nitrogen concentrations $> 10^{+19} cm^{-3}$ and oxygen concentrations

TABLE I

Effects of Imputities on Diffusion Lengths

Discharge Atmosphere	L_1(m)	W_o(m)	Impurity Concentrations (cm^{-3})				
			$O(X10^{19})$	$C(X10^{19})$	$N(X10^{19})$	$Cl(X10^{17})$	$F(X10^{18})$
SiH_4 (control)	0.52	1.47	1.7	2.0	0.03	1.2	----
0.1% SiF_4 in SiH_4	0.36	0.93	2.3	1.6	0.07	0.07	1.6
2% CH_4 in SiH_4	0.32	0.83	5.2	25.0	----	0.7	----
1% CF_4 in SiH_4	0.31	0.84	0.56	14.0	0.15	0.84	140
1% CO in SiH_4	0.31	1.32	8.7	6.9	0.06	0.36	----
0.07% SiH_2Cl_2 in SiH_4	0.30	1.56	0.3	0.3	0.15	17.0	----
0.2% $(SiH_3)_2O$ in SiH_4	0.18	0.33	15.0	0.8	1.0	1.6	----
1% C_2H_4 in SiH_4	0.15	0.77	0.25	42.0	0.26	0.08	0.3
2% N_2 in SiH_4	0.12	0.23	3.8	0.2	24.0	0.4	----

NOTE: L_1 is the diffusion length measured after exposure to 1 sun illumination for 48 hours, and W_o is related to the space-charge width in the dark (5).

172

$> 10^{+20} cm^{-3}$ caused additional metastable centers (about $10^{+17} cm^{-3}$) to be created after light soaking. However, this work clearly showed that metastable centers are still present in relatively pure a-Si:H films (total impurities $< 3 \times 10^{+18} cm^{-3}$), and these centers must be associated with complexes other than impurities.

Crandall et al. (54) used DLTS techniques to demonstrate that carbon impurities can introduce a metastable center with an activation energy of about 0.4 eV. Carlson et al. (5) showed that the presence of about 10^{+20} carbon atoms per cm^3 in the i layer of an a-Si:H p-i-n solar cell caused significantly more light-induced degradation than for a cell containing about 10X less carbon.

CONCLUSIONS

It is clear that the electronic properties of a-Si:H films can be strongly influenced by impurities. In some cases, impurities such as phosphorus, oxygen and nitrogen introduce donor-like states that move the Fermi level toward the conduction band resulting in increased dark conductivity and photoconductivity. Similarly, impurities such as boron and chlorine create acceptor-like states that move the Fermi level toward the valence band and for small concentrations tend to reduce both the dark conductivity and photoconductivity. Impurities such as oxygen and nitrogen also create deep gap states that act as recombination centers leading to reduced diffusion lengths in solar cells.

However, most of the impurities in a-Si:H do not produce gap states either because of hydrogen passivation or valence accommodation. In the case of oxygen or nitrogen, only about 1 in 10^{+4} atoms appears to generate a state in the gap. For good quality a-Si:H, the doping efficiency of boron is estimated to be about 1% (56) and that of phosphorus about 10% (57) so that these atoms are much more efficient in altering the electronic properties when present as impurities. Since good quality a-Si:H has a density of gap states of about $10^{+15} cm^{-3}$, one must keep the concentration of impurities such as oxygen and nitrogen below about $10^{+19} cm^{-3}$ in order to assure the measurement of intrinsic properties. In the case of boron or phosphorus impurities, the level must be kept below about $10^{+16} cm^{-3}$ to insure undoped material.

While there is considerable evidence that impurities such as oxygen, nitrogen, and carbon can create additional metastable centers, they do not appear to play an important role when the impurity concentration falls below about 10^{+19}cm^{-3}. The type of metastable defect associated with impurities may be similar to that in present in pure a-Si:H. One model for the metastable center in the latter case involves the breaking of a weak Si-Si bond by recombination and the movement of a nearly hydrogen atom to that site (58). Impurities could induce weak Si-Si bonds by locally distorting the a-Si:H matrix or weak Si-impurity bonds may be present in some cases. In either case, the weak bond could be ruptured by recombination or by trapping a hole, and the movement of a neighboring hydrogen atom to that site would result in a reconstructed complex with two dangling bonds.

REFERENCES

1. R. C. Chittick, J. H. Alexander and H. F. Sterling, J. Electrochem. Soc. 116, 77 (1969).
2. G. J. Clark, C. W. White, D. D. Allred, B. R. Appleton, C. W. Magee and D. E. Carlson, Appl. Phys. Lett. 31, 582 (1977).
3. J. C. Zesch, R. A. Lujan and V. R. Deline, J. Non-Cryst. Solids 35 & 36, 273 (1980).
4. C. W. Magee and D. E. Carlson, Solar Cells 2, 365 (1980).
5. D. E. Carlson, A. Catalano, R. V. D'Aiello, C. R. Dickson and R. S. Oswald, Conf. Record of 17th IEEE Photovoltaic Specialist Conf., May 1-4, 1984, Kissimmee, Florida, p. 330.
6. C. C. Tsai, J. C. Knights and M. J. Thompson, J. Non-Cryst. Solids 66, 45 (1984).
7. C. R. Dickson, private communication.
8. P. E. Vanier, A. E. Delahoy and R. W. Griffith, J. Appl. Phys. 52, 5235 (1981).
9. G. Lucovsky et al., Final Report, SERI/STR-211-2475, October 1984.
10. C. C. Tsai, J. C. Knights, R. A. Lujan, B. Wacker, B. L. Stafford, and M. J. Thompson, J. Non-Cryst. Solids 59 & 60, 731 (1983).
11. R. R. Corderman and P. E. Vanier, J. Appl. Phys. 54, 3987 (1983).
12. D. E. Carlson, A. Catalano, R. V. D'Aiello, C. R. Dickson and R. S. Oswald, AIP Conf. Proc. No. 120 (AIP, NY, 1984) p. 234.
13. M. Akhton and A. E. Delahoy, private communication.
14. A. Gallagher and J. Scott, private communication.
15. A. E. Delahoy and R. W. Griffith, Conf. Record of 15th IEEE Photovoltaic Specialist Conf., 1981, p. 704.
16. Y. Kuwano, M. Ohnishi, S. Tsuda, Y. Nakashima and N. Nakumura, Jap. J. Appl. Phys. 21, 413 (1982).
17. N. Fukada, T. Imura, A. Hiraki, Y. Tawada, K. Tsuge, H. Okamoto and Y. Hamakawa, 3rd Photovoltaic Sci. & Eng. Conf., Kyoto, Japan, May 1982, p. 99.
18. J. A. Naber, C. E. Mallon and R. E. Leadon, "Radiation Damage and Defects in Semiconductors," The Institue of Physics, London, C1973, p. 26.
19. Y. Tokumaru, H. Okushi, T. Masui and T. Abe, Jap. J. Appl. Phys. 21, L443 (1982).
20. P. M. Mooney, L. J. Cheng, M. Suli, J. D. Gerson and J. W. Corbett, Phys. Rev. B15, 3836 (1977).
21. I. Weinberg and C. K. Swartz, Conf. Record of 15th IEEE Photovoltaic Specialists Conf. 1981, p. 490.
22. L. C. Kimerling and J. L. Benton, Appl. Phys. Lett. 39, 410 (1981).
23. H. Lefevre, Appl. Phys. 22, 15 (1980).
24. G. D. Watkins and J. W. Corbett, Phys. Rev. 121, 1001 (1961).

25. G. S. Oehrlein, D. J. Challou, A. E. Jaworowski and J. W. Corbett, Phys. Lett. 86A, 117 (1981).

26. H. J. Stein, "Radiation Effects in Semiconductors," eds. J. Corbett and G. Watkins, Gordon & Breach Science Publ. London, C1971.

27. W. Scott and C. E. Jones, J. Appl. Phys. 50, 7258 (1979).

28. U. S. Vavilon, B. N. Mukashev and A. V. Spitsyn, "Radiation Damage and Defects in Semiconductors," The Institute of Physics, London, C1973, p. 284.

29. B. Ross, "Lifetime Factor in Silicon," ASTM Special Technical Publication 712, p. 22.

30. V. A. Singh, C. Weigel, J. W. Corbett and L. M. Roth, Phys. Stat. Sol. (b) 81, 637 (1977).

31. S. J. Pearton, Phys. Stat. Sold. (a) 72, K73 (1982).

32. J. I. Pankove, D. E. Carlson, J. E. Berkeyheiser and R. O. Wance, Phys. Rev. Lett. 51, 2224 (1983).

33. J. I. Pankove, R. O. Wance and J. E. Berkeyheiser, Appl. Phys. Lett. 45, 1100 (1984).

34. S. J. Pearton and A. J. Tavendale, J. Appl. Phys. 54, 820 (1983).

35. D. E. Carlson, R. W. Smith, C. W. Magee and P. J. Zanzucchi, Philos. Mag. B45, 51 (1982).

36. Y. Tawada, K. Tsuge, M. Kondo, H. Okamoto and Y. Hamakawa, J. Appl. Phys. 53, 5273 (1982).

37. R. W. Griffith, F. J. Kampas, P. E. Vanier and M. D. Hirsch, J. Non-Cryst. Solids 35 & 36, 391 (1980).

38. G. J. Smith and W. J. Milne, Philos. Mag. B47, 419 (1983).

39. T. Noguchi, S. Usui, A. Sawada, Y. Kunoh and M. Kikuchi, Jap. J. Appl. Phys. 21, L485 (1982).

40. S. M. Pietruszko, K. L. Narasimhan and S. Guha, Philos. Mag. B43, 357 (1981).

41. C. C. Tsai, M. Stutzmann and W. B. Jackson, AIP Conf. Proc. No. 120 (AIP, NY, 1984) p. 242.

42. N. Nakamura, S. Tsuda, T. Takahama, N. Nishikumi, K. Watanable, M. Ohnishi and Y. Kuwano, ibid, p. 303.

43. W. M. Pontuschka, W. E. Carlos, P. C. Taylor and R. W. Griffith, Phys. Rev. B25, 4362 (1982).

44. G. Lucovsky and S. Y. Lin, AIP Conf. Proc. No. 120 (AIP, NY, 1984) p. 55.

45. G. Lucovsky, J. Yang, S. S. Chao, J. E. Tyler and W. Czubatyj, Phys. Rev. B28, 3234 (1983).

46. W. Paul et al., Tech. Rpt. No.1 , SERI/XW-1-9358-1-3, July 1 - Sept. 30, 1981.

47. W. Y. Ching, J. Non-Cryst. Solids 35 & 36, 61 (1980).

48. D. A. Anderson and W. E. Spear, Philos. Mag. 36, 695 (1977).

49. P. E. Vanier, A. E. Delahoy and R. W. Griffith, AIP Conf. Proc. No. 73 (AIP, NY, 1981), p.227.

50. W. E. Spear, H. L. Steemers, P. G. LeComber and R. A. Gibson, Philos. Mag. B50, L33 (1984).

51. G. A. Swartz, J. Appl. Phys. 53, 712 (1982).

52. R. S. Crandall, Phys. Rev. B24, 7457 (1981).

53. D. E. Carlson, J. Vac. Sci. Technol. 20, 290 (1982).

54. R. S. Crandall, D. E. Carlson, A. Catalano, and H. A. Weakliem, Appl. Phys. Lett. 44, 200 (1984).

55. P. C. Taylor and W. D. Ohlsen, Solar Cells 9, 113 (1983).

56. B. W. Faughnan and J. J. Hanak, Appl. Phys. Lett. 42, 722 (1983).

57. B. vonRoedern, L. Ley, M. Cardona and F. W. Smith, Philos. Mag. B40, 433 (1979).

58. H. Dersch, J. Stuke and J. Beichler, Appl. Phys. Lett. 38, 456 (1981).

BONDING IN AMORPHOUS SEMICONDUCTORS; BEYOND THE 8-N RULE

John Robertson

Central Electricity Research Laboratories
Leatherhead
Surrey, UK

Atomic coordinations in amorphous semiconductors are reviewed in terms of octet and non-octet bonding. Particular attention is paid to the possibility of three-centre bonding in amorphous boron-silicon and gallium-arsenic alloys, to defects and to the doping mechanism in amorphous silicon.

INTRODUCTION

In 1967 Mott[1] noted that the absence of topological constraints in a random network allowed each atom to satisfy its valence requirements locally, by adopting a coordination of 8-N, where N is the element's column in the Periodic Table. This accounted for the intrinsic semiconduction of STAG (Se, Te, As, Ge) glasses and the general insensitivity of amorphous semiconductors to impurities, because if every atom follows the 8-N rule, all its valence electrons are assigned to bonds and therefore cannot cause doping.

The 8-N rule has provided a valuable basis with which to describe bonding in amorphous semiconductors and network glasses[2]. More accurately, such bonding should be called octet bonding, as each atom completes its valence octet and this then includes such tetrahedrally-bonded systems as GaAs, in both crystalline (c-) and amorphous (a-) phases. Covalent octet systems are essentially a network of two-centre, two-electron (2c-2e) bonds, supplemented sometimes by non-bonding orbitals. In this paper we consider some more unusual bonding situations, the multi-centre bonding in electron-deficient and electron-rich systems (e.g. a-Si,B and a-GeTe) which are definitely non-octet, and also the bonding at defects and impurities, much of which can be rationalized into an octet framework.

THE THREE-CENTRE BOND

The axial (σ symmetric) 2c-2e bond is the main building block of random networks. It consists of two orbitals on adjacent atoms interacting with each other. The symmetric combination is stabilized with respect to the unbonded state and is called the bonding, σ or σ_g state, while the antisymmetric antibonding, σ^* or σ_u state is destabilised compared to the

unbonded situation. The stability is highest when the σ state is doubly
occupied and the σ* state is empty, giving a 2c-2e bond.

The three-centre bond involves three atoms[3-5]. The bond can either
by symmetric or antisymmetric with respect to some middle atom or some
mirror plane. We first consider the molecular orbitals of a symmetric
bond using the two B-H-B bridge bonds in diborane B_2H_6 as an example.
By forming sp^3 hybrids on each B, we can treat each B-H-B bridge
separately. Each bridge now consists of a sp^3 hybrid from each B and the
s orbital on the central H atom. The isotropy of the H s orbital
defines the bond as symmetric. The BHB bridge is also called an open
bond because the H bond angle exceeds 60°. We consider forming the
bridge orbitals in two steps (Fig. 1a). First we form symmetric and
antisymmetric combination of the B hybrids

$$\phi_+ = (\phi_1 + \phi_2)/\sqrt{2} \tag{1}$$

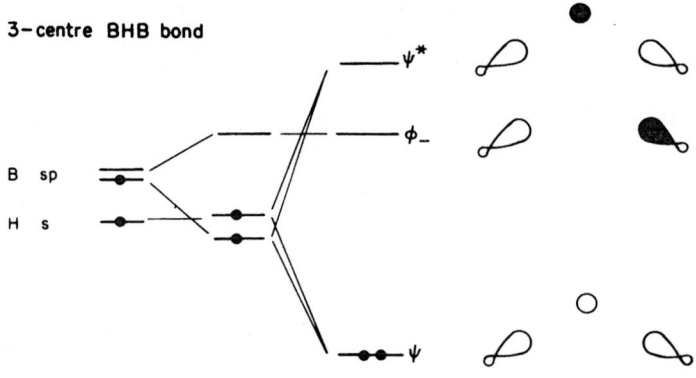

3-centre BHB bond

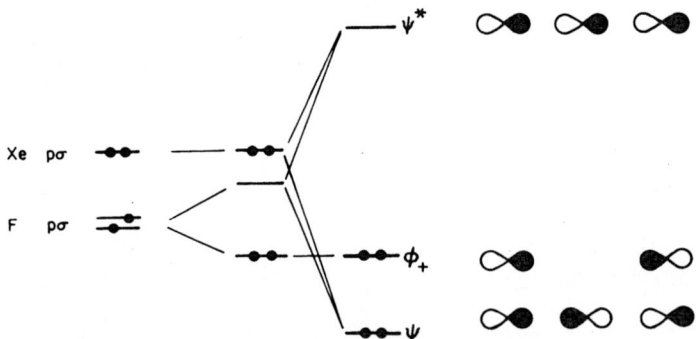

3-centre bond in XeF_2

Fig. 1. Orbitals in 3-centre bonds

$$\phi_- = (\phi_1 - \phi_2)/\sqrt{2} \tag{2}$$

These are lowered and raised respectively compared to the free hybrid energy by the small B-B interaction. The hydrogen s orbital ϕ_h only interacts with ϕ_+ and forms bonding and antibonding combinations

$$\Psi = u\,\phi_+ + v\,\phi_h \tag{3}$$

$$\Psi^* = v\,\phi_+ - u\,\phi_h \tag{4}$$

where $u^2 + v^2 = 1$. These are now strongly lowered or raised compared to ϕ_+ by the stronger B-H interactions. The ϕ_- orbital remains unperturbed by B-H interactions. Hence, the combination of B-B and B-H interactions opens up a large gap between one strongly stabilised state, Ψ, and the two others ϕ_- and Ψ^*. The lower state is symmetric for all three pairwise interactions, as seen from the orbital schematic in Fig. 1(a). Clearly, the most stable closed configuration is if the Ψ state is doubly occupied and the others empty, giving a 3c-2e bond. This indeed is found in B_2H_6, with each B contributing half an electron and the hydrogen contributing one electron to each bridge. A variant of the 3c symmetric bond is the closed bond, consisting of three borons at the corners of a triangle, each directing a symmetric orbital (e.g. a sp^3 hybrid) towards the triangle's centre. This configuration has D_{3h} symmetry. The Ψ state now has a_1 symmetry and consists of a symmetric combination of all three hybrids. This orbital remains strongly stabilised and leaves two destabilised, degenerate states of e symmetry. Again, double occupation gives the only stable configuration.

A typical antisymmetric 3c bond consists of a p orbital on a central site and σ orbitals (s, p or d) on the outer sites. Again, we form ϕ_+ and ϕ_- combinations (1), (2). It is now the ϕ_- state which interacts with the central orbital ϕ_x to form the Ψ and Ψ^* states

$$\Psi = u\,\phi_- + v\,\phi_x \tag{5}$$

$$\Psi^* = v\,\phi_- - u\,\phi_x \tag{6}$$

The ϕ_+ orbital is left relatively non-bonding. This configuration opens up a gap between two predominantly bonding orbitals and one fully antibonding orbital. Clearly, its most stable configuration is when both the Ψ and ϕ_+ orbitals are occupied by a total of four electrons, as in Fig. 1(b), giving a 3c-4e bond. The hypervalent noble gas molecule XeF_2 is used as an example in Fig. 1(b). As the molecule is linear only the $p\sigma$ states are shown in Fig. 1(b), all the $p\pi$ states are doubly occupied and non-bonding and are consequently omitted.

Note the relative separation of the final states in the symmetric and antisymmetric 3c-bonds. The stabilisation of only one orbital in the symmetric closed bond suits a 3c-2e configuration and electron-deficient systems (those with an electron/orbital ratio less than unity), and is a typical example of sub-octet bonding. Conversely, the stabilisation of two orbitals in the antisymmetric bond suits a 3c-4e configuration and electron-rich systems. It is a typical example of hypervalent or super-octet bonding. Thus, symmetric bonds are sometimes found for $\bar{N} < 4$ and antisymmetric bonds may be found for $\bar{N} > 4$.

The bonding configurations of group III elements are more complicated than those of the electron-rich elements. The simplest covalent bonding scheme for boron envisages an sp^2 hybridized configuration forming planar-3-fold coordinated sites, denoted $B_3{}^\circ$, where the superscript denotes the formal charge. This configuration occurs in glassy B_2O_3 and B_2S_3, but it is not the most common configuration in boron chemistry. This is because the $B_3{}^\circ$ site possesses an empty $p\pi$ orbital which is still available for bonding. Thus boron tends to maximise its bonding energy by using this orbital in either 3-centre bonds or dative bonds.

The configuration of boron in many of its molecules, such as boranes or carboranes, conforms to the following limiting condition – all its $2s,p$ orbitals are used in bond formation in either 3c-2e or 2c-2e bonds. As the two electrons of a 3c-2e bond are spread over three sites, fractional charges are involved. At a BHB bridge, the two electrons are formally distributed between the sites as 1/2B, 1H, 1/2B, and at a closed BBB bond the occupation is 2/3 per site. Consequently 3-centre bonds tend to cluster in order to retain integer charges. The possible combinations of 2- and 3- centre bonds for a particular borane are found with the aid of Lipscomb's styx numbers[3,4]. Consider a borane of composition $B_pH_{p+q}{}^c$ where c = charge. It always has at least p terminal (2c-2e) B-H bonds. Additionally let it possess

s   3c BHB bonds
t   3c BBB bonds
y   2c B-B bonds   (7)
x   additional 2c B-H bonds

If all orbitals are used in bonding, then by equating the number of orbitals and electrons available[4]

$$
\begin{aligned}
x &= q - s + c \\
t &= p - s + c \\
y &= 1/2(s - x + 3c)
\end{aligned}
$$
(8)

Positive integer solutions for t, y, x are sought for trial values of s. Typical examples are worked through by Wade[3].

Boron is a common p-type dopant in amorphous hydrogenated silicon (a-Si:H)[6]. It is well known that the doping efficiency is low because most dopant atoms adopt a non-doping or "alloying" configuration rather than the active substitional configuration. The alloying configuration of group V elements such as phosphorus is $P_3{}^\circ$, whereas boron could adopt either a $B_3{}^\circ$ or 3-centre bonded configuration. We noted that isolated 3c-2e bonds involve fraction charges, so such boron configurations in a-Si:H would involve either charge separation or clusters. The possible cluster configurations can be found by adapting the styx method to a random network. A cluster containing the boron sites is removed from network by cutting only 2c-2e bonds. These are then saturated with hydrogens (Fig. 2) to give a "siborane" $Si_rB_{p-r}H_{p+q}{}^c$. The styx equations (8) can then be used, if they are modified for the higher electron occupations by replacing the charge c by (c-r) and if the bond definitions (7) are generalised to include mixed Si,B configurations. The following small configurations are found to be possible

$\equiv$B-H-Si$\equiv$ (9)

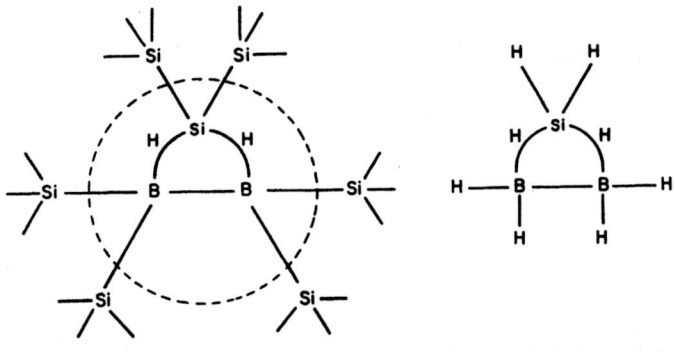

Boron cluster in a network   Equivalent Si-boron molecule

Fig. 2. Construction of a cluster to represent a boron-containing region in a-Si:H.

$$=B\!\!\overset{H}{\underset{H}{\bigcirc}}\!\!Si^+= \qquad\qquad (10)$$

$$=B\!\!\overset{H}{\underset{H}{\bigcirc}}\!\!B= \qquad\qquad (11)$$

$$=B\!\!\overset{H\ \ Si^+=}{\underset{H\ \ Si=}{\big\langle\big|}} \qquad\qquad (12)$$

Experimentally, the evidence for any 3-centre boron bonding in a-Si:H is inconclusive. A major signature of 3-centre bonds is the appearance of a broad vibrational band at a lower frequency that the corresponding 2-centre bond. Ovshinsky and Adler[7] first proposed the possibility of 3-centre boron bonding and Tsai[3] interpreted the vibration spectrum of B-rich a-Si,B:H alloys as evidence in favour of B-H-B bridges. However, Shen and Cardona[9] found no evidence of such bridges in their lower B-content alloys. The boron coordination can also be deduced from the spin-spin relaxation times in nuclear magnetic resonance (NMR)[10-12]. Two types of sites were detected, but both were 3-fold coordinated, presumably $B\equiv$ and $HB=$. Furthermore, little evidence of B-B clustering was found, which we noted often accompanies 3-centre bonding. If future work confirms that few 3-centre bonded B sites exist in a-Si:H, this would set Si-B alloys apart from related boron compounds such as the carboranes in which 3-centre bonds are predominant. The reason for their absence is not obvious. The B-H-Si bridge (9) requires only the presence of hydrogen and no other form of clustering. It is possible that as 3-centre bonds are formed by the overlap of three orbitals, they are favoured if each orbital has a similar size, which requires the outer atoms to be from the same row of the Periodic Table.

Dative bonding causes group III elements to be 4-fold coordinated. Characteristic examples are the III-V compounds like GaAs. These are also tetrahedrally coordinated when amorphous. Interestingly, 3-centre bonding has been observed in the vibration spectra of hydrogenated a-GaAs and assigned to Ga-H-Ga bridges[13]. Two hydrogen-related peaks are observed in their valence band density of plates (DOS) in photoemission, at -7 eV and -15 eV[14]. The -15 eV peak was assigned to non-bridging As-H sites, in accord with the As s-like character of the lowest valence band of GaAs, and the -7 eV peak to Ga-H states[14]. The Ga-H-Ga bridges could occur singly as:

$$\equiv Ga-H-Ga\equiv \tag{13}$$

or $=Ga-H-Ga=$ $\qquad\qquad$ (14)

or in diborane-like bridge-pairs:

$$=Ga-H_2-Ga= \tag{15}$$

Tight-binding cluster calculations[15] find that hydrogen pairs like (15) must introduce a twin-peaked structure in the density of states, at perhaps -6 and -10 eV, in conflict with photoemission. The single bridge is found to give a single H-related peak at around the observed position, only if the Ga sites remain four-fold coordinated overall, as in (13). Wang et al.[13] proposed that the bridges were bent. While the photoemission data and the tight-binding calculation do not rule this out, a linear bridge would be favoured energetically.

The presence of hydrogen bridges in a-GaAs:H but not a-Si,B:H is unusual. Wang et al[13] suggested that if the ionicity of Ga-As bonds gave GaAs an effective Ga^+As^- configuration, then on average, each Ga contributes 1/2e to each of its four surrounding bonds. This allows Ga-H-Ga bridges to possess the necessary two electrons without invoking clustering. However, this argument is erroneous. The possibility of local stoichiometry deviations is only included by taking atoms to be initially neutral. Then, each 4-fold Ga site contributes 3/4e to its surrounding bonds and each 4-fold As site contributes 5/4e to each of its bonds. Clusters then appear and an appropriate styx formula is needed. If we also allow the possibility of =AsH, $\equiv$AsH and $\equiv$As sites, we find the simplest configuration to be $Ga-H_2-Ga$ (15), in disagreement with the photoemission data.

ELECTRON-RICH ELEMENTS

In amorphous semiconductors the bonding configurations of elements with $N \geqslant 4$ generally follow the octer rule, e.g. Ge 4, As 3, Se 2 etc, There are a few occasions in which these elements exhibit higher coordinations. One of the best examples of such "hypervalent" bonding is the pentavalent phosphorus sites in oxide and sulphide glasses, represented as $\equiv P=O$ and $\equiv P=S$. The shorter P-O bond length to the terminal site (1.39 Å versus 1.62 Å) is evidence of the higher bond order. Hypervalent configurations can be represented in the (2c-2e) format if excited P 3d orbitals participate, in this case forming a pdπ bond to the terminal site.

Hypervalent bonding can also be described within a multi-centre scheme. Consider the linear molecule XeF_2 again. A (2c-2e) format could be retained by constructing two linear pd^2 hybrids at Xe, using an excited d state. Alternatively, a (3c-4e) format can be used, as in Fig. 1(b), and is now preferred by most workers for such molecules[16-18]. Hypervalent multi-centred bonding is sometimes also referred to as resonant bonding. Noting that the non-bonding orbital ϕ_+ is localized entirely on the fluorines, which are therefore negatively charged, XeF_2 can be represented as the resonance of a single 2c-2e bond between two positions:

$$F - Xe^+ \ F^- \ \rightleftharpoons \ F^- \ {}^+Xe - F \qquad (16)$$

Hypervalent or resonant bonding is found in many heavier crystalline semiconductors, such as As, Se or GeTe. The inter-chain bonding of Se (or inter-layer bonding of As and GeTe) is often described as the resonance of the covalent chain bonds into the inter-chain positions[19]. Alternatively, asymmetric 3c-4e bonds can be constructed from the intra-chain bond and a $p\pi$ lone pair on an adjacent chain.

Resonant bonding is rare in the amorphous state because it requires the alignment of p orbitals on three adjacent sites[2,19], as seen from Fig. 1(b). The best examples of this change are the IV-VI semiconductors GeSe and GeTe. While c-GeSe and c-GeTe have the resonantly-bonded orthorhombic or rhombohedral structures, the $a\text{-}Ge_x Se_{1-x}$ and $a\text{-}Ge_x Te_{1-x}$ alloys are 4:2 coordinated for most of their composition range in accord with the octet rule. Thus, resonant bonding is lost[2,20], although remnants are found under some conditions at $x = 0.5$[21,22].

DEFECTS IN CHALCOGENIDES

The original motivation for the 8-N rule of Mott[1] was the absence of a doping effect in STAG glasses. However, it is now believed that doping effects are small in such chalcogenide glasses because of the presence of special defects called valence alternation pairs (VAPs). The VAPs of a-Se are formed by the equilibrium

$$2 \ Se_2{}^\circ \ \rightleftharpoons \ Se_3{}^+ + Se_1{}^- \qquad (11)$$

These centres have a negative correlation energy (U) and therefore can pin E_F near midgap, even in the presence of carrier-generating impurities[23,24]. It is clear that the coordinations of each VAP defect are in accord with octet bonding as they are isoelectronic with As and Cl respectively. Their paramagnetic configuration $Se_1{}^\circ$ does not follow the octet rule. This site also decays exothermically

$$2 \ Se_1{}^\circ \ \rightarrow \ Se_3{}^+ + Se_1{}^- \qquad (12)$$

releasing an energy of U.

DEFECTS AND DOPING IN a-Si:H

The only defect so far definitely identified in a-Si:H by electron spin resonance is the silicon dangling bond, $Si_3{}^\circ$.

The observation of doping of a-Si:H by phosphorus and other elements[6] appears to contradict the octet rule. The absence of topological constraints was expected to allow all phosphorus atoms to enter $P_3{}^\circ$ alloying sites, while substitutional $P_4{}^\circ$ sites would break the octet rule. The doping mechanism is now understood to involve a mixture of 3- and 4-fold sites[25-27]. Most phosphorus atoms indeed enter $P_3{}^\circ$ sites and only a small fraction enter P_4 sites. The P_4 site is allowed by the octet rule if it is permanently ionized as $P_4{}^+$ (Fig. 3). To achieve this, the electron is transferred to a dangling bond, giving $Si_3{}^-$, which therefore must be present in equal numbers for compensation. This keeps the reaction

$$P_3{}^\circ + Si_4{}^\circ \ \rightleftharpoons \ P_4{}^+ + Si_3{}^- \qquad (13)$$

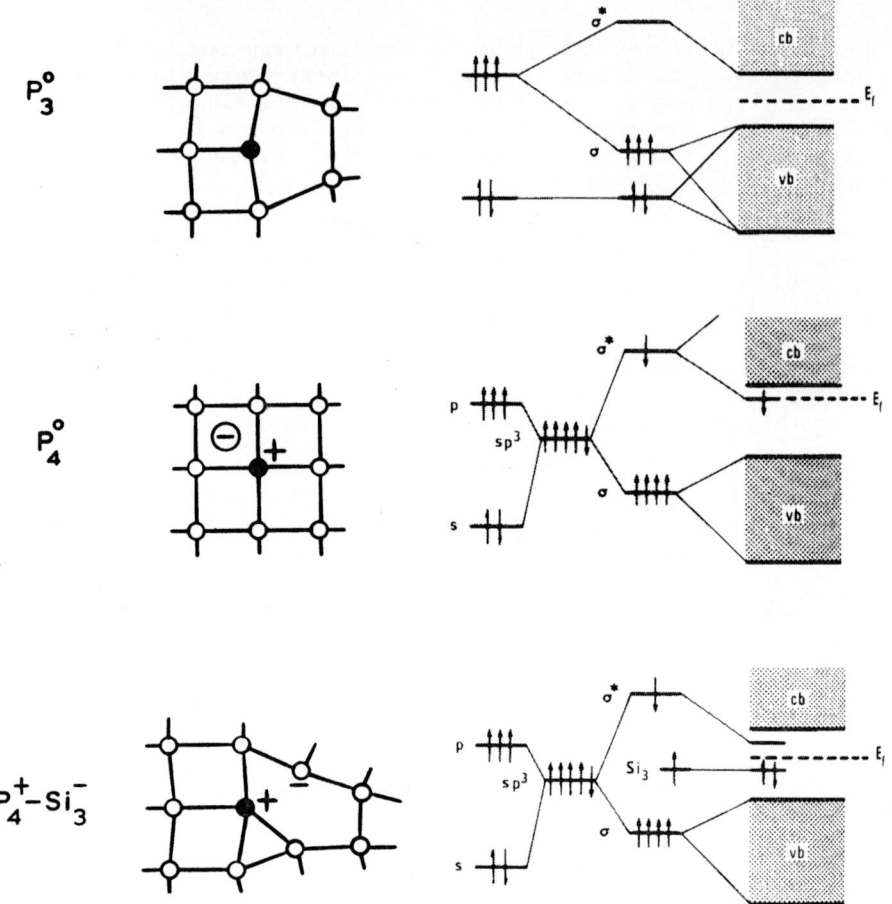

Fig. 3. Bonding configurations of Si:P.

(a) passive 3-fold coordinated P_3° site, as in a-Si:H

(b) doping simple substitutional P_4° site, as in C-Si

(c) doping P_4^+-Si_3^- complex present in a-Si:H
(these sites need not be adjacent)

in chemical equilibrium during deposition. Thus doping is only achieved
at the expense of introducing additional dangling bond centres, and these
are indeed seen experimentally.

In this model of doping there is a limit to how close E_F can approach
the conduction band edge E_c. The occupation of Si_3^- states around midgap
pushes E_F into the upper part of the gap (Fig. 3). However, if E_F were
to lie in the donor levels, P_4° states would result, violating the octet
rule. No such states are seen by ESR. As the donor levels lie below
E_c, E_c-E_F always exceeds about 0.15 eV, and experimentally[6] the conductivity
never reaches semi-metallic values, even at the highest doping levels.

From being a challenge to the octet rule, substitional doping of
a-Si:H now provides additional support. Firstly, the alloying site P_3°
is octet and is the ground state of P. The P_4^+-Si_3^- combination is also
octet and is the first excited state of P in a Si host. Indeed the site

184

pair can be recognised as a VAP by comparing equations (13) and (11). Finally, $P_4°$ is the only non-octet site. It is unstable in a-Si:H and only occurs in crystalline Si because both the others are topologically forbidden. Thus, the stable defects in amorphous chalcogenides, pnictides and doped amorphous silicon are all octet centres[26,27].

Si-H-Si bridges have been considered as possible negative U centres[28] in intrinsic a-Si:H, but so far the characteristic ESR signature of their paramagnetic configuration has not been observed. Possible Si-H-Si and $Si-H_2-Si$ bridge configurations have also been studied in relation to H_2 evolution and the saturation of weak Si-Si bonds[29].

One of the major outstanding questions in a-Si:H is to identify the mechanisms involved in the Staebler-Wronski effect and the associated metastable defects[30-38]. The effect is a bulk, intrinsic effect[35,37] involving a change in the density of midgap states[34,36] and the creation of $Si_3°$ centres[32-34]. The effect is often suggested to be caused by the rupture of weak Si-Si bonds and be followed by a reaction with hydrogen. This is because the effect anneals away with an activation energy similar to that of H diffusion. Two recent models suggested that the broken bond configuration was stabilised by the formation of SiHSi 3-centre bridges[37,38]. However, the model of Mosley et al.[38] invokes a SiHSi$^-$ bridge. We saw from Fig. 1 that a H-bridge must be symmetric and is only stable when electron-deficient, i.e. for positively charged Si. An electron-rich bridge is stable only if antisymmetric, with a p orbital on its central site as in Fig. 1(b) but H has no such p orbital. Stutzmann et al.[37] also invoke a 3-centre SiHSi bond, but again the suggested configuration appears too electron-rich to be stable.

REFERENCES

1. Mott, N.F., Adv. Phys., 16, 49 (1967).
2. Robertson, J., Adv. Phys., 32, 361 (1983).
3. Wade, K., "Electron Deficient Compounds", Nelson, London (1971).
4. Lipscomb, W.N., "Boron Hydrides", Benjamin, New York (1963).
5. Mutterlies, E.C. (Ed.), "Chemistry of Boron and its Compounds", Wiley, New York (1967).
6. Spear, W.E., LeComber, P.G., Solid State Commun., 17, 1193 (1975).
7. Ovshinsky, S.R., Adler, D., Contemp. Phys., 19, 109 (1978).
8. Tsai, C.C., Phys. Rev. B, 19, 2041 (1979).
9. Shen, S.C., Cardona, M., Phys. Rev. B, 23, 5322 (1981).
10. Greenbaum, S.G., Carlos, W.E., Taylor, P.C., Solid State Commun., 43, 663 (1983).
11. Greenbaum, S.G., Carlos, W.E., Taylor, P.C., Physica, 117B, 886 (1983).
12. Greenbaum, S.G., Carlos, W.E., Taylor, P.C., J. App. Phys., 56, 1874 (1984).
13. Wang, Z.P., Ley, L., Cardona, M., Phys. Rev. B, 26, 3249 (1982).
14. Karcher, R., Wang, Z.P., Ley, L., J. Non-Cryst. Solids, 59, 629 (1983) and private communication.
15. Robertson, J., unpublished work.
16. Rundle, R.E., J. Am. Chem. Soc., 85, 112 (1963).
17. Dewar, M.J.S., Healy, E., Organometallics, 1, 1705 (1982).
18. Hay, P.J., J. Am. Chem. Soc., 99, 1003 (1977).
19. Lucovsky, G., White, R.M., Phys. Rev. B, 8, 660 (1976).
20. O'Reilly, E.P., Robertson, J., Kelly, M.J., Solid State Commun., 38, 565 (1981).
21. Takahashi, T., Sakurai, A., Sagawa, T., Solid State Commun., 44, 723 (1982).
22. Takahashi, T., Sayawa, T., J. Non Cryst. Solids, 59, 879 (1983).

23. Street, R.A., Mott, N.F., Phys. Rev. Let., $\underline{35}$, 1293 (1975).
24. Kastner, M., Adler, D., Fritzsche, H., Phys. Rev. Let., $\underline{37}$, 1504 (1976).
25. Street, R.A., Phys. Rev. Let., $\underline{49}$, 1187 (1982).
26. Robertson, J., J. Phys., C$\underline{17}$, L349 (1984).
27. Robertson, J., Phys. Rev. B, to be published.
28. Fisch, R., Licciardello, D.C., Phys. Rev. Let., $\underline{41}$, 889 (1978).
29. Eberhardt, M.E., Johnson, K.H., Adler, D., Phys. Rev. B, $\underline{26}$, 3138 (1982).
30. Staebler, D.L., Wronski, C.R., App. Phys. Let., $\underline{31}$, 292 (1977).
31. Staebler, D.L., Crandall, R.S., Williams, R., App. Phys. Let., $\underline{39}$, 733 (1981).
32. Dersch, H., Stuke, J., Beichler, J., App. Phys. Let., $\underline{30}$, 456 (1981).
33. Lee, C., Ohlson, W.D., Taylor, P.C., Ullal, H.S., Ceasar, G.P., Phys. Rev. B, $\underline{31}$, 100 (1985).
34. Street, R.A., App. Phys. Let., $\underline{42}$, 507 (1983).
35. Cohen, J.D., Lang, D.V., Harbison, J.P., Sergent, A.M., Solar Cells, $\underline{9}$, 119 (1983).
36. Okushi, H., Miyagawa, M., Tokumaru, Y., Yamasaki, S., Oheda, H., Tanaka, H., App. Phys. Let., $\underline{42}$, 895 (1983).
37. Stutzmann, M., Jackson, W.B., Tsai, C.C., App. Phys. Let., $\underline{45}$, 1075 (1984).
38. Mosley, L.E., Paesler, M.A., Shimizu, I., Phil. Mag. B, $\underline{51}$, Lxx (1985).

186

WEAK BONDS IN AMORPHOUS SEMICONDUCTORS

Tatsuo Shimizu,[*]Nobuhiko Ishii and Minoru Kumeda

Department of Electronics, Kanazawa University
Kanazawa 920, Japan
[*]Department of Electrical Engineering and
Electronic Engineering, Fukui Institute of
Technol., Fukui 910, Japan

INTRODUCTION

In this paper, we would like to point out the important role played by the presence of weak bonds in various kinds of amorphous semiconductors. Most atoms have a normal structure bonding (NSB) configuration even in amorphous semiconductors. Only a small fraction of atoms have a deviant electronic configuration (DEC), that is, coordination number different from NSB. Besides such atoms, it is possible that atoms with a largely prolonged bond are present. So far several proposals for the existence of such weak bonds in amorphous semiconductors have been done. One of the authors (T.S.) suggested that defects in chalcogenide glasses are not three-fold coordinated chalcogens but positively and negatively charged dangling bonds with the positively charged ones interacting with neighboring lone-pair, making weak bonds.[1] In hydrogenated amorphous silicon (a-Si:H), the ESR signals with $g = 2.0043$ and $g = 2.013$ were ascribed to unpaired spins in the antibonding and bonding states of weak Si-Si bonds, respectively.[2,3] Very recently, we attributed the ESR hyperfine lines (ESR-HFL) observed in P-doped a-Si:H[4,5] to an electron in the antibonding state of a weak Si-P bond, and the change of the Si-P bond length according to its charge state was suggested.[6]

A wide range of phenomena observed in various amorphous semiconductors such as tetrahedrally bonded ones, chalcogenides and group-V ones can be explained by assuming the presence of weak bonds in which the bond length changes with their charge states in many cases. The existence of such weak bonds and the change of the bond length with charge state depend on the flexibility of the atomic structure, and they are considered to be characteristic phenomena in covalenty bonded amorphous materials. Several examples will be presented in following sections.

P-RELATED DEFECTS IN P-DOPED a-Si:H

ESR-HFL's have been observed by Stutzmann et al. in a-Si:H and a-Ge:H doped with P (doublet HFL)[4] as well as in a-Si:H doped with As (quartet HFL).[7] The doublet ESR-HFL has also been observed in P-doped a-Si:H by photo-induced absorption-detected ESR measurement by Hirabayashi et al. who suggest that this HFL is originated from unpaired electrons on two-fold

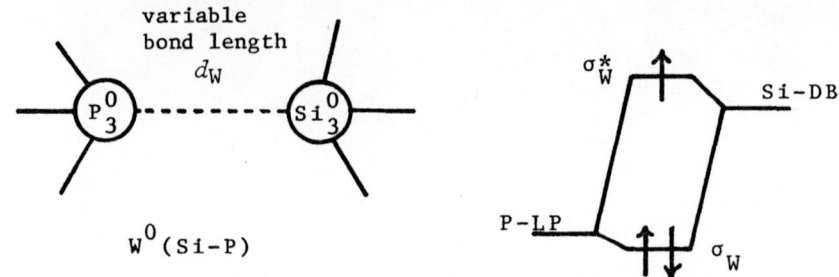

Fig. 1. A schematic representation of a Si-P weak bond (dashed line). An unpaired electron in the antibonding state σ_W^* of this weak bond is the origin of the doublet HFL.

coordinated P atoms.[5]

We calculated the electronic states of various defects containing P atoms by using the extended Hückel theory.[6] Then we estimated the strength of the hyperfine interaction (HFI) between an unpaired electron in the defect state and a P nucleus. As a result, we concluded that the origin of the doublet HFL observed in P-doped a-Si:H is an unpaired electron in the antibonding state for the weak Si-P bond between a three-fold coordinated Si atom and a three-fold coordinated P atom (W^0(Si-P)) as shown in Fig.1.[6]

In what follows, we discuss in more details the results calculated for a 26-atom cluster containing W^0(Si-P) with its variable d_W.

The calculated isotropic HFI (IHFI) between the unpaired electron in σ_W^*(Si-P) and the P nucleus increases and the energy level of σ_W^*(Si-P) moves upward with decreasing d_W. $d_W = 3.35$Å gives the closest agreement between the strength of the calculated IHFI (252 G) and the observed splitting of the HFL (245 G).[4]

σ_W^*(Si-P) with unpaired electrons locate in the range of $U(\sigma_W^*$(Si-P)) below the Fermi level, as shown by the shaded part in Fig.2(a). The state density for the ESR spin with $g = 2.0043$ is not shown in Fig.2(a) for simplicity. The shaded part shown in Fig.2(a) originates from W^0(Si-P) which exists in well-annealed film and contributes to the doublet HFL.

The similar results are expected to be obtained for W^0(Ge-P) because the difference between chemical properties of Si and Ge atom is little. In addition, the case for W^0(Si-As) should also be similar to that for W^0(Si-P).

Next we discuss the change induced by illumination. The charged defects are mainly created in P-doped a-Si:H by the Street's doping mechanism:[8]

$$P_3^0 \longrightarrow P_4^+ + D^-. \tag{1}$$

where P_4^+ and P_3^0 represent a positively charged four-fold coordinated P atom and a neutral three-fold coordinated P atom, and D^- is a negatively charged Si dangling bond (Si-DB). When P_4^+ captures an electron, one of the Si-P bonds can be lengthened to lower the total energy if its environment is flexible enough. This structural change is schematically shown in Fig.3. If the structure is completely flexible, P_4^+ trapping an electron, that is, P_4^0, changes its structure toward more stable configuration, that is, $P_3^0 + D^0$. On the other hand, if the structure is completely rigid, P_4^0 can not change its structure and releases an electron in order to revert to the stable state, that is, P_4^+. P_4^0 and the pair of P_3^0 and D^0 are two extreme cases of W^0(Si-P).

In our weak bond model, the following reactions are expected to occur

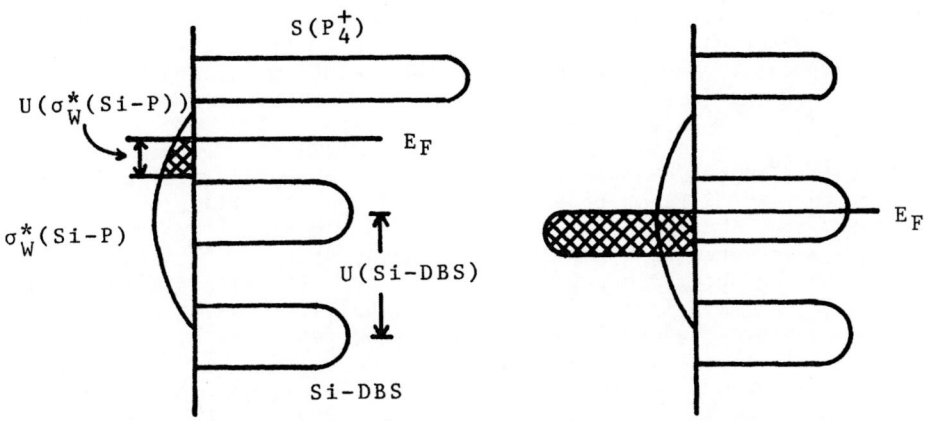

(a) Before Illumination   (b) After Illumination

Fig. 2. Schematic energy diagrams for P-doped a-Si:H before and after illumination. The energy of the antibonding state of neutral weak bonds (W^0(Si-P)) locate in the shaded area. $U(\sigma_W^*$(Si-P)) and U(Si-DBS) mean the electron correlation energy for the antibonding state of the Si-P weak bond and for the Si dangling bond state, respectively.

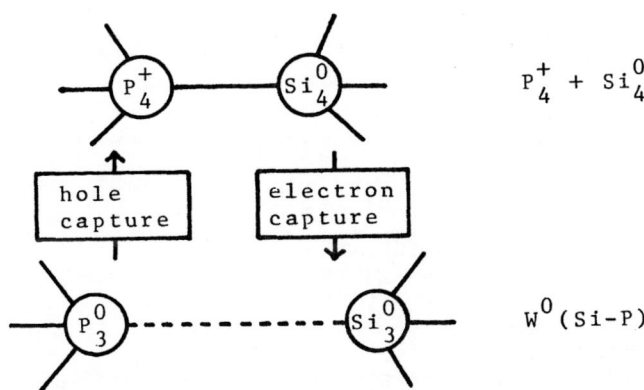

Fig. 3. Change of a bond length of a defect with its charge state. The dashed line means a weak bond.

by capturing photo-created electrons and holes:[6]

$$P_4^+ + \text{electron} \longrightarrow W^0(\text{Si-P}) \tag{2}$$

and

$$D^- + \text{hole} \longrightarrow D^0. \tag{3}$$

By these reactions, the intensity of the HFL will increase and the ESR signal with $g = 2.0055$ will appear. The density of the gap states after prolonged illumination is shown in Fig.2(b). In addition, the decrease in the density of D^- with an increase in that of D^0 moves the Fermi level toward the center of the mobility gap as shown in Fig.2(b).

By subsequent annealing, the following reaction proceeds:[6]

$$W^0(\text{Si-P}) + D^0 \longrightarrow P_4^+ + D^-, \tag{4}$$

and the photo-induced changes are recovered as shown in Fig.2(a). Here we assume that $W^0(\text{Si-P})$ which exists in well-annealed films has less flexible environment than $W^0(\text{Si-P})$ created by illumination. Thus, although the former can have partners with different charge states, $W^+(\text{Si-P})$ and $W^-(\text{Si-P})$, the latter can not have them because $W^+(\text{Si-P})$ and $W^-(\text{Si-P})$ are changed to P_4^+ and a pair of P_3^0 and D^-, respectively. With this assumption, the ESR spins increased by illumination can be recovered by annealing to its original level before illumination.

These reactions are considered to be a possible origin of the photo-induced changes in Si-based amorphous semiconductors.[9] We will discuss them in the next section by generalizing our weak bond model discussed above.

WEAK BOND MODEL FOR PHOTO-INDUCED CHANGES IN Si-BASED AMORPHOUS SEMICONDUCTORS

We generalize Eqs.(1)-(4) for the case of P-doped a-Si:H in order to discuss the case of X-doped a-Si:H (a-Si$_{1-y}$X$_y$:H) for X = group-V or VI element as follows:

$$X_n^0 \longrightarrow X_{n+1}^+ + D^-, \tag{5}$$

$$X_{n+1}^+ + \text{electron} \longrightarrow W^0(\text{Si-X}), \tag{6}$$

$$D^- + \text{hole} \longrightarrow D^0, \tag{7}$$

and

$$W^0(\text{Si-X}) + D^0 \longrightarrow X_{n+1}^+ + D^-. \tag{8}$$

Here n = 3 for X = group-V element and n = 2 for X = group-VI element according to the 8-N rule. The potential barrier exists between the right and left hand sides in these equations. The heights of them depend on the energy difference between the related states and the structural flexibility around the defects.

Some of $W^0(\text{Si-X})$ and D^0 created by the reactions of Eqs.(6) and (7) are expected to remain after the cessation of prolonged illumination because of the presence of the barrier for the reaction of Eq.(8). The decrease in the density of D^- with an increase in the density of D^0 makes the Fermi level move toward the center of the mobility gap as shown in Fig.2 which is considered to be the origin of the Staebler-Wronski effect (SWE).[9]

The following fact should be noted: When the electronegativity of the atom X ($x(X)$) is considerably larger than $x(\text{Si})$, the Si-X bond polarizes, that is, the bonding and antibonding states for the Si-X bond are mainly constituted of the atomic orbitals of the X atom and the Si atom, respectively. If the Si-X bonds in X_{n+1}^0 becomes longer, a mixing of the atomic orbitals of the X atom into $\sigma_W^*(\text{Si-X})$ should become smaller and the difference between $\sigma_W^*(\text{Si-X})$ and the Si dangling bond state (Si-DBS) should become negligible. As a result, if $x(X)$ is considerably larger than $x(\text{Si})$, $W^0(\text{Si-X})$ should effectively be equal to a pair of D^0 and X_n^0. For example, in the case of N or O impurities, $W^0(\text{Si-N})$ or $W^0(\text{Si-O})$ is regarded as a pair of D^0 and N_3^0 or D^0 and O_2^0, respectively, because $x(N)$ (= 3.0) or $x(O)$ (= 3.5) is considerably larger than $x(\text{Si})$ (= 1.41).[10] In addition, the creation rate of the pair of D^- and X_{n+1}^+ by the reaction of Eq.(5) is

190

expected to be small when $x(X)$ is large, because the creation of X_{n+1}^+ becomes energetically unfavorable. In fact, P atoms act as dopants more effectively than N and O atoms.

The SWE originating from X = group-III (for example, B) or II element can be explained by the same manner as that from X = group-V or VI element by interchanging 'positively' with 'negatively', 'hole' with 'electron', and 'σ_W^*' with 'σ_W'. Here, σ_W is the bonding state for $W^0(Si-X)$. In such cases, n = 2 for X = group-II element and n = 3 for X = group-III element.

The undoped a-Si:H should contain a certain amount of O and N as impurities. Especially, it is difficult to suppress the amount of O below 2×10^{18} cm^{-3}.[11] Thus, we regard the undoped a-Si:H as O-doped a-Si:H. O_3^+ and D^- are created by the reaction of Eq.(5). The fact that the ESR spin density with g = 2.0055 observed in lightly B-doped a-Si:H is several times as large as that in undoped a-Si:H[12] is consistent with the presence of D^- in undoped a-Si:H. $W^0(Si-O)$ and D^0 created by the reactions of Eqs.(6) and (7), respectively, are the origin of the SWE for the undoped a-Si:H. The fact that the density of photo-induced ESR spins does not scale with the O concentration below about 10^{20} cm^{-3} [13] does not necessarily deny the explanation that the SWE is due to O impurities.

Yamazaki et al. also emphasized the influence of the O contamination on the SWE.[14] In the model proposed by Yamazaki et al., the presence of the O-H bond is essentially important. On the other hand, in the present model, O_3^+ without an O-H bond is important for the SWE rather than that with an O-H bond, because $W^0(Si-O)$ with an O-H bond is considered to be recovered more easily than that without an O-H bond after illumination.

WEAK BOND MODEL FOR PHOTO-INDUCED ESR IN AMORPHOUS P AND As

By the photo-induced ESR measurement, two distinct ESR centers have been observed in a-P by Shanabrook et al.[15] These two ESR centers denoted by C_1 and C_2 are characterized by wave functions which are localized mainly on a single P atom and localized on two equivalent P atoms, respectively. The ESR signal for the C_1 centers resembles the dark-cold ESR signal. The C_1 and C_2 centers are increased by prolonged illumination at low temperature. The density of C_2 centers decreases rapidly above 80 K after the cessation of illumination while that of C_1 centers decreases more gradually. Bishop et al. have also been obtained similar results in a-As.[16] In what follows, we mainly discuss about a-P since our discussion for a-As is similar to that for a-P.

The P_4^+ and P_2^- centers are expected to exist in a-P.[17] If P_2^0 exists in a-P, the DB orbital of P_2^0 weakly interacts with a lone pair orbital (LPO) of P_3^0 around this P_2^0. Such a weak bond is denoted by W_1^0 as shown in Fig.4(a). Then, if the structure of a-P is sufficiently flexible, two W_1^0 should change into P_4^+ and P_2^-, respectively, by transfering an electron from one W_1^0 to the other W_1^0. The structure of the real a-P should be composed of the regions with various structural flexibilities. Accordingly, P_4^+, P_2^- and W_1^0 should coexist in well-annealed a-P. W_1^0 existing in well-annealed a-P can scarcely relax its structure due to a less structural flexibility even if its charge state changes. The weak antibonding state originating from W_1^0, which is occupied by an unpaired electron, is expected to be more localized on P_2^0 than on P_3^0 because the energy of the DB orbital of P_2^0 is higher than that of the LPO of P_3^0 as shown in Fig.4(a). Accordingly, we attribute the ESR centers C_1 to W_1^0.

Next, we discuss the photo-induced changes of the ESR signal in a-P. The following reactions proceed in the region with flexible structure by prolonged illumination:

$$P_4^+ + \text{electron} \longrightarrow W_1^0. \tag{9}$$

and

$$P_2^- + (\ P_3^0 \text{ around this } P_2^-\ ) + \text{hole} \longrightarrow W_1^0. \tag{10}$$

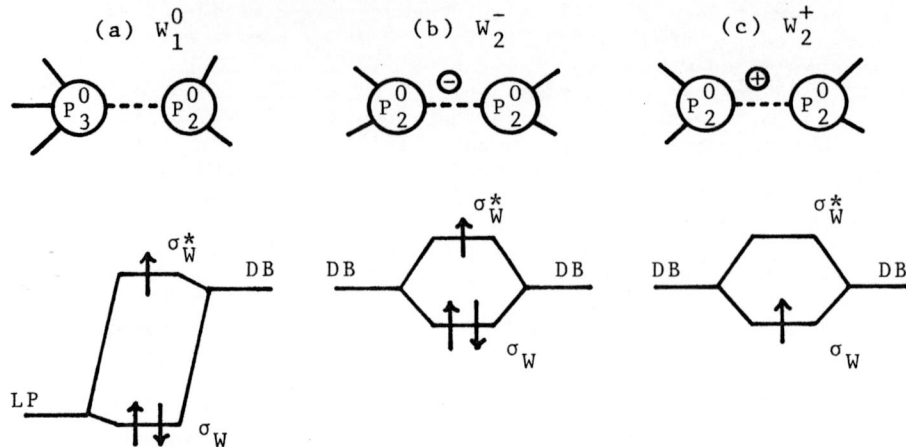

Fig. 4. Defects which are the origin of ESR signals in a-P and
their electronic configuration : (a) is a neutral weak
bond W^0, between a three-fold coordinated P atom and
a two-fold coordinated P atom, (b) and (c) are weak
bonds between two two-fold coordinated P atoms
capturing an electron and a hole, respectively.

In addition, the following reactions are expected to proceed: If a normal
bond between two P_3^0 traps an electron or a hole, this normal bond becomes
weak in order to lower the total energy. These weak bonds are denoted by
W_2^- and W_2^+, respectively, as shown in Fig.4(b) and (c). Here, we assumed
that the top of the valence band is constituted of the bonding state. Such
weak bonds are similar to charged weak Si-Si bonds in a-Si:H.

Unpaired electrons in σ_W^* for W_2^- and in σ_W for W_2^+ should localize on
two equivalent atoms because of the symmetry of these defects. In addition,
W_2^+ and W_2^- are expected to be unstable in comparison with W_1^0 because the
structures of W_2^+ and W_2^- can continuously change into the normal configuration. From these facts, we attribute the photo-induced ESR signal from
C_2 centers to a superposition of the ESR signals from W_2^+ and W_2^-.
Thus the weak bond model can reasonably explain the photo-induced changes
of the ESR signal in a-P or a-As.

Ge-S GLASSES

Figure 5 shows the compositional dependence of the g-value of the
ESR signal in $Ge_{1-y}S_y$ glasses for $y < 0.67$ and its change by illumination.[18,19] These ESR signals have so far been ascribed to Ge dangling
bonds (Ge-DB).[20] In order to explain the observed results in Fig.5, we
assume the existence of two kinds of ESR centers, that is, a neutral Ge-DB
(D^0) and a neutral weak Ge-S bond (W^0(Ge-S)). The latter is D^0 weakly
interacting with the LPO of the surrounding S_2^0. The antibonding state
(σ_W^*(Ge-S)) for W^0(Ge-S) has an unpaired electron and contributes to the ESR.
The g-shift for the defect state Φ_d mainly localized on the A atom
can approximately be expressed as follows:[21]

$$\Delta g_{ij} \sim 2\zeta_A \left\{ \Sigma_v \frac{<\Phi_d|l_A^i|\Phi_v><\Phi_v|l_A^j|\Phi_d>}{E_d - E_v} - \Sigma_c \frac{<\Phi_d|l_A^i|\Phi_c><\Phi_c|l_A^j|\Phi_d>}{E_c - E_d} \right\}, \quad (11)$$

where ζ_A is the spin-orbit coupling constant of the A atom, l_A^i is the i-th
component of the angular momentum around the A atom, E_d, E_v and E_c are

192

energy eigenvalues belonging to the state Φ_d, Φ_v and Φ_c, and Σ_v and Σ_c mean the summations over the valence and conduction band states, respectively. By using Eq.(11), we can qualitatively discuss the g-value for the defect state in terms of the energies E_d and E_v. First, the g-value of Ge-DBS or σ_W^*(Ge-S) should decrease with an increase in the number of S atoms bonding to the Ge atom of each defect, because the energy of the bonding state for the Ge-S bond is lower than that for the Ge-Ge bond, resulting in a decrease in E_v. Second, the g-value of σ_W^*(Ge-S) should be smaller than that of the Ge-DBS, because σ_W^*(Ge-S) locates higher in energy than the Ge-DBS does.

The change of the g-value with the S content can be expained as follows: The ESR center in a-Ge are D^0. With an increase in the S content, the number of S atoms bonding to D^0 and W^0(Ge-S) increases and a ratio of the density of W^0(Ge-S) to that of D^0 also increases. Accordingly, the g-value decreases with an increase in the S content y in the region with $0 < y < 0.5$. In $Ge_{1-y}S_y$ glasses with $0.5 < y < 0.6$, most of the ESR centers should be W^0(Ge-S) and the g-value becomes minimum. When the S content further increases, the structural flexibility increases and the fraction of W^0(Ge-S) in ESR centers decreases again by the following reaction;

$$2W^0(\text{Ge-S}) \longrightarrow S_3^+ + D^-, \qquad (12)$$

and the fraction of D^0 increases in comparison with that of W^0(Ge-S). Most of the ESR centers should be D^0 in the GeS_2 glass. Accordingly, the g-value increases with an increase in the S content y in the region with $0.6 < y < 0.67$. Although the g-value of the Ge-DBS might decrease with an increase in the number of S atoms bonding to D^0, this effect could be negligible because the range of y is small ($0.6 < y < 0.67$).

Next, we discuss the photo-induced change of the ESR signals. g-values observed after prolonged illumination are greater than those before illumination as shown in Fig.5. Both the charged defects, S_3^+ and D^-, which are

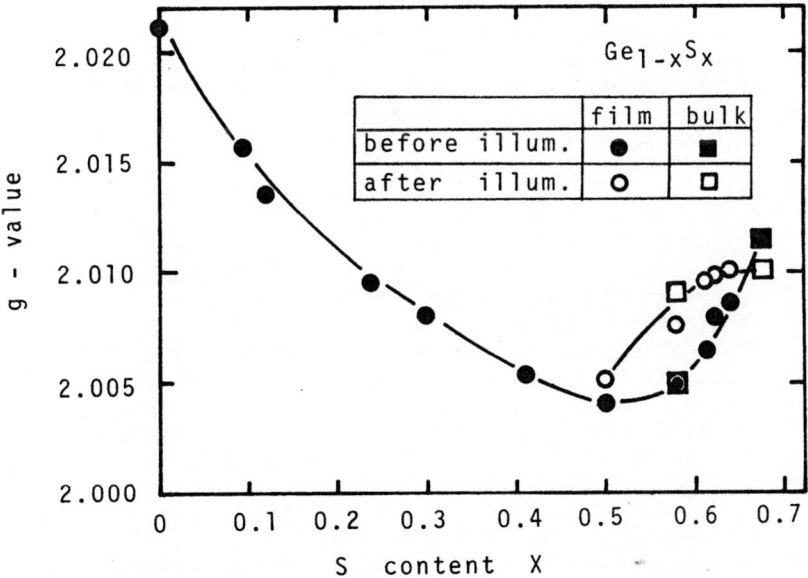

Fig. 5. Compositional dependence of the g-value of the ESR signal in $Ge_{1-y}S_y$ glasses before (closed symbols) and after (open symbols) illumination. Squares are for melt-quenched bulk glasses and circles are for evaporated films.

present before illumination, are changed to D^0 by illumination, because the environment of S_3^+ is expected to be more flexible than that of $W^0(Ge-S)$. The g-value for D^0 is greater than that for $W^0(Ge-S)$ as mentioned above, resulting in the increase in the g-value after illumination. Thus, the changes of the g-value by the change of the composition and by prolonged illumination in $Ge_{1-y}S_y$ glasses can qualitatively be explained by assuming the existence of a weak Ge-S bond.

CONCLUDING REMARKS

As stated in previous sections, the existence of weak bonds plays an important role in amorphous semiconductors. There are two types of weak bonds; one is a weakened bond between normal structural bonding atoms, for example a weak Si-Si bond, W_2^+ and W_2^- shown in Fig. 4(b) and (c) etc., and the other is a dangling bond interacting with a surrounding LPO or an empty orbital, for example a weak Si-P bond shown in Fig.1, W_1^0 shown in Fig.4(a), a weak Ge-S bond etc. The latter type of weak bonds exists when the atom with a LPO or an empty orbital is present. It is a characteristic property of a weak bond that the bond length changes with its charge state.

A pair of P_4^+ and D^- in Eq.(1) is equivalent to what is called a valence alternation pair.[22] A valence alteration pair is usually known to be present in amorphous semiconductors with a constituent of group-V or VI element. However, it should be noticed that a similar defect can be present in those with a constituent of group-II or III element if we replace "electron" with "hole". One such example besides B_4^- + D^+ is a doubly negatively charged four-fold coordinated Mn, $i.e.$ Mn_4^{2-}, in chalcogenide glasses.[23]

REFERENCES

1. T. Shimizu, Defect States in Chalcogenide Glasses, Jpn. J. Appl. Phys. 17:463 (1978).
2. S. Hasegawa, T. Kasajima and T. Shimizu, ESR in Doped CVD Amorphous Silicon Films, Philos. Mag. B43:149 (1981).
3. N. Ishii, M. Kumeda and T. Shimizu, The g-values of Defects in Amorphous C, Si and Ge, Jpn. J. Appl. Phys. 20:L673 (1981), ibid. 20:L920 (1981).
4. M. Stutzmann and J. Stuke, New Paramagnetic States in Amorphous Silicon and Germanium, J. Non-Cryst. Solids 66:145 (1984).
5. I. Hirabayashi, K. Morigaki, S. Yamasaki and K. Tanaka, Photoinduced Absorption and Photoinduced Absorption-Detected ESR in P-doped a-Si:H, in: "Optical Effects in Amorphous Semiconductors (AIP Conf. Proc. No. 120)", P. C. Taylor and S. G. Bishop, ed., AIP, New York, (1984) p.8.
6. N. Ishii, M. Kumeda and T. Shimizu, P-Related Defects in P-Doped a-Si:H, Solid State Commun. in press.
7. M. Stutzmann, private communication.
8. R. A. Street, Doping and the Fermi Energy in Amorphous Silicon, Phys. Rev. Lett. 49:1187 (1982).
9. D. L. Staebler and C. R. Wronski, Optically Induced Conductivity Changes in Discharge Produced Hydrogenated Amorphous Silicon, J. Appl. Phys. 51:3262 (1980)
10. J. C. Phillips, "Bonds and Bands in Semiconductors", Academic Press, New York and London (1973).
11. C. C. Tsai, J. C. Knights and M. J. Thompson, Clean a-Si:H Prepared in a UHV System, J. Non-Cryst. Solids 66:45 (1984).

12. H. Dersch, J. Stuke and J. Beichler, Electron Spin
    Resonance of Doped Glow-Discharge Amorphous Silicon,
    Phys. Stat. Sol. (b) 105:265 (1981).
13. C. C. Tsai, M. Stutzmann and W. B. Jackson,
    The Staebler-Wronski Effect in Undoped a-Si:H :
    Its Intrinsic Nature and the Influence of Impurities,
    in: "Optical Effects in Amorphous Semiconductors
    (AIP Conf. Proc. No. 120)", P. C. Taylor and S. G.
    Bishop, ed., AIP, New York (1984) p.242.
14. S. Yamazaki, T. Shiraishi and D. Adler, Isomerization
    Model for Photo-induced Effects in a-Si:H,
    J. Non-Cryst. Solids 68:167 (1984).
15. B. V. Shanabrook and P. C. Taylor, Optically and
    X-Ray-Induced Paramagnetism in Amorphous Phosphorus,
    Phys. Rev. B28:1239 (1983).
16. S. G. Bishop, U. Strom and P. C. Taylor, Optically
    Induced Localized Paramagnetic States in Amorphous As,
    Solid State Commun. 18:573 (1976).
17. M. Kastner and H. Fritzsche, Defect Chemistry of Lone-Pair
    Semiconductors, Philos. Mag. B37:199 (1978).
18. T. Shimizu, M. Kumeda and M. Ishikawa, Effects of
    Alloying Chalcogen Atoms into Amorphous Germanium
    Films, J. Non-Cryst. Solids 33:1 (1979).
19. M. Kumeda, G. Kawachi and T. Shimizu, unpublished.
20. I. Watanabe, M. Ishikawa and T. Shimizu, On the Property
    and Origin of Paramagnetic Defects in Amorphous Ge-S
    and Ge-S-Ag, J. Phys. Soc. Jpn. 45:1603 (1978).
21. A. J. Stone, Gange Invariance of the g Tensor,
    Proc. Royal Soc. 271:424 (1963).
22. M. Kastner, D. Adler and H. Fritzsche, Valence-Alternation
    Model for Localized Gap State in Lone-Pair Semiconductors,
    Phys. Rev. Lett. 37:1504 (1976).
23. M. Kumeda, Y. Jinno, M. Suzuki and T. Shimizu,
    Influence of Mn Impurity in Te-As-Ge-Si Glasses,
    Jpn. J. Appl. Phys. 15:201 (1976).

DANGLING BOND DEFECTS IN a-Si,Ge ALLOYS: A THEORETICAL STUDY USING THE

TIGHT-BINDING METHOD

S.Y. Lin and G. Lucovsky

Department of Physics
North Carolina State University
Raleigh NC 27695-8202

ABSTRACT

This paper presents a theoretical study of Si and Ge atom dangling
bond defects in a-Si,Ge alloys. We use a tight-binding Hamiltonian, and
a structural model based on a cluster Bethe Lattice. The central cluster
contains the Si or Ge atom with the dangling bond and all of the
possible configurations of nearest neighbor Si and Ge atoms. These
clusters are terminated by a Bethe Lattice consisting of virtual atoms
having the average properties at the particular alloy composition. We
employ self energies and interaction parameters previously used for
calculations of the band structures of crystalline Si and Ge and
formulated using an sp^3s* basis set. We find that for a given set of Si
and/or Ge nearest neighbors at each alloy composition, the Ge atom
dangling bond state lies deeper in the gap than the corresponding Si
atom dangling bond state. However, the spread in defect energies for a
Si or Ge atom dangling bond with different nearest neighbors is greater
than the separation between the statistically averaged Si and Ge
dangling bond state energies. This makes it difficult to compare the
results with experimental data. Nevertheless, the calculation
suggests that the explanation for a difference in dark conductivity
activation energies from about 0.8 eV to 0.65 eV in alloys with
approximately equal Si and Ge atom concentrations, but grown by
different deposition techniques, is related to a shift in Fermi level
pinning from Ge to Si atom dangling bond states. The calculated spread
in dangling bond energies and overlap between the manifold of Si and Ge
states also explains the simultaneous occurrence of Si and Ge atom

dangling bonds in a-Si,Ge:H alloys with Ge concentrations ranging to about sixty atomic percent.

I INTRODUCTION

There has been considerable interest in using a-Si,Ge:H alloys as the absorbing intrinsic layers in photovoltaic devices. The band-gaps in this alloy system can be compositionally tuned from about 1.1 eV (a-Ge:H) to 1.9 eV (a-Si:H)[1] and in this way one can design single and multilayer photovolatic devices that optimize the utilization of the solar spectrum. To date highly efficient photovoltaic devices have not been fabricated using a-Si,Ge:H i-layers with band-gaps less than about 1.5 eV, primarily because of a substantial increase in the number of mid-band gap defect states in the a-Si,Ge:H alloys relative to the number of similar defect states in photovoltaic quality a-Si:H[2-6]. The extent of the degradation is dependent on the particular deposition technique, glow discharge (GD) decomposition of silane and germane mixtures[2,3], or reactive diode (DS)[4] or magnetron (MS) sputtering[5,6].

Studies of the electronic properties of alloy films grown by the GD method have generally identified a large increase in the number of Ge atom defect states (singly occupied dangling bonds) as the Ge atom fraction is increased beyond about forty atomic percent[2,3,7]. This observation derives primarily from electron spin resonance (ESR) studies[7], and is consistent with other measurements of mid-band-gap state sensitive properties such as the photoconductivity and luminescence[2,3]. The ESR measurements also indicate the simultaneous existence of singly occupied Si and Ge dangling bond defect states. It has also been observed that the dark conductivity activation energies in GD films can display values in excess of half of the optical band-gap. For example, the Harvard group has reported dark conductivity activation energies of about 0.80 to 0.85 eV in GD alloys with optical band-gaps of about 1.4 eV and Ge concentrations of about fifty atomic percent Ge[8]. These films display relatively poor photoconductivity and photoluminescence[2,3]. In contrast films grown by dual magnetron sputtering of Si and Ge from separate targets, with comparable Ge atom concentrations and optical band-gaps, display substantially higher photoconductivity, with dark activation energies slightly less than one half of the band-gap[5,6]. For example, a value of 0.63 eV is reported in a film with an optical band-gap of 1.43 eV. Studies of the sub-band tail absorption, and space charge limited currents in these MS films indicate mid-gap defect concentrations in the range of $2-3 \times 10^{16}/cm^3$, or about a factor of seven to ten higher than in the best a-Si:H films. In addition, and in contrast to the GD films which show a strong preferential attachment of hydrogen to Si atom sites as the Ge atom concentration is increased, the MS films show essentially constant ratios for the relative SiH and GeH concentrations (SiH/Si and GeH/Ge, respectively) over the entire alloy range[5,6]. We present a theoretical analysis of dangling bond defects in a-Si,Ge alloys that will attempt to explain the differences in the conductivity activation energies between the GD and MS films in terms of Fermi level pinning by defect states associated with different alloy constituents. The results of this analysis also explain the simultaneous occurrence of both Si and Ge atom singly occupied dangling bond states (ESR results, Ref. 7).

In previous publications we have addressed the question of dangling bond defect state energies within the pseudo-gap of a-Si alloys[9,10]. We have found that an sp^3s* basis set[11] using tight-binding self energies and interaction parameters that have been fit to the band structure of crystalline Si give an excellent description of the valence band and band-gap of a-Si. This work emphasized shifts in the dangling bond states that were driven by changes in the near neighbor chemistry, and considered oxygen, nitrogen and carbon nearest neighbors to the silicon atom with the dangling bond[9,10]. This paper represents a continuation of these studies. In particular we have found that similar tight-binding parameters derived from a fit to crystalline Ge give an excellent description of the valence band and band-gap of a-Ge. We have explored the properties of Si and Ge atom dangling bond defects in a-Si and a-Ge, and in the binary alloy series a-Si,Ge, and have found that for all alloy compositions, the Ge atom dangling bond state, for a particular arrangement of Si and/or Ge nearest neighbors, is deeper in the gap than the corresponding Si atom dangling bond state. We find that the spread in energies for both Si and Ge dangling bond states, with different second neighbor bonding configurations, is greater than the difference between the statistically averaged Si and Ge dangling bond energies. Nevertheless, the difference in the calculated average energies of the respective Si and Ge atom dangling bond states is comparable to the difference in activation energies quoted above.

This paper is organized into four sections: (II) the method of computation; (III) the results of the computations; (IV) the application of the computation results to the interpretation of available experimental data; and (V) a summary of the important results and conclusions. In section IV, we also discuss the extension of these results to other alloy systems, and in particular point out some of the problems in applying the same method to a-Si,Sn and a-Si,C alloys.

II THE TIGHT-BINDING CALCULATION

The calculations we present are based on the empirical tight-binding formalism that has previously been applied to both crystalline and amorphous tetrahedrally bonded solids[12]. The Hamiltonian for the system can be written in the following form:

$$H = \sum_i E_i |i\rangle\langle i| + \sum_{ij} V_{ij} |i\rangle\langle j| \qquad (1)$$

where E_i represents the interaction between orbitals on the same atom, the so-called self energy terms, and V_{ij} represents the interaction between orbitals on neighboring atoms, i and j. There have been numerous studies of c-Si and c-Ge by the tight-binding method. In these instances, the E_i and V_{ij} parameters have been determined by either fitting the band structures, or from pseudo-potential calculations[12,13]. Dow and coworkers[11] have shown that an excellent fit to both c-Si and c-Ge can be achieved using an sp^3s* basis set. We have shown that the E_i and V_{ij} determined from this fit also yield a good description of the valence band and band-gap of a-Si. This calculation for a-Si is based on a Bethe Lattice structural model. The calculated band-gap of a-Si is 1.8 eV compared to a measured band-gap of about 1.6 eV.

Table I Tight-Binding Parameters sp^3s* Model

Material	Self Energies (eV)			Interaction Parameters (eV)				
	E_s	E_p	E_{s*}	$V_{sp\sigma}$	$V_{pp\sigma}$	$V_{pp\pi}$	$V_{ss\sigma}$	$V_{s*p\sigma}$
Si	-4.20	1.72	6.69	2.48	2.72	-0.72	-2.08	2.33
Ge	-5.88	1.61	6.39	2.37	2.85	-0.82	-1.70	2.26

Table I includes the E_i and V_{ij} terms. We have restricted the number of orbital terms to include only nearest neighbor interactions. Figure 1 gives the average electronic density of states (DOS), as well as the s and p-state local DOS (LDOS) for the Si Bethe Lattice. Table I also includes the corresponding E_i and V_{ij} obtained for c-Ge, and Figure 1 includes the corresponding DOS and LDOS functions calculated using the

Ge Bethe Lattice structure. We obtain a band–gap of 1.4 eV for Ge compared to a measured band–gap of about 1.1 eV, so that both gaps are over–estimated by the same amount, about 0.3 eV. We now discuss the Bethe Lattice structural model in order to set the framework for the defect state calculations in the binary alloys.

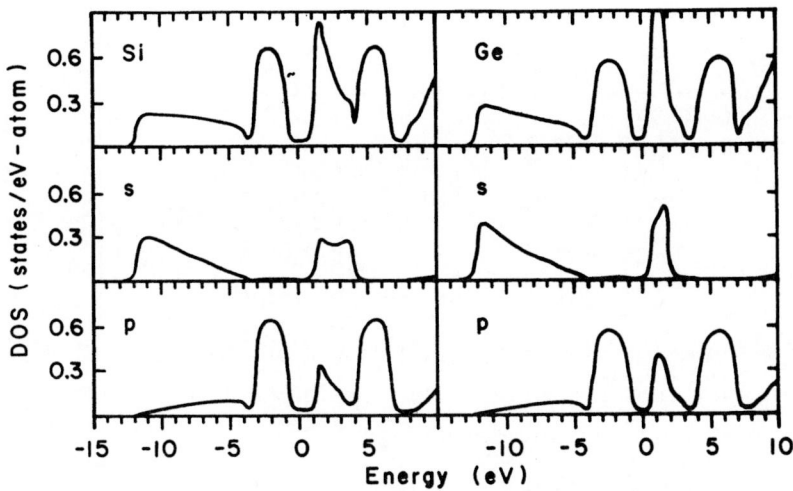

Figure 1 Density of state function (DOS) for (a) a–Si and (b) a–Ge.

The Bethe Lattice for a Group IV elemental solid is an infinite, aperiodic structure in which each atom has an ideal tetrahedral coordination, and in which there are no rings of bonded atoms. The electronic DOS function can be solved exactly for this structure, and it has been shown that this DOS function resembles what is obtained from a large continuous random network model. The results shown in Fig. 1 were derived for this type of structural model, and yield good agreement with the valence band and band–gap of the amorphous materials. The study of defect configurations is based on an extension of this model in which a local cluster containing the defect configuration of interest is terminated by Bethe Lattices. This type of structural model has been applied to studies of defect states and is called the Cluster Bethe Lattice (CBL) method[9,10,14].

Figure 2(a) gives a schematic representation of the Bethe Lattice, and Fig. 2(b) the way that it is used to terminate a defect bonding structure in the CBL method. We have shown these structures for all Si atoms; similar diagrams apply to Ge. In this paper we emphasize defect configurations with Si and Ge atom dangling bonds. We are interested in the LDOS functions for the Si atom with the dangling bond and its immediate neighbors. We describe the atomic configurations in terms of shells of atoms. Referring to Fig. 2(b), the first shell contains the only the Si atom with the dangling bond defect. This Si atom is then bonded to the three atoms that make up the second shell. For a Si atom dangling bond in a-Si, these are all Si atoms, whereas for a Si atom dangling bond in an a-Si,Ge alloy there are four possible configurations of Si and/or Ge atoms. We discuss the particular combinations a little

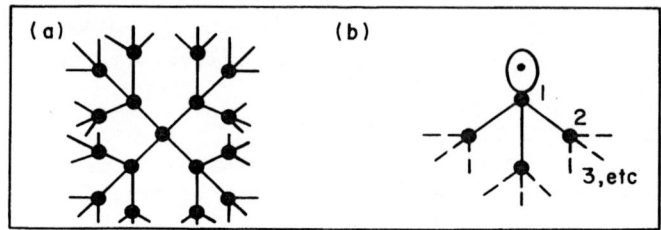

Figure 2 Schematic representation of (a) Bethe Lattice, (b) defect cluster containing a dangling bond. The dashed lines represent the connections between the defect cluster to the Bethe Lattice.

later on. The three atoms in the second shell make nine bonds to atoms of the third shell. We have found it adequate to terminate the defect cluster for a dangling bond at the end of the second shell and attach a Bethe Lattice to each of these nine bonding sites. Other defect configurations may require additional shells of atoms before their terminal bonds are connected to Bethe Lattices[14]. The CBL method gives a good description of the local electronic properties of the atom with the bonding defect and its immediate neighbors; i.e., the atoms of the first and second shells[9,10,14].

The structural model we use for the Si and Ge atom dangling bonds is shown in Fig. 3. Consider first a Si atom dangling bond, and note that parallel considerations hold for a Ge atom with a dangling bond. The atom in shell one is the Si atom. One of its sp^3 orbitals is singly occupied and this is the neutral dangling bond. The other three orbitals are terminated by atoms of the host material. If this material is a-Si as in Fig. 3(a), then all of the atoms in the second shell are Si atoms, and these in turn are terminated by Si Bethe Lattices. If the Si atom with the dangling bond is an impurity or alloy atom in a-Ge, then all of the atoms of the second shell are Ge atoms (as in Fig. 3(b)), which in turn are then connected to Ge Bethe Lattices. Finally, in the Si atom with the dangling bond is in an a-Si,Ge alloy, then the atoms of the second shell, and the Bethe Lattice may be either Si or Ge atoms. We make the following approximation to the real alloy structure. We consider all possible combinations of atoms in the second shell, three Si, two Si and one Ge, one Si and two Ge and three Ge (see Fig. 3(c)). For the third shell we use virtual atoms, denoted by the symbol V, that have average

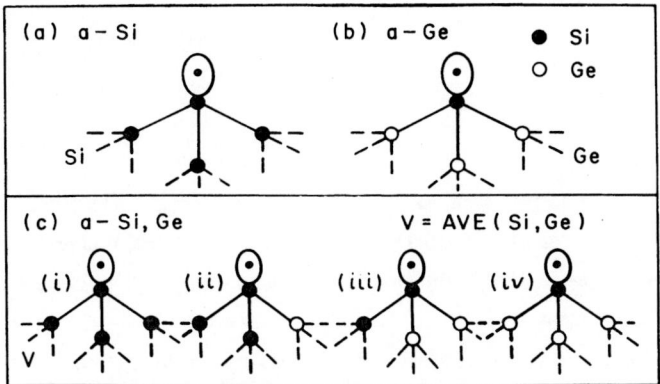

Figure 3 Si atom dangling bond defect clusters in (a) a-Si, (b) a-Ge and (c) a-Si,Ge alloys. For (c) ther are four different second shell arrangements which are terminated by virtual atom, V, Bethe Lattices.

properties of the particular alloy composition. If the alloy composition is a-$Si_xGe_{(1-x)}$, then the self energies and interaction parameters for these atoms are given by:

$$E_i(V) = xE_i(Si) + (1-x)E_i(Ge) \qquad (2)$$
$$V_{ij}(V) = xV_{ij}(Si-Si) + (1-x)V_{ij}(Ge-Ge) \qquad (3)$$

This is equivalent to defining a virtual atom Hamiltonian by:

$$H(V) = xH(Si) + (1-x)H(Ge) \qquad (4)$$

The configurations of second shell atoms vary as function of alloy composition. The probability of a given configuration being present can be treated statistically. If x and (1-x) are the alloy compositions of Si and Ge atoms respectively, then these probabilities (P(nSimGe)) are given by the following terms in a binomial expansion[15].

$$P(nSimGe) = C(n,m)x^ny^m \qquad (5)$$

where n = 0,1,2,3 and m = 3 - n, and C(n,m) is a binomial coefficient defined from the expansion of $(x + y)^3$; i.e.,

$$(x + y)^3 = x^3 + 3x^2y + 3xy^2 + y^3 \qquad (6)$$

so that C(3,0) = 1, C(2,1) = 3, C(1,2) = 3 and C(0,3) = 1.

The methods for calculating the various DOS functions for Cluster Bethe Lattice configurations are well documented[12-14] and we shall not present them here. Instead we present the results of our calculations.

III THE RESULTS

Figure 4 contains the results of our calculation of the DOS functions for a Si dangling bond in a-Si. We include the LDOS functions for the atom with the dangling bond, and its nearest and next nearest Si neighbors. The atoms labelled Si(2) represent the first shell of neighbors, and the atoms labelled Si(3) represent the first shell of atoms of the Si Bethe Lattices. Similar results apply to a-Ge. Note here that the dangling bond state that we have examined is the neutral site, the so-called D^o center. We have not attempted to describe the properties of the associated charged defect centers, D^+ and D^-.

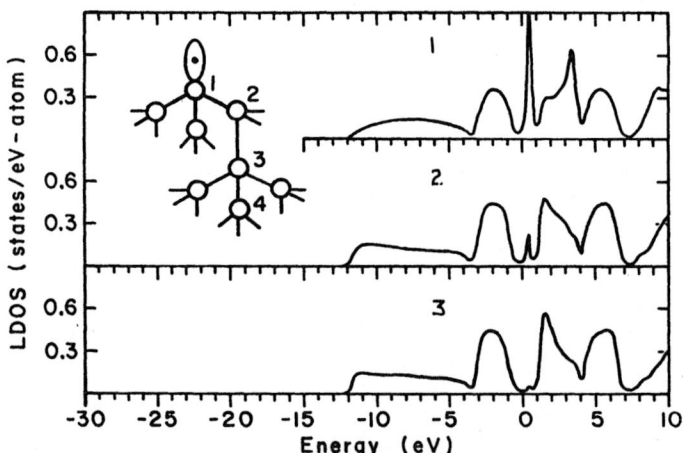

Figure 4 LDOS functions for a-Si. The defect configuration is shown. in the diagram, and identifies the atoms for which the LDOS functions are presented.

In both a-Si and a-Ge host materials, the energy of the Si atom dangling bond is higher in the gap than the energy of the Ge dangling bond. This difference is about 0.1 eV in a-Si, and is reduced by about a factor of two in a-Ge. The differences in Si and Ge atom defect energies (for the same nearest neighbor bonding groups) vary between these limits in the alloys with intermediate Si and Ge atom compositions.

We have shown in previous papers[9,10] that one of the factors that contributes to the shift of a dangling bond defect state induced by changing the near neighbor chemistry is the relative bond strength of the atoms that are back-bonded to the Si atom with the dangling bond defect. The dangling bond state has an anti-bonding character and to a first approximation, its energy in the gap is determined by the bonding states of that atom; the greater the bonding state energy, the larger the separation between bonding and anti-bonding states. In the case of the atoms from the first row of the periodic table, C, N and O, this bond strength scales directly with the relative electronegativities of these atoms[10]. In the case of the Si and Ge atom dangling bond defects discussed above, the difference in the energies of the Si and Ge atom dangling bond states derives directly from the relative energies of Si-Si (53 kcal/mol), Ge-Ge (45 kcal/mol) and Si-Ge(49 kcal/mol) bonds, and can not be simply related to the relative electronegativities of these two atoms.

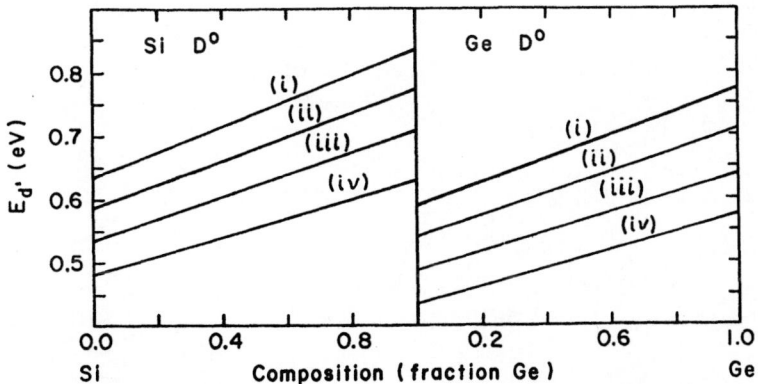

Figure 5 Normalized energies for Si and Ge atom dangling bonds. The designations (i)-(iv) refer to local bonding groups in Figure 3(c).

Figure 5 gives the results of our calculation for the a–Si,Ge alloy series. For each alloy composition other than x = 0.0 and 1.0, there are four possible next nearest neighbor bonding environments to the atom with the dangling bond. These are shown in Fig. 3(c). Figure 5 contains the calculated energy for each of these local bonding arrangements. For each of these local bonding groups, the energy of the Si atom dangling bond is higher in the gap than the Ge atom dangling bond. The average energy difference is about 0.1 eV. We have plotted the energy of the dangling bond state in normalized units, E_d, defined by the relationship given below:

$$E_{d'} = \frac{(E_d - E_{vb})}{(E_{cb} - E_{vb})} \qquad (7)$$

The energies of the Si and Ge atom dangling bond states are represented by two sets of four parallel lines. Note that the slopes of the Ge and Si lines are just about the same. Since the band–gap decreases in going from a–Si to a–Ge, this means that the actual defect state energies for Si and Ge dangling bonds get closer together in going from a–Si to a–Ge.

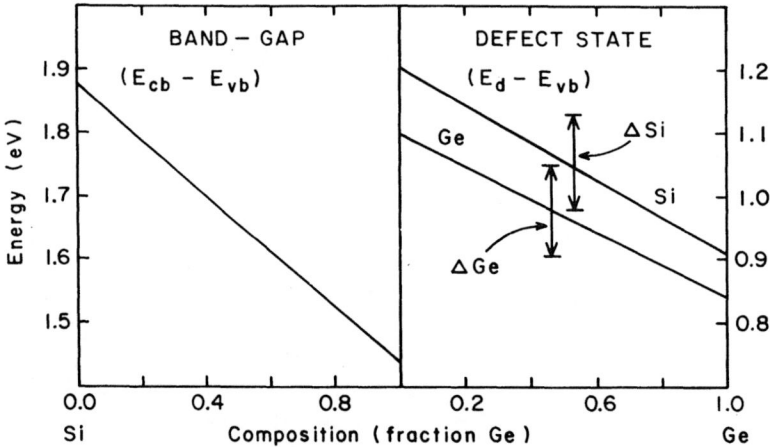

Figure 6 Average energy gap, and statistically weighted Si and Ge atom dangling bond energies as function of alloy composition.

Figure 6 gives a plot of the statistically averaged dangling bond energies as a function of the alloy composition. This average defect state energy takes into account the discrete energies shown in Fig. 5

and weights these according to the probability of the occurrence of a particular configuration involving the atom with the dangling bond and its three nearest neighbors. This figure also give the energy dependence of the average energy gap. For this presentation we have referenced the defect energy to the valence band edge;i.e., we have plotted the defect energy, $(E_d - E_{vb})$, and the band-gap, $(E_{cb} - E_{vb})$, as functions of the alloy composition. The average band gap shows a linear dependence on composition as is expected from the virtual atom approximation. The statistically averaged dangling bond energies also show linear relationships with the Si atom dangling bond state higher in energy by about 0.1 eV than the corresponding Ge atom dangling bond state for all alloy compositions. The separation in energy between the Si and Ge states decreases in going from a-Si to a-Ge. Figure 6 also includes the spread in energy for the four different local bonding environments associated repspectively with the Si and Ge atom dangling bond states. This spread is about 0.35 eV and is greater than the separation between the averaged defect state energies. This means that there is an energy region within the pseudo-gap in which there can be both singly occupied Si and Ge dangling bond states. The gap state distribution will depend on the relative number of defect states associated with each of the constituent atoms. For example, it the Ge atom defect concentration exceeds the Si defect concentration by a significant factor, then the dominant defects will be Si D^+ states and Ge D^0 and D^- centers.

IV DISCUSSION

There are two aspects of the experimental data that we wish to highlight: (1) the first relates to the different behavior of a-Si,Ge:H alloys with about fifty atomic percent Ge as formed by the glow discharge process and by magnetron sputtering; and (2) the second relates to the relative number of Ge atom dangling bonds in alloys prepared by the glow discharge process.

a-Si,Ge:H alloys grown by the GD and MS processes display band-gaps that are about the same; however, other properties of these alloys can differ appreciably. For example, the activation energy for the dark conductivity is a little less than one half of the band gap in the MS alloy, 0.63 eV with a Tauc's band gap of 1.43 eV[5,6], and has a value of more than one half of the band gap in the GD alloys, 0.8 eV with a Tauc's band gap of 1.45 eV[8]. The MS alloy discussed above, shows a relatively high photoconductivity (excited by photons from a He-Ne laser source), while the photoconductivity in the GD sample is generally about a factor of 20 less. Additional measurements on the MS sample indicate: (1) that the sub-band tail absorption is about of factor of 7 to 10 higher than in the best intrinsic a-Si:H alloys, and (2) that the trap density at the Fermi Level, as determined from measurements of the onset of space charge limited current flow, is about 2-3 x 10^{16}/cm^3, or about a factor of 5 to 7 more than in intrinsic a-Si:H. The photoconductivity measurements, sub-band tail absorption and trap densities at the Fermi level are consistent with each other;.i.e, they all are in accord with deep trap and tail state densities about a factor of 5 to 10 higher than that of the best intrinsic a-Si:H. In samples that are prepared by GD, it has been shown that the ESR signal for alloys with fifty atomic percent Ge is dominated by Ge dangling bond states[7], and that this ESR signal corresponds to a deep trap density in excess of 10^{17}/cm^3.

The picture that then emerges from these experimental results is that the Fermi level is pinned by Ge atom dangling bond states in the GD films, and that the energy of these neutral Ge dangling bond states is about 0.80 to 0.85 eV below the conduction band edge. In contrast, the Fermi level in the MS material is assumed to be pinned by Si atom dangling bond states. This places the the Si dangling bond state about 0.65 eV below the conduction band edge. This means that the Ge atom dangling bond state is deeper in the gap by about 0.15 eV than the Si dangling bond state. This difference in dangling bond state energies is consistent with the predictions of the calculation presented above, where for a fifty atomic percent Ge alloy, the calculation yields a difference of about 0.1 eV between the two dangling bond states (see Figures 5 and 6). The interpretation is however clouded by the fact that there is a considerable spread in the Si and Ge atom dangling bond states due the differences in local bonding arrangements. In this context, one should exercise some caution in interpreting the experimental results. It is likely that at each alloy composition there are both Si and Ge defect centers that are active in transport and recombination processes. In addition, there must also be pairs of charged centers as well. We have not addressed this problem here, so that we cannot comment on the energies of charged centers with respect to the neutral centers, or on the interactions that might occur because of near neighbor charged defect pairs. The ESR results discussed below also indicate an energy overlap in the position of the Si and Ge neutral dangling bonds states.

In a very recent paper[17], the Harvard group presented the results of a comprehensive study of the properties of a-Si,Ge:H alloys produced by the GD method. Based on their studies of the alloy dependence of the band-gap, and a variety of other electronic and photoelectronic properties, they have proposed a a model for the density of states structure within the pseudo-gap. They propose that the singly occupied Ge dangling bond state is higher in the gap than the singly occupied Si dangling bond state. This assignment is exactly the opposite of what our calculations have shown. The calculated relative positions of the Si and Ge atom dangling bond states are closely coupled to the values of the self energies used in the tight-binding calculations. The values we have chosen for the end member materials, a-Si and a-Ge, are the same values that have been used by Dow and his coworkers to calculate the band structures of crystalline Si and Ge, respectively[11]. The energies of Si and Ge atom core states in a-Si,Ge alloys have been studied by X-Ray Photoelectron Spectroscopy (XPS)[18]. The energies of these states show at most only very small shifts as a function of the alloy composition. Therefore, on the basis of the XPS results, it was concluded that there was very little charge transfer between Si and Ge atoms of the alloy, and that the electonegativities of the two elements did not differ by more than about 0.1, the difference assigned by Pauling scale on the basis of thermochemistry[19]. The parameterization of our tight-binding model is consistent with the XPS results, in the sense that we have assumed that the Ge and Si atom contributions to the self energy terms do not change with alloy composition. In particular, we have used this assumption in our calculation of the relative energies of Si and Ge atom dangling bond states in the end member materials, a-Si and a-Ge. In both instances we find that the Si atom dangling bond state is higher in the gap than the Ge dangling bond state. The model we have used to calculate the defect states for intermediate alloy compositions also preserves this relationship between the relative energies of the Si and Ge atom dangling bond states. The model proposed by the Harvard group is in disagreement with the predictions of this calculation, and therefore with the values assigned to the self enegy terms. On the basis of these arguments, we believe that the model proposed by the Harvard group needs reexamination and further study. In particular we suggest the obvious, that spectroscopic measurements be undertaken to pin down the relative positions of Si and Ge D^{o} states. Since we have shown that the manifolds of Si-atom and Ge-atom dangling bond states display considerable overlap within the pseudo-gap, considerable care must be exercised in the interpretation of any spectrocsopic measurements.

In principle the calculations we have performed can be extended to other group IV amorphous alloy systems, in particular a-Si,C and a-Si,Sn. In practice these calculations are made difficult by the nature of the local bonding of the alloy atoms C and Sn as their respective concentrations are increased. It has been shown that in both GD and DS a-Si,C alloys that the local bonding of C changes from a tetrahedral environment at low C concentrations to a trigonal or three fold-coordinated environment at high C concentrations[20]. This means that one cannot simply track dangling bond defect energies in a-Si,C alloys using a structural model similar the one that we have applied to a-Si,Ge alloys and in which both atomic species are assumed to retain a tetrahedral environment for all possible alloy compositions. In the a-Si,Sn system a similar problem may arise. In this case, about ten atomic percent Sn drives the alloy band-gap to zero[21]. Moreover the end member material Sn is a metal rather than a semiconductor or insulator, and the local atomic coordination is non-tetrahedral. In addition Sn displays a divalent character in a majority of its compounds. We therefore anticipate non-tetrahedral bonding of Sn in the a-Si,Sn system, for some composition range, and/or for some deposition schemes. It may however be possible to treat both these alloy systems , a-Si,C and a-Si,Sn, over limited composition ranges in which both atoms retain a tetrahedral coordination. We expect these ranges to be to about ten atomic percent Sn, and to about fifty atomic percent C. Our calculations of dangling bond defect state energies indicate that in a-Si, the C dangling bond state is higher in the gap than the Si dangling bond state[9,10], and that the Sn dangling bond state is lower in the gap than the Si dangling bond state. We will attempt to track these energies over the limited range of alloy compositions mentioned above in a later study.

IV SUMMARY

We have presented a theoretical analysis of the properties of Si and Ge atom neutral dangling bonds for the alloy system a-Si,Ge. We have used a tight-binding formalism, based on an sp^3s* basis set, and a structural model based on the cluster Bethe Lattice method. We have found that for all alloy compositions, and similar nearest neighbor bonding configurations, that the Ge defect state is deeper in the gap than the Si defect state. The separation in these energies decreases as the Ge concentration is increased. In addition, we have found that the spread in the defect state energies for Si (and Ge) atoms with dangling bonds with different nearest neighbor configurations is greater than the difference between the corresponding Si and Ge states themselves. We have also neglected charged defect pair which will occur in the alloys simply because of the differences in the energies of the various D^o centers. These factors then imply that experimentally determined activation energies, and deep trap energies can not be unambiguously assigned to Si and Ge atoms on the basis of their respective energies alone.

ACKNOWLEDGMENTS

This work is supported under ONR Contract N00014-79-C-0133 and SERI Subcontract XB-2-02065-1.

REFERENCES

1. J. Chevallier, H. Wieder, A. Onton and C.R. Guarnieri, Solid State Commun. 24, 867 (1976).
2. B. von Roedern, D.K. Paul, J. Blake, R.W. Collins, G. Moddel and W. Paul, Phys. Rev. B 25, 7678 (1982).
3. W. Paul, D.K. Paul, B. von Roedern, J. Blake and S. Oguz, Phys. Rev. Lett. 46, 1016 (1981).
4. S.Z. Weirz, M. Gomez, J.A. Muir, O. Resto, R. Perez, Y. Goldstein and B. Abeles, Appl. Phys. Lett. 44, 634 (1984).
5. R.A. Rudder, J.W. Cook, Jr. and G. Lucovsky, Appl. Phys. Lett. 45, 887 (1984).
6. R.A. Rudder, J.W. Cook, Jr. and G. Lucovsky, J. Vac. Sci. A., May-June (1985), in press.
7. A. Morimoto, T. Muira, M. Kumeda and T. Shimuzu, Jpn. J. Appl. Phys. 20, L833 (1981).
8. W. Paul in U.S. Department of Energy Report, SERI/CP-211-2167 (1983).
9. S.Y. Lin, G. Lucovsky and W.B. Pollard, J. Non-Cryst. Solids 66, 291 (1984).
10. G. Lucovsky and S.Y. Lin, AIP Conf. Proc. 120, 55 (1984).
11. P. Vogl, P.H. Hjamarson and J.D. Dow, J. Phys. Chem. Solids 44, 365 (1983).
12. J.D. Joannopoulos, J. Non-Cryst. Solids 32, 241 (1979).
13. D.J. Chadi and M.L. Cohen, Phys. Stat. Solidi 68, 405 (1975).
14. D. Allan and J.D. Joannopoulos, Topics in Appl. Phys. 56, 5 (1984).
15. H.R. Phillip, J. Non-Cryst. Solids 8-10, 627 (1972).
16. R.A. Rudder, J.W. Cook, Jr. and G. Lucovsky, Appl. Phys. Lett. 43, 871 (1983).
16. K.D. MacKenzie, J.R. Eggert, D.J. Leopold, Y.M. Li, S. Lin and W. Paul, Phys. Rev. B 31, 2198 (1985).
17. G. Lucovsky, S.S. Chao, J.E. Tyler, G. De Maggio, J. Vac. Sci. Tech. 21, 838 (1983).
18. A.L. Allerd, J. Inorg. Nucl. Chem. 17, 215 (1961).
20. T. Shimada, Y. Katayama and K.F. Komatsubara, J. Appl. Phys. 50, 5530 (1979).
21. H. Ituzaki, N. Fujita, T. Igarashi and H. Hitotsuyanagi, J. Non-Cryst. Solids 59 & 60, 589 (1983).

STUDY OF DISORDER IN FLASH-EVAPORATED AMORPHOUS InP FILMS

Adriana Gheorghiu and Marie-Luce Thèye

Laboratoire d'Optique des Solides
Unité Associée au CNRS 040-781
Université P. et M. Curie
4 Place Jussieu, 75230 Paris Cédex 05, France

INTRODUCTION

The similarities between the structure and the electronic properties of amorphous III-V compounds and their elemental counterparts, amorphous Ge and Si, have early been emphasized [1,2]. However, due to the presence of two constituents, and to the partial ionic character of the bonding, both the type of disorder and the nature of defects are likely to be different in the amorphous compounds. The possibility of chemical disorder must in particular be comtemplated. Amorphous InP is a good candidate for studying these problems. Its high ionicity (f_i = 0.421 according to Phillips' scale [3]) should energetically favour heteropolar bonding, and oppose the formation of bonds between atoms of the same constituent, or wrong bonds. However, real samples are obtained by atom-by-atom deposition processes, under conditions far from equilibrium, which may be responsible for a certain amount of chemical disorder, accompanying structural disorder. The effects of such chemical disorder should be strongly apparent, both in the structure, because of the large difference in atom size between the two constituents, and in the electronic properties, due to important potential fluctuations related to the ionic contribution to the bonding. The purpose of this paper is to investigate the disorder in flash-evaporated a-InP films, and to compare the results to those previously reported for amorphous Ge and other amorphous III-V compounds.

EXPERIMENT

The method of film preparation must be specified for such a study, since it is expected to play a crucial role in the degree of disorder of the samples. Our samples are thin films (300 to 1000 Å thick), deposited under high vacuum (10^{-8} Torr range) by flash-evaporation of crystalline powder onto silica substrates at room temperature. The best evaporation conditions for obtaining nearly-stoichiometric films, as controlled by α - particle back-scattering, are : crucible temperature $\simeq$ 1450°C, powder grain size $\simeq$ 160 -250 µm, deposition rate 0,5 Å/sec (for a crucible / substrate distance of about 40 cm). Unlike co-sputtered films which presented an excess of In [4], these flash-evaporated films tend to be richer in P; we limited ourselves to films close to stoichiometry within 5%. The samples were studied

213

as-deposited, and after annealing at 200-250°C, temperatures at which they reach their limit metastable state. They start to crystallize at about 350°C.

The film structure was investigated by electron diffraction experiments on a Philips EM 300 electron microscope (accelerating voltage of 100 keV) without velocity filter. The method used to get rid of the inelastically scattered intensity, as well as the data analysis procedure, have been reported elsewhere [5,6]. Reliable data were obtained up to $k = 4\pi \sin \emptyset / \lambda \simeq 11 \text{ Å}^{-1}$.

The complex dielectric constant $\tilde{\epsilon} = \epsilon_1 + i \epsilon_2$ was deduced from 0.5 to 6.2 eV from accurate transmission and reflexion measurements performed on a Cary 17D spectrophotometer. Exact thin film formulas taking into account multiple reflections in the substrate were used, and the thickness was determined independently by an X-ray interference technique. The d.c. electrical conductivity was measured as a function of temperature with a Keithley 616 electrometer, using thin-wire thermo-couples directly soldered to the substrate.

CHARACTERIZATION OF THE DISORDER

Figure 1a shows the interference functions i(k) deduced from the diffraction data for as-deposited and annealed a-InP. The two salient features are, on the one hand the rapid damping of the oscillations with increasing k, on the other hand the absence of the characteristic shoulder on the high-k side of the third peak, observed for most amorphous tetracoordinated semiconductors [1]. The first observation indicates a large spread of the first-neighbour distances, the second one a loss of correlation between first and second neighbours, due to variations in bond angles. These conclusions are confirmed by the corresponding radial distribution functions W(r) (figure 1b); although

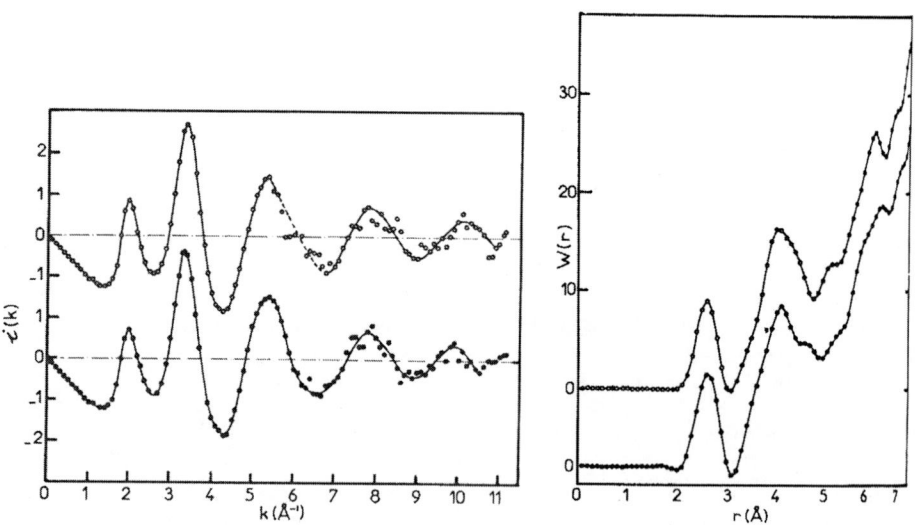

Fig.1. Interference functions i(k) (a) and radial distribution functions W(r) (b) for as-deposited (solid circles) and annealed (open circles) a-InP.

the diffraction data are limited to rather small k values, the first peaks of the RDF are well resolved because of the rapid damping of the i(k) oscillations. The first neighbour peak is quite broad, and centred at an average distance $r_1{}^a = 2.60$ Å significantly larger than the crystalline distance $r_1{}^c = 2.54$ Å. This suggests the presence of some proportion of wrong bonds, differing in length from the normal heteropolar In-P bond. The second neighbour peak is also broadened, markedly asymmetric, and centred at a distance $r_2{}^a$ smaller than the value expected from regular tetrahedral bonding : $r_2{}^a / r_1{}^a \simeq 1.57$ compared to $r_2{}^c / r_1{}^c = 1.633$. This indicates large fluctuations in bond angles, with an average value $\bar{\theta}$ smaller than the regular tetrahedron value. This disorder in bond angles can be explained by a larger flexibility of the bonds which, as their ionic character increases, loose part of their directional stability [3], and probably also by some relaxation of the bonding configuration at dangling bond and wrong bond sites [7]. As for the third peak which can clearly be observed in the RDF of other amorphous tetracoordinated semiconductors, and which is related to a peak in the distribution of three-bond neighbours [8], it is here almost burried into the background. This confirms that a-InP is much more disordered than its parent materials. Annealing produces little change in the structure of the films. The only detectable effect on the RDF is a slight shift to lower r values and a narrowing of the first neighbour peak, which suggests a moderate decrease of the proportion of wrong bonds.

The assumption of a certain amount of chemical disorder in our a-InP films has been tested by ESCA studies of the core level spectra of each constituent [9]. The resolution was 0.1 eV and the binding energies were referred to the Fermi energy. The results clearly show a shift of the In core levels towards lower binding energies, and a shift of the P core levels towards higher binding energies, in a-InP compared to c-InP. The effect cannot be explained by a change in location of the Fermi level, since the shifts occur in opposite directions for the two constituents. For both In and P, the core levels are shifted towards their positions in the pure elements. This can be interpreted as due to the presence of wrong bonds. However, no broadening of the core level distributions, which would be expected from the fluctuations of the Coulombic field resulting from a random distribution of charges, is detected within experimental uncertainty. The observed shift without broadening suggests a rearrangement of the charges tending to fix the core levels at some intermediate energy, characteristic of the amount of disorder.

Wrong bonds in amorphous III-V compounds can have essentially two origins. They can be introduced in an otherwise chemically ordered network by the formation of odd-membered rings. They can alternatively be a consequence of short-range composition fluctuations due to the process of film deposition, some chemical disorder accompanying structural disorder. Although the distinction between the two types of continuous random network models with [10,11] and without [12] odd-membered rings is quite difficult to make, the structure data reported above suggest that the a-InP network is better described by a model without odd-membered rings. The main argument, based on a recently proposed criterion, is that the ratios k_i/k_1 of the positions of the successive ith peaks of the interference functions to that of the first peak are very close to the values computed for the unrelaxed Connell-Temkin model [6]. Besides, certain features of the radial distribution functions, in particular the asymmetry of the second peak, are more compatible with such a model. Therefore, wrong bonds in a-InP are likely to be related to chemical disorder due to the deposition conditions, rather than to be introduced by the network topology.

The type of disorder and the nature and amount of defects determine the density of electronic states, both in the valence and conduction bands and within the pseudo-gap. We discuss in the following some features of the optical and transport properties which are particularly sensitive to disorder and defects and can help to their study. The main parameters are summarized in table I.

The ϵ_1 and ϵ_2 spectra of a-InP exhibit the characteristic shapes of all amorphous tetracoordinated semiconductors. We concentrate here on the value of the static dielectric constant $\epsilon_1(0)$, obtained by extrapolation of the ϵ_1 data at low energies. For as-deposited films, $\epsilon_1(0)$ is significantly larger than the crystalline value (table I); the relative increase is of the order of 45%, while it is only of 14% in the Ge case for example. It is well-known that $\epsilon_1(0)$ is related to the average gap E_G, which can be considered as a measure of the average bond strength in both the amorphous and crystalline materials, through [3] : $\epsilon_1(0) = 1 + (\hbar w_p)^2 / (E_G)^2$, where $\hbar w_p$ is the plasma energy of the valence electrons. Apart from density effects on the $\hbar w_p$ value, the variation of $\epsilon_1(0)$ from the crystalline to the amorphous material essentially reflects a change of the average gap E_G : here, an important decrease. For a partly ionic compound like InP, E_G includes both a covalent part E_h and an ionic part C [3] : $E_G^2 = E_h^2 + C^2$. E_h, like in the purely covalent case of Ge, is determined by the first neighbour environment. It will be modified by the presence of both dangling bonds and wrong bonds. Dangling bonds reduce the average coordination number C_1, and consequently E_h through [14] : $E_h \propto C_1^2$. Wrong bonds change the nature of the bonding; this effect can be accounted for by considering only the modification of the average bond length [15] : $E_h \propto d^{-2.5}$. Since the average first neighbour distance r_1^a is found to increase in a-InP with respect to c-InP, both dangling bonds and wrong bonds will contribute to a decrease of E_h. As for C, it depends on the potential difference between the two constituents, and may be perturbed by chemical disorder. In any case, the $\epsilon_1(0)$ value confirms the existence of a large proportion of coordination defects in as-deposited a-InP. Annealing produces a rather large decrease of $\epsilon_1(0)$ (table I). This is somewhat surprising, because the distribution of first neighbours changes very little, although in the

Table I. Values of the static dielectric constant $\epsilon_1(0)$, of the optical gap E_o, of the inverse slope of the exponential absorption edge E_e and of the conductivity activation energy E_σ for as-deposited and annealed amorphous InP; the values of $\epsilon_1(0)$ and E_o for crystalline InP are also indicated.

material	$\epsilon_1(0)$	E_o(eV)	E_e(eV)	E_σ (eV)
a-InP as-dep.	13.7 + 0.2	0.90 + 0.03	0.13 + 0.01	–
a-InP ann	11.3 + 0.1	1.20 + 0.01	0.09 + 0.01	0.55 + 0.01
c-InP	9.52	1.34	–	–

correct direction since the average first neighbour distance decreases slightly. The observed decrease of $\epsilon_1(0)$, or increase of E_G, could therefore be explained by an appreciable reduction of the number of dangling bonds , more than by the elimination of defects of the wrong bond type.

The optical gap E_O, determined by extrapolation of the low-energy side of the ϵ_2 spectrum according to the law : $(\hbar\omega)^2 \epsilon_2 = A (\hbar\omega - E_O)^2$, is smaller than the minimum direct gap of c-InP for both as-deposited and annealed a-InP (table I). In the present status of our knowledge, such a comparison bring no information on the effects of disorder. We focuss here on the large increase of E_O upon annealing, which is about twice that observed for a-Ge for example. This increase of E_O is accompanied by a shift of the whole ϵ_2 spectrum towards high energies by about the same amount. This indicates important network rearrangements, which tend to widen the gap between the valence and conduction bands. Since the distribution of the first and second neighbours is little modified by annealing, these rearrangements must concern more remote neighbours, via changes in the dihedral angle distribution and the ring statistics. These elements of disorder indeed seem to play a crucial role on the density of states, especially near the band extrema [16].

The optical absorption coefficient α in the edge region follows a roughly exponential behaviour : $\alpha \propto \exp(\hbar\omega /E_e)$ in the $5.10^3 - 5.10^2$ cm^{-1} range for as-deposited and annealed a-InP, as shown in figure 2. The edge is rather broad for as-deposited films, but sharpens upon annealing (table I). The effect is similar to the one observed for a-Ge [17]. In this case, the absorption tail is ascribed to localized states in band tails, which are created by disorder and shrink while the optical gap increases when the disorder is reduced by annealing; the defect states introduced by the dangling bonds are located deeper in the pseudo-gap. One can tentatively apply the same interpretation to a-InP. However, a-InP presents a high degree of disorder and a large amount of structural and chemical defects; besides, InP has a marked ionic character. One would therefore expect large band tails and a high density of defect states throughout the pseudogap. The optical data do not confirm this prediction. The behaviour of a-InP strongly contrasts with that of similarly prepared a-GaP [18], which is also believed to present some chemical disorder. In this case, as shown

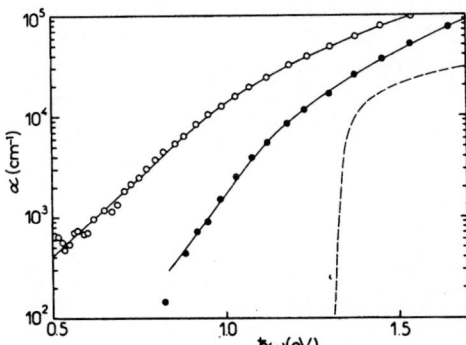

Fig.2. Logarithm of the optical absorption coefficient α versus energy $\hbar\omega$ for as-deposited (open circles) and annealed (solid circles) a-InP; the absorption edge of c-InP is indicated by the discontinuous line.

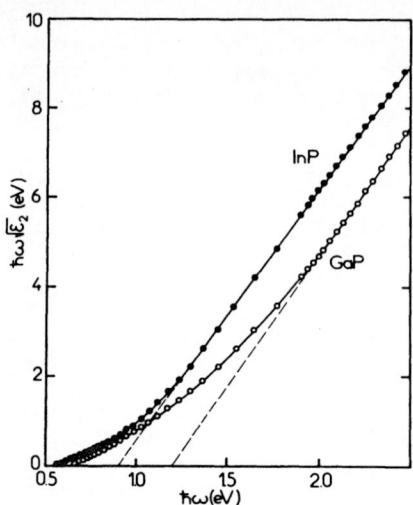

Fig.3.　　$\hbar\omega(\epsilon_2)^{1/2}$　　versus energy $\hbar\omega$ in the absorption edge
region for flash-evaporated, as-deposited a-InP (solid circles)
and a-GaP (open circles) films.

on figure 3, large absorption tails are observed, indicating a large
density of extra-states both near the band edges and deeper in the
pseudo-gap. This comparison emphasizes that the modifications of the
band edges by the disorder [16] as well as the electronic states associated
with the various types of defects [7] may be quite different depending
on the material considered.

Additional information on the density of states in the pseudo-
gap can be gained from the behaviour of the electrical d.c. conductivity.
Figure 4 shows the　　σ (1/T) curves for different a-InP films. At
high temperatures for annealed films, σ has an activated behaviour
indicating conduction in extended states, with an activation energy
E_σ of the order of half the optical gap (table I). Below room
temperature, σ follows a $T^{-1/4}$ behaviour like in the a-Ge case [19],
but the T_0 slope is larger (> 10^9K), varies somewhat with deposition
conditions, and increases further upon annealing. For annealed films
and for some as-deposited ones, the σ data are better described by
a $T^{-1/2}$ law. This behaviour, which has also been reported for other
flash-evaporated a-III-V compounds except precisely a-InP [20], indicates
that hopping to nearest neighbours in localized states away from the
Fermi level, dominates over variable-range hopping near E_F. This suggests
that the density of states at the Fermi level is rather small, and
that defect states must be located closer to the band edges.

CONCLUSION

From the data discussed in this paper, it can be concluded that
our flash-evaporated a-InP films have a highly disordered network and
certainly contain a large proportion of defects of the dangling-bond
and wrong-bond types. If the average bond strength is found to be much
smaller than in the crystal, one does not observe a high density of
localized states, either at the band edges or in the pseudo-gap. This
is somewhat contrary to what is expected, especially in view of the
high ionicity of this compound. More work is needed, both theoretically
and experimentally, to understand the problem of amorphous compounds.

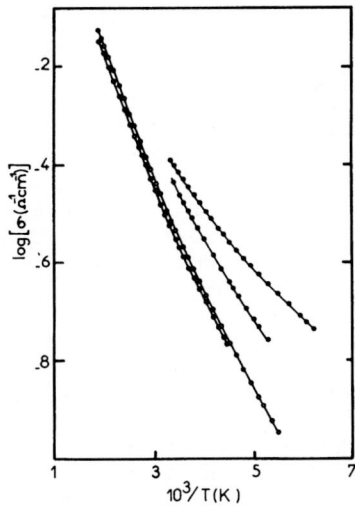

Fig.4. Logarithm of the conductivity σ versus reciprocal temperature 1/T for as-deposited (solid circles) and annealed (open circles) a-InP films.

ACKNOWLEDGEMENTS

The authors like to thank M. Ouchène, T. Rappeneau and S. Fisson for their participation in the experiments, C. Sénémaud for the ESCA studies, J. Dupin for the composition measurements and J. Dixmier for valuable comments.

REFERENCES

1. N.J. Shevchik and W. Paul, The structure of tetrahedrally coordinated amorphous semiconductors, J. Non-Cryst. Sol. 13 : 1 (1973/74)
2. N.J. Shevchik, Photoelectron spectroscopy of amorphous Group IV and III-V semiconductors, in : "Tetrahedrally bonded amorphous semiconductors", A.I.P. Conf. Proc. N°20, M.H. Brodsky, S.Kirkpatrick and D. Weaire, eds., American Institute of Physics, New-York (1974).
3. J.C. Phillips, "Bonds and bands in semiconductors", Academic Press, New-York (1973).
4. N.J. Shevchik, J. Tejeda and M. Cardona, Densities of valence states of amorphous and crystalline III-V and II-VI semiconductors, Phys. Rev. B9 : 2627 (1974).
5. M. Gandais, M.L. Thèye, S. Fisson and J. Boissonade, Structure studies by electron diffraction on amorphous Ge films, Phys. Stat. Sol (b) 58 : 601 (1973).
6. J. Dixmier, A. Gheorghiu and M.L. Thèye, Local order in amorphous III-V compounds $A_{1-x}B_x$ by electron diffraction, in relation with electronic properties, J. Phys. C. : Solid State Phys. 17 : 2271 (1984).
7. E.P. O'Reilly and J. Robertson, The electronic structure of defects in amorphous GaAs, Phil. Mag. Lett. B50 : L9 (1984).
8. R.J. Temkin, A theory of the structure of tetrahedrally bonded amorphous solids, J. Non-Cryst. Sol. 28 : 23 (1978).
9. M. Ouchene, C. Senemaud, E. Belin, A. Gheorghiu and M.L. Thèye, Influence of disorder on the electronic distribution of InP by X-ray and photoelectron spectroscopies, J. Non-Cryst. Sol. 59-60 : 625 (1983).

10. D.E. Polk and D.S. Boudreaux, Tetrahedrally coordinated random net-work structures, Phys. Rev. Lett. 31 : 92 (1973).
11. P. Steinhardt, R. Alben and D. Weaire, Relaxed continuous random network models, J. Non-Cryst. Sol. 15 : 199 (1974).
12. G.A.N. Connell and R.J. Temkin, Modelling the structure of amorphous tetrahedrally coordinated semiconductors, Phys. Rev. B9 : 5323 (1974).
13. J. Stuke and G. Zimmerer, Optical properties of amorphous III-V compounds, Phys. Stat. Sol. (b) 49 : 513 (1972).
14. J.C. Phillips, Electronic structure and optical spectra of amorphous semiconductors, Phys. Stat. Sol. (b) 44 : K1 (1971).
15. J.A. Van Vechten, Quantum dielectric theory of electronegativity in covalent systems : I. Electronic dielectric constant, Phys. Rev. 182 : 891 (1969).
16. J. Singh, Influence of disorder on the electronic structure of amorphous Si, Phys. Rev. B 23 : 4156 (1981).
17. M.L. Thèye, Influence of annealing on the optical properties of amorphous Ge films, Mat. Res. Bull. 6 : 103 (1971).
18. A. Gheorghiu and M.L. Thèye, Disorder effects and the optical properties of amorphous GaAs and GaP, Phil. Mag. B44 : 285 (1981).
19. M.L. Thèye, A. Gheorghiu, T. Rappeneau and A. Lewis, Transport properties of evaporated versus sputtered amorphous Ge films, J. Physique (Paris) 41 : 1173 (1980).
20. W. Beyer, "Thermokraft und leitfähigkeit von amorphen halbleitern mit tetraedrischer nahordnung", Görich and Weïershäuser, Marburg (1974).

GAP STATES IN HYDROGENATED AMORPHOUS SILICON: THE TRAPPED HOLE CENTRES

(THE A CENTRES)

K. Morigaki, H. Takenaka, I. Hirabayashi and M. Yoshida

Institute for Solid State Physics
University of Tokyo
Roppongi, Tokyo 106, Japan

INTRODUCTION

We have previously suggested the existence of trapped hole centres, which are designated as the A centres, as radiative recombination centres in hydrogenated amorphous silicon (a-Si:H) from the measurements of optically detected magnetic resonance (ODMR)[1-6], photoinduced absorption (PA)[7] and photoinduced absorption-detected ESR(PADESR)[8,9]. Furthermore, we have pointed-ed out that the A centres are correlated with the presence of hydrogen in a-Si:H films[10]. In this article, we review these results and present some additional evidence for the existence of the A centres that has been obtained from the time-resolved (TR) ODMR experiments[11,12]. We discuss the nature of the A centres on the basis of the results obtained by the ODMR and PA measurements. The correlation between our ODMR and PA experiments and other experiments concerning dual beam photoconductivity and thermally stimulated currents is also discussed.

EXPERIMENTAL

The TRODMR and steady state ODMR experiments were carried out at 2 K, using a microwave source consisting of a klystron (9V54) and a travelling wave tube (11W71) operating at 9.6 GHz. The microwave power was about 5 W at peak for microwave pulse and about 500 mW for chopped microwave at 1 kHz. Pulsed light was obtained from either an N_2-pumped pulsed dye laser or a CW argon ion laser with use of an acousto-optic modulator. The TRODMR signal was detected, using a cooled Ge detector with a response time of 300 ns and a two-channel boxcar integrator.

The photoinduced absorption measurement was carried out at 1.8 K, using a krypton ion laser as an excitation light, a tungsten lamp as a probe light, a single prism monochromator, a cooled PbS detector, and a lock-in amplifier. The PA-detected ESR measurement was carried out at 2 K, using the same microwave source as in the ODMR measurement. The detail has been described in ref. (8) and (9).

Samples used in the present experiment were prepared by glow-discharge decomposition of pure silane, using either a capacitively coupled system (for samples Nos. 519, 540 and 541) or a cylindrical diode system[13](for samples Nos. 927 and 933). Substrate temperatures were 300°C, 75°C, 120°C,

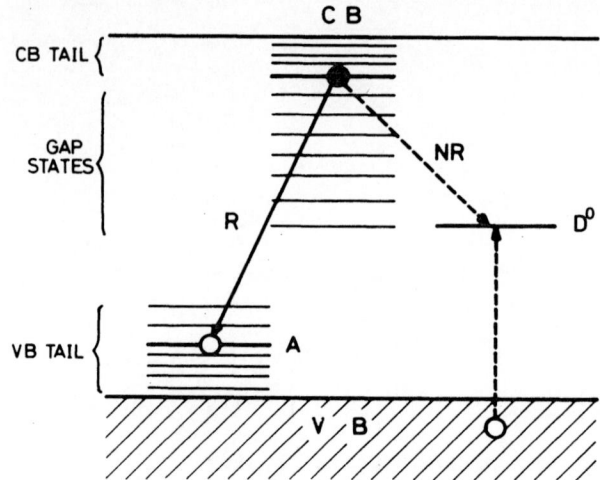

Fig. 1. Schematic diagram of recombination processes
and recombination centres in a-Si:H. R and
NR designate radiative and nonradiative recombination, respectively.

230°C and 230°C for samples Nos. 519, 540, 541, 927, and 933, respectively.
Spin density (dangling bond density) and hydrogen content (estimated from
IR measurements) of those samples are 7.0×10^{15}, 2.9×10^{18}, 2.5×10^{17},
1.6×10^{15}, and 3.0×10^{15} cm^{-3} and 7, 33, 28, 18.6, and 17 at.%, respectively.

RESULTS AND DISCUSSION

First, before describing the detail of experimental results, it would
be convenient to illustrate in Fig. 1 recombination processes involved in
the ODMR and PA measurements which are main topics of this article. As
shown in this figure, radiative recombination occurs between trapped electrons and trapped holes in the tail and gap states, as has been generally
accepted. The A centres we consider in this article are dominant trapped
hole centres. Nonradiative recombination between electrons and holes takes
place at the dangling bond centres (D^o), as having also general consensus.
At low temperatures, trapped electrons in the tail and gap states are first
captured by neutral dangling bonds for nonradiative recombination channel
in Fig. 1, and then holes recombine with these electrons at dangling bonds.
Trapped holes in the tail and gap states can move to the dangling bond sites
by hopping[*], but most dominant process seems to be capture of free holes by
negatively charged dangling bonds under optical excitation, because hopping
motion of trapped holes is very slow compared to free holes and the above
capture process has a large cross section for free holes created by optical
excitation.

---

[*] Dersch et al[14,15] suggested the importance of hole hopping in the tail and
gap states at low temperatures from spin-dependent photoconductivity measurements. Without assuming spin-dependent hole diffusion they did, their
results are also well understood, however, in terms of an alternative
model[16] that the light-induced ESR centres[17] with g = 2.0043 and 2.013
are not tail electrons and tail holes, but are nonradiative centres whose
defect model has already been speculated by us.

222

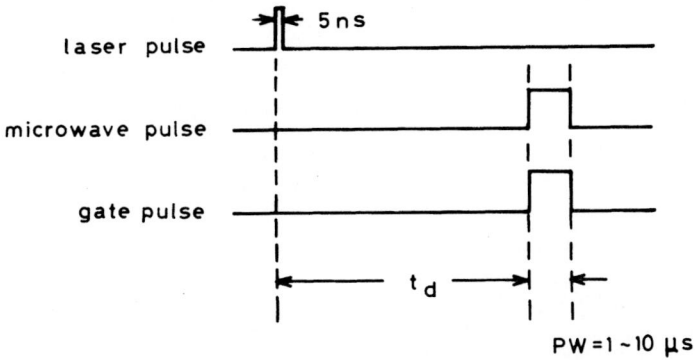

Fig. 2.   Timing of laser pulse, microwave pulse and
sampling gate for the measurement (A) (see text).

Optically Detected magnetic Resonance

An enhancing ODMR signal observed in a-Si:H under steady state excitation has been attributed to trapped hole centres, i.e., the A centres[1 - 6]. However, Street[18] interpreted this signal in terms of an exchange-coupled pair of a tail electron and a tail hole, because g-value for this signal,i.e. g = 2.0085, almost corresponds to an average value of those of two light-induced ESR (LESR) centres which have been identified as being a tail electron with g = 2.0045 and a tail hole with g = 2.011. We have already pointed out that the LESR centres have been observed as quenching ODMR signals, so that they are identified as being nonradiative centres. A model of these nonradiative centres has also been discussed in our recent paper[16]. The enhancing ODMR signal should be attributed to radiative centres which are identified as trapped hole centres (the A centres) in our model.

In the following, we present additional evidence that the enhancing ODMR signal is not due to an exchange-coupled pair of a tail electron and a tail hole. This was obtained from our recent TRODMR experiments[11,12]. In these experiments, two types of excitations were used, namely (A) a pulsed dye laser at 510 nm whose peak power and width were 220 kW/cm^2 and 5 ns, respectively and (B) a pulsed light of 200 μs width and 200 mW/cm^2 peak power formed,using an acousto-optic modulator, from argon ion laser at 514.5 nm. For the measurement(A), a microwave pulse of either 10 μs or 1 μs width, depending on the delay time, was applied at a given time, t_d, after the laser pulse was switched off. The sampling gate for observation of ODMR signals, whose width was the same as the microwave pulse, was opened at the same time as for the microwave pulse. Such a pulse sequence is shown in Fig. 2. The measurement (B) was performed in the same way as the measurement (A), except that a microwave pulse of 10 μs width and a sampling gate of 10 μs width were used. Fig. 3 shows the ODMR spectra observed for various delay time, monitoring total emitted light under excitation (A). Two significant features are noticed for the observed enhancing signal; first its peak position shifts towards high magnetic field corresponding to small g-value with decreasing t_d. Secondly, the ODMR spectrum observed for t_d = 50 μs appears to consist of two unresolved lines, i.e., a main line with g = 2.0041 and a shoulder line with g = 1.989. The g-value corresponding to the peak position of the enhancing line is shown as a function of t_d in Fig. 4, where the enhancing line was observed, monitoring either total emitted light or monochromatized emission light at 1.28 eV. This figure also includes the g-value of the enhancing line observed under exci-

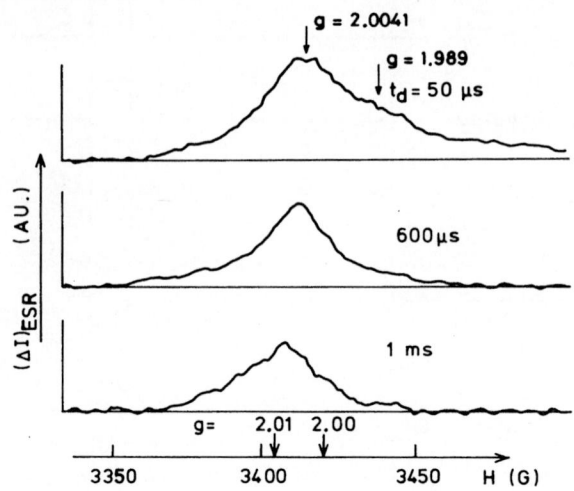

Fig. 3. TRODMR spectra observed at 9.6 GHz and 2 K
in a-Si:H No.933 for various microwave and
gate delay, t_d[11].

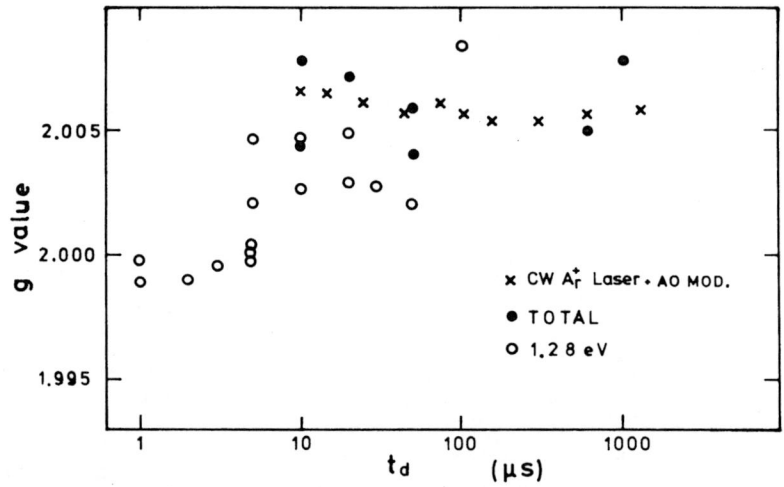

Fig. 4. g-value of the enhancing ODMR signal as a function
of t_d[11,12]. The signal was detected, monitoring
either total emitted light (closed circles) or monochromatized emission light at 1.28 eV (open circles)
under excitation (A). For excitation (B), CW argon
ion laser light was pulsed, using an acousto-optic
modulator. The signal was detected, monitoring
total emitted light.

tation (B), where a shoulder line seen in Fig. 3 was not observed for any t_d values. The g-values measured under excitation (B) are almost independent of t_d in contrast with those measured under excitation (A).

We discuss the above results in terms of a recombination model shown in Fig. 1. The shoulder line is attributed to trapped electron centres in the conduction band tail by considering its g-value. The reason that the shoulder line is not observed under pulsed excitation (B) and also under steady state excitation is given as follows: Trapped electron centre resonance affects the luminescence in two ways, namely enhancing the luminescence via the radiative recombination channel, R, as shown in Fig. 1 and quenching the luminescence via the nonradiative recombination channel, NR, as also shown in Fig. 1. Such competing nature of the trapped electron centre resonance turns out to suppress its ODMR signal in the steady state excitation and the pulsed excitation (B). Under the intense pulsed excitation (A), however, the NR channel is saturated as a result of rapid electron capture by dangling bond centres. During a certain time, i.e., about 100 μs after the intense pulsed excitation is switched off, this saturation continues and the radiative recombination is enhanced at the trapped electron centre resonance. Thus its ODMR signal can be observed as an enhancing line that appears in the shoulder of the A centre ODMR line. The shift of g-value of the A centre ODMR line towards its small g-value with decreasing t_d is interpreted in terms of exchange interaction between the A centres (the trapped hole centres) and the trapped electron centres. With decreasing t_d average exchange interaction becomes strong owing to shortening of average separation of trapped electrons and holes. This naturally accounts for the above experimental results obtained under the intense pulsed excitation. Under the pulsed excitation (B), such g-value shift with t_d is not observed owing to weak excitation. These results obviously show that the enhancing ODMR line observed under steady state excitation is not due to exchange-coupled pairs of trapped electrons and holes. As such, the above consideration provides additional evidence for the identification of the steady state enhancing ODMR signal as being due to the A centres.

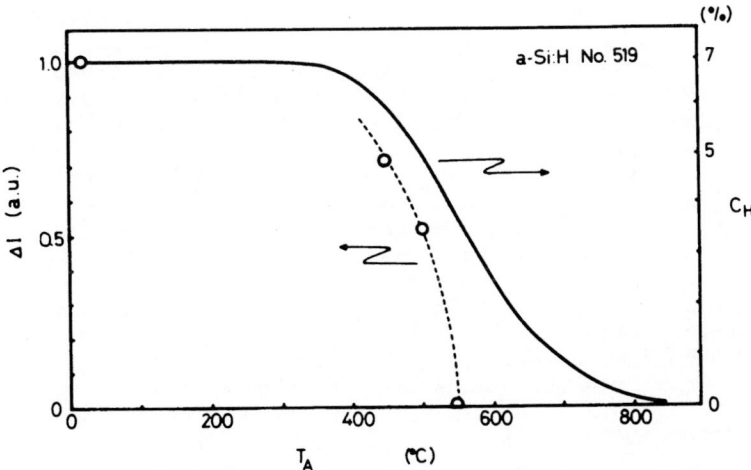

Fig. 5 Intensity of the A centre ODMR signal, ΔI, and hydrogen content in a-Si:H No.519 as a function of annealing temperature, T_A.

In the following, we discuss the correlation between the A centre ODMR signal and the presence of hydrogen in the film which has been suggested from the ODMR measurements on annealed a-Si:H samples . Fig. 5 shows the A centre ODMR signal intensity, i.e., change in the luminescence intensity at resonance, ΔI, as a function of annealing temperature, T_A. The hydrogen content was also estimated from hydrogen evolution experiments by Nitta and Hatano[19] on samples prepared at the same time as those used for the ODMR measurements. This result is also plotted as a function of T_A in Fig. 5, where the values of hydrogen content are normalized to the initial value estimated from the IR measurements. Two curves of the ODMR signal intensity and the hydrogen content against T_A are not fitted well with each other, but their decreasing tendency with increasing T_A is seen in Fig. 5. At T_A = 550°C, a large number of dangling bonds are created as a result of hydrogen effusion from the sample, so that nonradiative recombination at dangling bond centres is enhanced and the A centre ODMR signal intensity is rapidly decreased with increasing T_A. The above results indicate that the A centres are correlated with hydrogen in the sample. However, as clearly seen above, it should be pointed out that such correlation may arise, in part, through dangling bond density which varies with hydrogen content. The above correlation has also been suggested from the ODMR measurement on a-Si:F[20], in which the ODMR signal of the A centre, appearing to arise from the presence of hydrogen contained in the sample as contaminated impurities, was very weakly observed. Such correlation between the A centre and hydrogen is also shown from the PA measurement, as will be described in the next subsection. The nature of the A centre will be discussed on the basis of the ODMR, PA and PADESR measurements after the latter two types of measurements are mentioned.

Photoinduced Absorption

PA spectra in a-Si:H were analyzed in terms of the following relationship[7], $- \Delta T/T$ vs. $\hbar\omega$, where T is the transmitted intensity of probe light at photon energy of $\hbar\omega$.

$$- (\Delta T/T) \propto (\hbar\omega - E_{th})^{3/2}/(\hbar\omega)^3, \qquad (1)$$

where E_{th} is the threshold energy for photoionization of either trapped electrons or trapped holes into their respective band. This relationship was originally suggested by Lucovsky[21] for those deep levels in crystalline silicon which δ-potentials are associated with. The observed PA spectra for three samples Nos. 519, 541 and 540 are fitted well to this relationship, as shown in Fig. 6, where $(- \Delta T/T)^{2/3}(\hbar\omega)^2$ is plotted against $\hbar\omega$ and this plot gives E_{th} as its linear extraporation. As such, the threshold energy is estimated to be 0.25 eV, 0.24 eV, and 0.35 eV for samples Nos. 519, 541, and 540, respectively[7]. This photoinduced absorption has been attributed to excitation of trapped holes at the A centres into the valence band. Sample No. 540 containing a large number of dangling bonds with N_S = 2.9 X 10^{18} cm^{-3} , also exhibits photoinduced absorption due to electron excitation from negatively charged dangling bond centres, D$^-$, whose threshold energy is 0.75 eV, into the conduction band, as shown in Fig. 6.

The annealing effect on PA[22] was investigated for a-Si:H sample No. 927 containing dangling bonds of 1.6 X 10^{15}cm^{-3} . Fig. 7 (a) shows the PA intensity for the A centre as a function of T_A. The hydrogen content estimated from hydrogen evolution measurements[19] is also plotted against T_A in this figure, where its initial value was estimated from IR measurements. As shown in Fig. 7(b), the area of the theoretical fit using eq.(1) in the energy range between 0.5 eV and 1.5 eV is taken as the PA intensity for convenience. As in the ODMR measurement, this result indicates that the PA intensity for the A centre is well correlated with hydrogen content, except for $T_A \gtrsim$ 600°C. The deviation of the PA intensity from the hydrogen content vs. T_A curve is also accounted for in terms of enhancement of nonradiative

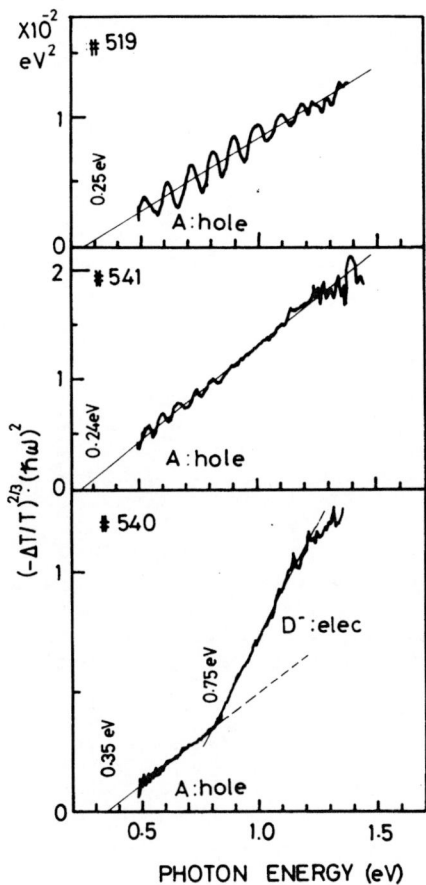

Fig. 6   Plots of the PA spectra, using eq. (1), for three a-Si:H samples at 1.8 K under excitation by krypton ion laser light at 530.9 nm with 0.13 W/cm^{-2}.[22] The threshold energies for hole excitation from the A centre and electron excitation from the D$^-$ centre are shown in the figure. Interference fringes are seen in the PA spectra.

recombination as a result of increase in the number of dangling bonds in a similar way to in the ODMR measurement.

The correlation between the A centre and hydrogen is also shown from the PA measurement in a-Si:F and a-Si:F,H as follows: Figs. 8(a) and (b) show the observed PA spectra for a-Si:F,H and a-Si:F samples, respectively. These samples were prepared at 450°C by glow-discharge decomposition of an intermediate-state active species of SiF$_2$ produced by chemical reaction between SiF$_4$ gas and solid silicon at 1150°C, adding H$_2$ gas for a-Si:F,H[23]. The PA spectrum of a-Si:F,H shown in Fig. 8(a) is due to hole excitation from the A centre, similarly to Fig. 7(b). On the other hand, this component is absent in the PA spectrum of a-Si:F shown in Fig. 8(b), which appears to be due to electron excitation from the D$^-$[22].

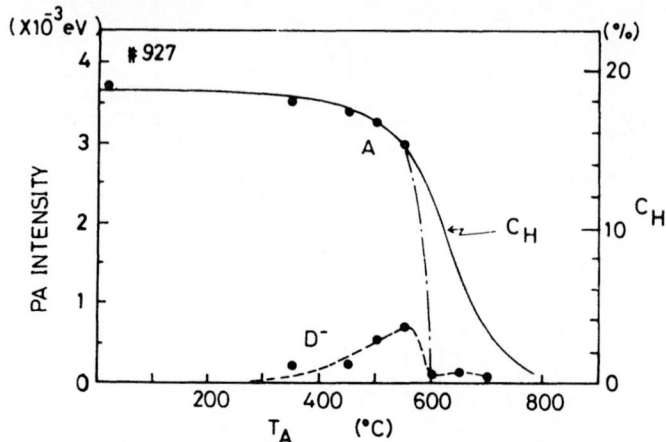

Fig. 7(a). Integrated PA intensity of two components, A and
D⁻, at 1.8 K and hydrogen content as a function
of annealing temperature, T_A[22]. For the definition of the integrated PA intensity, see text.

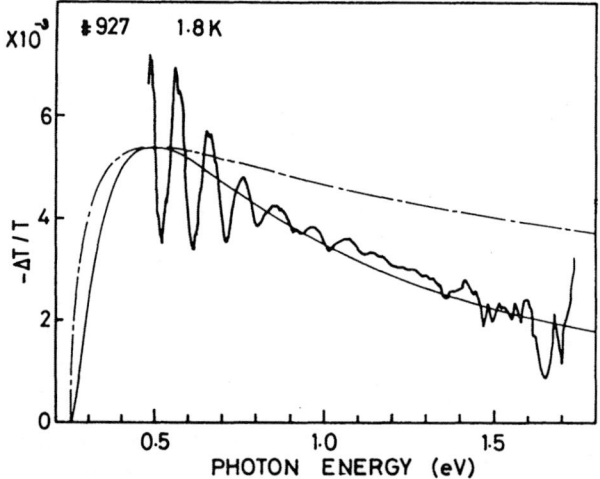

Fig. 7(b). PA spectrum observed at 1.8 K in a-Si:H No.927
(as deposited sample) under excitation by krypton
ion laser light at 568.2 nm with 0.45 W/cm² [7].
Two theoretical fits, using the O'Connor - Tauc
model and the Lucovsky model (eq.(1)) are shown
by dashed and solid curves, respectively (see ref.
(7)). Interference fringes are seen in the PA
spectrum.

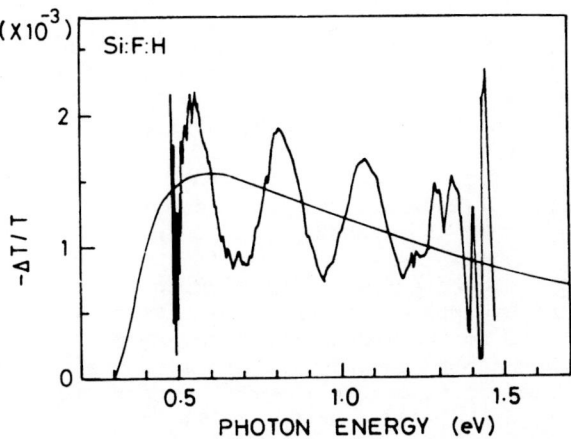

Fig. 8(a). PA spectrum observed at 1.8 K in a-Si:F,H under
excitation by krypton ion laser light at 530.9
nm with 0.43 W/cm^2 [22]. The theoretical fit using
eq. (1) is also shown by solid curve. Interference fringes are seen in the PA spectrum.

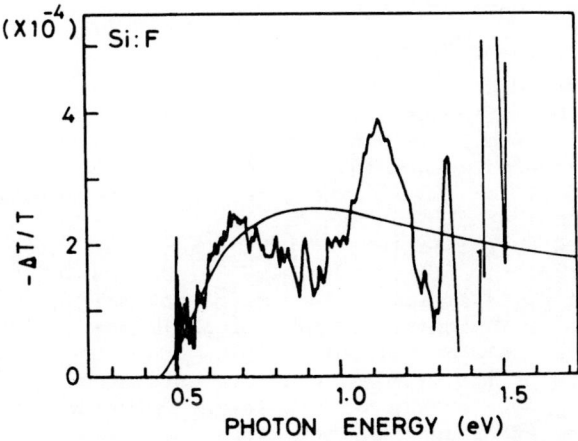

Fig. 8(b). PA spectrum observed at 1.8 K in a-Si:F under
excitation by krypton ion laser light at 530.9
nm with 0.43 W/cm^2 [22]. The theoretical fit using
eq.(1) is also shown by solid curve. Interference fringes are seen in the PA spectrum.

The identification of the PA spectrum with E_{th} = 0.24 - 0.35 eV as being due to hole excitation from the A centre into the valence band is also confirmed by the PADESR measurement[8,9], in which the PA intensity is decreased at the A centre resonance.

In the following, we discuss the nature of the A centre. The A centre which has been identified as the trapped hole centre from the ODMR, PA and PADESR measurements has the depth of its energy level, ranging at 0.24 - 0.35 eV and depending on the preparation condition. As mentioned above, this centre is correlated with the presence of hydrogen in the sample. From only such information, it is difficult to identify this centre with a certain specific defect. However, it would be worthy to note that if the A centre involves hydrogen as a constituent of the centre, the electron density on hydrogen should be small, because our ODMR measurements[20] on a-Si:F,H and a-Si:F,D indicate a similar broadening of the A centre line for both samples, so that hyperfine interaction with nuclei of H or D is not so strong as to contribute to the line broadening. One of the present authors[24] suggested a three-centre bond model[25] for the A centre from early ODMR measurements, although recent theoretical work[26] suggests that a single hydrogen three-centre bond is unlikely in a-Si:H. It is interesting to note, however, that the non-bonding state of the three-centre bond contributing to the A centre resonance has no electron density on the H site if H sits halfway between the two Si atoms[25]. This is consistent with the observation, as pointed out above.

In the following, we discuss briefly other experimental results related to the A centres. The observation of hole traps from thermally stimulated current (TSC) measurements has been reported by Vieux-Rochaz and Chenevas-Paule[27] and also by Yamaguchi[28,29], whose results indicate the trapping level depth of 0.3 eV and 0.22 eV, respectively. The former authors[27] concluded that the TSC peak associated with thermal excitation of carriers from the trap is due to holes, using Shottky diodes. Although, very recently, Ibuki and Fritzsche[30] and Misra et al[31] have interpreted the TSC peak in an alternative way that the peak is not necessarily associated with a localized level, the above estimate of the trap depth is almost consistent with that of the A centre.

From infrared quenching of photoconductivity observed in dual beam photoconductivity measurements, the depth of hole traps has been estimated by various authors as follows: Vanier and Griffith[32] have obtained the threshold energy for infrared quenching, depending on temperature, i.e., 0.41 eV for 100 K and 0.45 eV for 160 K. The threshold energy has also been reported by Persans[33], Fuhs et al[34] and Han[35] to be 0.6 eV at 100 K, 0.65 eV at 90 K and 0.4 eV at 130 K, respectively. Fuhs et al[34] estimated the thermal activation energy of the hole trap to be 0.35 eV from the thermal quenching of photoconductivity. A similar value of the thermal activation energy has also been obtained by Yamaguchi[36]. The threshold energy for infrared quenching seems to give the depth of hole traps. The depth of about 0.4 eV is almost consistent with the value of 0.35 eV, estimated from the PA measurements. The value of 0.6 eV seems inconsistent with our estimated values, but it is not clear at present why the threshold energy of infrared quenching of photoconductivity widely scatters in the range of 0.4 eV - 0.65 eV.*

---

* One of the possible reasons is given below: The A centre level may be broadened by disorder to a certain extent, i.e., 0.2 eV, so that occupation of this level by holes depends on the intensity of optical excitation. In this case, the trap depth should be estimated by a careful analysis as a function of excitation intensity. Our PA measurements give the same value for the depth of the A centre level within an experimental error ($\pm$ 0.02 eV) at excitation intensity ranging between 25 mW/cm^2 and 1.5 W/cm^2.

Although there is uncertainty in the value of the hole trap depth estimated from infrared quenching of photoconductivity, the hole trap responsible for this phenomenon seems to us identical with the A centre.

CONCLUSION

We have presented some evidence for the existence of trapped hole centres, designated as the A centres, in a-Si:H that has been obtained from tne ODMR, PA and PADESR measurements. The results obtained from the infrared quenching of photoconductivity and also from the TSC are also favorable for the existence of the A centres. The microscopic nature of the A centre is, however, unclear at present, although correlation between the A centre and hydrogen has been found in ODMR and PA measurements. Further investigations from the experimental side as well as the theoretical side are required to establish a conclusive microscopic model of the A centre.

Acknowledgments

We wish to thank Professor S. Nitta for invaluable discussions and him and Mr. A. Hatano for the hydrogen evolution measurements. We are also grateful to Mr. K. Fukui for providing us with samples Nos. 927 and 933 and also for informing us of their values of spin density and hydrogen content. We further acknowledge Professor H. Matsumura for providing us with samples of a-Si:F and a-Si:F,H. We are indebted to Dr. M. Yamaguchi and Mrs. Daxing Han for helpful discussions on the TSC measurements and the photoconductivity measurements, respectively.

REFERENCES

1. K. Morigaki, Y. Sano and I. Hirabayashi, Solid State Commun. 39:947 (1981).
2. K. Morigaki, J. Phys. Soc. Japan 50:2279 (1981).
3. K. Morigaki, Y. Sano and I. Hirabayashi, J. Phys. Soc. Japan 51:147 (1982).
4. K. Morigaki, Jpn J. Appl. Phys. 22:375 (1983).
5. K. Morigaki, Y. Sano and I. Hirabayashi, Optically Detected Magnetic Resonance in Hydrogenated Amorphous Silicon, in JARECT Vol. 6, Amorphous Semiconductor Technologies and Devices, Y. Hamakawa, ed., Ohmusha, Tokyo and North-Holland, Amsterdam (1983).
6. K. Morigaki, Optically Detected Magnetic Resonance, in Semiconductors and Semimetals Vol. 21, Hydrogenated Amorphous Silicon, Part C, J. I. Pankove, ed., Academic Press, New York (1984).
7. I. Hirabayashi and K. Morigaki, J. Non-Cryst. Solids 59 & 60:433 (1983).
8. I. Hirabayashi and K. Morigaki, Solid State Commun. 47:469 (1983).
9. I. Hirabayashi and K. Morigaki, J. Non-Cryst. Solids 59 & 60:133 (1983).
10. M. Yoshida and K. Morigaki, J. Non-Cryst. Solids 59 & 60:357 (1983).
11. H. Takenaka and K. Morigaki, to be published.
12. M. Yoshida and K. Morigaki, to be published.
13. S. Minomura, K. Tsuji, T. Hiraga, M. Matsumura, K. Fukui, T. Yoshida and Y. Okayasu, J. Non-Cryst. Solids 59 & 60:1119(1983).
14. H. Dersch, L. Schweitzer and J. Stuke, Phys. Rev. B28:4678 (1983).
15. H. Dersch and L. Schweitzer, Phil. Mag. 50:397 (1984).
16. K. Morigaki and M. Yoshida, Phil. Mag., in press, 1985.
17. R. A. Street and D. K. Biegelsen, Solid State Commun. 33:1159 (1980).
18. R. A. Street, Phys. Rev. B26:3588 (1982).
19. S. Nitta and A. Hatano, Tech. Digest 1st Int. Photovoltaic Sci. Eng. Conf., Kobe, p. 715 (1984).
20. M. Yoshida, K. Morigaki and H. Matsumura, Jpn J. Appl. Phys. 23:L593

(1984).

21. G. Lucovsky, Solid State Commun. 3:299 (1965).

22. I. Hirabayashi and K. Morigaki, to be published.

23. H. Matsumura and S. Furukawa, J. Non-Cryst. Solids 59 & 60:739 (1983).

24. K. Morigaki, B. C. Cavenett, P. Dawson, S. Nitta and K. Shimakawa, J. Non-Cryst. Solids 35 & 36:633 (1980).

25. R. Fisch and D. C. Licciardello, Phys. Rev. Lett. 41:889 (1978).

26. M. E. Eberhart, K. H. Johnson and D. Adler, Phys. Rev. B26:3138 (1982).

27. L. Vieux-Rochaz and A. Chenevas-Paule, J. Non-Cryst. Solids 35 & 36:737 (1980).

28. M. Yamaguchi, Jpn J. Appl. Phys. 21:L664 (1982).

29. M. Yamaguchi, J. Non-Cryst. Solids 59 & 60:425 (1983).

30. N. Ibuki and H. Fritzsche, J. Non-Cryst. Solids 66:231 (1984).

31. D. S. Misra, A. Kumer and S. C. Agarwal, Phys. Rev. B31:1047 (1985).

32. P. E. Vanier and R. W. Griffith, J. Appl. Phys. 53:3098 (1982).

33. P. D. Persans, Phil. Mag. B46:435 (1982).

34. W. Fuhs, H. M. Welsch and D. C. Booth, Phys. Stat. Sol. (b) 120:197 (1983).

35. D. Han, Chinese Phys. Lett. 1:65 (1984).

36. M. Yamaguchi, to be published.

232

ON THE NATURE OF GAP STATES IN HYDROGENATED AMORPHOUS SILICON ALLOYS

Subhendu Guha

Energy Conversion Devices, Inc.
1675 West Maple Road
Troy, Michigan 48084

INTRODUCTION

Study of localized states within the mobility gap of hydrogenated amorphous silicon alloys (a-Si:H) has received a great deal of attention in recent years. A simple energy distribution for the gap states was first suggested by Mott[1] where, in addition to the band tails caused by disorder, as proposed by Cohen, Fritzsche and Ovshinsky[2], there were states in the gap due to defects in the material. A specific defect center, namely a dangling bond, was proposed and it was suggested, depending on the occupancy, it may give rise to two bands separated by an appropriate correlation energy. Spear[3] suggested that there could be other defects, e.g., divacancies, giving rise to states in the gap. Since a-Si:H alloys contain many impurities at the level of $10^{18} cm^{-3}$ or higher[4], these impurities may form complexes with the inherent defects and give rise to a wide distribution of gap states.

The gap states act as traps and recombination centers and thereby play an important role in the performance of amorphous silicon solar cells. A variety of techniques has, therefore, been developed for studying the density and distribution of these states[5]. In this paper, I briefly review the nature of the gap states present in a-Si:H alloys. I restrict the discussion to the defect-induced deep states only; the band-tail states[6] arising from disorder will not be discussed in this paper.

INTRINSIC DEFECTS

The most common defect in a-Si:H alloys is the isolated dangling bond. ESR measurements show the presence of paramagnetic states with a g-value of 2.0055 in evaporated, sputtered or glow-discharge hydrogenated a-Si:H alloys. In good quality materials[7], the number of such spins is less than $10^{16} cm^{-3}$.

There is some controversy regarding the energy location of the dangling bond-induced state in the mobility gap. Using photo-thermal deflection measurements, Amer and Jackson[8] have observed a peak in the gap state distribution about 1.25 eV below the conduction band, E_c, in undoped samples. Based on a correlation between the spin density and the sub-band gap absorption corresponding to this peak in a number of samples prepared under different conditions, they conclude that these states are associated with singly-occupied dangling bonds. In phosphorus-doped samples, the peak shifts upwards appearing about 0.9 eV below E_c. This was identified as the state due to doubly-occupied dangling bonds. The interpretation, however, has been questioned by Adler[9] who has argued that since the Fermi level in the undoped sample is above the purported position of the doubly-occupied state, one should not have seen the singly-occupied centers in the undoped material. It should be mentioned that the determination of structures in gap state distribution from sub-band gap absorption measurements is not unambiguous[10]. Since absorption is governed by a convolution of the initial and final densities of states for the optical transitions, small structures in the density of states can not be discerned from these measurements. Independent experiments will be necessary to fix the energy locations of the dangling bond centers.

In addition to the isolated bonds, there could be two-fold coordinated Si atoms in its different charge states[11] giving rise to states in the gap. Isolated vacancies are unstable in crystalline silicon above 140K, but divacancies are stable at room temperature and may be present in a-Si:H alloys. a-Si:H alloys are also characterized by micro-voids, tissues and colunnar growths even in good quality materials.[12a] This implies presence of internal surfaces within the material. In crystalline silicon, surface states create a wide distribution of states within the gap[12b] arising from dangling bonds, strained bonds, re-constructed bonds and other complexes. It is, therefore, very likely that these micro-voids in a-Si:H alloys also cause a wide distribution of gap states.

DEFECTS INVOLVING HYDROGEN

Incorporation of hydrogen in general reduces the gap state distribution in a-Si:H films. In addition to removing the dangling bond states, singly bonded hydrogen also reduces states at the top of the valence band which are now replaced by deeper Si-H bonding states[12c]. The anti-bonding states of Si-H are located at the conduction band edge or above and hence do not affect transport properties. In addition to Si-H bonds, there are always dihydrides, polyhydrides and other bonding configurations of hydrogen present, even in good quality material[13]. It is well known that the transport and photovoltaic properties of a-Si:H alloys are significantly affected in the presence of the higher hydrides. Evidently, new gap states are created, but a spectroscopic determination of the energy location of these states in the gap has not yet been carried out.

DEFECTS INVOLVING IMPURITIES

Phosphorus

Phosphorus is the most common n-type dopant in a-Si:H alloy and addition of P has been shown to move the Fermi level over a wide energy range[14]. It is therefore likely that P goes in as a substitutional dopant entering the network in the tetrahedral (sp^3) configuration. The doping efficiency, however, is low and it is obvious that a significant fraction of the P atoms present in the material take up other configurations. The normal structure bonding for P is p^3 bonding and it is likely that three-fold coordinated P is present in phosphorus-doped a-Si:H. In addition, P can form various complexes with native defects.

234

In P-doped crystalline silicon, it is known from electron irradiation experiments[15] that P-vacancy complexes are located about 0.4 eV below conduction bond. P-related defect states, however, cover a wide energy range in a-Si:H alloys. Optical absorption measurements[16] show that incorporation of P creates new states at the valence band edge making the slope of the tail less steep. A shoulder in the density of states about 0.4 eV above valence band has been reported[17] for heavily doped (~8%) samples. The mid-gap states also increases with P-doping as evidenced from DLTS[18] and sub-band gap absorption measurements[8].

Boron

In a way, B behaves similarly to P causing substitutional doping. Boron chemistry, however, is very complex and there are many interesting configurations in which B can enter the structure. NMR experiments[19] show the presence of large amounts of three-fold coordinated B atoms. Three-center bonds with a bridging hydrogen atom[20] and other boron-vacancy complexes are the other possibilities which may introduce states in the gap. Incorporation of B may also weaken the structure of a-Si:H alloys[21]. This, in turn, could cause some of the intrinsic defects, that have already been discussed, to appear.

Experimental results on B-doped materials show presence of new states over a wide energy range. As in the case of P, the valence band tail widens and the mid-gap states increase[8,16].

Other Impurities

In a typical glow-discharge deposited a-Si:H film[4], the common impurities are oxygen, carbon and nitrogen with concentrations in the range 10^{18}-10^{20}cm^{-3}. In addition, impurities like S, Cl are also present. In crystalline silicon, these impurities take up various configurations in the lattice giving rise to a large number of deep states distributed within the gap[15,22]. There is evidence to believe that a similar situation exists in a-Si:H alloys.

Carlson et al.[23] have studied the effect of the addition of small amounts of impurities on the diffusion length of a-Si:H alloys. Trace impurities ($<10^{20}$cm^{-3}) of O, C and N were found to decrease the magnitude of the diffusion length. Since diffusion length is a measure of the minority carrier lifetime which in turn is dominated by the states below the Fermi level. These impurities, therefore, cause states below the mid-gap. There are other reports[24-26] of impurity related gap states indicating that gap states are created over a wide energy range. Pietruszko et al.[25] find a nitrogen related deep state about 0.4 eV below the conduction band edge. Nakamura et al.[24] also find new states associated with oxygen and nitrogen above the mid-gap. Interestingly enough, nitrogen behaves as a shallow donor in sputtered a-Si:H alloys[28].

Theoretical calculations have been made to obtain the energies of dangling bond defect states in a-Si:H alloys containing N, O and C. It was shown that the energy of the dangling bond state increases as the electronegativity of the neighboring atom of Si increases from Si to C, N and O. The defects associated with N or O will, therefore, give rise to states above the mid-gap.

Impurities like Li, Na, K, etc., also affect the electronic properties of a-Si:H alloys. Li is known to go interstitially in a-Si:H alloys and dopes the material n-type[30,31]. Doping efficiency is, however, small and it is believed that as in the case of crystalline Si,

Li forms many complexes with other defects in the material giving rise to new defect states. Doping effect has also been observed with incorporation of heavier alkali ions. Ion implantation of Na, K, Rb and Cs has been shown[32] to shift the Fermi level towards the conduction band.

LIGHT-INDUCED DEFECTS

Light soaking is known to change the properties of a-Si:H alloys[32]. The changes are reversible; annealing at temperatures above 150°C are found to restore the original properties. It is generally believed that recombination of excess carriers cause breaking of weak bonds which give rise to states in the gap[33]. One of the dominant states created is the isolated dangling bond. Based on spin resonance[7] measurements, it has been shown that the dangling bond density goes up by about a factor of 5 in good quality a-Si:H after prolonged light soaking. There is evidence[34-40], however, that in addition to dangling bonds, other defect states are created. Guha et al.[34,35] have shown that light soaking gives rise to states both below and above the mid-gap. Han and Fritzsche[36] have observed that the annealing rates of the defects that affect the photoconductivity and the sub-band gap absorption are different, indicating that there is more than one defect center present. In fact, the annealing rate is found to depend on the temperature at which the defects are created[37] and also on the nature of the dopants[38]. Light-induced defects in B-doped samples[39], for example, anneal out at temperatures below 100°C. This again proves that there are different types of defects that are created after light exposure.

CONCLUDING REMARKS

a-Si:H alloys are characterized by a large number of defects which give rise to states in the gap. A variety of defects exists, both intrinsic and extrinsic in nature. The predominant intrinsic defect is the isolated dangling bond. The extrinsic defects are associated with hydrogen and other impurities. In material prepared at low substrate temperature or high rf power, the isolated dangling bonds are the dominant defects and govern the recombination process. In high quality material, on the other hand, the dangling bond density is small and recombination is dominated by other defects which give rise to gap states over a wide range of energy[41,42].

ACKNOWLEDGMENT

We are indebted to Sir Nevill Mott who has been a source of inspiration to all of us working in the area of amorphous semiconductors.

The author is grateful to D. Adler, S.C. Agarwal, W. den Boer, H. Fritzsche, M. Hack, S.J. Hudgens, S.R. Ovshinsky, and J.S. Payson for interesting discussions. The work is supported by the Standard Oil Company (Ohio).

REFERENCES

1. N.F. Mott, Phil. Mag. 26, 505 (1972).
2. M.H. Cohen, H. Fritzsche and S.R. Ovshinsky, Phys. Rev. Lett., 22, 1065 (1969).

3. W. Spear, in Proc. Int. Conf. Amorphous and Liquid Semiconductors, Garmisch, 1973, eds., J. Stuke and W. Brenig, Taylor & Francis, 1974, p.1.
4. C. Magee and D.E. Carlson, Solar Cells, 2, 365 (1980).
5. From a recent review, see S. Guha, Sol. Energy Mat., 8, 269 (1982).
6. For a recent review, see T. Tiedje in Semiconductors and Semimetals, Vol. 21B, ed. J. Pankove, Academic Press, 1984, p. 207.
7. M.H. Dersch, J. Stuke and J. Beichler, Appl. Phys. Lett., 38, 456 (1981).
8. N.M. Amer and W.B. Jackson, App. Phys A, 32, 141 (1983).
9. D. Adler, in Optical Effects in Amorphous Semiconductors, ed., P.C. Taylor and S.G. Bishop, American Institute of Physics, 1984, p. 70.
10. J.S. Payson and S. Guha, Phys. Rev. B, 1985 (In press).
11. D. Adler, Solar Energy Mat., 8, 53 (1982).
12. (a) J.C. Knights, J. Non-Cryst. Solids, 35-36, 159 (1980); R.C. Ross, A.G. Johncock and A.R. Chan, J. Non-Cryst. Solids, 66, 81 (1984); (b) L.P. Wagner and W.E. Spicer, Phys. Rev. B, 9, 1512 (1974); N.M. Johnson, D.K. Biegelsen, M.D. Moyer, S.T. Chang, E.H. Poindexter and P.J. Caplan, Appl. Phys. Lett., 43, 563 (1983); (c) B. von Roedern, L. Ley and M. Cardona, Phys. Rev. Lett., 39, 1576 (1977).
13. H. Fritzsche, Sol. Energy Mat., 3, 447 (1980); S.R. Ovshinsky, J. Non-Cryst. Solids, 32, 17 (1974).
14. W.E. Spear and P.G. LeComber, Phil. Mag. 33, 935 (1976).
15. J.W. Corbett, Electron Radiation Damage in Semiconductors and Metals, Academic Press, New York and London (1977).
16. G.D. Cody, in Semiconductors and Semimetals, vol. 21 B, ed., J. Pankove, Academic Press, 1984, p. 11; C.R. Wronski, B. Abeles, T.Tiedje and G.D. Cody, Sol. State Comm., 44, 1423 (1982).
17. B. von Roedern and G. Model , Sol. State Comm., 35, 467 (1980).
18. J.D. Cohen and D.V. Lang, Phys. Rev. B, 25, 5321 (1982).
19. S.G. Grunbaum, W.E. Carlos and P.C. Taylor, Sol. State Comm., 43, 663 (1982).
20. S.R. Ovshinsky, in Amorphous and Liquid Semiconductors, ed., W.E. Spear (CICL, Univ. of Edinburgh, 1977), p. 519.
21. K. Chen and H. Fritzsche, Solar Energy Mat., 8, 205 (1982).
22. See, for example, A.G. Milnes, Deep Impurities in Semiconductors, Wiley, New York (1973).
23. D.E. Carlson, A. Catalano, R.V. D'Aiello, C.R. Dickson and R.S. Oswald, in Optical Effects in Amorphous Semiconductors, ed., P.C. Taylor and S.G. Bishop, American Institute of Physics, 1984, p. 234.
24. N. Nakamura, S. Tsuda, T. Takahama, M. Nishikuni, K. Watamabe, M. Ohnishi and Y Kuwano, ibid., p. 303.
25. S.M. Pietruszko, K.L. Narasimhan and S. Guha, Phil. Mag. B, 43, 357 (1981).
26. R.S. Crandall, Phys. Rev. B, 24, 7457 (1981).
27. A.E. Delahoy and R.W. Griffith, J. Appl. Phys. 52, 6337 (1981).
28. J. Baixeras, D. Menearaglia and P. Andro, Phil. Mag. B, 37, 403 (1978).
29. G. Lucovsky and S.Y. Lin, in Optical Effects in Amorphous Semiconductors, ed., P.C. Taylor and S.G. Bishop, American Institute of Physics, New York, 1984, p. 55.
30. W. Beyer and R. Fischer, Appl. Phys. Lett., 31, 850 (1970).
31. R.V. Navkhandewala, K.L. Narasimhan and S. Guha, J. de Physique, C-4, 803 (1981).
32. D.L. Staebler and C.R. Wronski, Appl. Phys. Lett., 31, 292 (1977).
33. For a recent review, see S. Guha in Amorphous Materials and Devices, ed., D. Adler, B.V. Schwartz and M.C. Steele, Plenum, 1985, (to appear).
34. S. Guha, C.-Y. Huang and S.J. Hudgens, Phys. Rev. B, 29, 5995 (1984).

35. S. Guha, C.-Y. Huang, S.J. Hudgens and J.S. Payson, J. Non-Cryst. Solids, 66, 65 (1984).
36. D. Han and H. Fritzsche, J. Non-Cryst. Solids, 59-60, 397 (1983).
37. S. Guha, C.-Y. Huang and S.J. Hudgens, Appl. Phys. Lett., 45, 50 (1984).
38. J. Jang, T.M. Kim, J.K. Hyun, J.H. Yoon and C. Lu, J. Non-Cryst. Solids, 59-60, 429 (1983).
39. W. den Boer and S. Guha, J. Appl. Phys. (to appear).
40. W. Kruhler, H. Pfleiderer, R. Plattner and W. Stetter, in Optical Effects in Amorphous Semiconductors, ed., P.C. Taylor and S.G. Bishop, American Institute of Physics, New York, 1984, p. 311.
41. M. Hack, S. Guha and M. Shur, Phys. Rev. B, 30, 6991 (1984).
42. S. Guha and M. Hack (to be published).

GAP STATES IN PHOSPHORUS-DOPED a-Si:H

Kazunobu Tanaka, Hideyo Okushi and Satoshi Yamasaki

Electrotechnical Laboratory

1-1-4 Umezono, Sakura-mura, Niihari-gun, Ibaraki 305, Japan

The density-of-state distribution of gap states and their relevant defect structures in phosphorus-doped a-Si:H have been investigated by ICTS (isothermal capacitance transient spectroscopy), PAS (photoacoustic spectroscopy) and ESR measurements. Two broad peaks are observed in the gap-state profile ; D^- at 0.52 eV below E_c, and $*D^-$ (D^- coupled with P^+_4) at 0.6 – 0.8 eV above E_v. ESR hyperfine spectra of the specimen deposited at room temperature suggest the existence of deep unpaired-spin centers coupled with P nuclei. Photo-induced effect is discussed in terms of $*D^- - D^-$ transformation and P-Si weak bond formation.

INTRODUCTION

In chalcogenide glasses, as has been initially pointed out by Street and Mott,[1] all of the dangling bonds are considered to be positively or negatively charged (in other words, empty or doubly-occupied), since an effective electron correlation energy at a dangling bond becomes negative with the assistance of a polaronic lattice distortion as well as a change in the coordination number of constituent atoms. This originates from p-like lone-pair electrons on chalcogen atoms and also from a low average coordination number m in the range of $2 \leq m < 3$.[2,3]

On the other hand, in tetrahedrally-bonded amorphous semiconductors like H-free a-Si, an effective correlation energy at a dangling bond is positive due to the lack of structural flexibility ($m = 4$), leaving the dangling bonds all in singly-occupied states (D^0). However, structural constraints can be partially relaxed by hydrogenation even in a-Si. For instance, a-Si$_{85}$H$_{15}$ results in the average coordination number of $m = 3.55$ according to the Phillips's definition.[3] Furthermore, several photo-induced phenomena have been observed in a-Si:H independent of whether doped or undoped,[4,5]

Table 1. Properties of P-doped a-Si:H used in the present work

Doping level (PH_3/SiH_4 vppm)	T_s(C)*	E_{opt}(eV)	$E_\sigma(= E_c - E_F)$(eV)
10	300	1.71	0.3
10^2	300	1.75	0.22
10^3	300	1.72	0.19
10^4	300	1.70	0.19
10^4	RT	2.00	> 0.5**

*   substrate temperature during deposition
** estimated from $\sigma_{RT} = 10^{-10}\Omega^{-1}cm^{-1}$ and ESR signal with $g = 2.0050$

suggesting that other different-natured defect states, in addition to singly-occupied dangling bonds (D^0), are expected to be formed in a-Si:H.

In this paper, firstly, we describe the gap-state profiles of P-doped a-Si:H obtained from several variations of ICTS (dark and photo) and PAS. Secondly, we present our recent data on ESR hyperfine structures associated with P atoms. Finally, we describe a reversible change in the gap-state profile induced by successive cycles of light soaking and thermal annealing, and try to understand its dynamics using defect models in terms of a charge-coupled dangling bond *D^- (D^- coupled with P_4^+) and a relevant Si-P weak bond.

SAMPLE PREPARATION AND EXPERIMENTAL PROCEDURES

Sample preparation

Samples used in this work were deposited from the glow-discharge plasma of SiH_4/PH_3 mixture on a glass substrate or an n^+ crystalline Si with a flow rate of 5 SCCM and a gas pressure of 50 mTorr. Basic properties of these samples are listed in Table 1.

Some samples were exposed to the repeated cycles of band-gap illumination (500-W xenon lamp with ir-cut filter, 4 hrs) and thermal annealing at 150 C, and after each process, ESR, PAS and ICTS measurements were performed on the samples.

ICTS (Isothermal Capacitance Transient Spectroscopy)

ICTS measurements were carried out on Schottky barrier diodes of Pt/P-doped a-Si:H / n^+c-Si structure. In the ICTS measurement, a transient capacitance $C(t)$ of the reverse-bias junction is measured, which is induced by applying four different modes of perturbation to the diode ; (1) a bias-voltage pulse V_p ($= -V_R$) in dark (dark ICTS), (2) a step function light under a constant reverse bias V_R (photo-ICTS : light-ON mode), (3) a light pulse under a constant V_R (photo-ICTS : light-OFF mode) and (4) a bias-voltage pulse under illumination (photo-ICTS : under-illumination mode).[6-9]

For each mode, ICTS signal $S(t)$ is defined as

$$S(t) = t df(t)/dt \quad \text{with} \quad f(t) = c^2(t) - c^2(\infty), \tag{1}$$

where t is the elapsed time after each perturbation is applied. For a system of continuously-distributed trap levels, the relation between $S(t)$ and a density-of-state distribution $N(E)$ for electron traps in the dark ICTS is given by

$$N(E) = - (1/kTB)S(t) \tag{2}$$

and

$$E_c - E = kT \ln\{\nu_n(E)t\}, \tag{3}$$

where $\nu_n(E) = N_c\sigma_n(E)v_{th}$ is the attempt-to-escape frequency of electron, $\sigma_n(E)$ the electron-capture cross section, v_{th} the electron thermal velocity, N_c the effective density of states in the conduction band and E_c the mobility edge of the conduction band. The energy dependence of $\nu_n(E)$ is obtained directly from the voltage-pulse-width (W_p) dependence of $S(t,W_p)$ expressed as

$$S(t,W_p) = S(t,\infty)\{1 - \exp[-W_p/\tau(E(t))]\}, \tag{4}$$

where $\nu(E(t)) = 1/n\sigma_n(E(t))v_{th}$, $S(t,\infty)$ the value of $S(t,W_p)$ at $W_p = \infty$ and n the density of free electrons. Then, $N(E)$ above the midgap is determined experimentally from eqs.(1), (2) and (3).

$N(E)$ in the energy range between the midgap and the valence-band mobility edge E_v is obtained mainly from photo-ICTS.[9] The time constant t corresponds to the hole-emission time if the hole-capture process is assumed to be dominant in the light-ON mode ; t is related with the energy E above E_v through the relation expressed as

$$E - E_v = kT \ln\{\nu_p t\} \quad \text{with} \quad \nu_p = N_v\sigma_p v_{th}, \tag{5}$$

where N_v is the effective density of states in the valence band, σ_p the hole-capture cross section. Then, $S(t)$ of the light-OFF mode gives $N(E)$ in the valence-band side, since ν_p or σ_p is experimentally determined from the light-ON spectra.

Theoretical and experimental details on ICTS involving the physical meaning of B in eq.(2) are given in separate papers.[7,8]

PAS (Photoacoustic Spectroscopy)

Self-supporting thick films were used for PAS measurements.[10,11] PAS signals of P-doped a-Si:H films were traced using a PAR model 6001 photoacoustic spectrometer provided with a 1-kW xenon arc at a chopping frequency of 40 Hz. The optical absorption coefficient $\alpha(h\nu)$ of the thin film sample is directly obtained from the normalized PAS signal as a function of photon energy without any information on the thermal parameters of the specimen, details of which were described earlier.[10]

ESR

The ESR hyperfine spectra were collected in the X-band using a JEOL JES-FE1XG and a BRUKER ER 200 D-SRC ESR spectrometer with the modulation frequency of 100KHz. P-doped a-Si:H deposited at T_s = room temperature as well as the samples at T_s = 300 C were subjected to ESR measurements in the temperature range between 4K and the room temperature.

RESULTS AND DISCUSSION

Density-of-state distribution determined from ICTS

Figure 1 shows the ICTS spectra $S(t)$ of P-doped a-Si:H for different doping levels of P atoms. Each spectrum was obtained in the time range from 10^{-3} to 10^4 sec under V_R = -1 V, V_p = 1 V and W_p = 10 msec. As shown in the figure, a single bump is observed on $S(t)$ and the peak height as well as the peak position of the bump varies with the doping level of P atoms.[7,8]

In order to convert the $S(t)$ spectra in Fig.1 into the density-of-state distribution N(E) the energy dependence of the attempt-to-escape frequency $\nu_n(E)$ of the specimens was determined from the measurements of W_p dependence of $S(t, W_p)$ using eq.(4). As described in previous papers,[6,8] $\nu_n(E)$ and $\sigma_n(E)$ decrease exponentially with an increase in the energy depth from E_c in the range $10^8 \text{ sec}^{-1} < \nu_n(E) < 10^{10} \text{ sec}^{-1}$ and $10^{-19}\text{cm}^2 < \sigma_n(E) < 10^{-17} \text{ cm}^2$, respectively. This exponential dependence of $\sigma_n(E)$ on E as well as its magnitude is quite conceivable if the multiphonon emission with a weak coupling predominates in the electron-capture process at those gap states.[6,8]

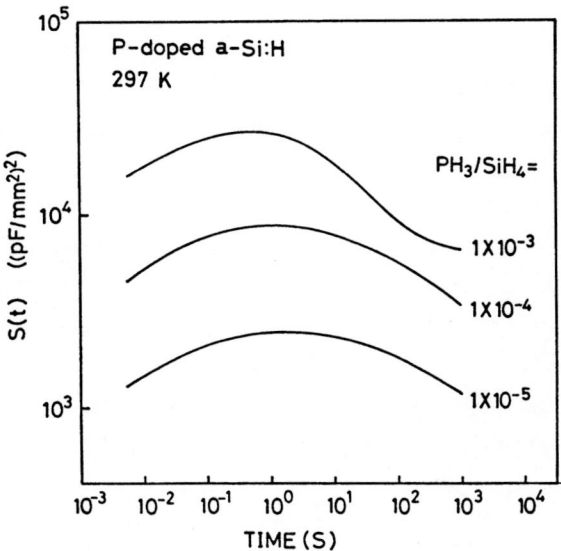

Fig.1. Dark ICTS spectra S(t) of P-doped a-Si:H for different doping levels.

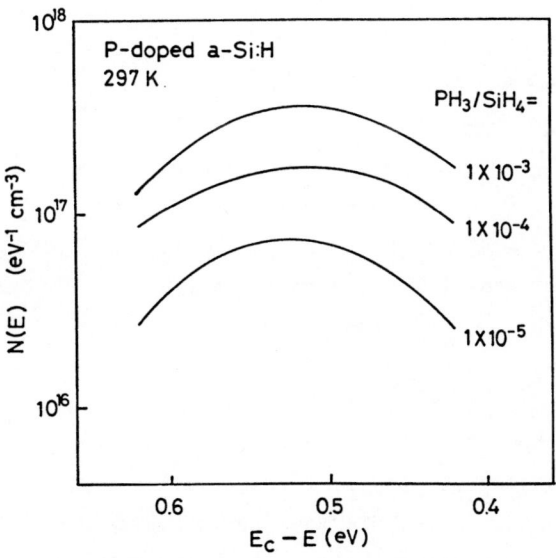

Fig.2. Density-of-state distributions N(E) of D⁻ states in P-doped a-Si:H for different doping levels.

Figure 2 shows the density-of-state distribution $N(E)$ of each specimen, being obtained from the results of $S(t)$ and $\nu_n(E)$ using eqs.(2) and (3). The spectrum shows a bump structure with a peak located at around 0.52 $\pm$ 0.02 eV below E_c independent of doping levels. The bump structure has been attributed to the doubly-occupied dangling-bond states (D^-), since its intensity increases in proportion to the square root of a doping level of PH_3/SiH_4.

Lang et al. studied the gap states profiles using a rate-window DLTS (Deep Level Transient Spectroscopy) on a back-to-back diode of P-doped a-Si:H, and determined the energy distribution of the density of localized states having a bump structure located at around 0.85 eV below E_c.[12] Cohen et al. have identified those states with the doubly-occupied-dangling-bond states (D^-) through the direct observation of the change in the dark absorption ESR signal.[13] However, obviously, there is non-negligible inconsequence on the energy location of D^- states between DLTS groups and our result. This apparent discrepancy simply originates from a difference in the estimation of the magnitude of $\nu(E)$ between those groups. Lang et al. used $\nu_n = 10^{13}$ sec^{-1} and Johnson obtained $\nu_n = 10^{12}$ sec^{-1}, assuming that the capture cross section of gap states is independent of both temperature and energy. Those values of ν_n are by several orders of magnitude larger than that which we determined from the direct measurement.

Morigaki et al. have shown through ODMR and PADESR that non-radiative recombination occurs via D^- located at around 0.6 eV below E_c.[15] Modulation photocurrent experiment done by Oheda et al. has given a similar energy location (0.6 eV).[16] Spear et al., recently also, have derived the energy associated with D^- from their excess carrier transport experiments, and obtained 0.6 eV.[17] Those results are compatible with our energy scaling within experimental margin of error.

Figure 3 shows $N(E)$ in the energy range of 0.6 - 0.9 eV above E_v of the P-doped a-Si:H specimens for different doping levels of P atoms. The results shown in the figure were determined from the photo-ICTS measurements with three different modes using eq.4. A shoulder structure is observed at around 0.6 eV above E_v, and its intensity increases with a doping level. The hole-capture cross section σ_p of these states was roughly estimated from the light-ON mode of photo-ICTS to be at least larger than 10^{-16} cm^2, suggesting that the states act as large hole-trap centers and, in contrast to D^- at 0.52 eV below E_c, a strong electron-lattice coupling takes place at these centers. On the other hand, the density of these states at 0.6 - 0.9 eV above E_v as well as D^- states at 0.52 eV below E_c increases in proportion to the square root of the concentration of P atoms, which is strongly associated with the

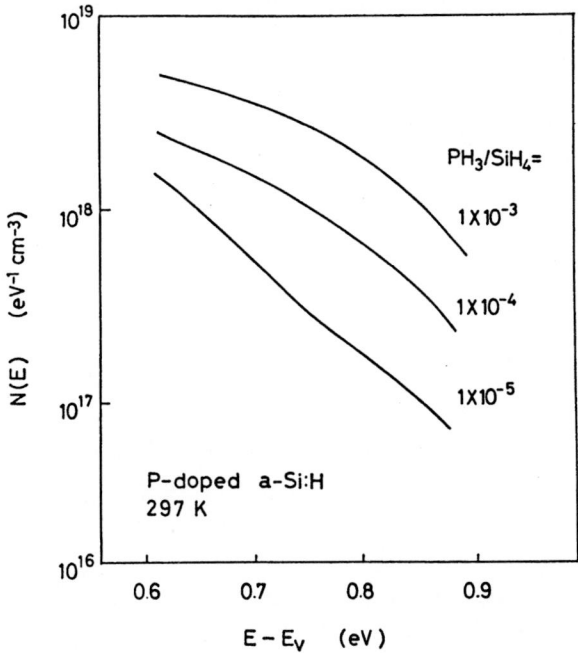

Fig.3. Density-of-state distributions N(E) near
the valence band tail of P-doped a-Si:H.
The data were obtained from photo-ICTS.

Street's prediction on the mass action law ($P_3^0 \rightleftharpoons P_4^+ + D^-$) based on 8-N
rule.[18] Furthermore, a total density of holes captured at the states at 0.6
- 0.8 eV above E_v nearly coincides with that of LESR reported by Street et
al.[19] Thus, we speculate that the nature of those states in Fig.3 is also
essentially like doubly-occupied dangling bonds, however, whose energy is
stabilized by a strong coupling with P_4^+ states. These charge-coupled D^-
states are denoted by *D^-. It is noted that, in contrast to D^-, the energy
location of *D^- is uncertain by about 0.2 - 0.3 eV because the precise
measurement of the magnitude of σ_p as well as its energy dependence has not
yet been made.

Figure 4 shows N(E) of P-doped a-Si:H($PH_3/SiH_4 = 3 \times 10^{-4}$) before and
after the light soaking. As shown in the figure, an increase in the density
of D^- states and a decrease in *D^- states in a higher energy side ($E - E_v >$
0.8 eV) are simultaneously induced by the band-gap illumination. This
complementary change of D^- and *D^- states is reversed by thermal annealing at
150 C, suggesting a reversible defect transformation between D^- and *D^-. It
should be noted that the density of *D^- states in a lower energy side ($E - E_v$
< 0.8 eV) remains unchanged or rather increases after the light soaking.

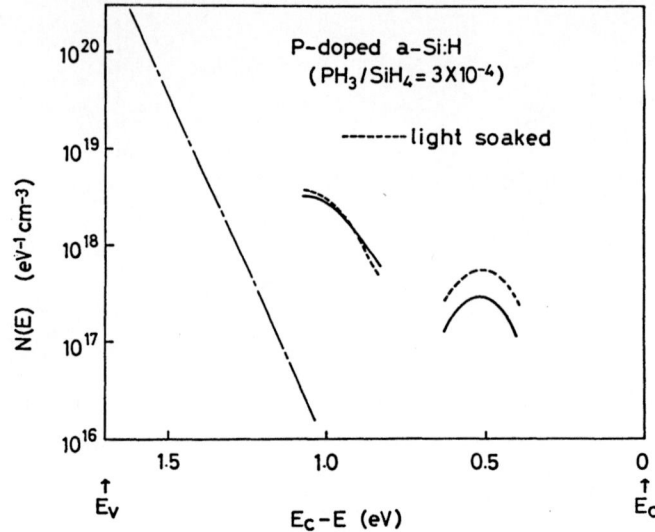

Fig.4. Density-of-state distributions N(E) of P-doped
a-Si:H before (solid line) and after (dashed
line) the light soaking.

Density-of-state distribution determined from optical absorption (PAS)

    Figure 5 shows the optical absorption spectra extending down to $\alpha = 1$
cm^{-1} of the P-doped a-Si:H for different doping levels of P atoms.   The
spectrum traced after the light soaking is also shown.   From the results in
the figure,   the gap-state profiles N(E) were derived by theoretical
simulation under the assumption that the matrix elements are kept constant
for transitions between extended states in one band and localized states in
the other one as well as transitions between extended states in both bands.

    Figure 6 shows thus-determined density-of-state distribution N(E) of P-
doped a-Si:H. The valence band tail drops exponentially with a slope of E_e =
0.06 eV in the relation $N(E) \propto exp(-E/E_e)$,   while the conduction band tail
falls off more sharply.   A broad peak appears in N(E) at around 0.7 eV above
E_v and its height increases with a doping level of P.   The density as well as
the energy location of these states is close to that of $*D^-$ states deduced by
photo-ICTS shown in Fig.3.   As shown in the figure,   the density of those
states increases after the light soaking and returns to the initial value by
thermal annealing.   This effect may correspond to the change observed in a
lower energy side of N(E) in Fig.4, although a direct comparison between N(E)

246

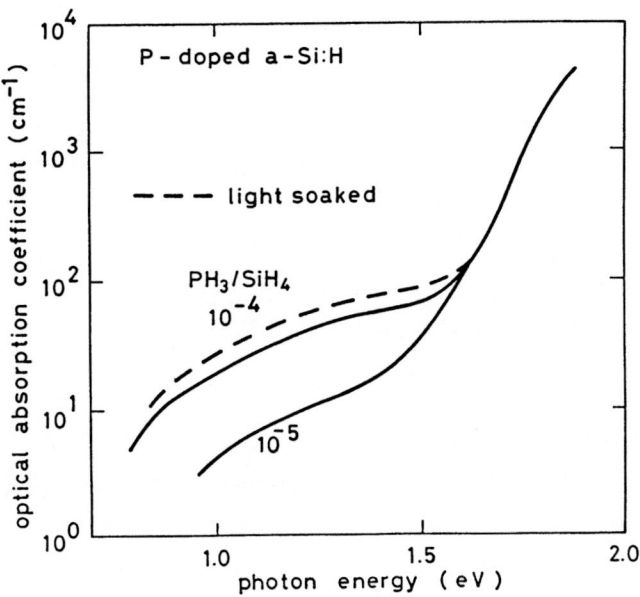

Fig.5. Optical absorption spectra below the optical gap of P-doped a-Si:H for different doping levels.

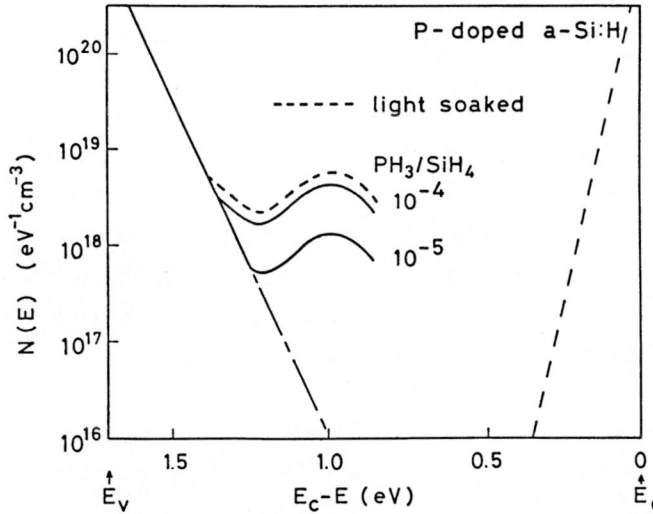

Fig.6. Density-of-state distributions N(E) of P-doped a-Si:H for different doping levels determined by the simulation of the data of Fig.5. The result for the light-soaked specimen is also shown (dashed line).

from photo-ICTS and that from PAS might be dangerous due to the uncertainty of the exact energy location of the hole-transport region[20] in ICTS and/or the Franck-Condon factor in PAS.

Johnson has deduced the 0.8-eV peak in N(E) from the constant-capacitance DLTS, and discussed it in relation to the below-gap absorption spectra determined by PDS.[14] However, as is seen in Figs.3 and 6, the below-gap absorption should not be ascribed to D^- but probably to $*D^-$ states. The contribution of D^- states to the below-gap absorption of P-doped a-Si:H could be negligibly small judging from the magnitude of their density around one order of magnitude lower than that of $*D^-$.

ESR hyperfine spectra

Recently, Stutzmann and Stuke observed new paramagnetic centers in P-doped a-Si:H through ESR hyperfine spectra measurements.[21] Hirabayashi et al. and ourselves also detected similar hyperfine structures in ODMR spectra.[22] Both works were done using highly-P-doped specimen deposited at T_s = 250 C, so the doping level is high enough to shift the Fermi energy close to the conduction band mobility edge.

Figure 7 shows the ESR hyperfine spectra observed at 40 K in P-doped a-Si:H (PH_3/SiH_4 = 1 %), deposited at T_s = 300 C and T_s = room temperature, respectively. In both cases a strong resonance line as well as a doublet with an approximately symmetric shape with respect to each strong central resonance is observed, although the central resonances are at different positions between both samples ; g = 2.0050 (T_s = room temperature) and g = 2.0044 (T_s = 300 C). It should be noted that similar hyperfine structures appear in both samples in spite of a considerable difference in the Fermi energy between two, suggesting that these paramagnetic centers exhibiting hyperfine interaction with a nucleus of spin I = 1/2 are energetically distributed in a broad energy range probably down to the midgap. Actually, the spin density of the hyperfine lines shows a very weak dependence on the measuring temperature and remains observable even at the room temperature. Another independent experiments on P-doped samples for a variety of doping levels clearly show that the hyperfine ESR spin density increases with the phosphorus concentration. Details of these experiments will be published elsewhere.

It is speculated from the above results that the hyperfine signal can be ascribed to P_4^o (neutral donors) states, as Stutzmann and Street have suggested quite recently.[23] However, the electronic configuration of P_4^o associated with the hyperfine spectra might be considerably modified, probably forming a Si-P weak bond, thereby its electronic energy being

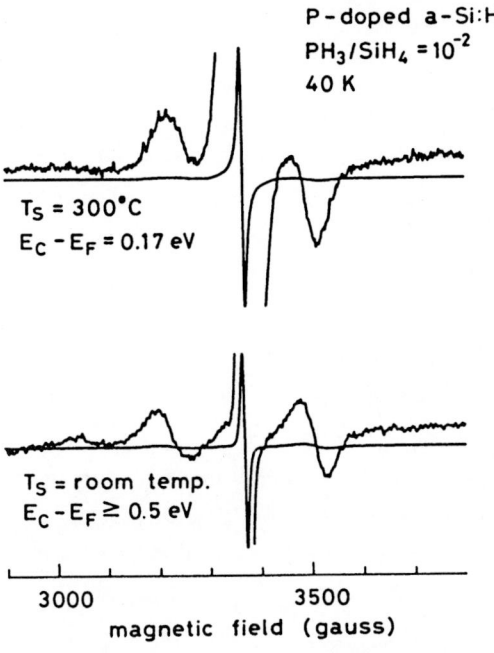

Fig.7. ESR hyperfine spectra observed at 40 K in P-doped a-Si:H ($PH_3/SiH_4 = 10^{-2}$), deposited at $T_s = 300°C$ (top) and $T_s = $ room temperature (bottom).

stabilized deep enough for explaining the present result (Fig.7). This concept is close to the model presented by Ishii et al.[24]

Defect Model

The ICTS results clearly show that at least two different defect states are involved in P-doped a-Si:H ; D^- at 0.52 eV below E_c (Fig.2) and $*D^-$ at around 0.6 - 0.9 eV above E_v with a higher density of states than D^- (Fig.3). PAS data also suggest the presence of the gap states probably associated with $*D^-$ (Fig.6), although some ambiguity exists in the energy location of $*D^-$ depending on the estimation of σ_p in photo-ICTS as well as the Franck-Condon factor in PAS. In our model, as described previously, $*D^-$ is characterized as the doubly-occupied dangling bond which is strongly coupled with the four-fold coordinated P_4^+ atom. This results in a stabler energy of $*D^-$ with respect to D^- because of the Coulombic interaction with P_4^+, and possibly, the lattice distortion around a pair of $*D^-$ and P_4^+. Then an effective electron-correlation energy U_{eff} at $*D^-$ is expressed as

$$U_{eff} = U_c - e^2/\varepsilon r - W, \hspace{3cm} (7)$$

where U_c is the pure correlation energy, e the electron charge, ε the dielectric constant r the $*D^- - P_4^+$ separation and W a polaronic relaxation energy. U_{eff} could be negative depending on the magnitude of W. A similar discussion on negative U_{eff} at the dangling bond state coupled with a charged state has been led by Robertson.[25]

Concerning U_{eff} at D^- as described earlier,[9] the temperature and the energy dependences of the electron-capture cross section σ_n indicate that the multiphonon emission process with a weak electron-lattice coupling predominates at the capture process, which means $W = 0$ and a positive U_{eff}. Another possible origin of the difference in U_{eff} between $*D^-$ and D^- may be a difference in electronic configuration of a Si dangling bond as discussed by Adler;[26] energetically stable doubly-occupied states $*D^-$ may correspond to $T_3^- (s^2p^3)$ while D^- to $T_3^0(sp^3)+e$.

The band-gap illumination creates excess $*D^-$ states (see Figs.4 and 6), but simultaneously, $*D^-$ is partially transformed to D^- (Fig.4). This $*D^- - D^-$ transformation is shown schematically in Fig.8 and corresponds to the change in charged states ($P_4^+ \rightarrow P_4^0$) and the atomic distortion.

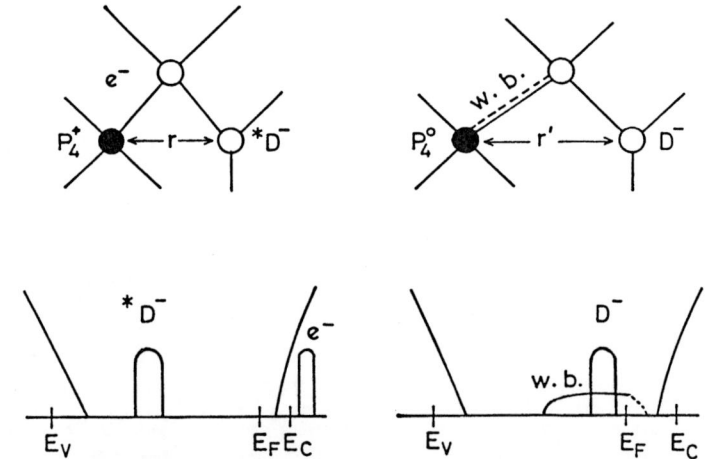

Fig.8. A schematic diagram of $*D^- - D^-$ transformation accompanying the formation of P_4^0 with a Si-P weak bond and their corresponding energy levels. P-doped a-Si:H involves all of the defects, and the band-gap illumination induces both additional $*D^-$ creation and $*D^- - D^-$ transformation.

The process is initiated by photo-excitation of electron-hole pairs. Holes are captured by $*D^-$ states and subsequently recombine with electrons. Judging from the large hole-capture cross section at $*D^-$ states ($\sigma_p > 10^{-16}$ cm^2), as mentioned above, an excess energy of photo-excited carriers is released to the lattice system surrounding $*D^-$ through a strong electron-lattice coupling. This may create Si-P weak bonds depending on local topological constraints produced by a variety of structural randomness. As shown in Fig.8, an electron, once trapped at P_4^+ having a Si-P weak bond, is energetically stabilized as the anti-bonding state of the weak Si-P bond (probably lying below E_F), therefore, the P_4^0 state remains neutral even at the room temperature, causing $*D^--D^-$ transformation as well as an increase of the hyperfine ESR signals. It appears that a bond length of weak Si-P bonds takes different values depending on their different local constraints, thereby producing a broad energy distribution of those paramagnetic states giving ESR hyperfine structures (see Fig.8). The weak bonds are restored to their original normal bondings by thermal annealing, and then P_4^0 are ionized to P_4^+ again.

It should be noted that the formation of Si-P weak bonds does not necessarily occur on every P_4^0 configuration, rather, many of them may remain as P_4^+ with normal bondings even after light-soaking.

The above reversible photo-induced change in local configurations is considered to originate from topological flexibility of a-Si:H realized by a decrease of the average coordinates number ($m = 3.55$) with respect to H-free a-Si. It is compatible with our discussion on the strong electron-lattice coupling at the hole-capture process at $*D^-$ states, and resulting possibility of negative U_{eff}.

The present model is quite tentative, and the current state of understanding of structural origins of defects in a-Si:H is still incomplete. However, the mechanism of the $*D^--D^-$ transformation accompanying the formation of Si-P weak bonds explains at least the present experimental observations in a unified manner, and seems to give insight to the general understanding of the behaviour of dangling bonds in a-Si:H associated with various degradation phenomena. For instance, the reversible change of electrical properties of the p-i-n diode induced by a current soaking and thermal annealing is well interpreted by the above mechanism, since the $*D^--D^-$ transformation is based on the hole-capture process at $*D^-$ states irrespective of whether holes are provided by optical excitation or current injection.

The present concept of the charge-coupled $*D^-$ can be applied to the undoped a-Si:H, because it is well known that various impurities are unintentionally incorporated into a-Si:H network during the film deposition process and may produce positively-charged defects instead of P_4^+. It should be noted that impurity levels still remain at 1 ppm for nitrogen, 8 ppm for carbon, and 40 ppm for oxygen even in the "clean material" prepared in a UHV plasma deposition system.[27]

SUMMARY

In summary, we determine the density-of-state distribution of localized states in P-doped a-Si:H using PAS and ICTS (dark as well as photo-) measurements. The bump peaked at 0.52 eV below E_c is attributed to D^- states and the other bump lying at 0.6 - 0.8 eV above E_v is to $*D^-$ (D^- intimately coupled with P_4^+), although the energy location of the latter is rather ambiguous. $*D^-$ states are described as doubly-occupied dangling-bond states which are energetically stabilized by interaction with P_4^+ and the polaronic lattice distortion. ESR hyperfine spectra observed in the P-doped specimens deposited at both T_s = 300 C and room temperature suggest that deep paramagnetic centers coupled with P nuclei are formed in the network. We present the model of $*D^-$–D^- transformation accompanying Si-P weak-bond formation for discussing the light-soaking effect.

Acknowledgments

The authors would like to express their thanks to Drs. M.Stutzmann and R.A.Street, and Profs. T.Shimizu and N.Ishii, for sending the transcripts of their works prior to publication.

REFERENCES

1. R.A.Street and N.F.Mott, States in the gap in glassy semiconductors, Phys. Rev. Lett. 35 : 1293 (1975).
2. M.Kastner, D.Adler and H.Fritzsche, Valence-alternation model for localized gap states in lone-pair semiconductors, Phys. Rev. Lett. 37 : 1504 (1976).
3. J.C.Phillips, Topology of covalent non-crystalline solids I:short-range order in chalcogenide alloys, J. Non-Cryst. Solids 39 : 153 (1979).
4. D.L.Staebler and C.R.Wronski, Reversible conductivity change in discharge-produced amorphous Si, Appl. Phys. Lett. 31 : 292 (1977).
5. M.H.Tanielian, N.B.Goodman and H.Fritzsche, Photo-creation of defects in plasma-deposited a-Si:H, J. de Phys. suppl. 42 : C4-375 (1981).
6. H.Okushi, T.Takahama, Y.Tokumaru, S.Yamasaki, H.Oheda and K.Tanaka, Temperature dependence of electron capture cross section of localized state in a-Si:H, Phys. Rev. B27 : 5184 (1983).
7. H.Okushi, A study of gap states in P-doped amorphous silicon by ICTS, Philos. Mag. (1985)(in press).
8. K.Tanaka and H.Okushi, Defect states and carrier capture processes in a-Si:H, J. Non-Cryst. Solids 66 : 205 (1984).

9. H.Okushi, M.Itoh. T.Okuno, Y.Hosokawa, S.Yamasaki and K.Tanaka, Gap states dynamics observed in light soaked P-doped a-Si:H, in AIP Conf. Proc. NO. 120 "Optical effects in amorphous semiconductors", P.C.Taylor and D.G.Bishop, ed., A.I.P. New York (1984).

10. S.Yamasaki, H.Okushi, A.Matsuda, H.Oheda, N.Hata and K.Tanaka, Determination of optical constants of thin films using photoacoustic spectroscopy, J. J. Appl. Phys. 20 : L665 (1981).

11. K.Tanaka and S.Yamasaki, PAS study of gap-state profiles of P-doped and undoped a-Si:H, Solar Energy Materials 8 : 277 (1982).

12. D.V.Lang. J.D.Cohen and J.P.Harbison, Measurement of the density of gap states in hydrogenated amorphous silicon by space charge spectroscopy, Phys. Rev. B25 : 5285 (1982).

13. J.D.Cohen, J.P.Harbison and K.W.Wecht, Identification of the dangling-bond state within the mobility gap of a-Si:H by depletion-width modulated ESR spectroscopy, Phys. Rev. Lett. 48 : 109 (1982).

14. N.M.Johnson, Measurement of deep levels in hydrogenated amorphous silicon by transient voltage spectroscopy, J. Appl. Phys. 51 : 327 (1983).

15. K.Morigaki, Y.Sano and I.Hirabayashi, Radiative and nonradiative recombination processes in hydrogenated amorphous silicon as elucidated by optically detected magnetic resonance, Solid State Commun. 39 : 947 (1981).

16. H.Oheda, S.Yamasaki, T.Yoshida, A.Matsuda, H.Okushi and K.Tanaka, Gap states distribution of undoped a-Si:H determined with phase-shift analysis of the modulated photocurrent, J. J. Appl. Phys. 21 : L440 (1982).

17. W.E.Spear, H.L.Steemers, P.G.LeComber and R.A.Gibson, Majority and minority carrier lifetimes in doped a-Si junctions and the energy of the dangling-bond state, Philos. Mag. B50 : L33 (1984).

18. R.A.Street, Doping and Fermi energy in amorphous silicon, Phys. Rev. Lett. 49 : 1187 (1982).

19. R.A.Street. D.K.Biegelsen and J.C.Knights, Defect states in doped and compensated a-Si:H, Phys. Rev. B24 : 969 (1981).

20. B.von Roedern, L.Ley, M.Cardona and F.W.Smith, Photoemission studies on in situ prepared hydrogenated amorphous silicon films, Philos. Mag. 40 : 433 (1979).

21. M.Stutzmann and J.Stuke, New paramagnetic states in amorphous silicon and germanium, J. Non-Cryst. Solids 66 : 145 (1984).

22. I.Hirabayashi, K.Morigaki, S.Yamasaki and K.Tanaka, Photoinduced absorption and photoinduced absorption-detected ESR, in Ref. 9.

23. M.Stutzmann and R.A.Street (to be published).

24. N.Ishii, M.Kumeda and T.Shimizu, P-related defects in P-doped a-Si:H, Solid State Commun. (1985)(in press).

25. J.Robertson, Unified theory of bonding at defects and dopants in amorphous semiconductors, in Ref.9.

26. D.Adler, Density of states in the gap of tetrahedrally bonded amorphous semiconductors, Phys. Rev. Lett. 41 : 1755 (1978).

27. C.C.Tsai, J.C.Knights and M.J.Thompson, 'Clean' a-Si:H prepared in a UHV system, J. Non-Cryst. Solids 66 : 45 (1984).

A TECHNIQUE FOR CALCULATING THE DENSITY OF ELECTRONIC STATES OF DISORDERED MATERIALS

T.M. Hayes

Xerox Palo Alto Research Center
Palo Alto, California 94304, USA

J.L. Beeby

University of Leicester
Leicester LE1 7RH, UK

A powerful new technique is presented by means of which the density of electronic states can be realistically calculated for a wide variety of disordered systems. The theory is based on partitioning the density of states into contributions associated with each atom's environment and on a parametric description of local atomic arrangements. Each contribution is approximated as a function of the local parameters and is linked to those from neighboring atoms by correlations between the respective sets of parameter values. Using a tetrahedrally-bonded amorphous semiconductor as an example, it is shown that the method is computationally tractable and gives realistic densities of states.

INTRODUCTION

One of the long-standing interests of Sir Nevill Mott has been the electronic properties of systems without long-range translational order (*e.g.*, see Mott and Davis 1971). The interesting physical properties of such systems and their technological importance give studies of them particular relevance at present. In the absence of translational periodicity, the powerful descriptive and theoretical techniques developed for the study of crystals cannot be used. There is no equivalent to the Bloch theorem! Hence existing theoretical treatments are restricted to simplified models for the potential, such as bonding or tight binding, and to cluster models of the structure (for reviews, see Kramer and Weaire 1979, Yonezawa and Cohen 1981). We present in this paper a new method with the power and flexibility to give accurate densities of electronic states for a wide variety of disordered systems.

The range of structural variation which a successful theory must be capable of treating is very wide: variable numbers and species of neighbors; variable neighbor distances and angular distributions; perhaps specific local complexes. The theory described in this paper is conceptually simple but nevertheless able to satisfy these demanding requirements. The formal treatment is algebraically complicated for the general case but the computational requirements are surprisingly modest. We will not go into all the necessary algebraic or numerical detail in this paper, but will refer the reader to earlier publications for complete discussions.

The theory requires that the one-electron potential can be adequately represented in the muffin-tin form of non-overlapping spheres of potential. This approximation can yield extremely accurate results in crystalline band-structure calculations and should therefore be a reasonable model for most disordered systems provided the potential within the spheres is chosen carefully.

The structural approximations follow from partitioning the density of states into contributions associated with each atom in the system. These contributions, though associated with individual atoms, depend on the positions of many atoms and are in no sense local densities of states. The contribution associated with an atom in a particular "local" configuration may, however, reasonably be expected to be largely independent of the exact positions of distant atoms, and hence to depend principally on parameters which specify the positions of nearby atoms—the parameters describing the "local" configuration (for a discussion, see Beeby 1983). The appropriate parameter set will usually include local coordinate axes and the near-neighbor distribution, but should obviously be as small as possible.

For a given system, with specific atomic positions, the set of exact integral equations coupling each contribution with those associated with neighboring atoms can be written in a form wherein each contribution is expressed as a function of its own set of parameter values. When these equations are averaged over the ensemble which describes the disordered system, the important structural information is the conditional probability for particular parameter values associated with neighboring atoms given specified parameter values associated with a central atom. The set of integral equations thus reduces to one equation with its kernel involving these conditional probabilities.

The local environment of each atom is retained in this treatment. No part of the system is omitted but distant neighbors are averaged more than near neighbors. The effect of this is that each local arrangement of atoms is embedded in an infinite network, the properties of which are self-consistently determined and matched to that local arrangement. Such an embedding is far beyond the capability of previous treatments in which the properties of the extended matrix inevitably affect the calculation, usually in an unphysical way.

In practice the theory can accommodate more structural information than is commonly available from experiments. Only near-neighbor distributions can be determined experimentally for most systems so that the theory will usually utilize additional structural information determined by physical modeling or computer energy-minimization methods. This theory is especially valuable in determining accurately the change in the density of states arising from postulated structural changes.

An elegant example of this treatment is provided by calculations for tetrahedrally-bonded amorphous semiconductors. In such systems, the nearest neighbors of each atom are very close in position to the tetrahedral configuration found in the diamond-cubic lattice. If only these atom positions are presumed to be important in detail, the only variable parameters to be specified are the Euler angles describing the orientation of the local tetrahedron. In the amorphous system, the tetrahedra around nearest neighbors are not in the relative orientations found in the crystal but instead are rotated from those orientations by dihedral angles, the distribution of which can in principle be obtained from model structures. The probability distribution for the dihedral angles then forms part of the kernel of the integral equation for the "local" contribution to the density of states. Structure thus enters this calculation in two ways—tetrahedra define the local structure and a dihedral-angle distribution specifies the positions of second nearest neighbors and beyond.

This theory can accommodate any degree of short-range order from complete disorder to crystalline order. In fact, it treats correctly the limit of translational periodicity. In the tetrahedrally-bonded semiconductor case just discussed, for example, the crystal corresponds to the special case in which the dihedral angles are fixed at values appropriate to a diamond-cubic lattice.

In the following, we present an expression for the density of states derived using multiple-scattering theory in which the effects of the atomic arrangements on electron propagation are isolated in a function F. Our structural model is invoked and the averaging technique applied to F. The resulting system of equations involves standard mathematical functions and can be solved in principle to any required accuracy. In order to reveal clearly the nature of these results, discussions in the remainder of the paper make use of the simplifying assumption that the dihedral-angle distribution is uniform.

GENERAL FORMULATION

Calculations of the electronic properties of disordered systems are often based upon multiple-scattering perturbation expansions involving non-overlapping muffin-tin potentials (*e.g.*, see Beeby and Edwards 1963, Beeby 1964). We have used such a method to obtain an expression for the density of electronic states in which the effects of the atomic arrangements are set out in particularly convenient form. Although this derivation has been carried out for a general muffin-tin potential (Beeby and Hayes 1985), the results presented here are restricted for simplicity to potentials which scatter only s waves.

Information about the electronic properties is conveniently expressed through the probability $\rho(\mathbf{k},E)$ for finding an electron with a given energy E and real momentum $\mathbf{k}$, often called the spectral density. It is more general than the density of states, to which it is related by

$$n(E) = (2\pi)^{-3} V \int dk \ \rho(k,E),\tag{1}$$

where V is the volume of the system. At a particular energy in a disordered system, there can be contributions from all momenta so that $\rho(k,E)$ spreads over regions of k,E space. $\rho(k,E)$ takes the place in disordered systems of the detailed Bloch wave functions of the perfect lattice.

The spectral density can be written in the general form (Beeby 1964):

$$\rho(k,E) = -(\pi V)^{-1} (E-k^2)^{-2} \ \text{Im} \langle T(k) \rangle,\tag{2}$$

where the Dirac brackets denote an ensemble average over the configurations defining the disordered system. Atomic units are used throughout (*i.e.*, $\hbar = m = e = 1$), except that energies are in rydbergs (1 ry = 13.6 eV). $T(k)$ is the diagonal part of the Fourier transform of the total scattering matrix for a single configuration:

$$T(k) = \int dr \int dr' \exp[-\textit{i}k \cdot (r-r')] \ T(r, r')\tag{3}$$

where

$$T = \Sigma_a t_a + \Sigma_{a,b \neq a} t_a G_0 t_b + \Sigma_{a,b \neq a, c \neq b} t_a G_0 t_b G_0 t_c + \dots\tag{4}$$

in operator notation. Here *a*, *b*, ... refer to the atoms positioned at R_a, R_b, ... and the scattering matrix for an individual atom at the origin is defined by

$$t(r,r') \equiv v(r) \delta(r-r') + v(r) \int dr'' \ G_0(r-r'') t(r'',r').\tag{5}$$

The muffin-tin atomic potential $v(r)$ is usually but not necessarily taken to be spherically symmetric. $G_0(r)$ is the free-particle propagator and is given by the Fourier transform of $(E-k^2)^{-1}$. Note that T, t, and G_0 are all implicit functions of E.

For spherically symmetric potentials, the t matrix is diagonal in angular momentum but the propagator from one atom site to another is not, giving rise to effects such as s–p hybridization. Though physically appropriate and important, these effects tend to obscure the analysis and understanding of the nature of our solution of these equations. In the following, therefore, the atomic t matrix is supposed to scatter only s waves. This approximation is not presumed to be accurate as a description of a semiconductor such as Si, though it does represent the system required to be solved after application of the transformation of Thorpe and Weaire (1971) to a very simple tight-binding model for amorphous Si. In addition, the range of energy is restricted to E < 0, corresponding to the occupied bands. Through a series of manipulations which are set out in detail by Beeby and Hayes (1985), the expression for $\rho(k,E)$ can be worked into the following form:

$$\rho(k,E) = -(4/V) (E-k^2)^{-2} [s_0(k)/s_0(i\mu)]^2 [-s_0(i\mu)/i\mu] \ \text{Im} \langle \Sigma_a F^a(i\mu,k) \rangle,\tag{6}$$

where $\mu^2 = -E$ and the product of the factors preceding the imaginary part operator is real. $s_0(q)$ is the Fourier transform of a particularly convenient representation of the atomic t matrix:

$$s_l(q) = \int dr \, j_l(qr) \, s_l(r) \tag{7}$$

where q can be k or $i\mu$,

$$s_l(r) = -i\mu \, v(r) \, j_l(i\mu r)$$
$$+ (2/\pi) P \int dk \, k^2 \int dr' \, r'^2 \, v(r) \, j_l(kr) \, (E - k^2)^{-1} \, j_l(kr') \, s_l(r') , \tag{8}$$

and P is the principal value operator.

All the information about the arrangements of the atoms which bears on n(E) is contained in the functions

$$F^a(i\mu,k) \equiv 1 + \Sigma_{b \neq a} \, t_0 \, G^{a,b} + \Sigma_{b \neq a, c \neq b} \, t_0 \, G^{a,b} \, t_0 \, G^{b,c} + \dots$$
$$= 1 + \Sigma_{b \neq a} \, t_0(i\mu, i\mu) \, G^{a,b}(i\mu, k) \, F^b(i\mu, k) , \tag{9}$$

where $t_0(i\mu, i\mu) = -s_0(i\mu)/i\mu$ is a Fourier transform of the t matrix. $G^{a,b}$, related to G_0, describes the propagation of a free electron from atom a to b, and will be specified precisely later. In defining F we have separated the structure dependent portion of $T(k)$ into terms, one for each specific atom a, each of which contains the effects of all those multiple scattering paths "beginning" at atom a. These are precisely the paths which dominate the influence of the structure in the vicinity of atom a on the electronic properties. It is the evaluation and averaging of this function for suitable physical models which is the primary concern of this paper.

Note that the treatment of atom arrangements in equation (6) is exact. We introduce in the next section an averaging procedure which leads to a tractable system of equations to be solved to obtain $\langle F^a \rangle$, and thereby n(E), for a disordered system.

AVERAGING OVER ATOM ARRANGEMENTS

An exact evaluation of F^a for an infinite system involves summing the infinite series given in the first line of equation (9) or, equivalently, solving the infinite set of equations represented by the second line. In a crystal, this otherwise impossible task is made tractable through translational periodicity, by virtue of which the set of distinct F^a can be reduced to a finite number (often only one). We have formulated an approximate procedure by which the set of practically distinct F^a can be reduced to a tractable set even in a disordered system (in which the set of formally distinct F^a is infinite in principle). This is accomplished by considering each F^a in turn, and averaging the expression for it (i.e., equation (9)) over a specified subset of the complete ensemble describing the disordered system, that subset characterized by particular values of the parameters describing the atom arrangements at short range. After this partial ensemble average, the remaining elements of the structure are contained in conditional probabilities relating the parameter values at near neighbor atoms with those at atom a. The resulting set of integral equations couples F's corresponding to different parameter values. After solving for these F's, the result is averaged over the distribution of parameter values at atom a, completing thereby the ensemble average. The success of this approach results from the freedom which one has to choose the set of parameters describing the environment of each atom and the remaining structural elements to be retained as conditional probabilities, which may be exploited to incorporate extensive short-range order.

Note that each distinct atom environment will be found in all orientations in a system without a macroscopic orientation, so that $\rho(\mathbf{k},E)$ must be independent of the direction of $\mathbf{k}$. It is then possible and convenient to set $\mathbf{k}$ parallel to the z direction in the laboratory coordinate frame when evaluating F^a. Accordingly, $\mathbf{k} = k\hat{z}$ and the integral over the direction of $\mathbf{k}$ in equation (1) yields 4π.

$G^{a,b}$ includes a factor of $\exp(-\mu|\mathbf{R}_b - \mathbf{R}_a|)$ for $E < 0$, where $\mu = (-E)^{1/2} > 0$. Our treatment in the remainder of this paper will be restricted to E sufficiently less than zero that we may neglect all $G^{a,b}$ except those where a and b are nearest neighbors of one another. While reminiscent of tight-binding approaches, this assumption is really much less restrictive since a <u>complete</u> basis set is always used for the expansion of eigenfunctions.

The first step in this theoretical approach is to choose the set of parameters with which each atom's environment is to be characterized. In the simple model used to illustrate our method, each atom has four nearest neighbors arranged in a perfect tetrahedron. For each atom, then, we can define a set of coordinate axes in which the associated tetrahedron has some standard orientation. The coordinate axes we choose for atom a are uniquely determined as follows: atom a is at the origin; the z axis points directly at one nearest neighbor; the vector to another nearest neighbor is in the xz plane with a component in the positive x direction. This choice is illustrated in Fig. 1(a). Let $R(0,a)$ be the operator which rotates the fixed laboratory coordinate axes into the set just defined for atom a. $R(0,a)$ is a function of the usual Euler

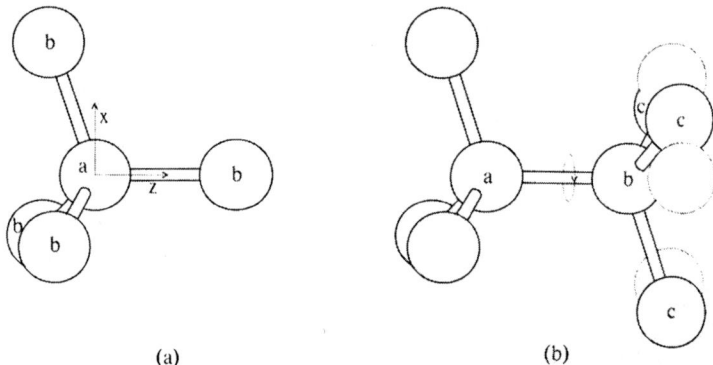

(a)   (b)

Fig. 1. Atom a is shown in (a) together with its nearest neighbors $\{b\}$. The standardized coordinate system describing the orientation of the environment of a is also shown, fixed by the orientation of the nearest neighbors of a. In (b), the averaging of the orientation of the coordinate system of b, a nearest neighbor of a, over a narrow distribution of dihedral angles so as to solve for the function $\langle F^a \rangle_a$ is illustrated schematically. The positions of the neighbors $\{c\}$ after a rotation about the ab axis are shown as faint outlines.

angles: α_a, β_a, γ_a. These angles are the only variable parameters needed to characterize an atom's environment in our simple model. This is a minimal set. In a more complicated model, the variable parameter set might also include the number and species of the neighbors, near neighbor distances or angular distribution, and so forth.

The second step is to perform the average over the ensemble of all atoms characterized by identical values of the local parameters. Given this particular local parameter set, the only correlations surviving are those between the orientation of the coordinate axes of each of the nearest neighbors to a, denoted by $\{b\}$, and that of a. These are represented by a conditional probability. Since a and $\{b\}$ are each surrounded by a perfect tetrahedron of nearest neighbors, the relative orientations can be characterized by a distribution of dihedral angles. Reasonable approximations for such distributions have been obtained from structural models (Connell 1974, Temkin 1974). The result of this average on equation (9) is

$$\langle F^a \rangle_a = 1 + \Sigma_{b \neq a} t_0 \langle G^{a,b} F^b \rangle_a , \tag{10}$$

where the dependence of these functions on $\mathbf{k}$ and on E through $i\mu$ has been suppressed for notational simplicity. The Dirac brackets in equation (10) denote that the coordinates axes of a have been held constant, implying that the positions of the nearest neighbors of a are fixed. This average is represented schematically in Fig. 1(b). In it, $\mathbf{R}_b - \mathbf{R}_a$ is held fixed but the positions of atoms $\{c\}$ are averaged over the distribution of dihedral angles. Note that $G^{a,b}$ depends on structure only through $\mathbf{R}_b - \mathbf{R}_a$, which is invariant during this particular partial ensemble average, and could have been removed from the Dirac brackets on the right-hand side of equation (10). We will not do so, however, so that we are able to develop the general techniques required to implement more complicated structural models.

In order to solve equation (10), it is necessary express the particular average of F^b found on the right-hand side as a function of the set of distinct $\langle F^a \rangle_a$ in the ensemble which characterizes the disordered system. This is an important step in which the correlations among parameter values associated with neighboring atoms are incorporated in the electronic properties. Let us first express $\langle F^a \rangle_a$ as an explicit function of the Euler angles which were held fixed during the partial ensemble average. A suitable basis is the set of D functions, each of which is a matrix element of the rotation operator between spherical harmonics in the laboratory coordinate system: $D_{lmm'}(0,a) \equiv \langle lm|R(0,a)|lm' \rangle$. The D_J, where $J = (l,m,m')$, form a complete orthogonal set in the space of the Euler angles (for a summary of their properties, see Rose 1957 or Brink and Satchler 1968). This gives

$$\langle F^a \rangle_a = \Sigma_J A_J D_J(0,a) , \tag{11}$$

with the coefficients A_J to be determined by means of equation (10). In the structural model being implemented here, the variable parameter set includes only the Euler angles specifying the orientation of the tetrahedron of nearest-neighbor atoms and the only correlations to be retained are specified through a dihedral-angle distribution. Accordingly, the F^b appearing in equation (10) is replaced by $\Sigma_J A_J$

$D_J(0,b)$, accounting thereby for the orientation of the environment of b, and subsequently subjected to an average over dihedral angle. In a more complicated structural model, F^b might have been expressed as a probability distribution of distinct $\langle F^a \rangle_a$, each subjected to an appropriate average over its Euler angles. In a disordered system, this latter average accounts for the greater disorder found at a b atom (when compared with the a atom) after the partial ensemble average holding fixed the values of the a-atom parameters. Our approach can also accommodate the crystalline limit, in which the disorder is not greater at a b atom, through a discrete distribution of Euler angles.

The propagator $G^{a,b}$ may be expanded similarly in D_J's. The only variable in $G^{a,b}$ which depends on the position of b is $\mathbf{R}_b - \mathbf{R}_a$, which is most conveniently expressed in the b coordinate system. If the nearest neighbors form a perfect tetrahedron, the four sets of Euler angles which rotate the coordinate system appropriate to one atom into that of a nearest neighbor are: $(\alpha,\beta,\gamma) = (0,\pi,\gamma_1)$, (π,β_0,γ_2), $(\pi/3,\beta_0,\gamma_3)$, $(-\pi/3,\beta_0,\gamma_4)$, where $\cos(\beta_0) = 1/3$. The only degrees of freedom are in the four angles γ_b, all of which vanish in the diamond-cubic structure. In the b coordinate system so obtained, $\mathbf{R}_b - \mathbf{R}_a$ is always in the $-z$ direction. Expanding $G^{a,b}$ in $Y_{lm}{}^*(\mathbf{R}_b - \mathbf{R}_a)$ in the laboratory coordinate system, and using the rotation matrices to express $Y_{lm}{}^*(\mathbf{R}_b - \mathbf{R}_a)$ in the b coordinate system, we obtain

$$G^{a,b} = \Sigma_{l,m} \, G_{lm} \, D_{lm0}(0,b) , \tag{12}$$

where

$$G_{lm} = - \, \delta_{m,0} \, (i)^{-l} \, (2l+1) \, i\mu \, j_l(kR) \, h_0{}^+(i\mu R) , \tag{13}$$

R is the magnitude of $\mathbf{R}_b - \mathbf{R}_a$, assumed to be the same for all nearest-neighbor bonds, and $h_0{}^+$ is an (outgoing) spherical Hankel function of the first kind.

With these definitions and approximations, the average on the right-hand side of equation (10) becomes

$$\langle G^{a,b} F^b \rangle_a = \Sigma_{l,j,n,\bar{n}} \, G_{l0} \, A_{jn\bar{n}} \, \langle D_{l00}(0,b) D_{jn\bar{n}}(0,b) \rangle_a . \tag{14}$$

The two D functions, having identical arguments, can be combined using the vector addition relation (Rose 1957). Finally, we use the additivity of the rotation operators, $R(0,b) = R(a,b)R(0,a)$, to express the resultant D function as:

$$D_{jn\bar{n}}(0,b) = \Sigma_{n'} \, D_{jnn'}(0,a) \, D_{jn'\bar{n}}(a,b) . \tag{15}$$

$D(0,a)$ is unaffected by the averaging process, while $D(a,b)$ is independent of the absolute orientation of the coordinate axes of a or b (depending only on the orientation of b relative to a). The only factor in $G^{a,b} F^b$ to be affected by the average $\langle \ \rangle_a$ is $D(a,b)$.

Using equations (11) through (15) in equation (10), multiplying the result by $D_{j\bar{n}n}{}^*(0,a)$, and integrating over all $(\alpha_a,\beta_a,\gamma_a)$ yields

$$A_{j0n} = \delta_{j,0} \, \delta_{n,0}$$
$$+ \, t_0 \, \Sigma_{l,j',n'} \, G_{l0} \, A_{j'0n'} \, C(lj'j;000) \, C(lj'j;0n'n') \, \langle \Sigma_{\{b\}} \, D_{jnn'}(a,b) \rangle_a , \tag{16}$$

where $C(l\ l'\ l'';\ m\ m'\ m'')$ is a Clebsch–Gordan coefficient (*e.g.*, see Rose 1957). Note that there is no coupling among $A_{j\bar{n}n}$ for different values of $\bar{n}$. Since the inhomogeneous term is non-zero only if $\bar{n}=0$, the terms $A_{j\bar{n}n}$ are zero unless $\bar{n}=0$, thus removing one dimension of the matrix equation. The sum over $b \neq a$ in equation (10) has become the sum over the four rotations to nearest-neighbor coordinate axes $\{b\}$ in equation (16). The arrangements of the atoms affect this equation only through the average in the last factor, which involves the dihedral-angle distribution.

Solving equation (16) will yield the coefficients A_J which determine $\langle F^a \rangle_a$ through equation (11). This function enters into $\rho(\mathbf{k},E)$ after being summed over all atoms a and averaged over the remainder of the complete ensemble describing the disordered system. These two operations result in an average over all orientations of the a coordinate axes, which will leave only the A_{000} term in the usual case of a disordered system without any macroscopic orientation, plus a sum over values of the other parameters which characterize the atom environments. There is only one such environment in the simple model considered here, so that the sum over a yields a factor of N, the total number of atoms. With these, equation (6) becomes

$$\int d\Omega_{\mathbf{k}}\ \rho(\mathbf{k},E) = -16\pi N/V\ (E-k^2)^{-2}\ [s_0(k)/s_0(i\mu)]^2\ [-s_0(i\mu)/i\mu]\ \text{Im}\ A_{000}\ . \quad (17)$$

The result is an averaged spectral density for clusters embedded in matching infinite tetrahedral networks. From the perspective of a calculation based on a specific cluster of atoms, we have terminated our central cluster with a self-consistently determined and, in some sense, ideal electron propagator. Although addressing in some respects the same issues as does the coherent potential approximation, our approach is much more sophisticated in that it preserves crucial elements of short-range order in a system without translational periodicity.

The remaining link in the derivation is the calculation of the imaginary part required in equation (17). The form of the solution can be seen readily if equation (16) is expressed in matrix form:

$$\mathbf{A} = \mathbf{e} + \mathbf{M} \cdot \mathbf{A} = (\mathbf{1} - \mathbf{M})^{-1} \cdot \mathbf{e}\ . \quad (18)$$

In this equation, $\mathbf{A}$ is a vector wherein each element is labeled by index $J = j,0,n$ coming from the expansion introduced in equation (11). $\mathbf{M}$ is a matrix with analogous indices expressing the coupling among the elements of $\mathbf{A}$, and $\mathbf{e}$ is a vector with elements $\delta_{J,0}$. This equation can be solved in terms of the eigenvalues, λ_i, and left and right eigenvectors, $\mathbf{u}_i^T$ and $\mathbf{u}_i$, of $\mathbf{M}$:

$$\mathbf{A} = \Sigma_i\ \mathbf{u}_i\ (1-\lambda_i)^{-1}\ \mathbf{u}_i^T \cdot \mathbf{e}\ . \quad (19)$$

The matrix $\mathbf{M}$ is not in general symmetric but its eigenvalues and eigenvectors are real so that Im A, and $\rho(\mathbf{k},E)$ by implication, is non-zero only for values of k and E such that $\lambda_i(k,E)=1$, corresponding to a pole in $\mathbf{A}$: Letting $E_i(k)$ be a value of E for which $\lambda_i(k,E)=1$, the desired quantity can be written as

$$\text{Im}\ A_{000} = \Sigma_i\ (\mathbf{e} \cdot \mathbf{u}_i)\ \pi\ \delta(E - E_i(k))\ (\partial\lambda_i/\partial E)^{-1}\ (\mathbf{u}_i^T \cdot \mathbf{e})\ . \quad (20)$$

In practice it is preferable to transform equation (16) into symmetric form before solving it numerically.

Numerical evaluation of the density of states is thus dominated by the need to solve equation (16) including enough terms to give the required precision. A great deal can be learned from simple models which illustrate the workings of the theory, and it is to this that the remainder of the paper is devoted.

EVALUATION OF STRUCTURE FACTORS

The structure affects equation (16) through the average in the last term. The D function may be written in terms of the Euler angles (Rose 1957) as

$$D_{jnn'}(\alpha,\beta,\gamma) = \exp(-in\alpha)\, d_{jnn'}(\beta) \exp(-in'\gamma). \tag{21}$$

The disorder in our model occurs in γ, so that the average in equation (16) may be expressed as

$$\langle \Sigma_i\, D_{jnn'}(\alpha_i,\beta_i,\gamma_i)\rangle = \Gamma_{n'}\, \Sigma_i\, D_{jnn'}(\alpha_i,\beta_i,0), \tag{22}$$

where the sum is over the four sets of Euler angles needed to rotate from the orientation of atom a to that of each of its nearest neighbors and

$$\Gamma_n \equiv \langle \exp(-in\gamma)\rangle = \int_0^{2\pi} d\gamma\, \Gamma(\gamma)\exp(-in\gamma). \tag{23}$$

$\Gamma(\gamma)$ is the distribution of dihedral angles measured relative to the diamond-cubic lattice configuration (that is, the usual definition of dihedral angle differs from ours by $\pm 60°$). Note that $\Gamma_n = \Gamma_{-n}$. The sum over i remaining in equation (22) can be performed explicitly in our model to yield

$$\langle \Sigma_i\, D_{jnn'}(\alpha_i,\beta_i,\gamma_i)\rangle = \delta_{n,3\times\text{integer}}\, \delta_{n',3\times\text{integer}}$$
$$\times (-1)^n\, \Gamma_{n'}\,[\,(-1)^j\, \delta_{n,-n'} + 3d_{jnn'}(\cos^{-1}(1/3))\,]. \tag{24}$$

The dihedral-angle distribution affects equation (16) only through $\Gamma_{n'}$. For the diamond-cubic structure, $\Gamma_{n'} = 1$. A special simplification occurs when γ is uniformly distributed: $\Gamma(\gamma) = (2\pi)^{-1}$ and $\Gamma_{n'} = \delta_{n'.0}$. Then the only terms of interest in equation (16) are those with $n = n' = 0$. The rotation matrix sum then takes the simple form

$$S_j \equiv \Sigma_i\, D_{j00}(0,\beta_i,0) = \Sigma_i\, P_j(\cos\theta_i) = (-1)^j + 3P_j(1/3), \tag{25}$$

in which the P_j are Legendre polynomials. The coefficients S_j reflect the local structure and will be referred to as structure factors. For the tetrahedral arrangement of neighboring atoms in this model, S_j vanishes for $j = 1, 2,$ and 5.

For a uniform dihedral-angle distribution, equation (16) is reduced to the form

$$A_{j00} = \delta_{j,0} + t_0\, S_j\, \Sigma_{l,j'}\, G_{l0}\,[C(lj'j;000)]^2\, A_{j'00}$$
$$= \delta_{j,0} + U(E)\, \Sigma_{j'}\, M'_{jj'}(k)\, A_{j'00}, \tag{26}$$

where all of the energy dependence of the matrix relating the components of A has been gathered into a scalar factor,

$$U(E) \equiv t_0(i\mu, i\mu) \, i\mu \, h_0^+(i\mu R) = t_0(i\mu, i\mu) \exp(-\mu R)/R , \qquad (27)$$

and

$$M'_{jj'}(k) = -S_j \sum_l (i)^{-1} (2l+1) [C(lj'j;000)]^2 j_l(kR) . \qquad (28)$$

Note that this matrix $\mathbf{M'}$ differs from the matrix $\mathbf{M}$ in equations (18) through (20) by a factor of $U(E)$. Equation (26) is straightforward to solve for any required number of j values. Its general analysis can be illustrated by considering the limit of a spherically averaged environment.

SPHERICAL LIMIT

If only the $j=0$ term is considered in equation (26), the effect is to perform a spherical average of the local structure. Then equation (26) has the solution

$$A_{000} = (1 - U(E) \, M'_{00}(k))^{-1} , \qquad (29)$$

where $M'_{00}(k) = -4j_0(kR)$. From equation (20),

$$\text{Im} \, A_{000} = \pi \, \delta(E - E_0(k)) \, (M'_{00}(k) \, \partial U/\partial E)^{-1}, \qquad (30)$$

where $E_0(k)$ are the values of E for which the denominator of equation (29) vanishes, (i.e., $U(E_0(k))^{-1} = M'_{00}(k)$). For a narrow s band and $E \ll 0$, it is a good approximation to write

$$U(E) \simeq W/(E - E_b) , \qquad (31)$$

where E_b is the bound-state energy of a single muffin-tin potential and W is a band width parameter. This is equivalent to the usual tight-binding approximation. Im A_{000} is then non-zero only when

$$E = E_b - 4W \sin(kR)/kR , \qquad (32)$$

showing the bottom of the band to be at $k=0$ with energy E_b-4W. This is the correct value when the number of nearest neighbors and the bond lengths are held fixed. The top of the band occurs near $kR=3\pi/2$ and has $E \simeq E_b+8W/3\pi$, to be compared with E_b+4W in the diamond-cubic lattice at $kR=3^{\frac{1}{2}}\pi$. The top of the band is reduced in energy relative to the crystal but occurs at approximately the same momentum. This lowering of the band edge is consistent with the frustration of phases in the disordered system. The band width is reduced to only ~60% of the value in the crystal due to the total elimination of short-range orientational order in the spherically symmetric limit (i.e., $j_{max}=0$). A similar effect has been seen in attempts to treat liquid transition metals using embedding methods (e.g., see Pendry 1980). Such a large reduction in band width is not observed experimentally, however, because even liquids are characterized by significant short-range orientational order. It will be seen in the next section that our approach has overcome this difficulty. The short-range order incorporated using non-zero but still small values of j_{max} is sufficient to increase the band width greatly.

When an explicit potential is used, the density of states in the spherical limit integrates numerically to unity. This is because equation (30) gives the proper solution for a structure for which $S_j = 4\delta_{j,0}$ even though that structure cannot be realized in a crystal. It is sufficient in the present context that the structure can be specified by structure coefficients. Truncating the set of equations (26) for any number of j values gives correctly normalized densities of states. On the other hand, the normalization will not be correct if the functions $s(k)$ and $s(i\mu)$ used in equation (17) do not correspond to a proper potential. It is thus important to use an $s(r)$ derived from equation (8) with an explicit $v(r)$. The model used in the next section is exactly of this kind.

DENSITY OF STATES

This section will illustrate the s-wave solutions for various choices of the number of j values, band width, and binding energy. The model potential used will be a square well of depth $-V_0$ and radius r_{SW}. The muffin-tin condition implies that $r_{SW} \leq R/2$. The energy range of interest is $-V_0 < E < 0$, in which case $\mu = (-E)^{\frac{1}{2}}$ and $\xi = (E + V_0)^{\frac{1}{2}}$ are real. Equation (8), which determines the t-matrix transforms, may be solved by writing $s_0(r) \propto R_0(r) v(r)$, where R_0 is a radial wavefunction of s symmetry, and determining the normalization using the inhomogeneous term. The resulting $s_0(r)$ is Fourier transformed to yield

$$s_0(i\mu) = (2i)^{-1} \{-1 + \exp(2\mu r_{SW})[1 - (\mu/\xi)\tan(\xi r_{SW})]/[1 + (\mu/\xi)\tan(\xi r_{SW})]\}, (33)$$

which has singularities at the bound states of the potential.

An analogous procedure leads to direct numerical evaluation of the required functions for any muffin-tin potential $v(r)$. It applies equally to other angular momenta. This means that the use of a more appropriate potential than the square well used in this work requires only routine numerical effort. There is little point in going beyond the square-well potential in the present discussion, however, because it serves to illustrate the range of behavior to be expected.

Using equation (20), the contributions to the density of states from the **A** vectors which solve equation (26) can be written as

$$\text{Im } A_{000} = \Sigma_i (\mathbf{e} \cdot \mathbf{u}_i) \pi \, \delta(E - E_i(k)) (\lambda_i(k) \, \partial U/\partial E)^{-1} (\mathbf{u}_i^T \cdot \mathbf{e}), \tag{34}$$

where $\lambda_i(k)U(E) = 1$ for $E = E_i(k)$ and $\mathbf{u}_i$, $\mathbf{u}_i^T$, and λ_i are the eigenvectors and eigenvalues of the matrix $\mathbf{M}'(k)$ defined by equation (28). The eigenvectors and eigenvalues can be found for each k, each eigenvalue giving a relationship between E and electron momentum k through $U(E_i(k))^{-1} = \lambda_i(k)$. The eigenvalues of $\mathbf{M}'(k)$ are shown as a function of k in Fig. 2 for $j_{max} = 0$, 6, and 10. The first of these is the solution in the spherical limit. $j_{max} = 6$ corresponds to four contributing values of j, while $j_{max} = 10$ corresponds to eight. As more j values are taken into account, the lines cover more and more of the k,E space. In the limit as $j_{max} \to \infty$, these lines will form a continuous distribution over much of the plane between some upper and lower limits. Note that there may be values of E for which there are no lines, leading to a zero density of states, or an energy gap. This behavior is properly expected for a disordered system. A tendency to model the areas of distribution expected in the complete solution can be seen even for $j_{max} = 10$.

The curves in Fig. 2 represent a great deal of information about the disordered system. They are properly compared in detail with $\rho(\mathbf{k},E)$ for a crystal, but only after the latter has been averaged over all orientations of the crystalline axes. In addition to the relation between E and k, it is possible to obtain information about the spatial dependence of an "averaged" electron eigenstate of energy E. Its properties are determined by the eigenvectors of $\mathbf{M}'$ at the intersections of the eigenvalue curves with a horizontal line corresponding to $U(E)^{-1}$. Of particular interest in this context, it can be seen from equation (34) that the contribution of each intersection to the density of states is proportional to $(\mathbf{e}\cdot\mathbf{u}_i)(\mathbf{u}_i^T\cdot\mathbf{e})$. The variation of this quantity is indicated in Fig. 2(d), wherein we have reproduced the curves of Fig. 2(c) as dotted lines. The solid lines represent eigenstates of $\mathbf{M}'$ for which $(\mathbf{e}\cdot\mathbf{u}_i)(\mathbf{u}_i^T\cdot\mathbf{e}) > 0.1$, where the eigenvectors of $\mathbf{M}'$ have been normalized so that $\Sigma_i\,(\mathbf{e}\cdot\mathbf{u}_i)(\mathbf{u}_i^T\cdot\mathbf{e})=1$. It is clear from this figure that the calculated density of states in this instance will be dominated by the highest and lowest eigenstates of $\mathbf{M}'$.

Further insight can be gained into the results of this theory by examining the densities of states which result from different values of the parameters. Densities of states are shown in Fig. 3(a) and (b) for $j_{max} = 6$ and 10, with the other parameters as specified in the figure caption. As desired, both distributions integrate to 2 electrons

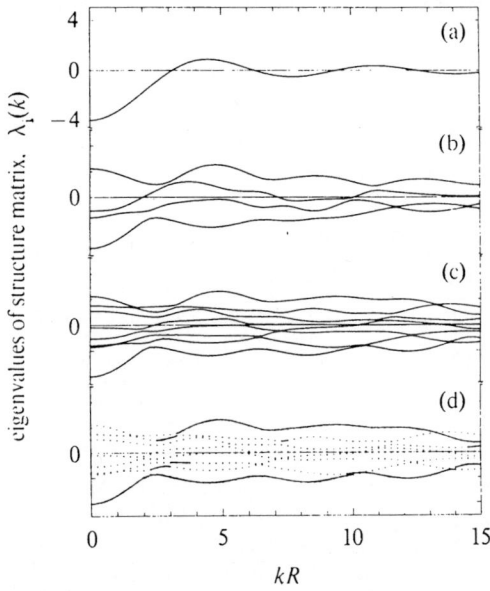

Fig. 2. The eigenvalues λ_i of the matrix $\mathbf{M}'(k)$ are shown as a function of kR, the product of electron momentum and the nearest-neighbor distance, in (a), (b), and (c) for the cases of $j_{max}=0$, 6, and 10, respectively. Through equation (34), they correspond to lines of non-zero strength in the spectral density function. The curves of (c) are reproduced as dotted lines in (d), except that solid lines are used to indicate the dominant contributions to the density of states as discussed in the text. Note the predominance of the highest and lowest eigenvalues of $\mathbf{M}'$.

when spin is taken into account. As j_{max} increases, the amount of structure increases because $\delta(E - E_i(k))$ gives a singular but integrable contribution to n(E) when $\partial E_i/\partial k = 0$. Thus, as is the case in a crystal, peaks in the density of states arise from nearly horizontal portions of the curves in Fig. 2. Increasing numbers of j values give more peaks of lesser weight. Note also that the upper limit of the band increases slightly on going from $j_{max} = 6$ to 10. There is little further change in the band as j_{max} increases further, suggesting a rapid convergence of the expansion in angular momentum of the atomic structure in this model. Apart from some fine structure between −5 and −3 eV, all of the features in Fig. 3(b) follow directly from the structure information incorporated in our model—that is, from the tetrahedral configuration of bonds and the dihedral-angle correlations.

The range of variation inherent in the square-well potential is illustrated in Fig. 3(c) and (d). In Fig. 3(c) is shown the n(E) which results from increasing the depth of the square well from 1.8 ry to 2.4. The band shrinks by nearly 2 eV and its bottom drops by nearly 6 eV. More interesting, the shape of n(E) changes significantly, with the features in the upper portion of the band being compressed much more than those in the lower portion. The n(E) which results from increasing the extent of the square well is shown in Fig. 3(d). Note that the band is broadened by ∼1.5 eV, preferentially in its lower portion. These examples suggest that the square-well

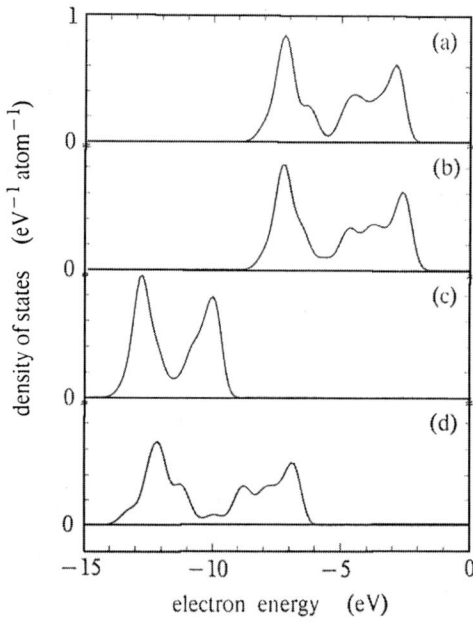

Fig. 3. The densities of states corresponding to $j_{max} = 6$ and 10 are shown for comparison as (a) and (b), respectively. The equations were evaluated using $V_0 = 1.8$ ry, $r_{SW} = 1.7768$ au ($= 0.4 \times R$), and $R = 4.442$ au. The parameters for (c) and (d) differ from those in (b) solely in that $V_0 = 2.4$ ry in (c) and $r_{SW} = 0.5 \times R$ in (d). Each n(E) has been broadened by convolution with a Gaussian of half-width 0.25 eV.

potential has enough flexibility to reproduce the qualitative aspects of most systems.

An important feature of this technique is its ability to yield correct results in the limiting case of translational periodicity. Equation (26) represents that limit for a linear chain given appropriate S_j, in which case the resulting density of states corresponds correctly to the well known result: $n(E) \propto \sin^{-1}[\pi(E - E_B)/(E_T - E_B)]$ in the tight-binding limit, where E_B and E_T are the bottom and top of the band, respectively. Equation (26) also gives the correct result for potentials more realistic than tight binding (Beeby and Hayes 1985).

Finally, equation (16) has been applied to interpret new and accurate x-ray photoemission spectra (XPS) of a-Si (Hayes, Allen, Beeby, and Oh 1985). The density of states was calculated for s-wave scattering potentials for two limiting structures of Si: diamond cubic; uniform dihedral-angle distribution. A procedure analogous to the transformation of Thorpe and Weaire (1971) was used to map these s-wave results onto the lower portion of the full s–p valence band. It was concluded that dihedral-angle disorder in a-Si is able to account fully for the principal observed differences between the XPS valence band spectra of c- and a-Si. Odd-membered rings, if they are indeed present in a-Si, cannot make a significant contribution to the XPS spectrum.

CONCLUSIONS

In this paper we have outlined the philosophy, formulation, and application of a method for calculating the density of electronic states of amorphous systems. While computationally tractable, it is powerful and flexible enough to allow inclusion of realistic structural information without recourse to arbitrary termination or embedding procedures. The detailed results presented for an amorphous semiconductor represent an interesting but structurally simple application. The method is readily applicable to all manner of disordered systems, however, including liquids, metallic glasses, and covalently-bonded glasses. Given its flexibility, it is hoped that this technique will indeed prove to be an advance towards the goal of a general theory of disordered systems.

Acknowledgments. One of the authors (JLB) is particularly indebted to Sir Nevill Mott for having pointed out in 1963 the problems associated with calculating the density of states of transition metal alloys. The resulting papers contributed to the development of the coherent potential approximation and led indirectly to the present work. The other (TMH) gratefully acknowledges the support of the Science and Engineering Research Council of the UK through a Senior Visiting Fellowship during the initial stages of this work.

REFERENCES

Beeby, J.L., 1964, The electronic structure of disordered systems, Proc. Roy. Soc. A, 279:82.

Beeby, J.L., 1983, The electronic structure of liquid metals, Phil. Mag. B, 48:L23.

Beeby, J.L., and Edwards, S.F., 1963, The electronic structure of liquid insulators, Proc. Roy. Soc. A, 274:395.

Beeby, J.L., and Hayes, T.M., 1985, A new method to calculate the electronic properties of disordered materials (to be published).

Brink, D.M., and Satchler, G.H., 1968, "Angular Momentum," 2nd ed., Oxford University Press, London.

Connell, G.A.N., 1974, Structures of amorphous tetrahedrally-bonded semiconductors, in: "Physics of Semiconductors," M.H. Pilkuhn, ed., B.G. Teubner, Stuttgart.

Hayes, T.M., Allen, J.W., Beeby, J.L., and Oh, S.-J., 1985, Structural origin of differences between the valence bands of crystalline and amorphous silicon (to be published).

Kramer, B., and Weaire, D., 1979, Theory of electronic states in amorphous semiconductors, in: "Amorphous Semiconductors," M.H. Brodsky, ed., Springer–Verlag, Berlin.

Mott, N.F., and Davis, E.A., 1971, "Electronic Processes in Non-Crystalline Materials," Oxford University Press, London.

Pendry, J.B., 1980, The electronic structure of liquids, J. Phys. C, 13:3357.

Rose, M.E., 1957, "Elementary Theory of Angular Momentum," John Wiley, New York.

Temkin, R.J., 1974, Dihedral-angle distribution of amorphous germanium, in: "Tetrahedrally Bonded Amorphous Semiconductors," M.H. Brodsky, S. Kirkpatrick, and D. Weaire, eds., American Institute of Physics, New York.

Thorpe, M.F., and Weaire, D., 1971, Electronic properties of an amorphous solid II. Further aspects of the theory, Phys. Rev. B, 4:3518.

Yonezawa, F., and Cohen, M.H., 1981, Theory of electronic properties of amorphous semiconductors, in: "Fundamental Physics of Amorphous Semiconductors," F. Yonezawa, ed., Springer–Verlag, Berlin.

THE OPTICAL THRESHOLD OF HYDROGENATED AMORPHOUS SILICON (*)

A. Frova and A. Selloni

Dipartimento di Fisica
Universita' di Roma I
Roma, Italy

INTRODUCTION

The optical behavior in the vicinity of the fundamental threshold is often the signature of a given material, leading to the determination of key parameters and physical properties. This is certainly true for amorphous semiconductors – and in particular for hydrogenated amorphous silicon and its alloys – whose optical investigation has been very intensive for a number of reasons. First, a-Si:H appears to be an important material for photoelectronic applications – solar cells, electrophotography, optoelectronic devices: spectral region of sensitivity, long-term stability under illumination, recombination kinetics, etcetera, are basic facts to be known. Second, the near-gap optical absorption is a direct diagnostical tool of the quality of the material. Third, due to the many available deposition techniques (glow-discharge, CVD, photoinduced deposition, diode and ion-beam sputtering, ion plating, evaporation, and others) and to their relatively advanced state of art, which permits a reasonable control over the results, amorphous silicon stands as a very good ground for the optical investigation of fundamental disorder effects, such as localization near the mobility edges. A number of reviews and books are available, which try to extract, from the harvest of published data on a-Si:H and related materials, unifying concepts.[1-5] The reader is referred to that literature – sometimes already of historical value – for a comprehensive summary of the state of research. In the present paper, we shall restrict ourselves to the discussion of a few aspects of special significance – namely the exponential absorption decay in the pseudogap (Urbach tail), the meaning of the so-called optical gap, and their relationship with disorder – be it thermal, structural, or compositional.

After a brief review of the experimental situation, we discuss the information – possibly not laboratory dependent – that can be extracted from the data. We shall base our arguments on some published a-Si:H results, but also on novel data obtained in our laboratory, including samples deposited by developmental techniques and amorphous silicon alloys. The latter ones are of great interest because they permit tailoring the degree of disorder

(*) This research has been supported as part of the GNSM Coordinated National Program, under grants from the Ministry of Education and CNR.

271

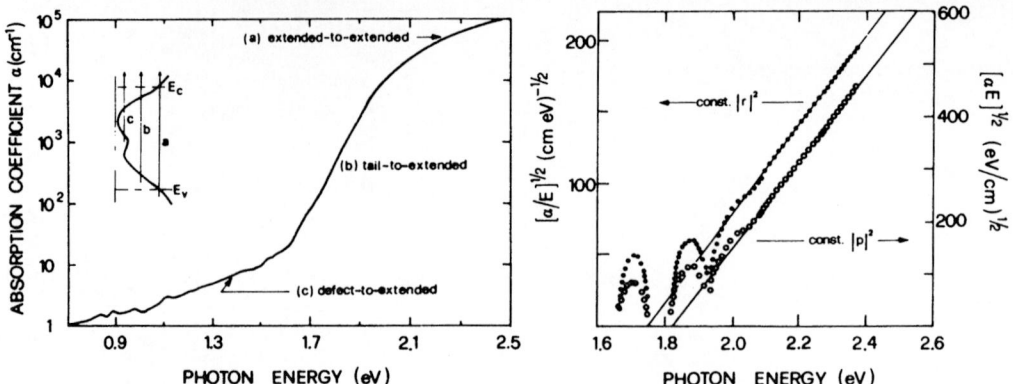

Fig. 1 The near-gap optical absorption coefficient of a-Si:H. The region above $\alpha \sim 10^4$ cm^{-1} is investigated by standard transmission,[7] the region below ~5x10^3 cm^{-1} by photothermal deflection spectroscopy (PDS).[8] The intermediate region is obtained by intrapolation. Interference fringes have been averaged out. The inset shows schematically the three types of transitions involved.

Fig. 2 Data of Fig.1 replotted - without fringe averaging - according to Eq.(3a), which assumes $|p|^2$=constant (Tauc's plot) and to Eq.(3b) for $|r|^2$=constant (Cody's plot).

over a wide range via changes in the composition. In the second part of the paper, we recall the theoretical understanding of the near-edge absorption behavior and introduce a model, which stems directly from the assumption of gaussian fluctuations in the potential, for quantitatively fitting the experimental spectra.

2. PRESENT EXPERIMENTAL KNOWLEDGE

The optical absorption spectrum of a glow-discharge a-Si:H film produced in our laboratory[6] is shown in Fig. 1. It is well accepted today that it comprises three regions: (a) the extended-to-extended state transition region, near and above the mobility gap (whose position cannot be determined from the spectra); (b) the valence band exponential tail-to-extended state region; (c) the defect-to-extended state region, involving mainly dangling-bond states D^0 or D^- in the lower half of the pseudogap. Discussion of the last region is beyond the scope of the present work.

(a) <u>Extended-to-extended state transition region</u> ($\alpha \gtrsim 10^4$ cm^{-1}). The absorption coefficient in this region is measured by standard thin-film transmission spectroscopy. Traditionally, α has been analized using the so-called Tauc plot:[10-14]

$$(\alpha nE)^{1/2} = \beta (E-E_g) \qquad\qquad (1a)$$

where E is the photon energy, E_g is the "Tauc optical gap" and β is a constant including the momentum matrix element squared $|p|^2$ and the density-of-state scale factor. Eq. (1a) descends from the relationship $\alpha = \epsilon_2 E/n\hbar c$ and the one-electron expression for ϵ_2 (for light polarization vector $\vec{n}$ and unit illuminated volume)[15]

$$\epsilon_2 (E) = 8(\pi e\hbar/m)^2 (1/E^2) \sum_{i,f} | (\vec{n}\cdot\vec{p})_{if} |^2 \delta(E_f-E_i -E) \quad , \qquad (2a)$$

under the following conditions: (i) for the amorphous material there exists

272

a region where the valence and conduction densities of states present a square root dependence on energy as for a virtual crystal where, however, the k-conservation rule is completely relaxed (nondirect transitions); (ii) $|p|^2$ is constant with energy. Since, in the experimentally accessible range of α , n varies only moderately, Eq. (1a) is usually replaced by

$$(\alpha E)^{1/2} = B(E - E_g) \qquad (3a)$$

with B constant. See Fig. 2 for the data of Fig. 1 replotted accordingly (circles). It appears that one can define an "optical gap" E_g by extrapolation to zero of the linear range. E_g marks approximately the transition between the virtual crystal behavior and the exponential tail, but it has no immediate physical meaning. In particular, it bears no obvious relationship with the properties of the real crystal. It is, however, a remarkably important parameter in determining the composition of the silicon alloys - including silicon hydride - and to some extent the degree of disorder. In this respect, one should hopefully correlate it with the mobility gap.

More recently, some authors[16,17] have pointed out that the Tauc plot is not the best choice possible. If data of the type shown in Fig. 2 are extended to a broader range - by use of very thin films or by reflectivity experiments - the linearity becomes poor. Data by Klazes et al.[17] are shown in Fig. 3 (full dots). The authors note that the same data, replotted as $(\alpha n E)^{1/3}$ vs. E, yield a better straight line (not shown). Physically this would imply a suitable energy dependence of $|p|^2$ or a linear distribution of the density of states, as shown by Davis and Mott.[18]

Theory[19] and recent inverse photoemission results by Jackson et al.,[20] however, suggest a different approach, recommended first by Cody et al.[21] For amorphous material and unpolarized light, Eq.(2a) can be also written as

$$\varepsilon_2 (E) = (8 \pi^2 e^2 /3) \sum_{i,f} | r_{if} |^2 \; \delta (E_f - E_i - E) \qquad (2b)$$

$|r_{if}|^2$ being the dipole matrix element squared. Jackson finds that, in the energy region of interest, it is $|r|^2$ in Eq.(2b) - rather than $|p|^2$ in Eq.(2a) - that is roughly constant with energy. A similar conclusion - although more qualitative - can be drawn from Ley's results.[22] As a consequence, a better linear behavior is obtained (see Fig. 3, circles) by plotting the

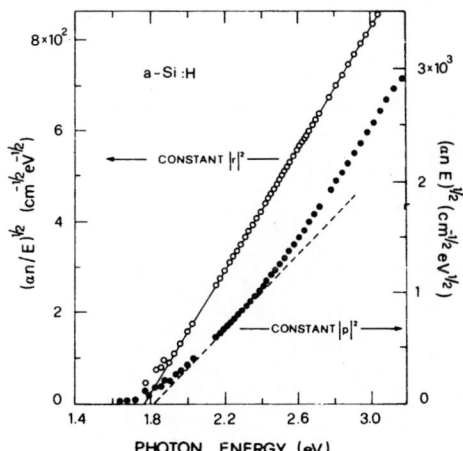

Fig. 3 Extended-range data by Klazes et al. shown in the constant $|p|^2$ approximation (full dots),[17] and replotted in the constant $|r|^2$ approximation (circles).

equation

$$(\alpha n/E)^{1/2} = \gamma (E - E_g) \qquad\qquad (1b)$$

or, for the near-gap region,

$$(\alpha /E)^{1/2} = C (E - E_g) . \qquad\qquad (3b)$$

This type of plot, which does not show a special improvement when the energy range is limited, does however result in a lower value of the optical gap (we shall hereafter refer to this as the "Cody gap"). In our samples, the difference is typically about 60 meV (Fig.2). A similar observation is reported by Jackson et al.[20] Although the $|r|^2$ approximation is clearly to be preferred, for the purpose of comparing different samples in the near-gap region, either choice is acceptable. However, one should bear in mind that the $(\alpha E)^{1/2}$ vs. E plot has the advantage of being found throughout the literature so far.

(b) <u>Exponential tail transition region</u> ($\alpha \lesssim 10^3$ cm^{-1}). The exponential absorption region can be described by the equation

$$\alpha(E) = \alpha_o \exp [(E - E_g)/E_o] , \qquad\qquad (4)$$

where E_o is the Urbach characteristic energy, and α_o is usually a function of it. The exponential region corresponds to transitions from the valence-band tail in the pseudogap, which is nearly twice as broad as the conduction tail,[23] to the extended states in the conduction band. It is now widely accepted[5,20,23,24] that these transitions are characterized by the same matrix elements as the extended-to-extended transitions, which means that the optical absorption spectra reflect the density-of-state distribution. E_o gives therefore a direct measure of the profile of the localized valence states.

The correct determination of E_o requires more sophisticated experimental techniques than straight transmission, because of the low optical density of the films. Recently developed methods are photothermal deflection spectroscopy (PDS),[9] and photoacoustic spectroscopy (PAS),[25] both providing an indirect determination of α via the effect of optical warming-up of the sample. The former is probably the most sensitive technique available. A widely used method, whenever applicable (e.g. not in short-lifetime materials such as the a-SiGe alloys), is spectral photoconductance.[26] If care is

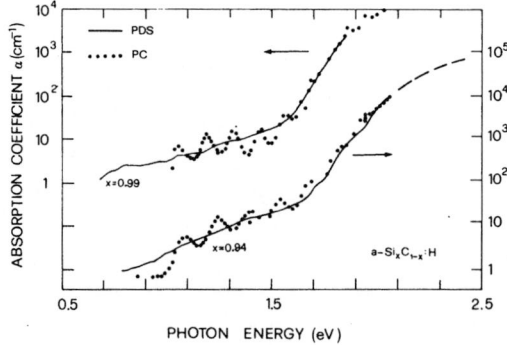

Fig. 4 Comparison of absorption coefficient as deduced from PDS [8] and photo-conductivity (interference fringes not averaged out in this case) in a-SiC:H, showing that the $\mu\tau$ product can be taken roughly constant throughout the spectral range.

taken to compensate for nonlinear dependence on illumination,[27] the method provides a good determination of low α levels, as long as the electron $\mu\tau$ product is constant with photon energy. This has proven true in a-Si:H,[28],[29] and a convincing evidence for a-SiC:H is provided by Fig.4. The discrepancy between the Urbach energy from PDS and PC measurements in all our samples never exceeds some 10%.

3. OPTICAL PARAMETERS AND DISORDER

We now discuss the correlation between the optical parameters E_o, E_g and B and the degree of disorder. The Urbach energy E_o gives a direct measure of disorder, much as Raman, X-ray or core-level spectroscopy. Cody et al.[30] have proposed a model where E_o is proportional to the sum of the mean square deviations of the atoms from their ideally ordered positions, assumed to have a gaussian distribution. This takes care of structural and thermal fluctuations. The model has been extended to include compositional disorder, brought about by hydrogenation[31] or alloying.[29] If we take the equivalent approach of gaussian potential fluctuations, i.e. the measure of the site disorder is given by the mean square deviation of the potential, the additivity of the variance leads to

$$E_O = K_O [W_X^2 + W_T^2 + W_H^2 + W_C^2] = K_O \sum_i W_i^2 \qquad (5)$$

with obvious meaning of the symbols. It has been reported [30],[31] that an increase in hydrogen content usually lowers E_o, suggesting that the healing effect (decrease in the structural term W_X^2) dominates over the increased compositional disorder (increase in W_H^2). In RF-sputtered material, Ceasar et al.[25] quote a hydrogen content of 19 atomic % for lowest E_o. This is not the case for other kinds of alloys where the term W_C^2 may become soon important enough, with increasing deviation from pure Si, to allow the other terms to be taken as constant. The one-to-one correspondence between E_o and the site potential fluctuations –and more generally the validity of Eq.(5) – are demonstrated by the data in Fig.5, referring to the a-Si$_x$C$_{1-x}$:H alloy at room temperature. This alloy is particularly interesting – as opposed, e.g., to a-SiGe:H – because the defect-to-extended absorption is not increased with respect to a-Si:H, which makes the determination of E_o very reliable. The curve has been calculated [29] by taking $K_O(W_X^2 + W_T^2 + W_H^2) = 54$ meV and $K_O W_C^2 = ax(1-x)$, as for the potential vari-

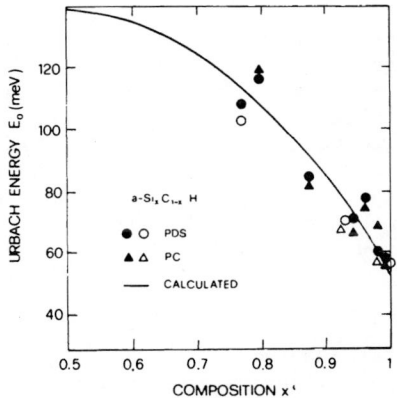

Fig. 5 Urbach energy E_o vs. composition x for the a-Si$_x$C$_{1-x}$ alloy. Full dots show PDS data [29] for sample grown in SiH$_4$+CH$_4$+H$_2$. The PDS data for samples grown in SiH$_4$+CH$_4$ (circles),[8] and all the PC data (triangles) are unpublished.

ance of an uncorrelated two-component alloy, with a= 337 meV. The interest of these results sits in the possibility, besides modelling the potential variance via changes in x, of theoretically estimating the site potential fluctuations, as discussed in Section 4.

The optical gap E_g , be it the Tauc or the Cody gap, is also somewhat affected by disorder. One can write, for its disorder-dependent part, an equation similar to Eq.(5) for E_o:[30]

$$E_x - E_g = K_g \sum_i W_i^2 \qquad\qquad (6)$$

where E_x would be the optical gap in the limit of zero disorder. It should therefore be a fixed number for a material of given composition and density. From Eqs.(5) and (6) we have

$$E_g = E_x - G E_o \quad \text{and} \quad G = K_g/K_o \quad . \qquad (7)$$

In a-Si:H, if the shift of E_g due to alloying with hydrogen could be neglected, Eq.(7) would provide an immediate measure of disorder, without a need for measurements of E_o . This approach has been taken by Cody and his coworkers.[5,30] By examining the dependence of E_g and E_o (measured by standard transmission) in a-Si:H on temperature and on hydrogen evolution treatments, they find G = 6.2 and suggest that Eq.(7) is of general validity for a-Si:H up to 20 atomic % hydrogen concentration.

This result, however, is questionable under various respects. A simple estimate of the gap opening due to hydrogenation, starting from the Si-H bonding-antibonding state separation,[32] leads to ~250 meV for 20% hydrogen. This is comparable with the overall change in gap, due to H evolution, reported by Cody et al.[30] Experimentally, the literature is rich with a-Si:H data where for nearly equal E_o the optical gap differs appreciably. For instance, heavily hydrogenated a-SiGe alloys may present a wider optical gap than a-SiC alloys of comparable Urbach energy E_o (see e.g. Fig.7 later). Since Ge lowers E_x , while C has the opposite effect, this in contradiction with Eq.(7). We will show in Section 5 that the G term in Eq.(7) is not quite a constant for varying E_o , and takes values close to 1.

The third optical parameter of importance is B in Eq.(3a), or C in Eq.(3b). There have been attempts to correlate it with disorder, but no definite correspondence has been proposed. Some results seem to suggest[33] - but the result is far from being confirmed on a general basis - that B decreases along with photoconductance. It is a fact rather that B is remarkably similar for good-quality samples produced by a variety of techniques, in a variety of laboratories, for a wide range of hydrogen concentrations and even for moderate alloying. Its value, found independent of temperature,[30] typically ranges between 750 and 850 $cm^{-1/2} eV^{-1/2}$, becoming smaller only in presence of severe structural changes, such as heavy alloying with C or N, or growth in shortage of H; or under serious disturbances, such as heavy doping or argon bombardment. These effects go far beyond disorder and imply a true redistribution of the states in the bands. These statements are better substantiated by close examination of the data of Figs.6 through 9.

Figs.6 and 7 show that adequately hydrogenated silicon or silicon alloys, no matter what the deposition method has been, behave almost identically. For germanium, an atom highly interchangeable with silicon, the slope B is constant all the way, and also for C and N (data not shown) the only effect observed is a gap widening, as long as silicon is strongly dominant in the alloy. For very large C or N concentrations, a decrease in slope is seen, probably associated with rearrangements in the nearest-neighbor confi-

276

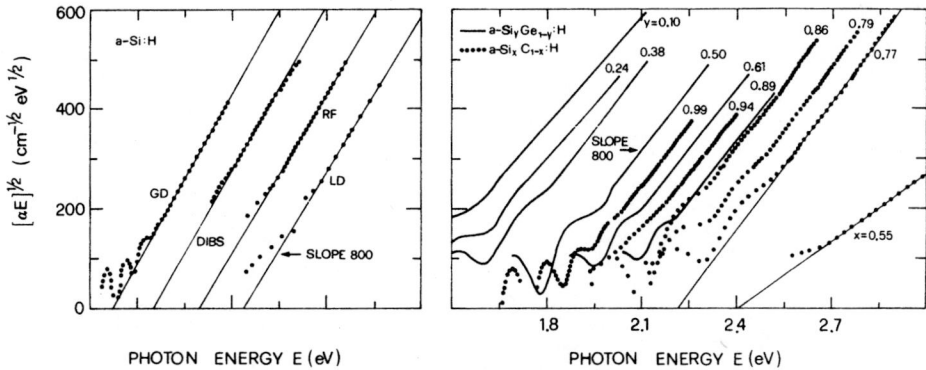

Fig. 6 Tauc's plot for a-Si:H produced by different deposition techniques, showing the uniformity of the B-value in Eq.(3a). GD=glow discharge in SiH , RF=R.F. diode sputtering, DIBS=dual ion-beam sputtering, LD=laser-induced deposition in SiH$_4$.[34] Curves are horizontally shifted for clarity.

Fig. 7 Tauc's plot for the most important a-Si alloys, grown by glow-discharge.[35] The large C-content data (x=0.55) are taken from the literature.[36]

guration, e.g. graphitic coordination in the C case.[37,38]

Quite instructive results are deduced from dual-ion beam sputtering deposited samples.[39] The DIBS technique is particularly useful because it permits careful and reproducible tailoring of the hydrogen content (by acting on the H-gun voltage and current). It also allows enhancing structural disorder by bombardment of the growing film by other species, e.g. argon. Fig.8(a) shows a few examples of the Tauc plot with and without argon bombardment. Fig.8(b) showing optical gap and slope parameter B obtained from a large number of curves of the same type, has an important meaning. Consider first the samples produced without Ar-bombardment (circles). Hydrogenation, while opening up the gap, makes the conduction edge steeper,[40] which is consistent with an increase of B. The monotonic rela-

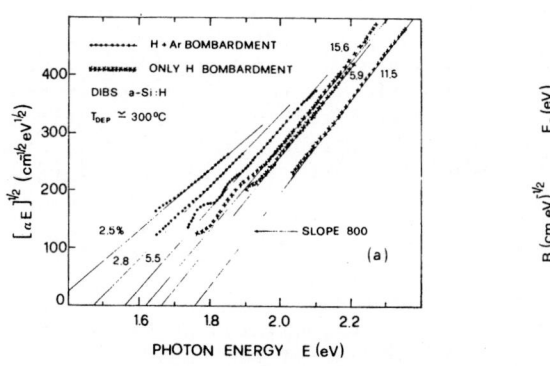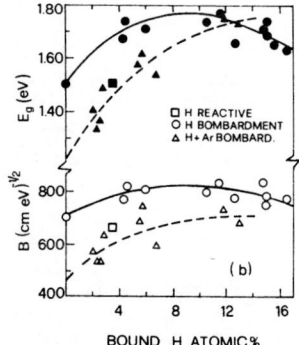

Fig.8 (a)Tauc's plot for DIBS a-Si grown with different hydrogen-gun voltage with or without simultaneous argon bombardment. The bound H-content is shown in atomic %. (b) Plots of E$_g$, B for the two cases above, vs. bound hydrogen concentration. Also shown is the behavior of a sample deposited by shutting off the H-gun and sputtering in H atmosphere (H reactive).

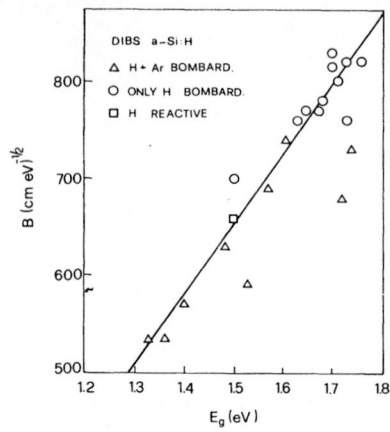

Fig. 9 Showing the linearity between optical
gap opening and amplitude factor B.

tionship between B and E_g (see Fig.9) is a reminder of the conservation of the total number of states. Both E_g and B present a maximal value for ~10 atomic % H concentration. Above that, consistent with our discussion following Eqs.(5) and (7), the dominant effect of H is to lower E_g (and B) via increased compositional disorder. On the low-H side the trend is instead guided by the "chemical" effect of alloying. The two competing effects result in an extended range of H-concentrations where E_g and B remain roughly constant. The change of ~200 meV from zero H to 10 atomic % is consistent with the theoretical estimate. This is not so (triangles) for samples grown under Ar-bombardment (energy 100-200 eV and fluence twenty times less than H from the same gun). Both E_g and B drop to much lower values for $C_H \rightarrow 0$. As the chemistry is the same, this proves that severe structural imperfections are now the dominant factor. The latter behavior is closer to that found e.g. for RF reactively sputtered or otherwise produced a-Si:H, indicating that a serious origin for disorder in a-Si:H may be undesired bombardment of the growing film by energetic species. In this respect, the DIBS technique is particularly promising.

In conclusion, when large B changes are observed, they should in general be related to strong structural damage and to more serious DOS changes than those merely due to disorder. When such effects are not dominant, there seems to be no easy and unambiguous way of linking B to disorder, particularly because its determination is affected to some extent by the uncertainty in the film thickness and by approximations in the refractive index, as well as by the choice made in drawing a straight line through the data points (particularly in the Tauc plot used so far in the literature).

4. THEORETICAL APPROACH TO THE URBACH TAIL

We now theoretically approach the problem of the exponential absorption edge which arises from transitions between localized states in the valence band and extended empty states. A calculation of the Urbach edge has been performed by Abe and Toyozawa[41] (AT) for a two-band tight-binding model with gaussian site-energy disorder, within the coherent-potential approximation. The basic parameter of this model is the ratio W/B, where W^2 is the variance of the gaussian distribution and B is the smaller (i.e. valence) bandwidth in the absence of disorder. For small to moderate disorder (W/B $\lesssim$ 0.6) the calculated spectrum exhibits an Urbach tail of width $E_0 \sim 0.5 \; W^2/B$, arising essentially from the exponential tails in the single-particle state den-

278

sities. The optical matrix element turns out to be almost constant within an energy interval 2 E_o. The exponential behavior of the density of states, in turn, is determined by an interplay of transfer energy and gaussian-distributed site energies. It occurs in the intermediate region where the wavefunctions change their character from extended (well inside the band) to localized (deep in the gaussian tail).

The AT results give some feeling about the origin of the Urbach tail and about the relationship between E_o and the microscopic parameters of the system. Any quantitative attempt to interpret the observed spectra in amorphous semiconductors in terms of such model, are however unsuccessful; in particular, they fail to describe the absorption coefficient above E_g. This is because the model is probably oversimplified: in particular neither intersite correlations nor topological disorder is included.

The problem of the disorder-induced low-energy tail in the density of states has been recently revisited by Soukoulis et al.[42](SCE). Their approach is similar to the one developed by Halperin and Lax,[43] where electronic states are localized by long wavelength potential fluctuations. But it differs in that it includes also small-scale potential fluctuations in the evaluation of the kinetic energy of localization. For the sake of clarity, we shall now sketch a simple derivation of the SCE results.

We consider a disordered system described by a one-electron Hamiltonian with a random potential V(r), which we assume to have gaussian fluctuations of correlation length L. Following a standard procedure,[44] we divide the macroscopic system of volume V into disjoint cells of volume λ^3 , with $\lambda \gg$ L, and denote n(E ,λ), the density of electron states in any of these cells. Our purpose is to evaluate the density of states N(E) of the system as: N(E) = ⟨n(E,λ_E)⟩, where ⟨...⟩ denotes the average over the cells, and λ_E is the value of λ determined by application of Lloyd and Best variational principle.[45] In order to have a simple expression for n(E , λ) , let us approximate V(r) by its average value $\bar{V}$ in the cell.[46] $\bar{V}$ is a gaussian-distributed random variable whose variance is inversely proportional to λ^3 and will thus be indicated as (L^3 /λ^3) W^2 . We may then write:

$$\langle n(E,\lambda) \rangle \propto (\frac{\lambda^3}{2W^2L^3})^{1/2} \int_{-\infty}^{E-\Delta E_\lambda} d\bar{V} (E-\bar{V}-\Delta E_\lambda)^{1/2} \exp(-\lambda^3 \bar{V}^2 /2W^2L^3) \quad (8)$$

where ΔE_λ is the kinetic energy of localization into the volume λ^3 . The usual expression for ΔE_λ is $\Delta E_\lambda = \chi/\lambda^2$, with $\chi = 3\pi^2 \hbar^2/2m$ in 3-D. This leads to a low-energy tail of the form N(E) $\propto$ exp $(-A|E|^{1/2}/ W^2)$, where A is a constant.[43] SCE, on the other hand, argue that in the presence of small-scale potential fluctuation the energy required to localize a state of energy E is:[47]

$$\Delta E_\lambda = \frac{B}{\lambda^3} \frac{1}{N(E)} g(\lambda,E) \quad (9)$$

where B is a constant and g(λ , E) is the dimensionless conductance. Since g(λ , E) $\sim 1/\pi^3$ when λ is small in comparison to the correlation length ξ(E) of wavefunction fluctuations,[48] there can be a range of energies where ΔE_λ is proportional to λ^{-3} , rather than λ^{-2} as usual. Using (9) in (8), the low-energy tail of N(E) takes the form:

$$N(E) \propto \exp \{-|E|/E_o\} \quad (10)$$

where $|E|$ is measured relative to the mobility edge E_c and E_o is given by[49]

$$E_O = \frac{\pi}{4} \frac{L^3}{\chi \xi_o} W^2 \quad (11)$$

ξ_o being the value of ξ(E) at $E \gg E_c$.
It is interesting to observe that the value of λ_E^3 obtained from the varia-

tional procedure[45] turns out to be inversely proportional to $|E|$:

$$\lambda_E{}^3 = \frac{2}{\pi} \frac{\chi \xi_0}{|E|} \tag{12}$$

Roughly speaking, this implies that the variance $W^2 L^3/\lambda^3$ of the potential fluctuations in Eq.(8) increases for larger $|E|$, i.e. for states lying deeper in the valence tail.

Eqs.(10)–(12) have of course a well defined range of validity. First, they require $\lambda_E \gg L$ and $\lambda_E \ll \xi(E)$. In addition, the potential fluctuation leading to the localized state of energy $|E|$ should be sufficiently isolated. From this, SCE estimate that (10) holds below an energy E_i such that $(E_C - E_i) \sim 3\text{-}6\ E_0$. In the very deep part of the tail, however – where λ_E given by (12) would become smaller than L – gaussian behavior occurs.

The microscopic parameters of the disordered system – ξ_0 , L and W^2 – entering the SCE expression for E_0 are in general difficult to evaluate. It is however useful to remind that in the limit of strong disorder and short potential correlation length ξ_0 becomes of the order of L, both being of the order of interatomic distances d.[42] Thus in this limit E_0 simplifies to $E_0 = \frac{\pi}{4}(L^2/\chi) W^2$. We may now use this expression, together with the experimental values of the Urbach energy E_0 , to estimate the value of $W_C{}^2$ – the potential variance associated with compositional disorder – in a-$Si_x C_{1-x}$. In Section 3 we found that the Urbach energy data of a-$Si_x C_{1-x}$ could be fitted introducing a compositional term $K_0 W_C{}^2 = a\ x\ (1-x)$, with a = 0.337 eV. In the limit $L \sim d$ (i.e. negligible intersite correlations), $W_C{}^2$ can indeed be expressed as $W_C{}^2 = x\ (1-x)\ (\Delta v)^2$, where Δv is the difference between the averaged atomic potentials of the two components of the alloy. This yields $(\Delta v)^2 = 12\pi\ a\ \hbar^2 /(2\ m\ L^2)$. We take L approximately equal to the Si-Si bond length (4.44 a.u.). We then find $\Delta v \sim 3$ eV. It is worth noting that this value is very close to the difference between the ionization potentials of the C and Si atoms. This seems a fair estimate of (Δv) for valence band states, consistently with the notion that E_0 is related to the width of the valence band tail. On the other hand, it gives an a posteriori justification of the $\xi_0 \sim L$ assumption.

5. MODEL CALCULATION OF NEAR GAP SPECTRA

Using the theoretical results of the preceding section, we now derive an expression for the absorption coefficient $\alpha(E)$. For complete relaxa-

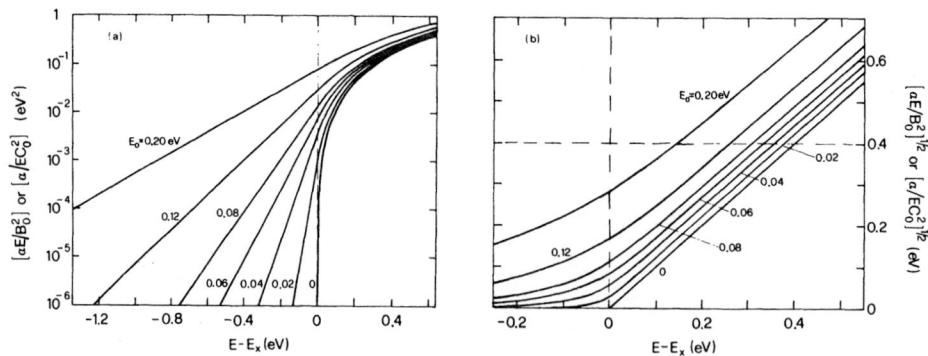

Fig. 10 Absorption coefficient in universal form, as calculated from Eqs(15) for different values of the Urbach energy E_0 , shown (a) on a logarithmic plot, (b) on a square root plot. $\alpha E/B_0{}^2$ and $\alpha/EC_0{}^2$ are for the constant $|p|^2$ and constant $|r|^2$ approximations respectively.

tion of the k-selection rule and constant momentum matrix element – which we take for historical reasons –, we have

$$\alpha(E) \propto \frac{1}{E} \int dE\, N_v(E)\, N_c(E + E) \qquad (13)$$

where N_v and N_c are the valence and conduction state densities respectively. From Eqs.(8) and (12) of the previous section we know that, for E in the region of tail states, $N(E)$ can be expressed as

$$N_v(E) \propto \int_{E-E_v^o}^{\infty} dy\, (E_v^o - E + y)^{1/2}\, e^{-y/E_o}$$

where E_v^o is the band edge in the absence of disorder. The above integral is in practice an exponential broadening of the DOS for the ideally ordered material. We ignore instead the conduction band tail, which is known to be narrower, and take $N_c(E) \propto (E - E_c^o)^{1/2}$. By a simple interchange of the order of integrations, Eq.(13) becomes then

$$\alpha(E) = \frac{B_o^2}{E} \int_{E_L}^{\infty} dy\, e^{-y/E_o}\, (E - E_x + y)^2 \qquad (14)$$

where $E_x = E_v^o - E_c^o$, B_o summarizes all constant factors and $E_1 = 0$ if $E - E_x \geqslant 0$, $E_L = E_x - E$ otherwise. We recall that B_o^2/E should be replaced by $C_o^2 E$ in the constant $|r|^2$ approach. In all the discussion that follows, however, it makes little or no difference if we stay in the constant $|r|^2$ or constant $|p|^2$ framework, due to the narrow range of photon energies considered. Although Eq.(14) can be theoretically justified only for a limited range of energies – i.e. in the interval corresponding to transitions from valence tail states to extended empty states –, we will show, by comparison with experiment, that this simple expression can be used also for transitions occurring above the optical gap.

The integral (14) is the same earlier introduced – however on a purely phenomenological basis – by Dunstan.[24] Analytical integration leads to:

$$[E \leqslant E_x], \qquad \alpha(E) = (2E_o^2 B_o^2/E)\, \exp[(E - E_x)/E_o] \qquad (15a)$$

$$[E \geqslant E_x], \qquad \alpha(E) = (B_o^2/E)[(E - E_x)^2 + 2E_o(E - E_x) + 2E_o^2]. \qquad (15b)$$

Eqs.(15) are plotted in Fig.10, using E_o as a parameter. For $E - E_x$ appreciably larger than E_o, Eq.(15b) can be approximated to

$$\alpha(E) = B(E - E_g)^2 \qquad . \qquad (16)$$

Calculated E_g and B/B_o as a function of E_o are shown in Fig.11. For practi-

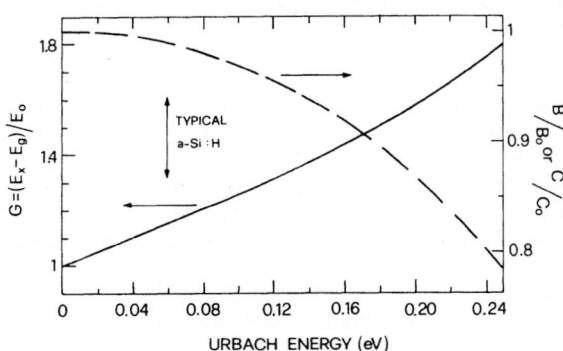

Fig. 11 Optical gap E_g and broadened amplitude factors B, C as obtained from the tangent to the curves of Fig.10(b) at $(\alpha E/B_o^2)^{1/2} = 0.4$ eV, which is at the center of the experimental range usually explored.

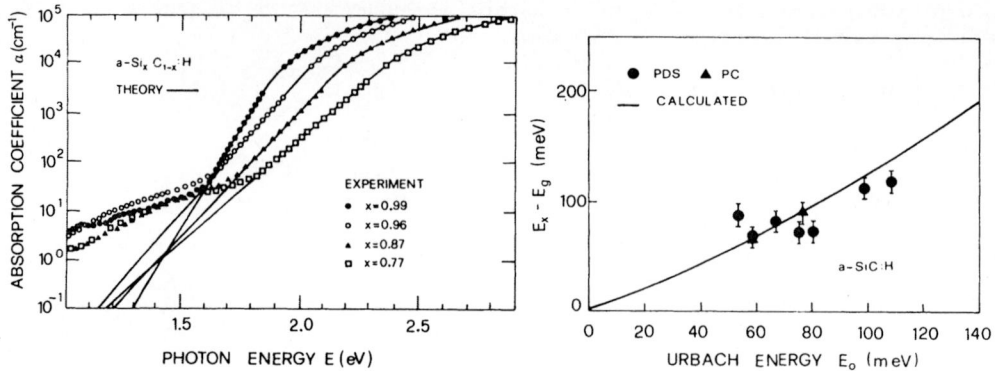

Fig. 12 Fitting of some a-Si$_x$C$_{1-x}$ samples in terms of Eqs.(15). The fitting
parameters are E$_o$, E$_x$ and B$_o$.

Fig. 13 Comparison between calculated E$_x$ -E$_g$ and experimental values from
fittings of the type shown in Fig.12.

cal E$_o$ -values, B/B$_o$ ∿1, as seen experimentally, and

$$E_g = E_x - G E_o \quad , \quad\quad\quad\quad\quad\quad (17)$$

G being a slowly varying function of E$_o$. In the range of interest for
a-Si:H, G ∿ 1.2.

We now rewrite Eq.(15a) in terms of E$_g$, so as to obtain a theoretical
expression for Eq.(4):

$$\alpha(E) = \alpha_o \quad \exp\left[(E - E_g)/E_o\right] \quad ,$$

where $\alpha_o E/B_o^2 = 0.6 E_o^2$. This result is identical to the one derived by
Cody with a simple band profile model.[5] We also note that, for G∿1.2, our
E$_x$ is not too different from Cody's estimate of the mobility gap. Jackson
et al.[20] arrive at similar numerical conclusions, noting that, in a sense,
the physically less justified Tauc plot is the one that yields the closest
E$_g$ -value to the mobility gap.

Eqs.(15) have proven capable of fitting, both below and above gap, all
the experimental spectra considered. We shall make the point by means of
the silicon-carbon alloy series, which we have already discussed in Fig.5.
This is because it offers a well-defined type of disorder and a wide range
of E$_o$ and E$_x$ -values. Four typical best fittings in the
extended-to-extended and tail-to-extended transition regions are shown in
Fig.12. The agreement below E$_g$ is granted, since E$_o$ is the starting parameter in generating each spectrum. The agreement above E$_g$ is instead critical
because this range of energies falls outside the exponential SCE range. In
Fig.13, we show E$_x$ -values, defined from the respective Tauc's observed
gaps, that were used for best fitting. Comparison with the calculated curve
(Eq.(17)), is quite satisfactory, considering that the determination of
E$_g$ is affected by some uncertainty and that an amplitude error in the Urbach
tail by PDS is undistinguishable from an energy shift of the spectra. It is
worth noting that our model also explains part of the E$_g$ vs. B trend exhibited by DIBS material in Fig.9, especially when grown under Ar-bombardment.
However, because of the nonnegligible "chemical" effect due to the changing
H concentration, we do not attempt a quantitative fitting, that might be
somewhat fortuitous.

We conclude therefore that Eqs.(15) are reasonably well applicable -

either in the constant $|p|^2$ or $|r|^2$ approach – over the whole near-gap region. This is so, as long as moderate disorder is involved, and severe rearrangements in the nearest-neighbor tetrahedral structure are not taking place.

CONCLUSIONS

In the present paper we have presented and discussed a number of experimental results – some of them unpublished – on the fundamental absorption edge of a-Si:H and a-SiC:H. We have also reviewed the theoretical understanding to date of the exponential (Urbach) broadening, as resulting from gaussian potential fluctuations. On this basis, we have introduced a model for the calculation of the absorption spectra, which accounts also for the near-gap extended-to-extended state transition region.

By investigating the exponential tail in a-SiC:H for varying alloy composition, we have deduced the relationship between the degree of disorder – expressed by the variance of potential fluctuations for a binary alloy – and the Urbach energy E_o, as well as the Tauc gap E_g. We have found that the latter shifts downwards, with respect to its limit value E_x for zero disorder, by slightly more than E_o, in very good agreement with our model estimate. In the silicon-hydrogen system this downward shift is the dominant effect in the E_g behavior for increasing C_H, but only at large H contents, against the "chemical" widening of E_x.

Another important parameter, the slope B in the Tauc plot (or C in the constant $|r|^2$ approximation) is examined and found to present – both theoretically and experimentally – too weak a dependence on disorder to be unambiguously used as reference figure, unless major alterations in the extended-state profile take place.

ACKNOWLEDGEMENTS

The authors are deeply indebted to the following people: P.Fiorini, A.Gregori and C.Coluzza for continued experimental assistance, S.Beretta for performing the calculations; N.M.Amer and A.Skumanich for the PDS results; M.Musci, R.Bilenchi, G.Fortunato, A.D'Amico and C.Giovannella for providing data, often unpublished; F.Evangelisti, C.Di Castro and D.Della Sala for very helpful discussions.

REFERENCES

1. J. Tauc, in "Optical Properties of Solids", p.279, F. Abeles ed., North-Holland Publ. Co., 1972; also in "Amorphous and Liquid Semiconductors", Chapter 4, J. Tauc ed., Plenum Press, 1974.
2. N.F. Mott and E.A. Davis, "Electronic Processes in Non-Crystalline Materials", 1979 Oxford University Press.
3. H. Fritzsche, Solar Energy Materials 3, 447 (1980).
4. G.A.N. Connell, in "Amorphous Semiconductors", Topics in Applied Physics vol.36, M.H. Brodsky ed., Springer-Verlag, 1979.
5. G.D. Cody, in "Semiconductors and Semimetals", Vol.21B, p.11, J. Pankove ed., Academic Press, 1984.
6. All glow-discharge samples from silane shown here are from P.Fiorini and his students.
7. We are indebted to A.Gregori for this kind of measurements.
8. We acknowledge the collaboration of A.Skumanich and N.M.Amer on this part of the work.
9. See, for instance, N.M. Amer and W.B. Jackson, in "Semiconductors and

Semimetals", Vol.21B, J. Pankove ed., Academic Press, 1984.

10. J. Tauc, R. Grigorovici and A. Vancu, phys. stat. sol. 15, 627 (1966); also Ref. 1.

11. M.H. Brodsky, R.S. Title, K. Weiser and G.D.Pettit, Phys. Rev. B1, 2632 (1970).

12. G.K.M. Thutupalli and G. Tomlin, J. Phys. C10, 467 (1977).

13. E.C. Freeman and W. Paul, Phys. Rev. B20, 716 (1979).

14. G.D. Cody, B. Abeles, C.R. Wronsky, B. Brooks, and W.A. Lanford, J. Noncryst. Solid, 35-36, 463 (1980).

15. See, for instance, L. Ley, in "The Physics of Hydrogenated Amorphous Silicon II", J.D. Joannopoulos and G. Lucovsky eds., Springer-Verlag, 1984, p.61.

16. V. Vorlicek, M. Zavetova, S.K. Pavlov and L. Pajasova, J. Noncryst. Solids 45, 289 (1981).

17. R.H. Klazes, M.H.L.M. van den Broek, J. Bezemer and S. Radelaar, Phil. Mag. B45, 377 (1982).

18. E.A. Davis and N.F. Mott, Phil. Mag. 22, 903 (1970).

19. J.D. Joannopoulos and M.L. Cohen, Phys. Rev. B7, 2644 (1973).

20. W.B. Jackson, S.M. Kelso, C.C. Tsai, J.W. Allen and S.-J. Oh, to be published.

21. G.D. Cody, B.G. Brooks and B. Abeles, Sol. En. Mat. 4, 231 (1982); see also Ref.5.

22. L. Ley, in Procs. of MRS Europe Conference, P. Pinard and S. Kalbitzer eds., Editions de Physique, 1984, p.451.

23. T. Tiedje, B. Abeles and J.M. Cebulka, Sol. St. Commun. 47, 493 (1983).

24. See, for instance, D.J. Dunstan, J. Phys. C 30, L419 (1982).

25. See, for instance, S. Yamasaki, N. Hata, T. Hishida, H. Oheda, A. Matsuda, H. Okushi and K. Tanaka, J. de Physique Suppl. 10, 297 (1981); G.P. Ceasar, K. Okumura, J. Lin, M.A. MacChonkin and L. Dudek, J. Noncryst. Solids, 59-60, 289 (1983).

26. C. Moddel, D.A. Anderson and W. Paul, Phys. Rev. B22, 1918 (1980).

27. M. Vanecek, J. Kocka, J. Stuchlik and A. Triska, Sol. St. Commun. 39, 1199 (1981); F. Evangelisti, P. Fiorini, G. Fortunato, A. Frova, C. Giovannella and R. Peruzzi, J. Noncryst. Solids 55, 191 (1983).

28. C.B. Roxlo, B. Abeles, C.R. Wronski, G.D. Cody and T. Tiedje, Sol. St. Commun. 47, 985 (1983).

29. A. Skumanich, A. Frova and N.M. Amer, Sol. St. Commun. (1985).

30. G.D. Cody, T. Tiedje, B. Abeles, B. Brooks and Y. Goldstein, Phys. Rev. Lett. 47, 1480 (1981).

31. J.P. Ferraton, A. Donnadieu, J.M. Berger, F. De Challe, S.P. Coulibaly and C. Ance, J. Noncryst. Solids 59-60, 313 (1983).

32. T. Hama, H.Okamoto, Y. Hamakawa and T. Matsubara, to be published; see also E.N. Economou and D.A. Papacostantopoulos, Phys. Rev. B 23, 2042 (1981), and J.A. Verges, Phys. Rev. Lett. 53, 2270 (1984).

33. See, for instance, I. Sakata, Y. Hayashi, M. Yamanaka and H. Karasawa, J. Appl. Phys. 52, 4334 (1981).

34. We are indebted to C.Coluzza and Petra Rudolf for the data in DIBS material, P. Fiorini for GD, M. Musci for LD, G. Fortunato and A. D'Amico for RF.

35. The SiGe data are due to C.Giovannella, the SiC data to A.Gregori.

36. R.S. Sussman and R. Ogden, Phil. Mag. 44, 137 (1981).

37. D.A. Anderson and W.E. Spear, Phil. Mag. 35, 1 (1977).

38. Y. Katayama, K. Usami and T. Shimada, Phil. Mag. B 43, 283 (1981).

39. C. Coluzza, D. Della Sala, G. Fortunato, S. Scaglione and A. Frova, J. Noncryst. Solids 59-60, 723 (1983); S. Scaglione, C. Coluzza, D. Della Sala, L. Mariucci and A. Frova, Thin Solid Films 120, 215 (1984).

40. W.B.Jackson, S.J.Oh, C.C.Tsai, and J.W.Allen, Phys. Rev. Lett. 53, 1481 (1984).

41. S. Abe and Y. Toyozawa, J. Phys. Soc. Japan 50, 2185 (1981).

42. C.M. Soukoulis, M.H. Cohen and E.N. Economon, Phys. Rev. Lett. 53, 616 (1984).

43. B.I. Halperin and M. Lax, Phys. Rev. 148, 722 (1966).

44. See, for instance, E.O. Kane, Phys. Rev. 131, 79 (1963).

45. P. Lloyd and P.R. Best, J. Phys. C 8, 3752 (1975).

46. J. Ziman, "Models of Disorder", Cambridge University Press, 1979.

47. D.J. Thouless, Phys. Rev. Lett. 39, 1167 (1977).

48. D. Vollhardt and P. Wolfe, Phys. Rev. Lett. 48, 699 (1982).

49. The expressions for E_0 and λ_E^3 given here differ by numerical factors from those reported in Ref.42. More precisely, our E_0 is larger by a factor (4/3) and λ_E^3 is larger by a factor 3. Our values are obtained by straight application of the Lloyd and Best maximization procedure[45] to Eq.(8), with $\Delta E_\lambda = 2X\xi_0/\pi\lambda^3$.

DENSITY OF STATES DISTRIBUTION AND

TRANSPORT PROPERTIES OF a-Ge:H

Harald Overhof

Fachbereich Physik
Universität Paderborn
4790 Paderborn, F.R.G

INTRODUCTION

While the electronic properties of hydrogenated amorphous silicon have been widely studied both theoretically and experimentally in the last years the electronic properties of the similar material a-Ge:H have received much less attention.This is of course mainly due to the fact that a-Si:H is considered superior to a-Ge:H for all technical applications.The larger band gap of the former material is also helpful to suppress contributions of ambipolar transport at elevated temperatures,a difficulty that is oftenly observed in a-Ge:H.

It is the aim of the present paper to show that in contrast the transport properties of amorphous germanium can be understood more easily than those of amorphous silicon.This is mainly due to the much larger density of midgap states in the former material as compared to the latter.While this larger density of states leads to a smaller sensitivity to doping (an effect which usually is considered as unfavourable) it also suppresses the relative sensitivity of the electronic properties to additional deep states created unvoluntarily by doping.

Ideally one would hope that the only action of dopants is the creation of additional donor (acceptor) states.These should be located closely enough to the band edges to ensure that the dopants are always ionized (and hence electrically active).If this where true the dark Fermi level could be moved by doping across the band gap from just below the donor level to just above the acceptor level.It turnes out,however,that this idealized picture oversimplifies the matter considerably in a-Si:H.There is now much evidence that doping by the various techniques also introduces additional states deep in the gap (see e.g. ref.1 and references therein).These states limit the range of control on the position of the dark Fermi level obtainable by doping which is unfavourable from the applications side of view.At the same time the variation of the density of states distribution with doping prohibits the interpretation of the electronic properties by a unique density of states model for all doping levels [2,3].We shall show in this paper that in the case of a Ge:H the overall density of deep midgap states is sufficiently large that the influence of additional states created by the doping process can be ignored to a first approximation.

The paper is organized as follows : In the next Section we shall review experimental data obtained by the Marburg group on dc transport properties and spin densities as a function of doping level and temperature In the third section we discuss the transport model employed in order to derive the Fermi level position with respect to the mobility edges from experimental transport data. Here we also discuss the admittedly unsolved controversial question as to which degree the electron - phonon coupling (which is mainly responsible for the temperature dependence of the optical band gap) alters the various transport properties. Combining spin density and transport data as a function of doping level and of temperature we derive a density of states distribution model for hydrogenated amorphous germanium. The last Section of this paper will be devoted to a comparison of the experimenetal data with the results obtained from this model.

EXPERIMENTAL DATA REVIEW

The experimental data to be reviewed here are taken from the literature and include dc transport[4,5] (conductivity,thermopower,and Hall mobility data) and spin resonance experiments[6,7]. The experimental data where obtained as a function of doping level and temperature on material prepared by the glow discharge process. As is common practice we shall specify the doping level by the volume ratio of the doping gas (PH_3 and B_2H_6, respectively) to germane prior to decomposition. It should be kept in mind that the samples used for the transport measurements where a few microns thick only and deposited on insulating substrates. In contrast the samples investigated in the spin resonance experiments had been deposited as rather thick films and peeled off the molybdenum substrate. We shall in the following assume that both types of films have identical electronic properties. As in the case of a Si:H we find that this assumtion does not lead to a contradiction.

Conductivity and thermopower

The conductivity data obtained by Hauschildt et al[4] on p -type and on n-type samples (similar data on n-type samples have been presented by Jones et al.[5])do not appear to be strictly activated in the temperature range covered. Instead the slope E_σ^* of an Arrhenius plot ($\ln \sigma$ vs 1/T) increases monotonically as one goes to higher temperatures. The same is true if e/k S (with e = $\pm|e|$ for holes and electrons, respectively and k Boltzmann's constant) is plotted versus inverse temperature ("Arrhenius plot of the thermopower"). Strikingly for highly doped samples at lower temperatures the thermopower is virtually constant. The absolute value of e/k S in this case is about 7 and hence too high to be ascribed to the heat of transport term alone. Note that a decrease in slope at elevated temperature commonly found in undoped or lightly doped a-Si:H is absent in the case of amorphous germanium.

The Q - function

As has been discussed at length previously (see e.g. ref.1) the temperature dependence of the dc transport data is predominantly governed by the position of the dark Fermi level with respect to the mobility edge. In particular for hydrogenated material with a relatively low density of deep midgap states the temperature dependence of the Fermi level cannot be ignored. In order to suppress this unknown quantity we discuss

$$Q (T) = \ln(\sigma \Omega \, cm) + e/k \, S (T) \qquad\qquad (3)$$

a quantity that does not depend on the position of the Fermi level at all provided we deal with unipolar conduction and can use nondegenerate statistics.Results for Q as a function of inverse temperature have been published by Hauschildt et.al.[4] and are reproduced in fig.1.In all curves Q appeares to be roughly activated with only small deviations from the simple law

$$Q (T) = Q_o - E_Q/kT \qquad\qquad (2)$$

For p-doped samples (light symbols) the slope of Q decreases slightly towards higher temperatures while for the undoped sample Q has a distinct maximum at higher temperatures.We shall show that these deviations from the simple law eq.2 are due to the onset of ambipolar transport which is detected readily in Q (T) .The transition from unipolar to ambipolar transport is more gradual for p-type samples than for n-type samples as will be shown below.Note that for the data shown in fig.1 the contribution of the minority carriers never exceeds 10 %. The very strong decrease of Q for the most lightly p-doped sample at lower temperatures is indicative of a contribution of hopping conduction far below the mobility edge (which again can be estimated to contribute 1/3 of the total current at most).

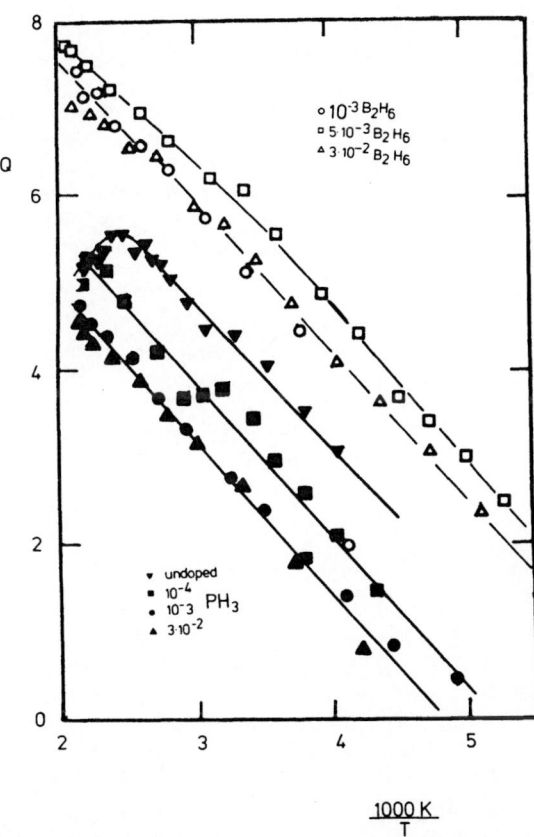

Fig.1.   Experimental data for Q vs inverse temperature by Hauschildt et. al.(ref.4)for a-Ge:H.Data for n-type samples are shown by full symbols while p-type samples are given by with light symbols.

In contrast to the case of amorphous silicon the magnitude of the slope, E_0, is virtually independent of the doping level while the intercept at $1/T = 0$,Q_0, for p-type samples systematically exceeds the corresponding value for n-type samples by about 2. We shall come back to this point in the next Section.

Spin densities

As in the case of a-Si:H there are three different spin resonance signals observed in a-Ge:H.The extensive work of the Marburg group [6,7] has shown that the different spin signals can be discriminated by the g-factor, the linewidth,and the saturation behaviour.Certainly the spin resonance signals are due to three different states in the gap.As in the case of amorphous silicon there is a spin resonance signal from states deep in the gap.These are interpreted as dangling bonds[6] while the other resonances originate from states near the valence and conduction band tails,respectively.We shall adress these states as dangling bonds or tail states following the interpretation of the Marburg group.In the next Section we shall exploit the fact that dangling bond states can be found in three different charge states depending on the Fermi level position:If singly occupied the state denoted by D^0 has a spin and is neutral.The unoccupied state,D^+ ,is positively charged and spinless as is the doubly occupied state,D^- ,which is negatively charged.For a given dangling bond the energy difference between the D^- state and the D^0 state is called the effective correlation energy,U.In comparison to the case of a-Si:H the dangling bond spin density decreases much faster with increasing doping level indicating that for a-Ge:H the density of additional dangling bonds created simultaneously with the doping process is not important.This is in part due to the fact that the spin density in undoped Ge is higher by an order of magnitude.We shall see that for our present analysis it is permissible to assume that the density of states in independent of the doping level.We do of course not claim that this is strictly true.In contrast we show that a single density of states distribution can be used to explain various transport properties and the spin density in doped and undoped samples.

Spin densities can be converted into a density of states distribution if the effective correlation energy U is known.In a-Ge:H the magnitude of U can be estimated from the temperature dependence of the spin density (if U is comparable to kT) or from a comparison of the optical spectra for doped and undoped samples (if U is larger than kT).We shall in this paper use the values estimated by the Marburg group.

TRANSPORT MODEL

The value of E_0

A finite value of E_0 has recently be explained by Fenz et.al.[8] as due to the dynamical electron phonon interaction which gives rise to a "soft" mobility edge.Detailed model calculations show,however,that values beyond 0.05 eV for E_0 can hardly be explained by this microscopic theory alone.Alternately we have proposed (see e.g.ref.1) that the presence of charged dopants or other charged centers must lead to a longranged electro-static potential[9] ("random potential").This longranged potential is caused by the inhomogeneity of a statistical distribution which is considered most homogeneous if random in space.In crystalline solids the presence of free carriers will completely screen this random potential except for the case of heavy doping at nearly complete compensation[10,11].The nonzero density of midgap states in amorphous semiconductors acts like a complete compensation keeping the density of free carriers small even at the highest doping levels.Hence the longranged potential caused by charged dopants will not be screened by free carriers (except at the highest doping levels). Instead screening is achieved by the redistribution of car-riers near the Fermi level caused by the random potential itself. It is evident that for moderately and highly doped a-Si:H films a considerable random potential must be present except for the case that the dopant charge and the charge neutralising the dopant are always closely correlated in space.

For moderately and highly doped a-Si:H we,therefore,are lead to conclude that the random potential is predominantly responsible for the rather high values of E_Q while for undoped material the electron-phonon coupling alone can explain the small values of E_Q.

The relatively high values for E_Q observed in undoped a-Ge:H can be explained quite naturally as due to a random potential.This potential,however,is caused by charged dangling bond states (the D^- states) rather then by charged dopants.If we combine an effective correlation energy $U = 0.1$ eV with a spin density of 10^{18} cm^{-3} we obtain a density of states that must exceed 10^{19} cm^{-3} eV^{-1}.The experiments indicate strongly that the width of the neutral dangling bond peak D^0 is several tenth of an eV wide. Hence for n-type (and for lightly doped p-type) material a considerable fraction of the dangling bond states must be negatively charged.Using the formulae given in ref.11 valid under the assumption that the charges are not clustered we obtain the values Δ =0.1 eV and l_{max} =50 Å.Here Δ is defined as the amplitude of the random potential via the distribution function $p(V)$ of this potential given by

$$p (V) = \exp (-(V/\Delta)^2) \qquad (3)$$

while l_{max} is the maximum length scale of the potential that can not be screened.Comparing with the formula

$$E_Q = 1.5 \, \Delta$$

derived from computer model calculations[12] we see that we can easily obtain values of E_Q that equal those found experimentally.As E_Q varies as the square root of the density of charges it is clear that for undoped and for n-type material the variation of E_Q will be quite small.Since in addition charged donor states are required to shift the Fermi energy towards the conduction band tail we do not understand that the observed variation of E_Q is virtually zero.

For the highly p-doped samples the above arguments can be reversed easily:in this case virtually no dangling bond states D^0 are left (no spins of the dangling bond type are observed) we must have the same random potential from D^+ states as we had from D^- states in highly n-doped samples. Hence the similarity of E_Q in highly p-doped and highly n-doped samples simply reflects the symmetry between electrons and holes.Note that for a-Si:H the situation is quite different: the spin density is smaller,the effective correlation energy is larger and the Fermi level for undoped samples is located in the minimum between the D^0 and the D^- peak .As a consequence the density of D^- states for undoped material is small and hence the variation of E_Q with doping is expected to be substantial.

Temperature dependent band edges

The observed temperature dependence of the optical band gap in non-metallic solids usually is explained by two different effects:the thermal lattice expansion and a contribution from the static electron-phonon coupling,the latter effect being most prominent in semiconductors.Naively one would like to treat the temperature dependence of the band edges in a fist approximation by inserting temperature dependent band edge energies into the transport formulae.Emin[13],however,has argued that this is not a correct procedure even in lowest order.Energies in quantum systems usually do not depend explicitly on temperature.Instead,energies are altered by a coupling to some other quantum system (the phonons in our case) whose occupancy depends on temperature.Hence temperature dependent energies are not single particle energies.Emin shows that for the carrier density in the conduction band one can use the standard formulae if $E_c(T)$ as derived from optical experiments is inserted as band edge energy.This,however,is not the quasiparticle energy of an extra electron in the conduction band. Assuming that the quasiparticle energy has to be used in the expression for the Peltier heat (and via Onsager's symmetry relations in the corresponding expression for the thermopower) one finds that in the formulae for the thermopower $E_c(T=0)$ enters rather than $E_c(T)$.

Unfortunately the transport properties of the coupled electron - phonon system can not been treated rigorously.It is,therefore,a plausible assumption only that the energy transported by a carrier in the conduction band is the quasiparticle energy containing both an electronic and a vibronic contribution.For crystalline semiconductors there seem to be no experiments which could help to decide the question experimentally:in good crystals the thermopower is dominated by the phonon drag contribution and the heat of transport term is not known to a sufficient accuracy.

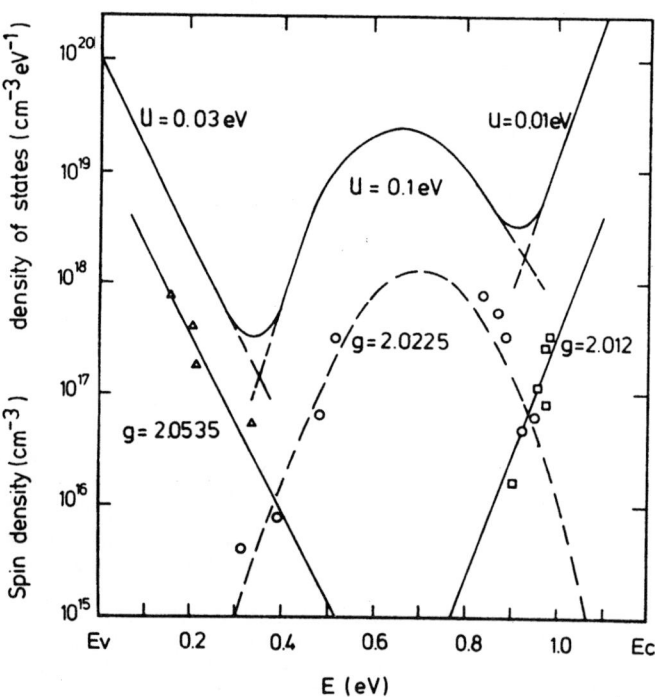

Fig.2 Spin densities of ref.6,7 and density of states distribution for a-Ge:H.The position of E_F is taken from the transport data (ref.4) The spin densities given by full lines are obtained from the density of states distribution for 80 K.

Friedman and Butcher[14] and recently Butcher[15] have questioned Emin's result.In their work ,however,the assumtion is made from the beginning that effect of temperature dependent band edges can be treated adequately by inserting the explicitely temperature dependent band edge energies into the standard transport formulae.This basic assumption,however,is not sub-stantiated and rather unlikely in the light of Emin's results.

We shall,therefore,use Emin's results in order to incorporate the temperature dependence of the band gap.As has been shown in our previous work[12] the quantity relevant for the transport coefficients is the shift of the band edges relative to the states where the Fermi level is located. Taking from optical data that the band gap,E_g is given by

$$E_g (T) = E_g^0 - 4 kT$$

and assuming that the dangling bond states midgap shift towards the va-lence states with about 3 kT we find from the formulae in ref.12 that Q_0

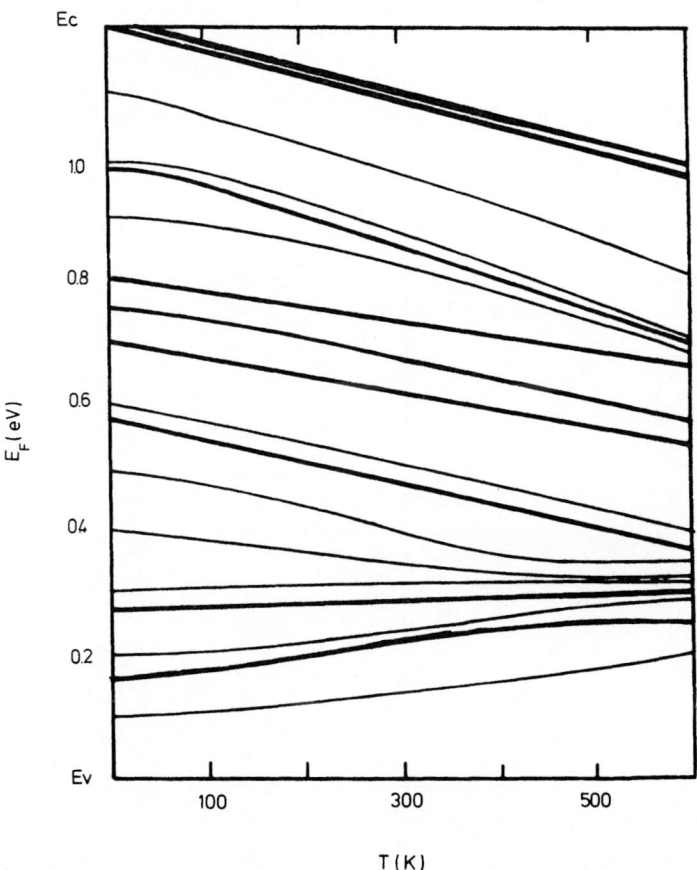

Fig.3 Temperature dependence of the Fermi energy calculated from the density of states distribution of fig.2.All energies are with respect to the valence band mobility edge.

for the p-type samples should be exceed the corresponding value for n-type samples by two if the properties of both mobility edges are identical.

Density of states distribution

We are now in a position to derive a density of states distribution model from the spin densities,assuming that we have no states that never show up in spin resonance (negative U states,e.g.).Combining the spin density with the effective correlation energy we obtain the density of states near the Fermi level.The position of E_F is obtained from the dc conductivity using

$$\ln (\sigma_0 / \sigma (T)) = (E_c (T) - E_F (T)) / kT \qquad (4)$$

where $\sigma_0 = 200 \, \Omega^{-1} \, cm^{-1}$ is used in agreement with the Q_0 data (if as usual the heat of transport term is set equal to unity).The resulting density of states distribution is shown in fig.2 in comparison with the spin densities found experimentally and obtained from the model.

RESULTS AND DISCUSSION

From a given density of states distribution it is relatively simple to calculate the temperature dependence of the Fermi energy if the initial position at T = 0 is known.From the neutrality condition the total number of electrons in the system must be temperature independent.This can be written as an integral over the density of states distribution function,g(E), and the Fermi distribution function $f_F(E , E_F, U, T)$ for states with a finite correlation energy as

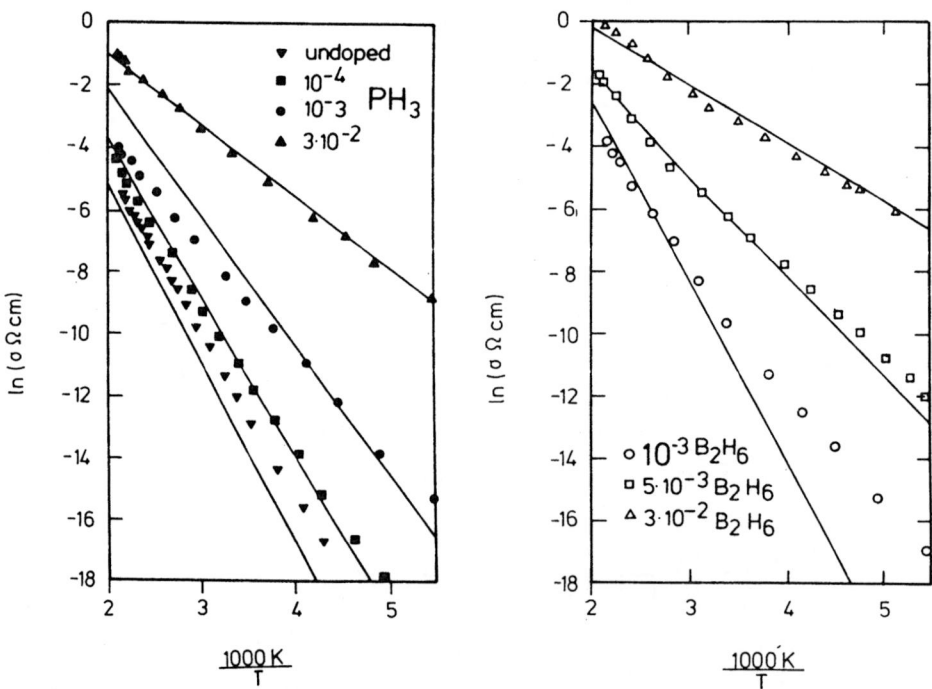

Fig. 4   Temperature dependence of the conductivity for n-type (left, full symbols) and p-type (right, light symbols) samples.

294

$$n = \int_{-\infty}^{\infty} g (E) f_F (E , E_F, U, T) dE \qquad (5)$$

which implicitely defines the temperature dependence of the Fermi energy. Results for a variety of different initial positions of E_F (at T = 0) are shown in fig.3.Here all energies are plotted with respect to the mobility edge of the valence band.The temperature shift of the conduction band is indicated by the double line.Full lines indicate the curves of E_F (T) which are used to interpret the conductivity and thermopover data by Hauschildt et.al [4].The asymmetry of the shift of the dangling bond states with respect to the band edges is clearly reflected in the E_F (T) curves:While for E_F (0) in the upper half of the gap the Fermi level at elevated temperatures can relatively freely move towards E_V,the opposite is not true.We expect that ambipolar transport will set in more rapidly with rising temperature for n-type samples than for p-type specimens.

Inserting the results for $E_F(T)$ from fig. 3 into our transport model it is easy to calculate the dc transport properties, taking Δ = 0.1 eV for all doping levels and using σ_0 = 200 $\Omega^{-1}cm^{-1}$ for both types of carriers. Fig. 4 shows a comparison of the results thus obtained with experimental data for the dc conductivity (a similar comparison can be made with the data of Jones et.al. (ref. 5). Fig. 5 shows a comparison of experimental

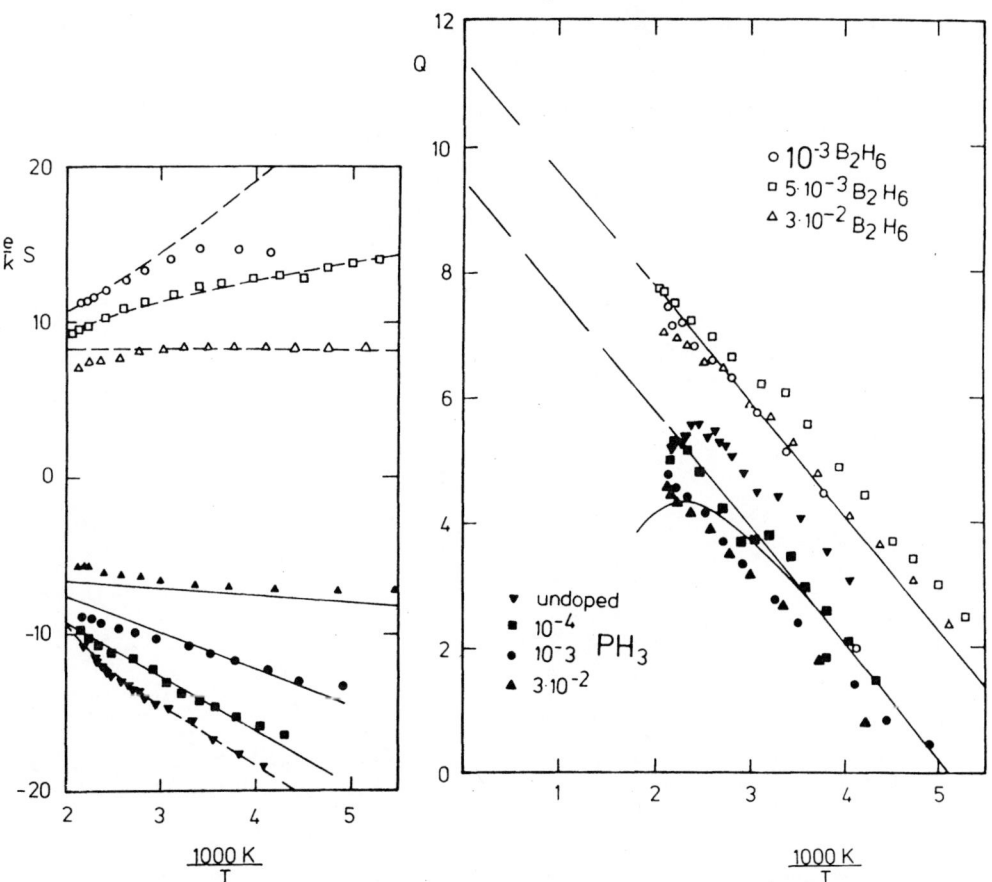

Fig. 5 Thermoelectric power and   Fig. 6 Q (T) for p-type (light
        symbols) and n-type (full symbols) samples.

data for the thermopower. The agreement of the experimental data with those obtained by our model using a single density of states distribution is excellent (note that a similar agreement for a-Si:H requires that the density of states model is substantially altered upon doping (ref. 2)).In particular the conspicuous constant value of the thermopower for the highly doped samples (both n-type and p-type) is reproduced, in fact even exaggerated, by the present calculation. For simplicity our model does not include any contributions from hopping. It, therefore, does not reproduce the bending over of e/k S at lower temperatures for the most lightly p-doped sample.

The Q-function calculated from our model is compared with that obtained from experimental data in fig. 6. Note the difference in Q_0 between p-type and n-type samples is produced in our model by the asymetry of the shift of the dangling bond peak with respect to the band edges. This difference which relies on the validity of Emin's results (see above) must not be taken as direct evidence for the validity of Emin's theory: agreement between experiment and theory is evidence for the validity of the theory if it can be shown that agreement cannot be obtained otherwise. In our case we cannot exclude that e.g. the heat of transport term for holes exceeds that for electrons by two - although we have no reasons to assume that this should be the case.

From the close agreement between thre results of the present transport model and the experimental data we may conclude that the density of states model distribution used is not unrealistic.It is then possible to calculate the gross doping efficiency (see,e.g.,ref.1)

$$\eta_D = \frac{\text{change in occupation of deep midgap states}}{\text{doping gas concentration}}$$

as a function of the doping level displayed in fig.7.As in the case of

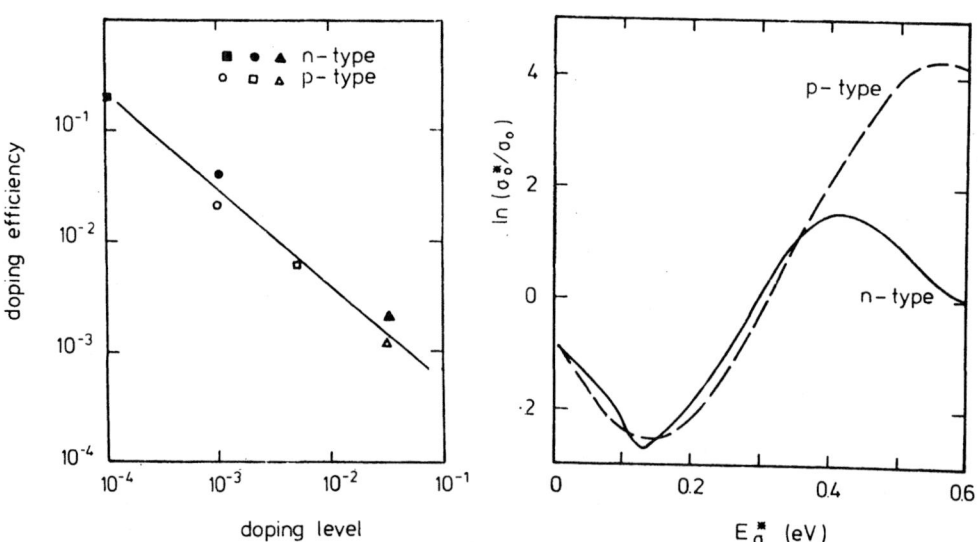

Fig. 7 Gross doping efficiency as a function of doping level

Fig. 8 Meyer-Neldel rule for n-type (full line) and p-type (broken line) a-Ge:H compared with experimental data.

296

a-Si:H η_D is not independent of the doping level (see ref. 1 for a discussion of η_D) but varies like $n_D^{-\epsilon}$ with ϵ between 1/2 and 1.

As in the case of a-Si:H the shift of the Fermi level with temperature must lead to a strong variation of the observed prefactor of the conductivity, σ_0^* with observed activation energy, E_{σ} , generally known as Meyer-Neldel-rule (ref. 16). We show the data of our model in fig. 8 evaluating the shift of E_F in the $300\,K \leq T \leq 400\,K$ temperature range (note from the Arrhenius plot of the conductivity that the activation energies depend on the temperature range). The experimental data roughly follow the pattern indicated by the results of our model as shown in fig.8.

In summary we have shown that the spin density data and the transport data of a Ge:H can be combined to give a density of states distribution that can in detail account for temperature and doping dependence of conductivity, thermopower, and spin densities. For the future it would be interesting to study in more experimental detail the transport properties of very lightly doped samples. These have been neglected in the past because the onset of ambipolar transport makes the evaluation of the thermopower quite difficult. At the present level of agreement between a theoretical model and the experimental data these experiments could help to establish that the D^0 peak shifts towards E_v in an asymmetric way.

A further interesting experiment could combine dc transport data with time-of-flight (TOF) experiments.If as shown in the present paper the longranged potential is significant we must expect that the TOF current traces are highly dispersive even above room temperature (see ref.17).

In many respects the electronic properties of a-Ge:H are similar to those of a-Si:H but in a complementary way.In a-Si:H the random potential is caused by charged dopants while in a-Ge:H it is produced by charged dangling bonds.While in the former material the amplitude of the potential depends strongly on the doping level,the opposite is true for the latter material.In a-Si:H the low density of deep states in undoped material is significantly increased upon doping the high density of dangling bond states in a-Ge:H can be considered as doping independent.

Acknowledgements

The author is indebted to Prof. Dr. J. Stuke for suggesting the problem and helpful discussions. He also expresses his gratitude to Dr. W. Beyer for a critical reading of the manuscript.

REFERENCES

1.   W. Beyer and H. Overhof,Doping Effects in a-Si:H, in:" Semiconductors and Semimetals ",vol. 21c,R.K. Willardson and A.C. Beer, eds., Academic Press, New York,1984, p. 258.
2.   H. Overhof and W. Beyer, Journ. Non-Cryst. Solids 35 & 36, 375, 1980.
3.   R. Tanaka and S. Yamasahi, Solar Energy Mat. 8, 277, 1982.
4.   D. Hausschildt,M.Stutzmann,J.Stuke,and H.Dersch,Solar Energy Mat 8, 319.1982
5.   D.I. Jones,W.E.Spear,P.G.LeComber,S.Li,and R.Martins,Phil.Mag.B39, 147,1979
6.   M. Stutzmann,J.Stuke,and H.Dersch,phys.stat.sol(b),115,141,1982
7.   M. Stutzmann and J.Stuke,solid State Comm.,47,635,1983
8.   P. Fenz, H. Müller,H. Overhof,and P.Thomas, Journ. Phys. C (in press).
9.   H. Overhof and W. Beyer, Phil. Mag. B43, 433, 1981.
10. J. Jäckle, Phil. Mag. B41,681,1980

11. Shklovskii,and A.L.Efros,Soviet Physics JEPT $\underline{33}$,468,1071
                              Soviet Physics JEPT $\underline{35}$,610,1972
12. H. Overhof and W. Beyer, Phil. Mag. $\underline{B47}$, 377, 1983.
13. D. Emin, Solid State Comm. $\underline{22}$, 409, 1977.
14. P.N. Butcher and L. Friedman, Journ. Phys. C $\underline{10}$, 3801, 1977.
15. P.N. Butcher, Phil. Mag. B50,L5,1984
16. W. Meyer and H. Neldel, Z. Techn. Phys. $\underline{120}$, 588, 1937.
17. H.Overhof,Journ.Non-Cryst.Solids $\underline{66}$,261,1984

IS THE DLTS DENSITY OF STATES FOR AMORPHOUS SILICON CORRECT?

J. D. Cohen
University of Oregon
Eugene, OR 97403

D. V. Lang
AT&T Bell Laboratories
Murray Hill, NJ 07974

ABSTRACT

A large number of different measurements which are sensitive to gap states in hydrogenated amorphous silicon have been analyzed over the past few years in an attempt to determine the generic form of the density of states function within the mobility gap of this material. These results tend to support one or the other of two quite different functions: a) the field-effect density of states, or b) the DLTS density of states. In this paper we review the evidence for the density of states function determined for a-Si:H by deep-level transient spectroscopy (DLTS). We argue that the strong and growing support for the DLTS picture is evidence that the two general forms for the density of states relate to different spatial regions of the specimen, e.g. interface vs. bulk properties.

INTRODUCTION

Understanding the electronic properties of hydrogenated amorphous silicon (a-Si:H) depends to a great degree upon having detailed knowledge of the energy distribution of localized states, $g(E)$, within the mobility gap. Because of this fact, and because so many measured properties are affected by aspects of the localized gap state distribution, a great many techniques have been applied to determine $g(E)$ for this material. Unfortunately, many researchers continue to disagree in a substantial way about the overall shape of $g(E)$ for a-Si:H. This disagreement centers on whether $g(E)$ shows a broad minimum in the center of the gap, or rather exhibits a peak near the center of the gap (due to the dangling bond state) and a substantial minimum about 0.5 eV below the conduction band mobility edge, E_c. Proponents of the first picture include most of the workers who have used the field effect technique,[1-3] researcher who performed some early low frequency capacitance voltage (C-V) measurements,[4,5] and recent proponents of the space charge limited currents (SCLC) technique.[6,7] The second picture was initially proposed by ourselves and coworkers on the basis of deep-level transient spectroscopy (DLTS) measurements,[8,9] and is now also shared by most of the groups studying electron spin resonance (ESR)[10,11] in this material as well as the workers employing photo-thermal deflection spectroscopy.[12]

In this paper we will review the evidence for this second viewpoint as obtained by a wide range of junction capacitance measurements utilizing Schottky barrier (or p^+n) junctions on doped and undoped amorphous silicon samples. We now believe that *all* such measurements can be shown to support the picture initially obtained from the DLTS studies. The methods to be discussed will include capacitance measurements as a function of temperature and frequency (C-T-ω), thermally stimulated capacitance (TSCAP), DLTS, junction photocapacitance studies, as well as a recent "drive-level capacitance profiling" technique. Since the methods mentioned include both dynamic transient as well as steady state capacitance measurements, we would argue that the evidence supporting this picture of g(E) for a-Si:H is very strong indeed. Thus while additional arguments may still be found to support the alternative picture of g(E) mentioned above, good reasons will then also have to be found to explain why all capacitance measurements yield a different conclusion. The more likely resolution of the controversy, we believe, lies in the fact that all of the supporting evidence for the first picture comes from measurements that have been demonstrated to be extremely sensitive to material properties at, or very near, interfaces on those samples. In contrast, the capacitance techniques (as well as the ESR and optical absorption measurements) are really showing us the bulk material properties.

CAPACITANCE STUDIES

Junction capacitance measurements are employed routinely to characterize and study crystalline semiconductors. If the space charge is dominated by the shallow donor or acceptor concentrations, the capacitance is largely independent of applied frequency or temperature providing that the dielectric relaxation time is short enough to allow charge to move into and out of the edge of the depletion region on the time scale of the measurement frequency. In contrast, the contribution to the capacitance due to deep levels varies markedly with temperature and frequency depending on the degree of participation of the deep levels to the total time varying charge distribution. While such contributions are small in most crystals, they can be very large in amorphous semiconductors where the concentration of deep levels dominates many of the material properties.

To understand this latter situation in detail, we need to examine how the total barrier charge is changed by a small change in the applied voltage, δV_A. Applying the identity

$$\frac{d}{dx}\left(x\,\frac{d\Psi}{dx}\right) = \frac{d\Psi}{dx} + x\,\frac{\rho(x)}{\epsilon}$$

to a Schottky barrier diode and integrating from $x = 0$ (at the semiconductor interface) to infinity (or far into the bulk where the band bending function Ψ and $d\Psi/dx = 0$) we obtain:

$$\frac{\delta Q}{\delta V_A} = C = \frac{\epsilon A}{\langle x \rangle} \tag{1}$$

where

$$Q = A \int_0^\infty \rho(x)\,dx \tag{2}$$

is the total barrier charge, A is the junction area, ϵ is the dielectric constant, and where $\langle x \rangle$ is the first moment of the time varying charge distribution, $\delta\rho$:

300

$$<x> = \frac{\int_0^\infty x\delta\rho(x)dx}{\int_0^\infty \delta\rho(x)dx} \qquad (3)$$

The above equation is quite generally valid. The difference between a low frequency of dc capacitance measurement and a high frequency ac measurement at frequency ω lies entirely in the difference in the fraction of the space charge region over which the charge density is able to change on the time scale $1/\omega$. This, in turn, is determined by limitations imposed by emission and capture times to the Fermi energy, E_F, which lies at increasingly deeper energies in the gap as one moves into the depletion region from the bulk semiconductor.

Figure 1 illustrates a low frequency case as well as a high frequency situation both for a crystalline semiconductor with a single discrete deep level and also for an amorphous semiconductor with a continuous large distribution of states. For the crystal, the charge density is essentially a constant equal to the net donor concentration so that $\delta\rho \neq 0$ only at the edge of the depletion region located at $x = W$. Thus $<x> = W$ and $C = \epsilon A/W$. A small increase in C will occur for low frequencies (or high temperatures) due to the deep level which crosses the Fermi energy closer to the junction interface because it will then contribute to $\delta\rho$ and shift $<x>$ slightly toward the interface (see Fig. 1(a)).

For the amorphous semiconductor at low frequency (Fig. 1(b)) $\delta\rho$ is nonzero throughout the depletion region and gives rise to a smaller $<x>$ and a larger capacitance for the same spatial extent of the space region as in Fig. 1(a). For the ac measurement the emission time limitation at the Fermi energy produces a "cutoff" to $\delta\rho$ which increases $<x>$ and decreases the capacitance.

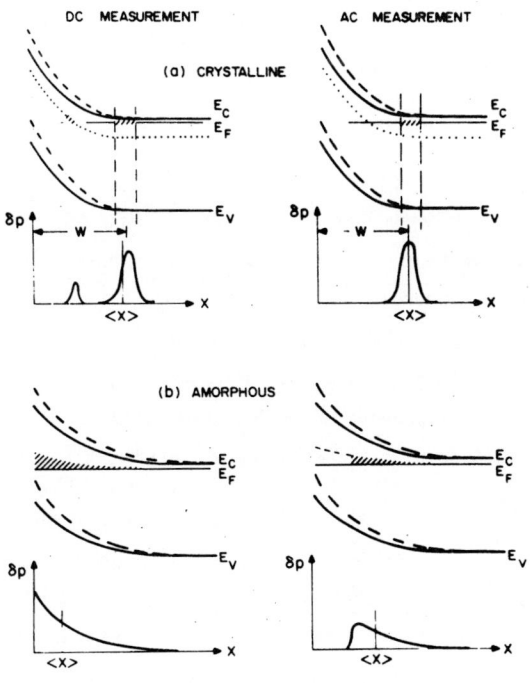

Fig. 1. Schematic diagram showing the qualitative differences between a dc and an ac capacitance measurement for (a) a crystalline semiconductor with a single discrete deep level and (b) an amorphous semiconductor with a continuous distribution of gap states. As discussed in the text, the measured capacitance in each case is inversely proportional to the first moment, $<x>$, of the alternating charge distribution.

The proper interpretation of capacitance is thus intimately connected to the relationship between the measurement time $1/\omega$ and the thermal emission process from deep gap states. The details of how the capacitance will vary with voltage, temperature, or frequency will depend on the energy distribution of localized states within the gap. Such measurements give information which may be used to deduce $g(E)$ in amorphous silicon samples. While the early measurements relied on studying the variations of junction capacitance versus applied voltage at fixed frequency (usually dc),[4,5] subsequent studies[13-15] looked at the dependence on frequency and temperature as well. This allowed both the spatial variation and the energy distribution of gap states to be examined independently.

Let us now consider the situation for moderate frequencies in more detail. In effect, the measurement time scale imposes an energy cutoff E_e such that states at E_F deeper than E_e from the closest mobility edge, E_c, cannot respond to the applied ac voltage. This energy cutoff is given by

$$E_e = k_B T \ln(\nu/\omega) \tag{4}$$

where ν is the exponential prefactor for majority carrier emission and capture. To determine the capacitance we thus determine the total change in barrier charge for a small change of voltage across the barrier interface subject to the condition that the occupation of states deeper than E_e cannot change. It has been shown,[16-18] for a spatially uniform sample, that this may be expressed in terms of the properties of the space charge region at the point where the Fermi level lies at the energy E_e below E_c; namely

$$C = \frac{\epsilon A}{x_e + (\epsilon F_e/\rho_e)} \tag{5}$$

Here x_e denotes the distance from the barrier interface at which E_F lies at $E_c - E_e$. The quantities F_e and ρ_e refer to the values of the electric field and charge density at the point x_e, respectively. The value ρ_e is given simply by the integral over the density of states:

$$\rho_e = \rho(E_e) = q \int_{E_F - E_e}^{E_F^o} g(E) dE \tag{6}$$

where E_F^o refers to the position of the Fermi level in the (neutral) bulk material and q is the electronic charge.

We expect Eq. (5) to be valid until we reach sufficiently high ω that the thermal emission time in the tail region of the depletion layer becomes comparable to dielectric relaxation times in the bulk. At this point the details of carrier transport through the film must be included for a proper understanding of the complex admittance of the device. This can be quite a difficult situation to analyze in detail although several attempts have been made.[19,20] However, one easily accessible piece of information that is obtained from observing the conditions corresponding to dielectric relaxation "freeze-out" is the activation energy of the bulk conductivity of the a-Si:H sample film.

With the above discussion in mind let us consider the experimental data of Fig. 2(a) in which capacitance and conductance vs. temperature curves are recorded at a series of measurement frequencies.[21] These data are for an n-type a-Si:H sample (100 vppm PH_3) which has been specially prepared to have a physically well-delineated area. The behavior exhibited by this sample is supris-

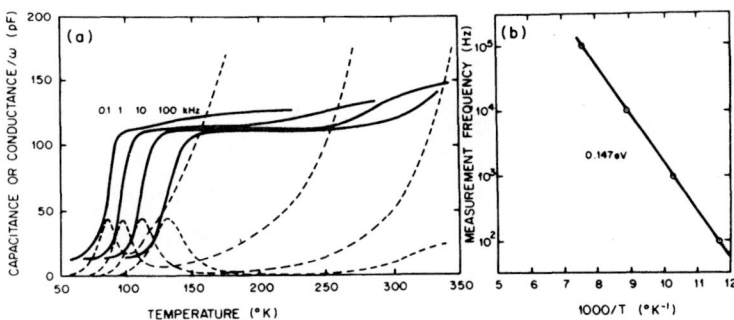

Fig. 2. (a) Measured capacitance and conductance at four ac frequencies vs. temperatures taken under 4 V reverse bias for an n-type a-Si:H sample in a cleaved chip configuration. (b) Semilog Arrhenius plot of the measurement frequency vs. T_0, the position of the conductance peak at each frequency. As explained in the text, the slope of the fitted lines gives the activation energy of the conductivity of the sample.

ingly similar to that for crystalline materials insofar as they indicate a fairly constant value of capacitance above the temperature for dielectric freeze-out. This indicates that g(E) has a significant shallow state concentration of roughly $1 \times 10^{17} \mathrm{cm}^{-3}$ but comparably few deep states over the energy range 0.4-0.6 eV below E_c (corresponding to the temperatures over which the capacitance is roughly constant). At high temperatures the capacitance begins to increase possibly indicating a more significant contribution from states deeper than about 0.8 eV.

To obtain the activation energy of the bulk conductivity from the dielectric freeze-out of these C-T-ω measurements one identifies the temperature, T_0, at which the capacitance abruptly increases from its low temperature value. This defines the temperature at which the RC time constant is comparable to ω^{-1}. Relating this resistance to the bulk resistivity, ρ, we have

$$\rho(T_o) \;-\; A/\omega Cd \qquad\qquad (7)$$

where d is on the order of the film thickness. Plotting the logarithm of each frequency vs. the corresponding values of $1/T_0$ gives the activation energy as shown in Fig. 2(b). This method for determining the conductivity is greatly immune to problems associated with the quality of ohmic contacts and contributions from surface conduction. It has been used to good advantage in several recent studies.[9,21]

In studies of this kind on n-type doped a-Si:H films, then, one obtains primarily the information that there are substantial numbers of shallow states (with $E_C - E_F < 0.4$ eV), few intermediate energy states (0.4 - 0.6 eV) and indications of a more significant number of deeper states (deeper than about 0.8 eV). Similar kinds of studies on undoped films[9,22] usually disclose only the density of states near E_F or the total numbers of states between E_F and midgap. Although early more detailed pictures of g(E) were presented on the basis of low frequency C-V measurements,[4,5] such results are now considered ambiguous due to the influence of interface states. More detailed pictures of g(E) are obtained using the more sophisticated methods described below. We will observe, however, that the general attributes of the above capacitance results are indeed consistent with these other measurements.

THERMALLY STIMULATED CAPACITANCE (TSCAP)

Capacitance measurements are a steady-state property of a p-n junction or Schottky barrier. As we discussed, the behavior of the capacitance as a function of the frequency and temperature of the measurement is somewhat complicated but can be readily understood in term of the dynamics of the states in the gap. We will now discuss capacitance transient measurements: TSCAP and (in the next section) DLTS. These thermally stimulated transient measurements are relatively easy to interpret since they exhibit the presence of deep gap states directly as simple traps.

The basic physics of TSCAP and DLTS is the same. The sample is first prepared in a meta-stable state with nonequilibrium carriers trapped in gap states within the space-charge layer. The capacitance is then recorded as a function of time as the system returns to equilibrium. The density of states can be determined from such capacitance transients measured at various temperatures. Figure 3 shows an example energy diagram for an n-type Schottky biased to observe a transient due to electron traps in the upper part of the gap. In this case the initial condition is set by maintaining the sample at zero bias long enough to fill states below the Fermi level, as shown in the top (zero bias) shaded region in Fig 3. The equilibrium state occupation is shown as the shaded region in the lower (reverse bias) part of the figure. The distribution in space and energy of the metastable trapped electrons which give rise to the capacitance transient is the difference between these two regions, as shown.

We should note here that the so-called isothermal capacitance transient spectroscopy (ICTS) method[27,28] can also be described by Figure 3. As we have pointed out elsewhere[29], this method is essentially the same as DLTS. Both measure the logarithmic derivative of the transient for various times and temperatures. The only difference between the two methods is the way in which the data is plotted, i.e. vs. log time in ICTS and vs. temperature in DLTS. This is an important point, since it has been claimed[27,28] that ICTS gives fundamentally different results from DLTS, namely, a peak in the density of states at 0.52 eV below the conduction band. As we shall discuss, this difference is due to measurements in two different time regimes of the transient, not due to different analysis methods.

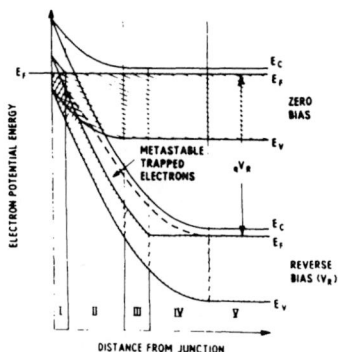

Fig. 3. Potential-energy (band-bending) diagram of an idealized barrier at both zero bias and an applied reverse bias. Regions of gap states which are filled under steady-state conditions are indicated for both biases by diagonal shading.

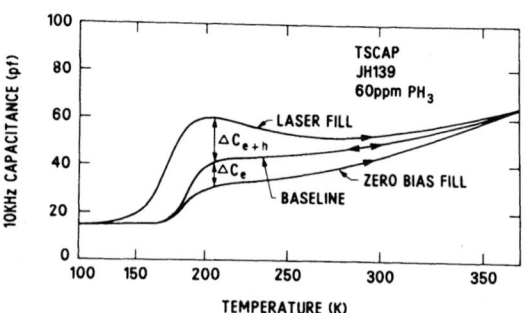

Fig. 4. Typical set of thermally stimulated capacitance (TSCAP) temperature scans for three different initial conditions.

304

Both TSCAP and DLTS give the well-known minimum in g(E) in the upper half of the gap. Even though the two techniques are superficially very similar, there are sufficient differences so that this independent verification of the same g(E) is significant. The major difference between the two techniques is in the time regime. TSCAP is very slow; it is sensitive to thermal emission times on the order of several seconds, whereas DLTS is typically performed in the msec regime. Another difference is that TSCAP is a "single-shot" procedure which depends strongly on the rate of the temperature scan, while DLTS involves repetitive transients observed as a function of temperature (independent of the scan rate). Finally, the DLTS signal in the ideal case is directly proportional to the density of states; the TSCAP signal is ideally the integral of DLTS. For this reason DLTS is usually used to obtain the shape of the density of state while the integral property of TSCAP is useful for setting the overall concentration scale.

A typical set of TSCAP data[9] is shown in Figure 4. The essence of TSCAP is that the traps are filled once at the beginning of a thermal scan and the measurement consists of observing the return of the capacitance to equilibrium as the temperature is scanned from some low temperature (typically <77K) to some high temperature (typically >350K). In Fig. 4 the middle curve (labeled "baseline") is the steady-state capacitance at this reverse bias. As shown in Figure 2, such a C(T) curve is frequency dependent. The capacitance measurement frequency for TSCAP (as well as for DLTS) is chosen to be high enough so the the states giving rise to the transient cannot follow the ac measurement voltage, i.e. capacitance transient measurement are carried out in the high-frequency limit, typically 10 kHz. The C(T) curve labeled "zero bias fill" in Fig. 4 corresponds to the conditions outlined above for Fig. 3.

At any given temperature the difference between the lower two curves in Fig. 4 is directly related to the integral of gap states between roughly mid-gap and an energy E below the conduction band, given by Eqn. 4 with $1/\omega \sim 10$ s. At the lowest temperature in Fig. 4 where this capacitance difference can reliably be measured (210K) this corresponds to 0.58eV. The fact that the two curves are parallel in this temperature range indicates a very low density of states at this energy. The fact that the curves converge as the temperature increases indicates that g(E) is increasing as E increases toward mid-gap, in agreement with the more exact determination from DLTS.

The upper capacitance curve in Fig. 4 was obtained by using above-gap laser illumination to saturate the density of states with a steady-state occupation of trapped electrons and holes. The difference between this curve and the "baseline" curve as a function of temperature is related to the density of states in the lower part of the gap. This aspect of g(E) is not particularly controversial, however, and will not be discussed at length. A detailed discussion of the procedure for obtaining g(E) has been given in refs. 9 and 17.

We have not attempted to extract the detailed shape of g(E) from TSCAP because of uncertainties resulting from the very long time constants involved in the thermal emission of the trapped carriers. In such a long time regime one must be concerned with phenomena other than thermal emission influencing the observed transient time constant. Indeed, we have recently shown[29,30] that leakage current can influence the data in this slow emission rate regime. Such leakage effects cause a spurious shift in the apparent shape of g(E) to lower energies. It is therefore all the more significant that the TSCAP data in Fig. 4 show a minimum in g(E) in the upper half of the gap in agreement with DLTS.

An additional, but little known, use for TSCAP measurements is to obtain information about the bandtail states within 0.3 eV of the conduction band[9]. It is by now well known that the states in this region are given by

$$g(E) = 10^{21} \exp{(-E/E_o)} \, cm^{-3}eV^{-1} \qquad (8)$$

with the characteristic energy parameter E_o in the range of 0.02 to 0.03 eV for most samples[31,32]. In ref. 9 we define the quantity N_s as the integral over g(E) from 0.5 eV below the conduction band to the Fermi level. We also show that N_s may be obtained directly from a TSCAP measurement, independent of the shape of g(E). N_s is proportional to the square of the capacitance of the lowest C(T) curve ("zero bias fill") in Fig. 4 at a temperature just above the dielectric freezeout temperature T_o (200K). If we assume a bandtail of the form in Eqn (8), we obtain values of E_o between 25 and 34 meV, in excellent agreement with drift mobility[31] and photoconductivity[32] measurements. More recently we have varied E_F^o and N_s via the Staebler-Wronski effect and directly confirmed Eqn (8) using only capacitance measurements[33]. Since N_s is an integral over the occupied states between 0.5 eV and the Fermi level (roughly 0.2 eV), it is very sensitive to states in the disputed region of the gap. The fact that the capacitance measurements agree so well with other determinations of the conduction band tail, therefore, supports the general picture of a deep minimum in g(E) at roughly 0.5 eV below the conduction band.

DEEP LEVEL TRANSIENT SPECTROSCOPY (DLTS)

The application of DLTS measurements to a-Si:H is straightforward. One uses the same experimental method applicable to crystals with only a few minor variations[9]. The main difficulty lies in the complicated nature of the capacitance in the presence of a large density of deep gap states distributed smoothly in energy, as discussed above. However, we have made a detailed numerical analysis of the problem[17] and have found that in lightly phosphorus-doped n-type a-Si:H one's intuition based on DLTS measurements of simpler systems[34,35] gives a reasonable zeroth order approximation to g(E). For the case we are primarily concerned with here--the 0.5 eV minimum--one does not really need the numerical analysis. Indeed, it is the very existence of this minimum which makes the analysis of DLTS in n-type a-Si:H so close to that expected for more normal semiconductors where the deep and shallow states are well separated in energy. As we discuss in ref. 17, the densities of states proposed in refs 1-7 would give rise to radically different DLTS signals than those which we have observed. The differences in g(E) which are currently disputed are not subtle effects which could be masked by the analytical machinery. In fact, as we describe below, the existence of a minimum at 0.5 eV can easily be seen in the raw data and is therefore one of the features of the density of states which is easiest to defend.

A typical set of DLTS spectra for hole- and electron-emission is shown in Figure 5 taken from ref. 9. Note especially the voltage pulse signal (which corresponds to the transient condition shown

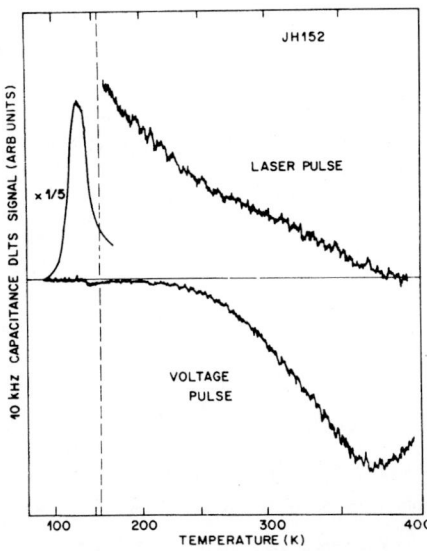

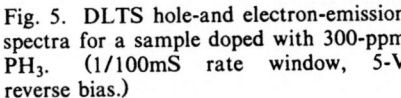

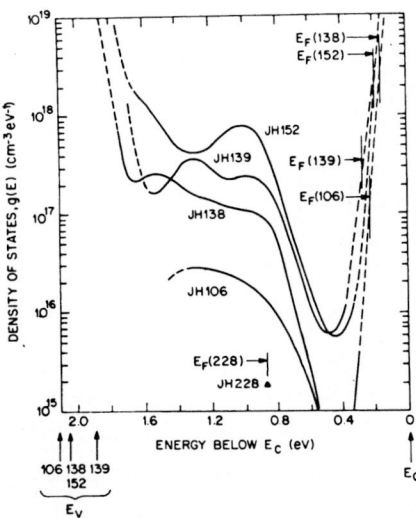

Fig. 5. DLTS hole-and electron-emission spectra for a sample doped with 300-ppm PH₃. (1/100mS rate window, 5-V reverse bias.)

Fig. 6. Summary of the DLTS density-of-states results for five different a-Si:H films. (ref. 9).

in Fig. 3). This negative signal is related quite closely to the actual density of states if the temperature scale is converted to an energy scale (measured down from the conduction band) using Eqn (4) with ω equal to the DLTS rate window. In Fig. 5 the minimum voltage pulse signal occurs at roughly 180K. From Eqn (4) with $\nu = 10^{13}\text{sec}^{-1}$ and the rate window of 10sec^{-1}, we have E = 0.43eV. One can see in Fig. 5 that the DLTS signal (and hence g(E)) is beginning to rise rapidly below 150K (i.e. below 0.36 eV). The results of our detailed quantitative analysis in Figure 6 show that this rise in g(E) seen in DLTS below 150K connects exactly with the band tails determined independently from TSCAP (Eqn (8)).

The only point in this discussion which needs to be justified further is the value of ν used in Eqn. (4). This quantity can be determined directly from an Arrhenius plot of DLTS spectra recorded with different rate windows, as shown in Figure 7. In this figure we have plotted both the temperature position of the DLTS voltage pulse peak (e.g. 370K in Fig. 5) and the position of a point a constant fraction up on the leading edge of this peak. Note that the latter is well behaved and has a prefactor $\nu = 10^{13}\text{sec}^{-1}$. The "normal" procedure of plotting the *peak* position, on the other hand, shows a well-defined kink and gives two different activation energies and prefactors. We have seen exactly analogous results in an entirely different system, namely, GaAs/SiN$_x$ interface states[30]. In that case we could show that the effect was due to leakage through the SiN$_x$ layer and that the true thermal emission data relevant to g(E) was at temperatures above the kink in the Arrhenius plot. One can show that in general the data at faster time constants is less subject to leakage effects[30]. Thus the DLTS peak data in Fig. 7 above the kink as well as the leading edge data both give a prefactor of 10^{13}sec^{-1}.

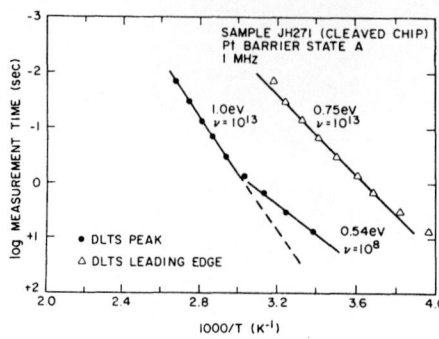

Fig. 7. Arrhenius plot of the apparent peak as well as the leading edge of the DLTS signal showing an anomalous energy and prefactor for peak data at transient time-constants longer than 1 sec.

The data in Figure 7 also explain the apparent discrepancy between DLTS and ICTS results[27,28]. As shown in ref. 29, the ICTS results in question were taken at the longer measurement times which we found to be subject to leakage in other systems and below the kink in Fig. 7. Note, however, that one may not see a kink or even an obvious leakage path and still have problems. The fundamental point is that the apparent peak in the DLTS electron emission signal in Fig. 5 is not necessarily a true peak in g(E) but rather the demarcation line in Fig. 3 between filled and empty states in the reverse-bias equilibrium state. As we discuss in refs. 9 and 7 this can be quite an arbitrary point which depends on the details of not only g(E) but also the device quality of the p-n junction, Schottky barrier, or (more realistically) MIS structure with an inadvertent interfacial layer of unknown composition between the a-Si:H under study and the metal contact or p^+ Si substrate. In the final analysis, perhaps the strongest verification of the DLTS prefactor of $10^{13} sec^{-1}$ in Fig. 7 is the close agreement between the DLTS and photocapacitance results to be discussed in the next section.

PHOTOCAPACITANCE

As an alternative to a measurement of the gap state charge distribution by thermal detrapping, as in TSCAP or DLTS, one may employ optical excitation to determine the density of gap states. Two such studies of this kind have been reported recently.[23,24] In our own studies, photocapacitance measurements were made at 10kHz for two n-type a-Si:H samples (100 vol. ppm PH_3 in SiH_4) at a temperature just above dielectric freeze-out for each sample. At these temperatures the *thermal* detrapping of gap state charge was determined to be negligible. Measurements were made at several values of applied bias using weak sub-band gap light of 1.0 - 2.0 μm obtained by a scanning monochromator and tungsten-halogen source. Before light exposure at each wavelength, the sample bias was reduced to zero and then restored to fill gap states in the depletion region, as shown in Figure 3. Exposing the sample to light then emptied states from E_c down to the corresponding optical energy. Light exposure was continued until no further capacitance change was observed (typically 1-10 minutes). Measurements of the capacitance, C, were taken using two nearby values of applied bias, V_A, so that a value of

$$N_{cv} = -\frac{C^3}{\epsilon q A^2} \left[\frac{dC}{dV_A} \right]^{-1} \qquad (9)$$

could be determined.

308

In carrying out these measurements[23] we determined that the presence of light could shift the quasi-Fermi level in the near depletion region of the barrier which caused the asymptotic value of the capacitance to change with light intensity and also to vary in a somewhat anomalous manner with applied bias. However, the value of N_{cv} defined above was largely insensitive to these problems and gave consistent values to within a factor of 1.5 over a range of applied bias from -4 to -8 volts. To circumvent the same problem another group[24] performed such measurements at very large values of applied bias (-39 volts) so that the anomalous near depletion region contribution could be neglected. This procedure, however, introduced the additional problem of field-induced injection of carriers. In more recent detailed modeling of the effects of quasi-Fermi level shifts on the measured capacitance we have found, however, that the degree of error in determining $g(E)$ from N_{cv} in the vicinity of -7 volts bias is actually slightly less than that introduced by analyzing capacitance directly at -39 volts bias. Representative data for the variation of N_{cv} versus optical energy obtained for one sample in the vicinity of both -4 and -7.5 volts applied bias are displayed in Fig. 8.

The densities of states deduced by the low temperature photocapacitance measurements for both samples are displayed in Fig. 9 along with densities of states determined for these samples by voltage-pulse DLTS. For the two samples there is good qualitative and quantitative agreement considering that sources of uncertainty can easily account for a factor of two in the absolute magnitude of $g(E)$ deduced by the two kinds of measurements. The shift to deeper energies and broadening of the midgap defect band for sample 1 may suggest that lattice-relaxation effects play a significant role in the bound-to-free transitions of this defect. Such an effect, however, is not evident in the data for sample 2. Studies to determine what conditions are responsible for such small energy shifts are still in progress.

The overall agreement between the photocapacitance derived density of states and that obtained by DLTS is quite striking. The same conclusion has been reached by the other group.[24] While photocapacitance has so far only been reported to determine the gap state distribution in the upper half of the mobility gap, these measurements can be extended to learn about states in lower half gap by employing above-bandgap laser pulse excitation as described for some of the thermal detrapping measurements of the previous Section.

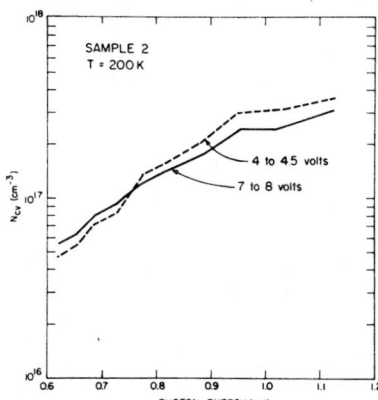

Fig. 8. Photocapacitance N_{cv} as given by Eq. (9) *vs.* photon energy in the vicinity of the two values of applied sample bias.

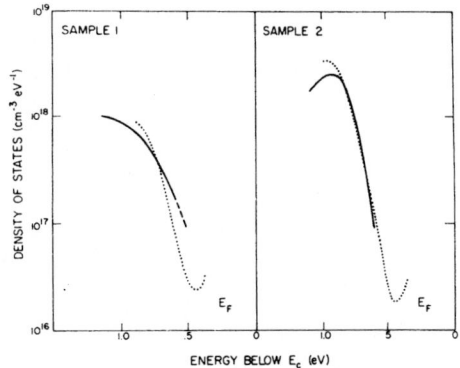

Fig. 9. Densities of states vs. optical energy derived from photocapacitance measurements (solid line) compared to the density of states *vs.* thermal energy derived from DLTS (dotted lines) for the same samples.

DRIVE-LEVEL CAPACITANCE MEASUREMENTS

In this section we will describe one alternative type of capacitance measurement by which the spatial variation of the gap state concentration in amorphous silicon can be studied.[25] In addition, the analysis required to yield gap state densities is extremely straightforward for this method compared with many of the other methods described above. Like DLTS, moreover, this method is also greatly immune to effects of surface and to anomalies near barrier interfaces which have been raised in connection with many of the other kinds of techniques. Finally, unlike DLTS and related techniques which generally require doped samples to yield sufficiently high conductivities, this method is more readily applicable to the study of undoped films and may be used to map out g(E) in such materials over a limited range of the mobility gap.

If we examine Eq. (5) we see that for uniform material the junction capacitance depends on the applied dc bias only through the term x_e. That is, increasing the bias will shift the space charge structure at and beyond x_e rigidly away from the barrier such that the quantities ρ_e and F_e will remain unchanged. By contrast, increasing the temperature (or decreasing ω) at fixed dc bias will increase E_e and thereby decrease x_e while also increasing ρ_e and F_e (See Fig. 10). Moreover, the spatial variation of the material in the region closer than x_e to the barrier does *not* affect the validity of this formula; indeed, sample attributes such as surface or near-interface anomalies lying closer to the barrier than x_e do *not* change either ρ_e or F_e and only modify the junction capacitance by the fact that they help determine the overall position of x_e.

Equation (5) was determined for the case of an infinitesimal oscillating voltage component of the applied bias. However if we instead examine the situation in which the p-p alternating voltage drive level, δV is not zero we obtain higher order corrections to the junction capacitance. Specifically

$$C = C_o + C_1 \, \delta V + C_2 (\delta V)^2 + \ldots \ldots \tag{10}$$

where C_o is given by Eq. 5 and

$$C_1 = -\frac{1}{2} \epsilon A \frac{\rho_e^2}{\left[\epsilon F_e + x_e \rho_e \right]^2} \tag{11}$$

We can utilize this variation with drive level amplitude to obtain the quantity

$$N_{DL} = -\frac{C_o^3}{2q\epsilon A^2 C_1} = \frac{\rho_e}{q} \tag{12}$$

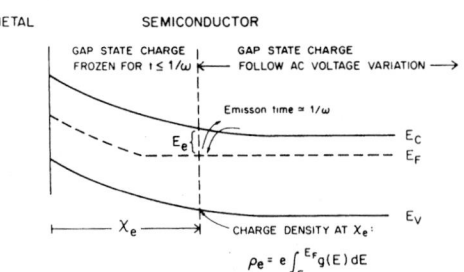

METAL   SEMICONDUCTOR

Fig. 10. Schematic diagram of the diode junction barrier to illustrate parameters defined in conjunction with Eq. (5) and relevant to the analysis of the drive level profiling method.

From the above discussion we deduce: (1) The quantity N_{DL} is independent of x_e and therefore also independent of surface states or near interface anomalies in the film; (2) the charge density N_{DL} is directly and simply related to the density of states in the material, $g(E)$, through Eq. 6; and (3) by choosing a measurement temperature and varying the applied voltage one can spatially profile either the shallow state densities or shallow plus deep state density.

To illustrate the drive-level profiling technique we consider two rf glow discharge a-Si:H samples.[25] Sample 1 is n-type doped (PH_3 at 100 vppm) and Sample 2 is undoped. To test this method for Sample 1, we first determined $g(E)$ from TSCAP and DLTS measurements in the usual manner. This density of states is shown in Fig. 11(b). The solid line portion of the curve is obtained from DLTS; the dotted line portion is inferred from the TSCAP data. In Fig. 11(a) we display the full variation of the 10 kHz drive-level capacitance vs. temperature at one value of applied bias. The 160K - 350K temperature range corresponds to a range of energy for E_e of roughly 0.3 to 0.6 eV.

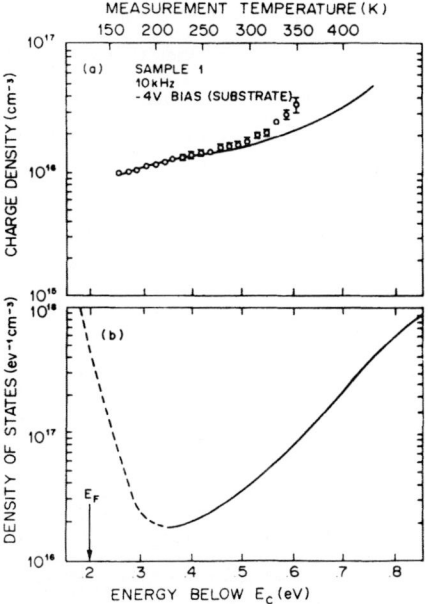

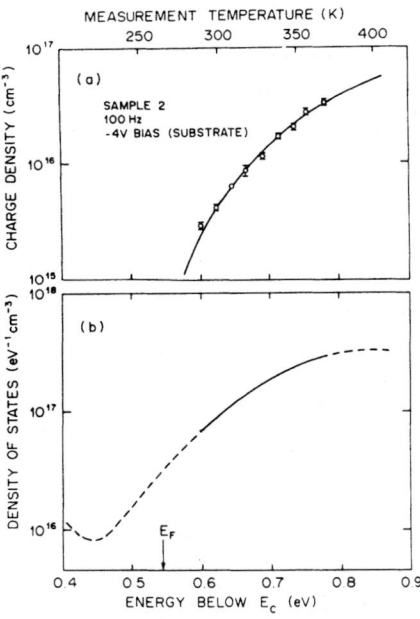

Fig. 11. (a) Variation of N_{DL} for sample 1 at -4 volts bias as a function of the measurement temperatures as given by the top scale with a corresponding value of E_e indicated below. The error bars indicate the range of values for N_{DL} between -3.5 and -4.5 volts applied bias. The solid line is derived by integrating the density of states shown in Fig. 11(b) below. (b) Density of states for sample 1 derived from DLTS measurements (solid line). The dotted lined portion of the curve is inferred from supplemental TSCAP measurements.

Fig. 12. (a) Variation of N_{DL} for sample 2 at -4 volt bias as a function of the measurement temperature. The error bars are obtained as in Fig. 11(a). The solid line is derived by integrating the density of states shown in Fig. 12(b) below. (b) Results of Gaussian-shaped density of states fit to data points of Fig. 12(a). Gaussian parameters are: Peak position at E_c - 0.85 eV; Width parameter of 140 meV; and Magnitude (at peak) of 3.3 x $10^{17}eV^{-1}cm^{-3}$. The minimum at E_c - 0.45 eV occurs if we *assume* an exponential bandtail from the conduction band with a characteristic energy of 35 meV.

311

The solid line in Fig. 11(a) is the integral of the DLTS density of states of Fig. 11(b). Agreement between this curve and the N_{DL} data, as predicted by Eq. 6, is seen to be quite good for energies shallower than E_c - 0.5 eV. This level of agreement may be taken as further evidence for the existence of the distinct minimum in g(E) near E_c - 0.4 eV. The difference between the two curves at deeper energies is likely due to sample area spreading[26] which would affect the two kinds of measurements in such a manner to account for the deviation observed. Alternatively, it might be due to differing values of the exponential prefactor for two such significantly different measurement times (100mS for DLTS versus 15 μS for N_{DL}).

Figure 12(a) shows the results of a 100 Hz drive-level density vs. temperature for Sample 2 (our undoped sample). Here the behavior appears quite different than that for Sample 1. In Fig. 12(b) we show a density of states which fits these data. In our fit we assumed a Gaussian shaped g(E) characterized by just three parameters: its position below E_c, its amplitude, and an energy width parameter. The solid line in Fig. 12(a) is obtained from the fit to g(E) and indicates the degree of agreement with the data. We believe that these results represent the most accurate determination of g(E) for an undoped sample from junction capacitance measurements. One should recognize the overall similarity of this g(E) to that previously determined for n-type doped a-Si:H samples. Several other undoped a-Si:H samples give densities of states quite similar to that shown in Fig. 12(b) but with up to an order of magnitude lower defect density.

Thus, we believe that the drive-level capacitance profiling technique offers many important advantages over traditional profiling techniques for semiconductors in which a substantial fraction of the depletion region charge is due to deep gap states. The technique is also applicable to the determination of the density of states in the upper half of the mobility gap in both doped and undoped samples.

CONCLUSIONS

In this paper we have reviewed the evidence for the DLTS density of localized states in the mobility gap of hydrogenated amorphous silicon. This generic function g(E) has been observed in both doped and undoped a-Si:H and is supported by five widely different types of capacitance measurements, as well as ESR and photo-thermal deflection spectroscopy. It is characterized by a peak near the center of the gap due to the dangling bond state and a substantial minimum about 0.5 eV below the conduction band mobility edge. This is to be contrasted with the so-called field-effect density of states which shows a broad minimum in the center of the gap.

We discussed the following measurements: (1) steady-state capacitance vs. temperature and frequency, (2) TSCAP, (3) DLTS, (4) photocapacitance, and (5) drive-level capacitance profiling. While these are all based on various junction capacitance measurements, they are actually quite distinct techniques. The only common point is Poisson's equation and a pn junction or Schottky barrier. Beyond this the five method differ widely--1 and 5 are steady-state capacitance measurements, 2 and 3 rely on thermally stimulated capacitance transients, and 4 is an optical emission method analogous to optical absorption spectroscopy. It is therefore noteworthy that ALL of these methods independently agree with the original DLTS density of states. It is also noteworthy that the major difference between the DLTS and field-effect densities of states--the deep minimum at 0.5 eV below E_c--is apparent in the various capacitance measurements without the rather complex numerical analysis we have typically used to obtain the most accurate form for the final $g(E)$.

We would argue on the basis of the results presented here that the evidence supporting the DLTS density of states for a-Si:H is very strong. Thus even if additional arguments are found to further support the field-effect picture, one must also explain why the capacitance, ESR, and optical absorption method all agree with the DLTS $g(E)$. We believe that the most likely resolution of this controversy lies in the fact that all of the supporting evidence for the field-effect picture comes from measurements which have been demonstrated to be extremely sensitive to material properties at, or very near, interfaces. The capacitance, ESR, and optical absorption measurements, on the other hand, are widely accepted as valid probes of bulk material properties. In fact, the evidence for the DLTS picture is now so strong that one should accept the persistent support for the two forms of $g(E)$ as prima facie evidence for different bulk and interfacial properties in a-Si:H.

ACKNOWLEDGEMENTS

The parts of this research carried out at the University of Oregon were supported by NSF Grant DMR-8207-437.

REFERENCES

1. W. E. Spear and P. G. LeComber, J. Non-Cryst. Solids *8/10*, 727 (1972).
2. A. Madan, P. G. LeComber, and W. E. Spear, J. Non-Cryst. Solids *20*, 239 (1976).
3. R. L. Weisfield, P. Viktorovitch, D. A. Anderson, and W. Paul, Appl. Phys. Lett. *39*, 263 (1981).
4. G. H. Dohler and M. Hirose, in *Amorphous and Liquid Semiconductors*, ed. by W. E. Spear (CICL, Univ. of Edinburgh, 1977), p. 372.
5. M. Hirose, T. Suzuki, and G. H. Dohler, Appl. Phys. Lett. *34*, 234 (1979).
6. W. den Boer, J. de Physique *42*, C4-451 (1981).
7. K. D. MacKenzie, P. G. LeComber, and W. E. Spear, Philos. Mag. *B46*, 377 (1982).
8. J. D. Cohen, D. V. Lang, and J. P. Harbison, Phys. Rev. Lett. *45*, 197 (1980).
9. D. V. Lang, J. D. Cohen, and J. P. Harbison, Phys. Rev. *B25*, 5285 (1982).
10. H. Dersch, J. Stuke, and J. Beichler, Phys. Status Solidi (b) *105*, 265 (1981).
11. D. K. Biegelsen and N. M. Johnson, in *Optical Effects in Amorphous Semiconductors*, ed. by P. C. Taylor and S. G. Biship, (AIP Conf. Proc., New York, 1984), p. 32.
12. W. B. Jackson, R. J. Nemanick, and N. M. Amer, Phys. Rev. *B27*, 4861 (1983).
13. P. Viktorovitch, D. Jousse, A. Chenevas-Paule, and L. Vieux-Rochas, Rev. Phys. Appl. *14*, 201 (1979).
14. J. Beichler, W. Fuhs, H. Mell, and H. M. Welsch, J. Non-Cryst. Solids *35-36*, 587 (1980).
15. D. Jousse and S. Deleonibus, J. Appl. Phys. *54*, 4001 (1983).
16. A. J. Snell, K. D. MacKenzie, P. G. LeComber, and W. E. Spear, Philos. Mag. *B40*, 1 (1979).
17. J. D. Cohen and D. V. Lang, Phys. Rev. *B25*, 5321 (1982).
18. I. W. Archibald and R. A. Abram, Philos. Mag. *B48*, 111 (1983).
19. P. Viktorovitch and G. Moddel, J. Appl. Phys. *51*, 4847 (1980).
20. T. Tiedje, C. R. Wronski, and J. M. Cebulka, J. Non-Cryst. Solids *35 & 36*, 743 (1980).
21. J. D. Cohen, D. V. Lang, J. P. Harbison, and A. M. Sergent, Solar Cells *9*, 119 (1983).
22. T. Tiedje, T. D. Moustakis, and J. M. Cebulka, Phys. Rev. *B23*, 5634 (1981).

23. A. V. Gelatos, J. D. Cohen, and J. P. Harbison, in *Optical Effects in Amorphous Semiconductors,* ed. by P. C. Taylor and S. G. Bishop, (AIP Conf. Proc., New York, 1984) p. 16.
24. N. M. Johnson and D. K. Biegelsen, in Proc. of the 17th International Conference on the Physics of Semiconductors, (San Francisco, 1984) to be published.
25. C. E. Michelson, A. V. Gelatos, J. D. Cohen, submitted to Applied Physics Letters.
26. D. V. Lang, J. D. Cohen, J. P. Harbison, and A. M. Sergent, Appl. Phys. Lett. *40,* 474 (1982).
27. H. Okushi, Y. Tokumaru, S. Yamasaki, H. Oheda, and K. Tanaka, Phys. Rev. *B25,* 4313 (1982).
28. H. Ikushi, T. Takahama, Y. Tokumaaru, S. Yamasaki, H. Oheda, and K. Tanaka, Phys. Rev. *B27,* 5184 (1983).
29. D. V. Lang, J. D. Cohen, J. P. Harbison, M. C. Chen, A. M. Sergent, J. Non-Cryst. Solids *66,* 217-222 (1984).
30. M. C. Chen, D. V. Lang, W. C. Dautremont-Smith, A. M. Sergent, and J. P. Harbison, Appl. Phys. Lett. *44,* 790 (1984).
31. T. Tiedje, Bull. Am. Phys. Soc. *26,* 330 (1981); G. D. Cody, T. Tiedje, B. Abeles, B. Brooks, and Y. Goldstein, Phys. Rev. Lett. *47,* 1480 (1981).
32. P. E. Vanier, A. E. Delahoy, and R. W. Griffith, in *Tetrahedrally Bonded Amorphous Semiconductors, Carefree, Arizona, 1981* ed. by R. A. Street, D. K. Biegelsen and J. C. Knights (American Institute of Physics, New York, 1981), p. 227.
33. J. D. Cohen, D. V. Lang, A. M. Sergent, and J. P. Harbison (to be published).
34. D. V. Lang, J. Appl. Phys. *45,* 3023 (1974).
35. D. V. Lang, in *Thermally Stimulated Relaxation in Solids,* Vol. 37 of *Topics in Applied Physics,* ed. by P. Brauhlich (Springer, Berlin, 1979), p. 93.

STAEBLER-WRONSKI EFFECT IN HYDROGENATED

AMORPHOUS SILICON

Richard S. Crandall

RCA Labs
Princeton, N.J.

ABSTRACT

A discussion of salient experimental observations of light induced
fatigue of hydrogenated amorphous silicon, Staebler-Wronski(S-W), effect
is presented. The aim is to formulate the boundaries within which any
model of the effect must lie. At present it is known that the S-W effect
involves thermally activated hole and electron trapping in deep traps.
The activation energies are in the vicinity of 1eV and 1.5eV for electron
and hole, respectively.

INTRODUCTION

During the past few years there has been considerable interest in
metastable defects in amorphous materials. These defects have the property that they can be formed by radiation of various energies and subsequently removed by annealing at moderate temperatures. Most investigations have been made on hydrogenated amorphous silicon (a-Si:H). Because
the first evidence for these defects in a-Si:H was obtained by Staebler
and Wronski (1977), the metastable effect has been termed the Staebler-
Wronski (S-W) effect. Although there is considerable scientific interest
in the S-W effect, a concern at present is the metastable degradation of
solar cells due to light-induced defects. Despite the remaining questions
as to the exact mechanism that is responsible for solar cell degradation,
there is no doubt that it is related to the S-W effect.

Staebler and Wronski observed that both photoconductivity and dark
conductivity decreased following intense illumination of a thin film of
a-Si:H at room temperature. The effect could be completely removed by
annealing the film above about 150C. Since then additional metastable
changes have been found in such properties of a-Si:H as electrical conductivity (Staebler and Wronski, 1977), photoconductivity (Staebler and
Wronski, 1980), luminescence (Pankove and Berkeyheiser, 1980), electron
paramagnetic resonance (Dersch et al., 1980; Hirabayashi et al., 1980;
Stutzmann et al., 1984), gap state density (Cohen et al., 1983 ; Beichler
and Mell, 1983 ; Huang et al., 1983; Grunwald et al., 1981; Stoica 1981),
sub-band-gap optical absorption (Amer et al., 1983), and solar cell
properties (Staebler et al., 1981). It has been demonstrated that these

metastable effects can be produced by current flow in the film as well as light (Staebler et al., 1981; Crandall, 1981; Crandall and Staebler, 1983). It seems clear that all the metastable effects in a-Si:H are intimately related. Various measurements (Crandall, 1981; Crandall and Staebler, 1983; Crandall et al., 1984 ; Jang et al., 1983) of the annealing rate have shown that the activation energies for this process fall into three groupings; 0.4eV, 1eV, and 1.5eV. The discussion that follows shall be restricted to those defects that fit the classification of 'classical' S-W defects. That is they can be formed by light or current flow and are stable at room temperature. The discussion thus precludes those metastable defects that are formed and studied at low temperature (Tomozane et al., 1983).

PROPERTIES OF THE METASTABLE DEFECTS

The metastable defects formed in a-Si:H possess varied and perhaps unique properties that must be explained by any model of the defects. Of course, the first property is that the defect is usually stable below about 100C. This means that, if the annealing of the defect can be described by simple escape kinetics, its activation energy must be on the order of 1eV. This limits the nature of the defect destruction to weak bond-breaking, thermal emission from a state below midgap, or escape over a high potential barrier.

Deep Level Transient Spectroscopy, DLTS, measurements have shown that the defects appear when they trap either an electron or hole. It is not clear, nor does it appear to be possible to distinguish, whether the defect is created during the process or nearly captures a free carrier from a band edge.

It is often assumed that light is necessary to produce the S-W effect. However, experiments have shown that this is not necessarily the case. The metastable defects can be populated with thermal electrons. This has been demonstrated by producing the defects by current flow alone (Staebler et al., 1981; Crandall, 1981; Crandall and Staebler, 1983).

Even though we, as well as others, have found metastable defects in a large variety of a-Si:H samples, there does not appear to be a significant correlation with sample preparation. There is mixed evidence regarding doping. Some find no effect with doping whereas others do. In two instances the defect density and properties could be correlated with impurities. Both oxygen (Crandall, 1981) and carbon (Crandall et al., 1984) are associated with metastable defects. However, to significantly increase the density of defects requires nearly percent impurity concentrations. Furthermore, defects associated with carbon have different properties from those associated with oxygen.

A characteristic feature of the S-W effect, not universally appreciated, is that the production of the metastable defects responsible for the effect is thermally activated. This is the main reason that the cross section for producing the effect is so low at room temperature. What is peculiar or remarkable is that the activation energy, E_p, for the defect production is nearly equal to the activation energy, E_e, for annealing the defect (Crandall, 1981). This relationship holds over a wide range of activation energies; from as low as 0.35eV to as high as 1.8eV.

The wide range of activation energies indicates that there is more than a single type of metastable defect associated with the S-W effect. However, this wide range reduces to three main groupings; 0.5eV, 1eV, and 1.5eV. The latter two energies can be associated with electrons and holes, respectively and the first has only been observed when carbon is in the material.

Accompanying this wide range of activation energies is a wide range of attempt-to-escape frequencies, ν, for the rate constant. The trend is that the lower the activation energy the lower ν. Values of ν range from $1 \times 10^{+4} s^{-1}$ to $7 \times 10^{+18} s^{-1}$; considerably different from the expected value of about a lattice vibration frequency.

To illustrate a few of the properties described above we outline DLTS measurements made at elevated temperature on a-Si:H Schottky barrier and p-i-n devices. These structures are particularly useful because capacitance DLTS can be used to determine the sign of the charge trapped to form the defect. Because the defects are charged, the depletion width and hence the capacitance will change following degradation. The recovery of the capacitance during annealing can be measured and analyzed to determine the activation energy for defect destruction as well as the attempt-to-escape frequency. The experimental details have been described elsewhere (Crandall, 1981; Crandall and Staebler, 1983).

The most convincing argument that light or current flow through an amorphous silicon film produce the same defect is to show that the same defect properties result in either case. One such property is the activation energy for annealing of the defect. This is a signature of the defect since different defects would have different activation energies.

Figure 1 shows the DLTS function S(t) which for this case is the difference in the capacitance at two different times t_1 and t_2 plotted versus the time after removal of the degradation. The time difference $t_2 - t_1$ is proportional to the time. The usefulness of this function is that it is strongly peaked at the characteristic time for the relaxation of the capacitance to its original value. This relaxation time is related to the emission energy by the well known relation

$$\tau^{-1} = \nu \, \exp(- \, E_e/kT)$$

where k is Boltzmanns' constant and T the absolute temperature. By repeating these measurements at other temperatures, an Arrhenius plot of the emission time versus reciprocal temperature can be constructed. From this plot, E_e as well as ν can be found.

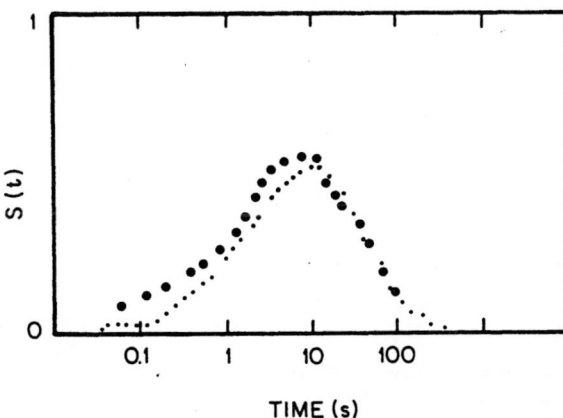

Fig. 1.   S(t) for both light and voltage pulse excitation of a p-i-n solar cell at 474K.   ooo -illumination with 690nm light for 1s   ...- 0.4s forward bias pulse. The sample was reverse biased at -5V for both measurements. The scales are not the same.

317

A detailed study of the sign of the capacitance change indicates that holes are trapped during degradation. Therefore, during annealing, holes are released from the bulk of the i layer causing the depletion width to widen. This produces a decrease in capacitance with time. The figure clearly shows that both current pulses and light pulses produce the same effect. There is good agreement between the position of the two S(t) shown in the figure. These measurements were repeated at different temperatures to construct an arrhenius plot yielding an emission energy E_e =1.8eV. This value is extraordinarily high. In fact it is larger than the bandgap energy.

Results similiar to those above have been obtained when electrons instead of holes are trapped. However, E_e is usually on the order of 1eV in contrast to the high value for the hole. When light is used to populate the defects both electrons and holes can be trapped. When current pulses are used, holes can be trapped only if they can be injected. For this reason it is difficult if not impossible to populate hole traps in an n-type Schottky barrier. In contrast both carriers can be trapped in a p-i-n structure because holes can be injected from the p layer.

It is difficult to observe hole trapping for two reasons. The first is that the high value of E_e means that high temperatures must be used for measurement so that the trapped hole can be emitted during the time scale of the measurement. The second is that the activation energy for hole trapping is higher than for electron trapping. Thus, at low temperature, electrons are trapped more readily than are holes, masking the effects of hole trapping.

A particularly simple experiment (Crandall and Staebler, 1983) demonstrates the similarity in activation energies for population and depopulation of the defects. It involves trapping and release of an electron in an n-type a-Si:H Schottky barrier device. In this experiment, conditions are such that electron trapping and release from metastable defects is the most likely process. Figure 2 is a plot of the DLTS function S(t) for an n-type Schottky barrier device reverse biased at 1V so

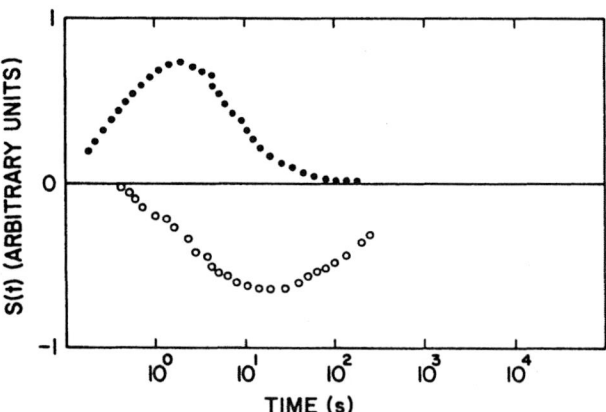

Fig. 2. DLTS function S(t) determined from transient capacitance measurements on an n-type Schottky barrier device at 500K. The positive S(t) in the upper half of the graph follows a reverse bias pulse. The sign of S(t) means an increase in net positive charge. The negative S(t) follows a forward bias pulse and shows a decrease in net positive charge.

318

that the depletion width extends an appreciable distance into the bulk. The capacitance probes effects that take place near the edge of the depletion width. Thus one measures a true bulk effect.

Figure 2 shows two distinct experiments. The one depicted by the negative S(t) in the lower half of the figure follows a short forward bias pulse. This pulse decreases the depletion width permitting electrons from the bulk to flow into the region that was originally depleted. These thermal electrons can be captured by the defect. At the end of the pulse defects containing electrons reduce the net positive charge resulting in a decrease in capacitance. As electrons are emitted from the defect they are swept out of the depletion width so that the capacitance increases and finally returns to its value before the pulse. This takes some hundreds of seconds. By repeating the measurements at different temperatures, the emission energy can be determined. For these measurements E_e=0.86eV. If one were to interpret this number in the usual manner it would indicate that the electron is trapped 0.86eV below the conduction band. If this were the case virtually, then we expect the rate constant for production of the defect to be temperature independent because an electron, if present in the conduction band, would simply fall into the defect. The second experiment shows the inconsistency of this interpretation.

The second experiment begins with the sample reverse biased as above. However, instead of a 'forward' bias pulse a 'reverse' bias pulse is applied to widen the depletion width. Defects normally occupied with electrons in the bulk low field region are now in the high field depletion region. Therefore, electrons thermally emitted to the band edge leave the depletion region during the pulse increasing the net positive charge in the depletion width. At the end of the pulse the depletion width capacitance is increased. Equilibrium is reestablished as conduction band electrons fall into the defects. If there were no barrier for their capture this would occur virtually instantaneously at this temperature. As can be seen from Fig. 2., the time constant for the trap filling is 2s. By repeating this type of measurement at different temperatures, the capture energy, E_p, can be determined. It is 0.98eV, somewhat higher than E_e.

From the above experiments we conclude the following:

(1) The emission and capture are thermally activated with E_e =.88E_p, a typical result for n type a-Si H.
(2) It is only necessary to have electrons present in the bulk to populate the defect. Light is not necessary. It is thermal energy that causes the electron to populate the defect.
(3) The defects are not produced by light. They are most probably present in equilibrium. In fact defects can be removed or depopulated by reverse bias. This is the origin of the reverse bias improvement of solar cells (Swartz, 1983).

The above experiment cannot, however, distinguish whether the defects are always present and can be populated or depopulated or that the defects are created thermally and stabilized by the presence of the electron.

Experiments similiar to those described above have been carried out on a large number of a-Si:H devices with the similiar results. These include p type and undoped material. The defects observed in these experiments all exhibit the above feature that the capture rate of the defect is thermally activated. More important, the higher the activation energy for annealing, the higher the activation energy for production.

The above experiment was for an n type device where only electron trapping is possible using current pulses to change the charge state. If a p-i-n device is used, or light is used so that holes are present in conducting states, it is possible to observe a hole trap with much the same properties as the electron trap. However, there are significant differences. The main one is that E_e is much larger for the hole. It is usually greater than 1.3eV. Again there does not seem to be much variation among sample type. However, we have not been able to make as detailed measurements of E_p for the hole trap as for the electron trap. We only know that it is quite high, certainly greater than 1eV.

Both hole and electron traps can be induced in the same sample of a-Si:H. As an example of this, results of a series of DLTS measurements on both p-i-n and n-i-p devices are shown in Fig. 3. Here the emission times for both hole and electron traps are shown as a function of inverse temperature. The data in Fig. 3 clearly show that both holes and electrons can be trapped in the same material and that the hole is always bound much tighter than the electron. This does not seem to depend on doping in any strong way.

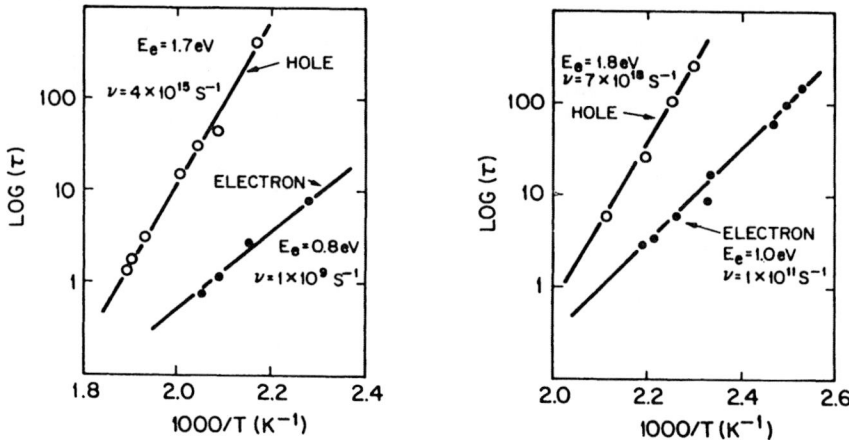

Fig. 3. Logarithm of the emission time for two different solar cell structures plotted as a function of inverse temperature. These data show both hole and electron traps in each sample.

The experiments described above were carried out on a-Si:H samples that exhibited only a moderate amount of degradation. Some samples degrade rapidly, even at room temperature. Most notably are those containing large amounts of carbon. With these it is possible to follow the kinetics of defect production over a long time span. At high temperature we usually find that degradation obeys first order kinetics with the number of defects that are produced increasing linearly with time until saturation when the steady state is reached. At low temperature the rate of degradation does not always follow first order kinetics. In fact the most common behavior is a degradation that increases sublinearly in time (Crandall and Staebler, 1983; Stutzmann et al, 1984). It is possible, over a limited time span to fit the data to a power law. However, if degradation is followed over a long enough time span then the data is much better characterized by a logarithmic dependence on time. An example of this is shown in Fig. 4 for a p-i-n solar cell that shows rapid degradation even at room temperature. The change in the number of recombination centers produced by forward bias current is plotted versus the logarithm of the time. The change in the number of recombination centers is proportional to the reciprocal of the drift length of the photocarriers (Crandall, 1982; Faughnan et. al., 1984). The drift length is proportional to the derivative of the short circuit photocurrent, J_{pc}, with respect to voltage, V. dJ_{pc}/dV is proportional to the in phase component of the ac admittance under illumination. For this example a forward bias current pulse degraded the cell. However, the same results were obtained using light to degrade the cell. It is obvious from the data that a logarithmic time dependence is obeyed over a large time interval. At short times, however, the degradation varies linearly with time.

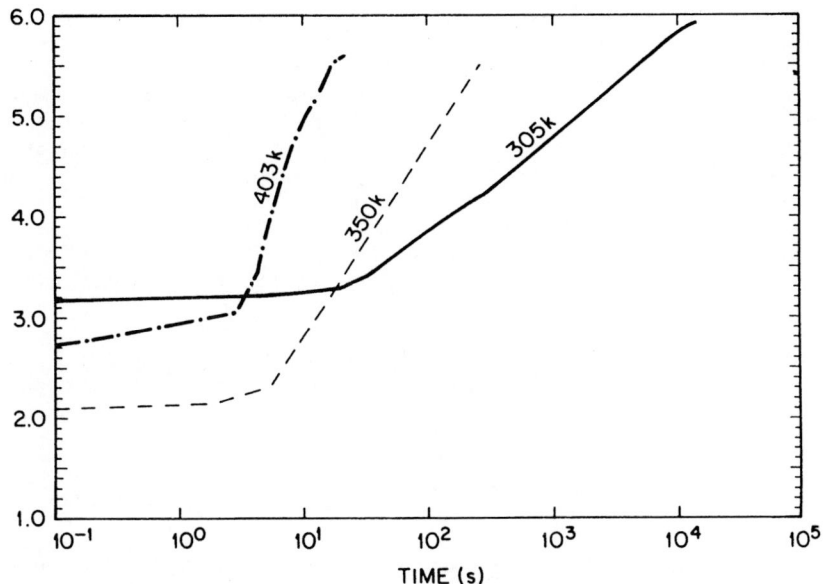

Fig. 4. Change in the density of recombination centers versus logarithm of the time of degradation at different temperatures for a p-i-n solar cell.

Figure 4 also shows the dramatic increase in the degradation at elevated temperature. A detailed investigation of the temperature dependence of the degradation at elevated temperature shows that it increases exponentially with increasing temperature (Crandall, 1981). This behavior appears to be a universal feature of the metastable effects. It or similiar behavior has been observed in other metastable effects (Pankove and Berkeyheiser, 1980; Jang et al., 1983). Even though degradation does not appear to saturate in time near room temperature, steady state conditions can be readily attained at elevated temperature. In this case the number of defects produced, in the steady state, increases linearly with light intensity until a saturation is obtained when all the defects are filled.

COMPARISON WITH MODELS

The surprising similarity between the activation energies for production and destruction must be explained by any model of the S-W effect. It places a severe restriction on the number of plausible models. In fact any model must explain why the activation energy for production is usually 10% larger than the activation energy for destruction.

Models that place the ground state of the defect within the energy gap have difficulty explaining a high activation energy for populating a state in the gap. There is evidence in crystalline material for strong electron-phonon interactions producing activated electron capture. However the activation energies are small, typically less than 0.1eV. If the state were at midgap then perhaps an unlikely process involving conduction and valence bands could be invoked. However, the large spread in measured activation energies, with most being larger than half the energy gap precludes this being a universal model.

An appealing model for the S-W effect involves creation of dangling bonds (Dersh et al., 1980 and 1983; Stutzmann et al., 1984). This idea stems from experiments that show an increase in the spin density associated with a singly occupied dangling bond. An increase in the density of occupied dangling bonds would also account for the decrease in the luminescence as well as an increase in the sub-band-gap optical absorption. However, there is always a question connected with these experiments. How does one insure that the apparent increase in the occupied dangling bond density is not caused by a shift in the the position of the dangling bond relative to the Fermi level. This can be a local effect. In fact, even though the Fermi level, as determined from the activation energy of the bulk conductivity may show little change (Dersch et al., 1983), the surface or regions near the surface that contribute little to the current can make large contributions to the spin signal.

It is not clear how or why the energy for production of a dangling bond should be so close to its energy of destruction. These energies include in the former case the energy gained in the electron capture and the energy lost during electron emission. To my knowledge there is not a model that has this feature.

Measurements (Lang et al., 1980; Beichler et al., 1983; Huang et al., 1983; Grunwald et al., 1981; Stoica 1981) of changes in the density of states near midgap, where the two principal states of the dangling bond are located present conflicting evidence. Some (Beichler et al., 1983; Huang et al., 1983; Grunwald et al., 1981) indicate an increase in the density of states in the upper half of the gap, indicating an increase in dangling bond density. DLTS measurements (Cohen et al., 1983), which give a direct measurement of bulk states, show no change in the density of states in the upper half of the band gap. They only show a movement of the Fermi level toward midgap. They do, however show an increase in states in the lower half of the gap. Some field effect measurements (Stoica, 1981) show a decrease in the gap state density above midgap whereas others (Goodman, 1982) show an increase.

In any measurement it is necessary, but often difficult to separate effects occurring near the free surface and the bulk. Field effect (Goodman 1982) and current transient DLTS (Beichler et al., 1983) weight effects near the surface most heavily, whereas capacitance transient DLTS (Crandall 1981; Cohen et al., 1983) weight bulk effects most strongly. This may be the reason for the above disagreements.

Staebler and Wronski postulated that the defects represented a metastable state most likely caused by the defect being surrounded by a high potential barrier. They reasoned that the state was populated when an electron hole-pair recombined. Since the energy of the light was greater than the band gap, sufficient energy was available to populate states with high barriers. This still seems like a reasonable model for the defect. It is consistent with my measurements insofaras the most reasonable explanation for the near symmetry of emission and capture energy would be the same potential barrier for each process. The ground state of the electron in the defect would lie near the conduction band. The small difference between emission and capture energy could arise from the charging energy. This is the energy necessary to add additional electrons to the defect. Since these defects are macroscopic, they can hold more than one electron. The logarithmic time dependence of the defect production shown in Fig. 4 can be explained in terms of charging energy(Crandall, 1980).

REFERENCES

Amer, N. M., Skumanich, A., and Jackson, W.B., 1983, Physica B&C, 117-118 B+C, pt.2:897.

Beichler, J., Mell, H., and Weber, K., 1983, J. Non-Cryst. Solids, 59 & 60: 257

Crandall, R. S., 1980, J. Electronic Matr., 9:713.

Crandall, R. S., 1981, Phys. Rev., B24:7457.

Crandall, R.S., 1982, J. Appl. Phys., 53:3350.

Crandall, R.S., Carlson, D.E., Catalano, A., and Weakliem, H.A., 1984, Appl. Phys. Lett., 44:200.

Crandall, R.S. and Staebler, D. L., 1983, Solar Cells, 9:63.

Cohen, J. D., Lang, D.V.,Harbison, J.P., and Sergent, A. M., 1983, Sol. Cells, 9:119.

Dersch, H., Stuke, J., and Beichler, J.,1980, Appl. Phys. Lett., 38 :456.

Dersch, H.,Schweitzer, L., and Stuke, J.,1983, Phys. Rev., B28: 4468.

Faughnan, B., Moore, A., Crandall, R., 1984, App. Phys. Lett., 44:613.

Goodman, N. B., 1982, Phil. Mag., B45:407.

Grunwald, M., Weber, K., Fuhs, W., and Thomas, P., 1981, J. Phys. (paris), Colloq. C4,42 (10,Suppl.):523.

Huang, C. Y., Guha, S, and Hudgens, S. J., 1983, Phys. Rev., B27:7460.

Hirabayashi, I, Morigaki, M., and Nitta, S.,1980, Jpn. J. Appl. Phys., 19:L357.

Jang, J., Kim, T.M., Hyun, J.K., Yoon, J.H., and C. Lee, C., 1983, J. Non-Cryst. Solids, 59&60:429.

Jang, J. and Lee, C.,1983, J. Appl. Phys., 54:3934.

Lang, D. L., Cohen, J. D., Harbison, R E, Sergent, P. P.,1983, App. Phys. Lett., 40:474.

Pankove, J. and Berkeyheiser, J. E., 1980, Appl. Phys. Lett., 38:456.

Staebler, D. L. and Wronski, C. R., 1977, Appl. Phys. Lett., 31:292.

Staebler, D. L.and Wronski, C. R., 1980, J. Appl. Phys., 51: 3262.

Staebler, D. L., Crandall, R.S., and Williams, R., 1981, Appl. Phys. Lett., 39:733.

Stoica, J., 1981, J. Phys. (paris), Colloq. C4,42 (10,Suppl.):407.

Stutzmann, M., Jackson, W.B., and Tsai, C.C.,1984, Appl. Phys. Lett., 45:1075.

Swartz, G., 1984, App. Phys. Lett., 44:697.

Tomozane, M., Hasegawa, F., Kawabe, M., and Nannichi, Y., 1983, Jpn. J. Appl. Phys., 21:L497.

HOPPING TRANSPORT IN TETRAHEDRALLY BONDED AMORPHOUS FILMS VIA STATES

NEAR THE FERMI LEVEL

P.N. Butcher[*], R.P. Ferrier[†], A.R. Long[†] and
S. Summerfield[*]

[*]Department of Physics, University of Warwick, Coventry
CV4 7AL, England

[†]Department of Natural Philosophy, University of Glasgow
Glasgow G12 8QQ, Scotland

ABSTRACT

The unified theory of ac and dc hopping conductivity involving
electron states near the Fermi level is reviewed and used to interpret
recent data for sputtered films of pure and hydrogenated a-Ge and a-Si.

1. INTRODUCTION

The electron states near the Fermi level in tetrahedrally bonded
amorphous films are localised[1]. Electron transport involving localised
states proceeds by the hopping mechanism originally introduced by Conwell[2]
and Mott[3] for impurity states in lightly-doped crystalline semiconductors.
A theory of hopping conductivity in these materials was developed in the
early 1960's by Miller and Abrahams[4] and Pollak and Geballe[5]. The extension
of the formalism to amorphous semiconductors had already been made possible
by Anderson[6] who drew attention to the intimate connection between disorder
and localisation in 1958. Anderson's arguments led Mott[7] to develop the
concept of variable range hopping between localised states near the Fermi
level in an amorphous semiconductor which implies a $T^{1/4}$ law in the dc
conductivity[7]. A corresponding theory of ac conductivity was given by
Austin and Mott[8].

These seminal ideas have led to extensive studies of hopping transport
in many different amorphous materials over the last decade. In this paper
we describe the current understanding of ac and dc conductivity involving
localised states near the Fermi level and use the theory to analyse recent
experimental data for sputtered films of pure and hydrogenated a-Ge and
a-Si. The theory is outlined in §2 and the data is discussed in §3.
Outstanding problems are reviewed in §4.

2. THEORY OF HOPPING CONDUCTIVITY

2.1. The Rate Equation Formalism

The treatments of hopping conductivity used in the interpretation of experimental data nearly all derive from the semiclassical theory of Miller and Abrahams[4].   We outline the simplest case in which the localised electron states are regarded as defining sites which may be occupied by just one electron.   Let f_m and R_{mn} be the occupation probability of site m and the hopping rate from m to n.   The f_m are determined by the rate equations

$$\frac{df_m}{dt} = \sum_n [f_n(1 - f_m)R_{nm} - f_m(1 - f_n)R_{mn}] \tag{2.1.}$$

when the applied electric field $\underline{E}$ is uniform, the potential energy of an electron on site m is $U_m = e\underline{E}.\underline{r}_m$ and its total energy is $\varepsilon_m + U_m$.   We suppose that the perturbed hopping rates satisfy the detailed balance relation

$$\frac{R_{mn}}{R_{nm}} = \exp[\beta(\varepsilon_m + U_m - \varepsilon_n - U_n)]$$

$$\simeq \exp[\beta(\varepsilon_m - \varepsilon_n)][1 + \beta(U_m - U_n)] \tag{2.2.}$$

to first order in $\underline{E}$, where $\beta = 1/k_BT$.   When $\underline{E} = 0$ the ratio R_{mn}/R_{nm} reduces to the Boltzmann factor required to maintain the f_m's at their thermal equilibrium values.

$$f_m^0 = [\tfrac{1}{2}\exp\{\beta(\varepsilon_m - \varepsilon_F)\} + 1]^{-1} \tag{2.3.}$$

where ε_F is the chemical potential.   When $\underline{E} \neq 0$ the ratio R_{mn}/R_{nm} determines the net particle flux from m to n and hence the Ohmic electrical conductivity of the localised electrons.

To calculate the conductivity we write $f_m = f_m^0 - (df_m^0/d\varepsilon_m)\phi_m$ in (2.1.), where ϕ_m is proportional to $\underline{E}$, and use (2.2.) to obtain the linearized rate equations[9]

$$C_m \frac{d}{dt} (V_m + Ex_m) = \sum_n g_{mn} (V_n - V_m) \tag{2.4.}$$

where, with the superscript "0" indicating $\underline{E} = 0$,

$$C_m = e^2 \beta df_m^0/d\varepsilon_m, \tag{2.5.}$$

$$g_{mn} = e^2 \beta f_m^0(1 - f_n^0) R_{mn}^0, \tag{2.6.}$$

and

$$V_m = -[Ex_m + e^{-1}\phi_m] \tag{2.7.}$$

The notation in (2.4.) has been chosen to emphasise that they are Kirchhoff's equations for the circuit shown in Fig. 1(a).

Let us suppose that $E = (E,0,0)$ is oriented in the x direction and has a time factor $\exp(-i\omega t)$. Then $d/dt = i\omega$ in (2.4.). Calculating the average current density across a plane perpendicular to the x axis in a cube of side L and dividing by E gives the conductivity formula

$$\sigma(\omega) = \frac{J}{E} = \frac{1}{2EL^3} \sum_{mn} g_{mn} (V_m - V_n)(x_n - x_m) \qquad (2.8.)$$

2.2. The Models

To complete the calculation of $\sigma(\omega)$ we require formulae for C_m, g_{mn}, $V_m - V_n$ and a description of the statistics of the system. The results obtained when R_{mn} is taken from Miller and Abrahams[4] define the "MA model". The energy dependences of g_{mn} and C_m come from Fermi and phonon number factors and the dependence of g_{mn} on $r = |\underline{r}_m - \underline{r}_n|$ is dominated by an overlap matric element of the form $R_0 (\alpha r)^\nu \exp(-2\alpha r)$ where R_0 is a characteristic hopping frequency, α is an asymptotic decay constant for the localised states and $\nu = 0$, 1.5 or 2 depending on the details of model. Considerable simplification may be achieved by using the asymptotic energy dependences originally introduced by Ambegaokar, Halperin and Langer [11]. Thus we obtain the "AHL Model" for which

$$C_m = e^2 \beta \exp[-\beta | \epsilon_m - \epsilon_F |] \qquad (2.9a.)$$

and

$$g_{mn} = g_0 \exp(-s) \qquad (2.9b.)$$

where

$$g_0 = e^2 \beta R_0 \qquad (2.10a.)$$

and

$$s = 2\alpha r + \nu \log(\alpha r) + \tfrac{1}{2} \beta [| \epsilon_m - \epsilon_F | + | \epsilon_n - \epsilon_F | + | \epsilon_m - \epsilon_n |] \qquad (2.10b.)$$

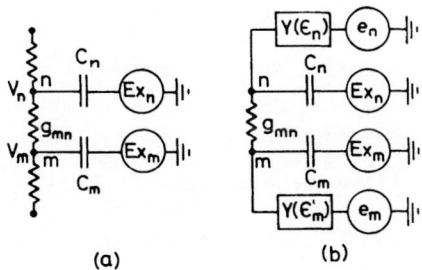

(a)   (b)

Fig. 1.   Equivalent circuits: (a) Miller and Abrahams[4]
(b) extended pair approximation[17].

All the calculations described in this paper assume that each site is equally likely to be anywhere inside the sample and that each site energy has an independent probability density proportional to a model-dependent density of states.

2.3. Ac Conductivity

The frequency dependence of the real part $\sigma_1(\omega)$ of $\sigma(\omega)$ is shown schematically in Fig. 2. To determine the saturated value for $\omega \gg R_0$ we replace d/dt by $-i\omega$ in (2.4.) and let $\omega \to \infty$. Then V_m must approach $-Ex_m$ to avoid a divergence on the left-hand side. The saturated value is obtained immediately by setting $V_m - V_n = -E(x_m - x_n)$ in (2.8.). However, this high-frequency asymptotic behaviour is inaccessible in conventional experiments which are confined to the region of low frequency response shown in Fig. 2.

A useful formula for $\sigma_1(\omega)$ which extends the above high-frequency result into the experimentally accessible frequency range is provided by the pair approximation[5]. To calculate $V_m - V_n$ from (2.4) we simply ignore all sites except m and n in Fig. 1(a). Solving the remaining two Kirchhoff equations for $V_m - V_n$ and substituting into (2.8.), we find that the contribution for each pair to $\sigma_1(\omega)$ is that of a Debye dipole with moment $e|x_m - x_n|$ and a relaxation time τ_{mn} equal to their RC time constant[10,12] when $\omega \ll R_0$ the result is dominated by pairs with $\tau_{mn} \sim \omega^{-1}$. System averaging produces power law behaviour: $\sigma_1(\omega) \sim \omega^s$, with s a little below one, as indicated by the straight line centre section in Fig. 2.

Power law behaviour is ubiquitous in ac experimental data and its prediction is the main success of the pair approximation which has been reviewed recently by Long[13]. Its chief failure is that the power law continues down to $\omega = 0$ (as indicated by the dashed straight line in Fig. 2) because the pair approximation makes no provision for the continuous passage of electrons through the system. A full solution of (2.4.) shows the low-frequency plateau indicated in Fig. 2.

The imaginary part $\sigma_2(\omega)$ of $\sigma(\omega)$ is determined from $\sigma_1(\omega)$ by the Kramers-Kronig relation. In broad terms[10], it rises linearly from zero when $\omega = 0$ to a wide peak in the neighbourhood of the inflexion point in $\sigma_1(\omega)$ and then falls off as ω^{-1}. Fine structure found in the low frequency regime is discussed in §3.3.

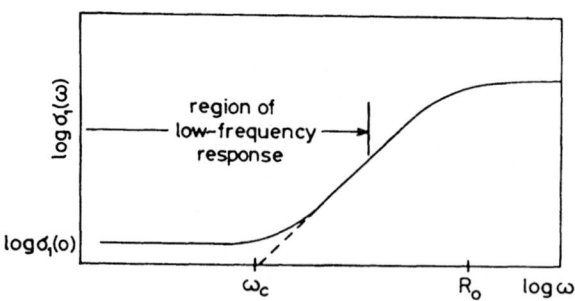

Fig. 2. Schematic diagram showing the behaviour of $\sigma_1(\omega)$.

2.4. Dc Conductivity

When $\omega = 0$ the left-hand side of (2.4.) vanishes and we are simply concerned with a random conductance network (see Fig. 1(a)). The value of $\sigma(0)$ is then controlled by an appropriate average, g_p, of g_{mn}. Equations (2.9.) and (2.10.) show that the random variables ε_m, ε_n and r appear exponentially in g_{mn} which is consequently spread over an enormous range. Ambegaokar et al.[11], Shklovskii and Efros[14] and Pollak[15] independently emphasised that g_p is therefore determined by the following percolation criterion. Take an infinite conductance network and set a conductance level G. Remove all conductances $g_{mn} < G$. If the remaining network contains infinitely long chains of conductances then reduce G and repeat. g_p is the critical value of G at which the last infinite chain breaks.

It is convenient to define a corresponding critical percolation exponent s_p by setting $g_{mn} = g_p$ in (2.9b.). The value of s_p may be determined numerically[16]. Alternatively we may summarise many investigations of percolation problems by the idea that the number N_p, of "bonds" (conductances) at the percolation threshold is determined primarily by the dimensionality of the problem. When s is given by equation (2.13.) the appropriate value of N_p is 2.1[16]. The equation for s_p is $2B/S' = N_p$ where B is the number of conductances per unit volume with $s < s_p$ and S' is the number of sites per unit volume with an energy close enough to ε_F to be involved in the critical percolation network[9].

Once $g_p = g_0 \exp(-s_p)$ has been calculated a very successful approximation to $\sigma(0)$ is obtained by replacing the summand in equation (2.8.) by[9]

$$g_{mn}(V_m - V_n)(x_n - x_m) \rightarrow g_{mn}E(x_n - x_m)^2 \qquad (2.11a.)$$

when $s > s_p$ and by

$$g_{mn}(V_m - V_n)(x_n - x_m) \rightarrow g_p^2 E(x_n - x_m)^2/g_{mn} \qquad (2.11b.)$$

when $s < s_p$. Results obtained in this way (after system averaging with a flat density of states) are given by the dashed curve in Fig. 3. They are in good agreement with the bold points which were obtained by direct numerical solution of Kirchhoff's equations for a large conductance network[17]. Fig. 3 also illustrates Mott's $T^{1/4}$ law[1,7] in the form[9]

$$\sigma(0) = \sigma_0 \exp[-(T_0/T)^{1/4}] \qquad (2.12.)$$

where

$$k_B T_0 = 40 N_p \alpha^3 / \pi N(\varepsilon_F) \qquad (2.13a.)$$

and

$$\sigma_0 = 40(g_0\alpha)N_p^2 T^{1/2}/T_0^{1/2} 3\pi \qquad (2.13b.)$$

with $N(\varepsilon_F)$ denoting the density of states at the Fermi level.

329

2.5. Unified Theory

In the last decade much experimental data has been analysed using the pair approximation to model ac conductivity and percolation theory for dc conductivity. A better interpretation of both types of data has now become possible with the development of a unified theory. We describe here the Extended Pair Approximation (EPA)[17]; a random walk treatment[18] yields similar results.

To calculate $V_m - V_n$ in (2.8.) we extend the pair approximation circuit described in §2.3 by adding additional elements to account for the rest of the network in an average way (see Fig. 1(b)). The values of e_m and e_n are determined by symmetry arguments to be $-Ex_m$ and $-Ex_n$. Then Kirchhoff's equations for the extended pair network immediately yield

$$V_m - V_n = - E(x_m - x_n)/g_{mn}Z_{mn} \qquad (2.14a.)$$

where

$$Z_{mn} = g_{mn}^{-1} + [Y(\varepsilon_m) - i\omega C_m]^{-1} + [Y(\varepsilon_n) - i\omega C_n]^{-1} \qquad (2.14b.)$$

The calculation of $\sigma(\omega)$ from (2.8.) is now trivial.

The pair approximation is regained by putting $Y(\varepsilon_m) = Y(\varepsilon_n) = 0$. To determine $Y(\varepsilon)$ in the EPA we use a renormalised mean field approximation to the average admittance seen looking out of site m. This yields the integral equation

$$Y(\varepsilon_m) = B^{-1} <\sum_k \{g_{mk}^{-1} + [Y(\varepsilon_k) - i\omega C_k]^{-1}\}^{-1}> \qquad (2.15.)$$

where the angular brackets signify an average over all stochastic variables except ε_m. The EPA automatically gives a $T^{1/4}$ law at low temperatures.

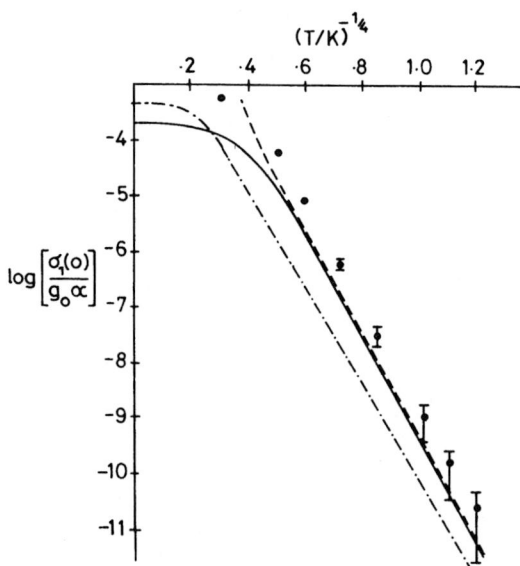

Fig. 3. Dc conductivities calculated as described in the text[17].

The value of B in (2.15.) is set equal to 4.4 so as to make T_0 agree with (2.13a.). The full line in Fig. 3 is the EPA result for $\sigma(0)$ for a rectangular density of states 10 meV wide. The dash-dot curve in Fig. 3 shows an analogous result calculated using random walk theory[18,19]. It differs from the EPA curve only because a different value of B was employed and because different approximations were used to evaluate the final formulae.

2.6. The Scaling Hypothesis

The most significant parameter involved in $\sigma_1(\omega)$ in the low-frequency regime is the frequency ω_c, shown in Fig. 2, at which the change from power law to dc behaviour occurs. A critical assessment of results calculated for many different models shows that ω_c is proportional to $\sigma_1(0)$. Then, for dimensional reasons, we must have $\omega_c = A\sigma_1(0)kT/e^2\alpha$ for the tunnelling models discussed in §2.2 where A is a dimensionless constant. Further inspection of the numerical results suggests that the scaling hypothesis[20]

$$\frac{\sigma_1(\omega)}{\sigma(0)} = f\left[\frac{\omega}{\omega_c}\right] = 1 + \left[\frac{\omega}{\omega_c}\right]^{0.725} \tag{2.16.}$$

is appropriate for typical parameter values in the low frequency regime.

To illustrate the success of (2.16.) in fitting numerical results it is convenient to make A explicit by writing $\omega/\omega_c = \tilde{\omega}/A$ where $\tilde{\omega} = \omega e^2\alpha/\sigma_1(0)k_BT$. The full curves in Fig. 4 are plots of $f(\omega/\omega_c)$ given by (2.16.), which have been fitted to the points calculated from the EPA by taking (from left to right) $\log_{10}A = 1.91, 2.96, 4.79$ and 5.71. The models considered are respectively AHL with $T \to \infty$, AHL with a constant density of states, MA with a Gaussian density of states and AHL with a Gaussian density of states. The span of $\sigma_1(0)$ covered by the numerical points is respectively 10, 5, 8 and 6 decades.

The simple scaling function given in (2.16.) implies that $\sigma_1(\omega) - \sigma(0)$ is proportional to $\omega^{0.725}$ for all ω. Deviations from this simple power law are not apparent in Fig. 3 but they are brought out by the more sophisticated plots used in §3.3. In particular, since $\sigma_1(\omega)$ is an even function of ω, $\sigma_1(\omega) - \sigma(0)$ is proportional to ω^2 when $\omega \to 0$. The change of the power of ω involved from 2 to 0.725 as ω increases through ω_c is responsible for the loss peaks discussed in Section 3.3[21].

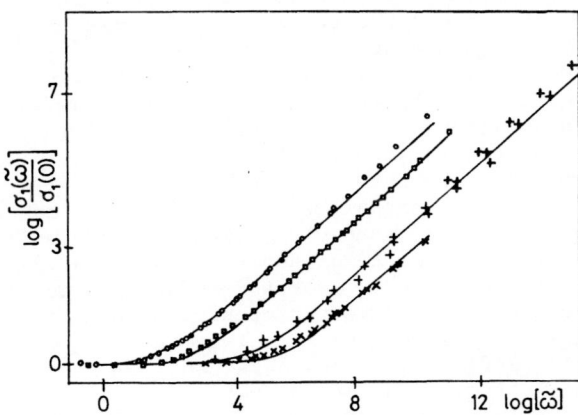

Fig. 4.   Tests of the scaling hypothesis for the models described in the text[20].

3. EXPERIMENTAL DATA ON HOPPING CONDUCTIVITY IN AMORPHOUS TETRAHEDRAL FILMS

3.1. Structural properties and electron spin resonance

In examining hopping transport in any material, it is important to know as much as possible about its structural properties and particularly about the nature of the states through which the transport is occurring. It is now widely accepted that the atoms in a-Si and a-Ge films are bound tetrahedrally into a continuous random network, in which the nearest neighbour distances and bond angles are close to those pertaining in the crystal. If the films are deposited by radio frequency (rf) sputtering on smooth, amorphous substrates then they generally show little large scale structure and electron micrographs taken of them have very little contrast[22,23]. As expected, contrast increases when films are deposited on rough substrates or at high substrate temperature, or both. Thermally evaporated films tend to show rather more growth structure, on the scale of tens of nanometers[24].

An important structural effect occurring in these films is a large amount of small angle scattering (SAS). The SAS in rf sputtered films arises from defects typically 1nm in extent[25], thought to be multi-vacancies or "voids". In rf sputtered material there is considerable evidence that these voids contain atoms of the inert sputtering gas at high density[25]. Voids are also present in evaporated material[24,26], but typically on a larger and much less reproducible scale.

One advantage of the rf sputtering technique is that by using it, active gases can be incorporated into the growing films in a reasonably reproducible manner. The most important of these impurities is hydrogen. The incorporation of hydrogen does not appear greatly to change any structural parameters of the films[23,22]; SAS recurs on a similar scale, consistent with a roughly equivalent incorporation of the passive sputtering gas. Hydrogenated tetrahedral films can also be prepared by decomposing the appropriate gaseous hydride, most frequently by striking an rf plasma within it [27]. Such films are not dissimilar structurally from rf sputtered films, though the scale of the SAS is rather larger[28]. Films can be laid down with great rapidity by the glow-discharge decomposition technique, and under unfavourable conditions they can show significant columnar growth structure[29].

An important technique for examining the localised states in amorphous films is that of electron spin resonance (esr). (For a review of early work see reference 30). Films deposited at room temperature in the absence of hydrogen contain typically around 10^{19} spins/cm^3; this number declines significantly if the deposition is made at elevated substrate temperature or hydrogen is incorporated. The g-values (2.022 for a-Ge, 2.0055 for a-Si) are identical to those observed for crushed crystalline material, and hence these spins are identified with dangling bonds, either distributed singly through the complete film, or associated with the surfaces of any voids present. These observations allow us to make important assertions about the defect sites in a-Si:H, namely that there is a finite density of them in the region of the fermi level and that they have positive correlation energy (unlike defects in amorphous chalcogenide films, which are thought to have negative effective correlation energy and therefore to be spin paired[31]. The low spin density observed in the glow discharge material [32] (10^{16} cm^{-3}) confirms the idea obtained from the measurement on hydrogenated sputtered films that one primary function of hydrogen is to passivate dangling bond states. One of our objectives in this paper is to examine whether this is the only effect that hydrogen has on them.

Of course, no information about the distribution of dangling bond states in space can be derived from measurements of the spin density alone but such information can be deduced from measurements of the esr linewidth, and we discuss this in §3.4.

In the remainder of this section, we concentrate on the properties of films prepared by rf sputtering. There are several reasons for this. Firstly, this preparation technique gives samples whose transport properties are reproducible (to ± 50%) over a wide range of temperature. Secondly, active gases like hydrogen can be introduced into the films in continuously increasing amounts, and their effects on transport thereby monitored. By using elevated preparation temperatures and large hydrogen concentrations, materials can be prepared which are not dissimilar from those produced by glow discharge decomposition. Thirdly, considerable structural information is available about materials prepared in this manner, which will be useful in interpreting the transport data.

3.2. Dc transport

Pure (i.e. unhydrogenated) amorphous tetrahedral films follow the Mott $T^{1/4}$ law (2.12.) over a wide range of temperatures (see Fig. 5). Typical parameters for films deposited by rf sputtering at room temperature

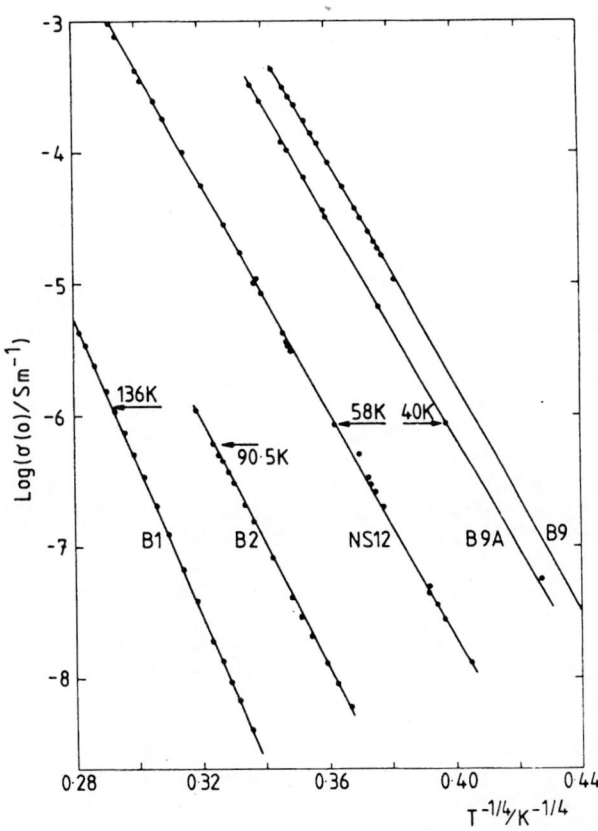

Fig. 5.   Dc conductivity of a-Ge samples[53] showing parallel shift with anneal (B9→B9A) and with hydrogenation ([H] - B9/0, NS12/17%, B2/25%, B1/33%-see also Fig.9).

Table 1. α^{-1} Values

Material[a]	a-Ge	a-Ge	a-Ge	a-Ge	a-Ge	a-Ge:H	a-Si	a-Si	a-Si	a-Si	a-Si:H
Method[b]	A	A	B	B	C	D	A	B	B	C	D
Value/nm	1.0	1.4	1.4 ±0.1	1.2 ±0.1	2.2 ±0.2	1.1	0.3	1.9	1.5	1.7 ±0.2	0.4 ±0.1
Reference	33	54	55	This	work		32	53	54	This work	32

a. Preparation techniques may vary.   b. Methods: A - 2-D Hopping,
    B - dc field dependence, C - ac fitting, D - esr.

are $T_0 \sim 10^8$K, with the conductivity at 77K (a convenient reference temperature) around 10^{-3} Sm^{-1} for a-Ge, and 5.10^{-5} Sm^{-1} for a-Si. Two parameters can be derived from the temperature dependence of the low field dc conductivity, σ_0 and T_0. However the simple variable range hopping model for a constant density of states (§2), involves three basic quantities $N(\varepsilon_0)$ and α^{-1}, which together determine T_0, and the rate constant R_0, which contributes with α^{-1}, to σ_0(2.13.). Hence to derive definite information about these quantities from the dc conductivity it is necessary to examine further data. Several methods have been proposed for doing this. Knotek, Pollak and coworkers[33] investigated transport along very thin a-Ge films prepared in ultra-high vacuum. This restricts the dimensionality of the hopping process, and causes a transition from a dc conductivity obeying the 3-D Mott relation (2.12.) to one following the relation

$$\sigma = \sigma_0' \exp\left(-(T_0'/T)^{1/3}\right) \qquad\qquad (3.1.)$$

characteristic of 2-D hopping[33]. From the parameters derived from the two regimes, values of α^{-1} may be deduced (see Table 1). α^{-1} may also be derived (independently of $N(\varepsilon_F)$ and R_0) by examining the electric field dependence of the dc conductivity. This has been analysed theoretically by several groups of workers [34,35,36]. Their conclusions as to the expected behaviour are broadly in agreement, though detailed numerical differences are found, which are not unexpected given the differing theoretical techniques employed. Some α^{-1} values determined experimentally for pure materials using this technique are given in Table 1. A further method for deriving α^{-1} is to examine the frequency dependence of the hopping conductivity; this is discussed in §3.3. Finally we note that, by generating dangling bonds under electron bombardment of glow discharge films, and studying the changes in esr spin density and hopping conductivity as these defects are annealed away, Stutzmann and Stuke[32] have also been able to derive α^{-1} values for hydrogenated a-Si and a-Ge. These values too are included in Table 1. Once the value of α^{-1} has been derived independently, then values of R_0 and $N(\varepsilon_F)$ can also be deduced. We summarise such an analysis for a variety of rf sputtered a-Si and a-Ge samples in §3.3.

One of the more perplexing aspects of dc hopping conduction in tetrahedral films is the "parallel shift", the tendency of σ_0 to change without any significant variation in T_0 when a sample is annealed[37] or when hydrogen is introduced[38]. Some data illustrating this effect for rf sputtered a-Ge and a-Ge:H samples are given in fig. 5. Note that the true parallel shift is only observed for low hydrogen concentrations; at higher hydrogenations an increase in T_0 is observed, as one would expect if the increase in hydrogen content were causing a decrease in the number of dangling bond states. The data for weakly hydrogenated a-Si films is much sparser than for a-Ge, but such as there is (§3.3.) suggests a much smaller parallel shift effect than for a-Ge.

3.3.   Frequency dependent hopping transport

The frequency dependent conductivity of amorphous tetrahedral semiconductors has been measured extensively over the last fifteen years or so (for a review, see reference 13).   Although the general trends in the early data were mutually consistent, numerically the observations varied quite considerably.   Recently however the Glasgow group has performed a detailed survey of frequency dependent loss in rf sputtered tetrahedral films[39,40].   These data show a high degree of internal consistency, and we use them to illustrate the arguments in this section.

As a first approximation to the ac response of the hopping network, we may use a conductivity of the form

$$\sigma_1 (\omega) = \sigma_1 (0) + \sigma_{pa} (\omega) \tag{3.2.}$$

where $\sigma_1(0)$ is the Mott $T^{1/4}$ conductivity given by (2.12.) and $\sigma_{pa} (\omega)$ is the ac response calculated using the pair approximation, as pioneered by Austin and Mott[8] and described in §2.4.   However it was realised on the basis of early data on evaporated materials that not only were the parameters $N(\varepsilon_F)$ and α^{-1} derived from this approximation in the dc and high frequency limits inconsistent[41], but also the pair approximation calculation did not predict accurately the observed temperature and frequency dependences in the dispersive region[42].   Early attempts to resolve this discrepancy[13] concentrated on modifying the mechanism for hopping in these materials from the pure electron tunnelling presumed in deriving the Austin/Mott relation. These attempts were empirically quite successful, but theoretically not fully convincing.   However the development of more accurate means of describing the frequency response of the hopping network by Movaghar and co-workers[18,19] and by the Warwick group (the EPA – see §2) showed that (3.2.) is a poor approximation to the total conductivity in the region where σ_1 becomes dispersive.   In the remainder of this section we use the EPA calculated for the AHL model (2.9, 2.10.) with a constant density of states to analyse the experimental data[43].

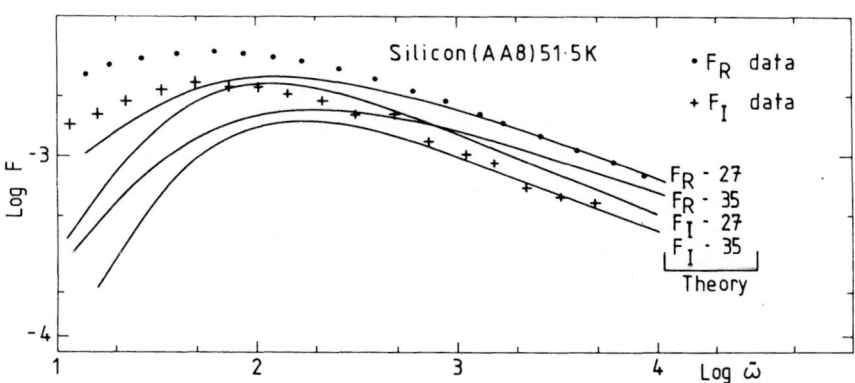

Fig. 6.   Frequency dependent loss against reduce frequency for an a-Si sample[43].   Functions plotted: $F_R = [\sigma_1(\omega)/\sigma_1(0) -1]/\tilde{\omega}$. $F_I = d[\sigma_2(\omega)/\sigma_1(0)\tilde{\omega}]/d[\ln\tilde{\omega}]$.   Theoretical curves plotted at $(T_0/T)^{1/4} = 27, 35$.

Provided that the comparison is made at a relatively high dc conductivity ($\sim 10^{-6}$ Sm^{-1} - see below for a discussion of this point) the EPA is in excellent agreement with the measurements of both real and imaginary parts of the loss for a wide range of a-Ge and a-Si samples. Fig. 6 shows data derived from both real and imaginary parts of the conductivity of an a-Si sample, plotted against the reduced frequency $\tilde{\omega}$ defined in §2.6. The relative magnitudes of the real and imaginary parts, and the loss peak observed near the transition from dispersive to non-dispersive behaviour, are both qualitatively fitted, though detailed differences between theory and experiment remain. Fig. 7 demonstrates, for the same sample, that, when plotted against $\tilde{\omega}$ the observed loss curves are to a first approximation independent of temperature over a considerable range. Finally in Fig. 8, we plot the total conductivity σ_1, normalised to the dc value at the same temperature $\sigma_1(0)$, against $\tilde{\omega}$ for a range of a-Ge and a-Si samples. In all cases, the correspondence between the theoretical curves and the observed data is strikingly good. It is equally good for all other samples measured to date, provided that there are no contact effects and that the dc conductivity is dominated by fermi level hopping.

As noted in §2.6, the reduced frequency $\tilde{\omega}$ for tunnelling is given by $e^2 \alpha\omega/\sigma_1(0)k_BT$. Thus the fit in Fig. 8 involves choosing the value of one unknown parameter, the localisation length α^{-1}. This is the additional technique for determining α^{-1} mentioned in §3.2. In Figs. 9a and 9b we show the variation of the localisation length with hydrogen incorporation (measured by the hydrogen concentration in the sputtering gas). α^{-1} values

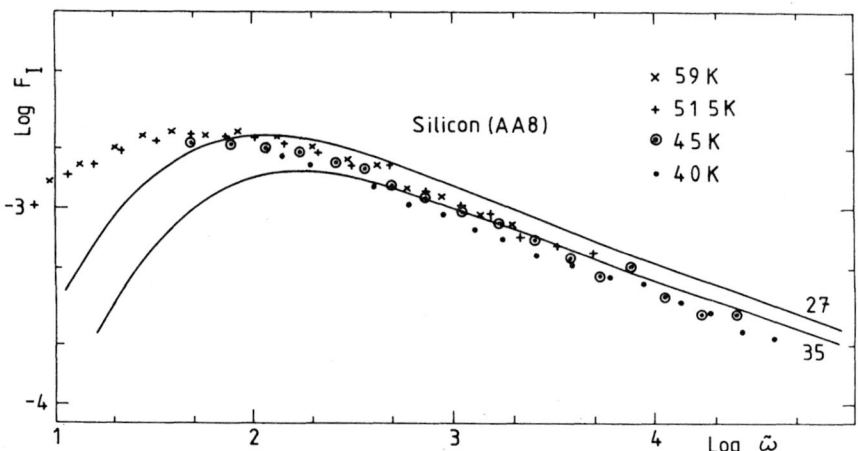

Fig. 7. F_I data at different temperatures for sample of Fig. 6[43]. Theoretical curves as Fig. 6.

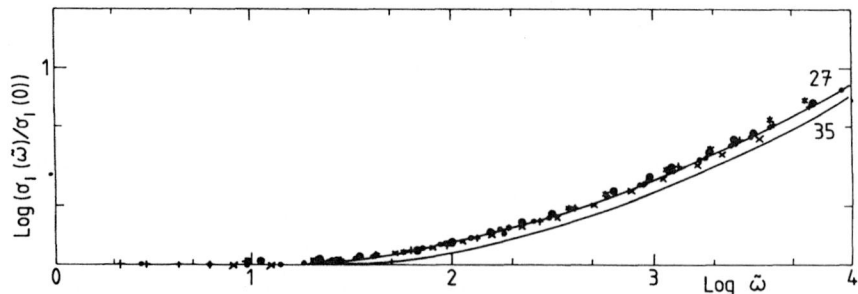

Fig. 8. Reduced conductivity data for 6 a-Ge and a-Si samples[43]. Theoretical curves as Fig. 6.

were determined in two ways, by fitting the ac data as in Fig. 8, and by examining the field dependence of the dc conductivity in the light of the theoretical analysis of Apsley and Hughes[35]. The α^{-1} derived by this technique is a weakly decreasing function of temperature and we have taken values at temperatures where the dc conductivity is around 10^{-6} Sm^{-1}. For the unhydrogenated films, the α^{-1} values derived by the two methods are in approximate agreement (and in agreement with most of the other reported values - see Table 1). However, as hydrogen is incorporated serious discrepancies develop between the results of the two techniques. Whereas α^{-1} determined from the ac data decreases markedly, that from the field effect shows a correspondingly dramatic increase. This result suggests that there remain inconsistencies in the formulation of the hopping problem in hydrogenated material.

In Fig. 9c we plot the hopping rate parameter derived for the same samples against hydrogen content. We have assumed that R_0 increases linearly with temperature, as suggested by single phonon calculations of the Miller-Abrahams type[4], and hence the ordinate is given in the form R_0/T. The "parallel shift" observed for amorphous germanium takes the form of a large drop in (R_0/T) as hydrogen is introduced. The data suggests however that such an effect is not present for a-Si, where the (R_0/T) values are of a similar order for all samples. Finally density of states values may be deduced using T_0 and ac values of α^{-1}. In distinct contradiction to the expected behaviour, the apparent density of states rises as hydrogen is incorporated. The reason for this is that the number evaluated from the parameters T_0 and α^{-1} is dominated by the rapid decrease in the latter. This too casts further doubt on direct applicability of the simple tunnelling model in the presence of hydrogen.

We noted above that the scaling model applies only at high temperatures, where the dc conductivity is around 10^{-6} Sm^{-1}. In Fig. 10 we plot, as a function of temperature, the conductivity of an a-Ge sample measured at 3kHz[40]. Below a temperature of 20K the observations deviate from the EPA prediction (which is also plotted in the figure). In this low temperature regime σ_1 becomes weakly temperature dependent (varying typically as $T^{0.4}$) but varies more rapidly with frequency than at higher temperatures (typically as $\omega^{0.9}$). All samples investigated to date, both of a-Ge and of a-Si, behave in a similar manner, although the transition point between high and low temperature regimes is at a higher temperature in hydrogenated samples. We believe[40] that this behaviour is strong evidence for an inhomogeneous distribution of defect states in these materials. At low temperatures the charge is trapped within defect

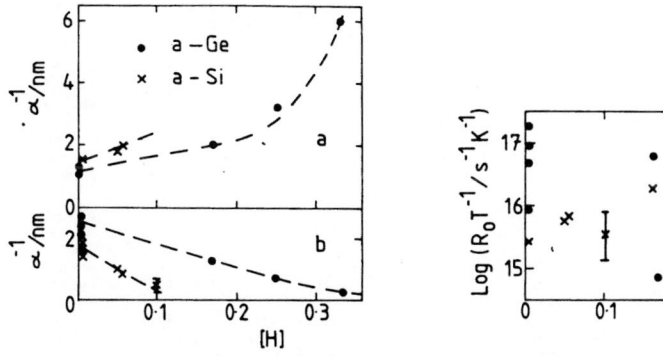

Fig. 9. Hopping parameters for various a-Ge and a-Si samples plotted as a function of hydrogen concentration in the sputtering gas. (a) α^{-1} calculated from dc field dependence (b) α^{-1} calculated from ac loss[43] (c) rate parameter[43].

clusters and the loss is due to the polarisation of such clusters.   At high temperatures the charge can hop from cluster to cluster and eventually percolate right through the film to give the dc current.   This view finds support from the investigations of the dependence of the loss on the exciting ac field[40,44], which suggests a scale for these clusters of typically 1nm. It is natural to associate them with the voids which are known to be present in these rf sputtered films and to have a similar scale (§3.1).

3.4.   Esr linewidth measurements.

Much information relevant to our discussion of transport properties in amorphous tetrahedral films can be obtained by examining the linewidth of the dangling bond esr signal as a function of frequency and temperature (for a review see reference 45).   The linewidth can be split into a zero temperature residual term $\Delta H_{pp}(0)$ and a temperature dependent term $\Delta H_{pp}(T)$. The former can itself be decomposed using measurements at different static fields[46] into a field dependent part due to the spectrum of g values present and a field independent contribution, arising in the main from unresolved hyperfine interaction.   Whereas for glow discharge a-Si films, the field dependence is adequately explained by neglecting exchange interactions, in evaporated films it is necessary to postulate an exchange constant of 1.4mT (equivalent to around 10^{-3}K) to explain the data.   Bachus et al note that this exchange interaction hardly changes with spin density.   Moreover they also observe that it is small compared with the values of 1T (1K) derived from susceptibility data[47].   They are therefore led to the conclusion that, in evaporated materials, the spins are strongly clustered, with the intracluster exchange interaction lying around 1T and the 1.4mT figure, determined from the linewidth data, associated with intercluster exchange.

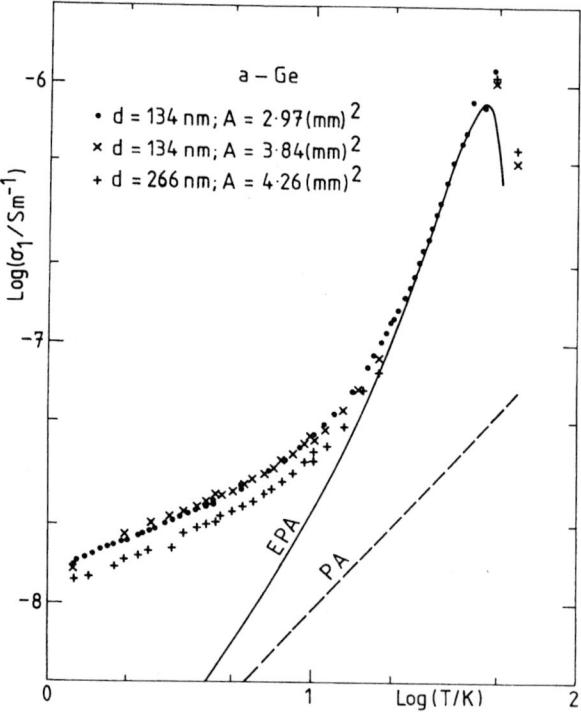

Fig. 10.   Ac conductivity at 3kHz for a-Ge sample B9A[40].   EPA denotes theoretical fit with same parameters as in Fig. 8. PA denotes pair approximation calculated with same parameters.

The temperature dependent linewidth, which is only found in materials showing Fermi-level hopping, is related to the dc conductivity by the relation

$$\Delta H_{pp}(T) = C\bigl(\sigma_1(0)\bigr)^m \tag{3.3.}$$

where C is a constant and m lies between 0.5 and 1. Movaghar, Schweitzer and Overhof[48,49] have shown that this relation arises naturally from hopping processes in which the electron spin is flipped, which determine the spin-lattice relaxation time and hence, in the appropriate limit, influence the linewidth. This provides important independent evidence for the basic correctness of the hopping picture we have described.

4. DISCUSSION

When the data of §3 is interpreted within the theoretical framework described in §2, we are able to make the following observations.

(i)   The ideas of hopping within a distribution of localised states close to the Fermi level enable us to understand qualitatively many aspects of the ac and dc transport in a-Si and a-Ge films at relatively high temperatures.

(ii)   However, the scaling hypothesis of §2.6 suggests that many hopping rates lead to similar forms for the frequency dependence of the conductivity, and hence that the excellent agreement between theory and experiment noted in Fig. 8 should not necessarily be taken to imply that charge transfer occurs by electron tunnelling alone. Agreement with the scaling curve reflects primarily that the transport is occurring in states which can be represented by a conductance network with a wide distribution of conductances. The truth of a particular model for the transition rates can only be established by a careful examination of the parameters derived by fitting the transport data, in conjunction with data derived from other sources.

(iii)   The values of α^{-1} derived from the EPA fits to the ac data for unhydrogenated a-Si and a-Ge are consistent with those from other sources. However, upon the introduction of hydrogen, serious inconsistencies arise. These are shown by our observation in §3.3 that, if the α^{-1} values are interpreted literally, the density of states derived from α^{-1} and T increases with hydrogen incorporation, instead of decreasing as expected. We doubt therefore whether the decrease in α^{-1} shown in Fig. 9b is a real effect. There are a number of possible reasons for this behaviour. Firstly, the fits of §3.3 assume a constant density of states at the Fermi level. It is possible that hydrogen, in addition to reducing the density of states, could lead to the Fermi level lying in a region of rapidly varying $N(\varepsilon)$. There is however no independent evidence for this, and we are inclined to believe that the explanation may well lie elsewhere. We suspect that in heavily hydrogenated material the coupling rates may not adequately be described by (2.9). These possibilities are currently being investigated further.

(iv)   One of the most mysterious features of the dc transport data is the "parallel shift" observed on annealing or hydrogenation. In §3.3 it was noted that this was reflected in a-Ge by a decrease in R_0/T upon the introduction of hydrogen, but that the equivalent effect appeared to be absent in a-Si. We do not know whether this observation is of fundamental importance, but are inclined to doubt it, as the parallel shift is still

339

observed when a-Si films are annealed. The observed values of R_0/T have always been regarded as large, the standard of comparison being the calculation of Miller and Abrahams[4] for impurity band conduction in c-Ge. However until we obtain a greater understanding of the wave functions associated with the deep localised states in the amorphous material, and hence are able to make a more appropriate calculation to compare with experiment, such criticisms might be regarded as premature.

(v)   One of the least considered aspects of hopping transport in a-Si and a-Ge is the role of inhomogeneities in the localised state system, and particularly the possibility that localised states might be clustered round voids. As we have seen, there is considerable evidence in favour of this picture, particularly from the low temperature ac loss data and from the esr linewidth results. Clustering appears to have little effect on the transport data in the high temperature, percolation, regime. However we may well need to look at details of the clustering to explain the insensitivity of T_0 to preparation conditions and the relationship between $N(\varepsilon_F)$ derived from the transport data and the spin density, N_s. The possibility also arises that changes in internal structure of the clusters on annealing may influence the parallel shift.

(vi)   One further effect which we have not considered in our analysis of the hopping data, is that of intrasite and intersite correlation. In an analysis based on their random walk theory of hopping transport, the Marburg Group[50] have calculated the effect of a constant positive intrasite correlation energy U on the dc conductivity. For $\beta U \gg 1$, they confirm their previous single electron results whereas in the opposite high temperature limit $\beta U \ll 1$ the conductivity declines from the extrapolated low temperature behaviour to follow a $T^{1/4}$ law with a T_0 twice the low temperature value. Given that U is currently estimated at around 0.1eV for a-Ge and 0.4eV for a-Si (admittedly for hydrogenated material[32]), the influence of the finite U on the dc conductivity should be detectable, at least in a-Ge, at temperatures below 300K. Many deviations from the $T^{1/4}$ law have been reported in the literature in this temperature range, but the experimental position is confused, and it is not clear that the effects of a finite U have ever been observed.

In the study of ac conductivity, the effects of a finite U are not difficult to predict, at least within the pair approximation. Particularly interesting are the effects predicted for negative U, but these are outside the scope of this review and the reader is referred to the literature[13] for a discussion. The effects of the intersite correlation energy E_{12} are also straightforward to discuss within the pair approximation. Pollak[51] was the first to note that, in the limit $\beta E_{12} \gg 1$, an ac conductivity independent of T is predicted for tunnelling in a constant density of states, compared with the direct proportionality to T predicted in the pair approximation[8]. Intersite correlation effects have not been included in the discussion of §2 and the excellent agreement between theory and experiment suggests that they are not significant in the high temperature range. However the low temperature loss regime (§3.3) is characterised by a temperature dependence around $T^{0.4}$, and it is probable, given the short distances associated with these processes (§3.4), that this weak power dependence reflects the presence of strong intersite correlation effects.

(vii) Finally, we note that, although in this paper we have restricted our discussion to amorphous tetrahedral semiconductors, the EPA has been applied with great success to other systems, notably doped c-Si[20] and c-GaAs[52], and polyacetylene[20].

340

ACKNOWLEDGEMENTS

We should like to acknowledge the contributions of many past and
present colleagues to the development of this work, particularly those of
Drs. N. Balkan, K.J. Hayden, W.R. Hogg and J.A. McInnes.

REFERENCES

1. N.F. Mott and E.A. Davis, "Electronic Processes in Non-Crystalline
    Materials", Clarendon Press, Oxford (1979).
2. E.M. Conwell, Impurity band conduction in germanium and silicon,
    Phys. Rev. 103:51 (1956).
3. N.F. Mott, On the transition to metallic conduction in semiconductors,
    Can. J. Phys. 34:1356 (1956).
4. A. Miller and S. Abrahams, Impurity conduction at low concentrations,
    Phys. Rev. 120:745 (1960).
5. M. Pollak and T.H. Geballe, Low-Frequency conductivity due to
    hopping processes in silicon, Phys. Rev. 122:1742 (1961).
6. P.W. Anderson, Absence of diffusion in certain random lattices,
    Phys. Rev. 109:1492 (1958).
7. N.F. Mott, Conduction in non-crystalline materials, Phil. Mag.
    19:835 (1969).
8. I.G. Austin and N.F. Mott, Polarons in crystalline and non-crystalline
    Materials, Adv. Phys. 18:41 (1969).
9. P.N. Butcher, K.F. Hayden and J.A. McInnes, Analytical formulae
    for dc hopping conductivity, Phil. Mag. 36:19 (1977).
10. J.A. McInnes, P.N. Butcher and J.D. Clark, Numerical calculations
    of ac hopping conductivity, Phil. Mag. B 41:1 (1980).
11. V. Ambegaokar, B.I. Halperin and J.S. Langer, Hopping conduction
    in disordered systems, Phys. Rev. B 4:2612 (1971).
12. P.N. Butcher, AC conductivity, in "Handbook on Semiconductors Vol. 1,
    W. Paul ed., North Holland, Amsterdam (1982).
13. A.R. Long, Frequency dependent loss in amorphous semiconductors,
    Adv. Phys. 31:553 (1982).
14. B.I. Shklovskii and A.L. Efros, Impurity band conductivity of
    compensated semiconductors, Sov. Phys. JETP 33:468 (1971).
15. M. Pollak, A percolation treatment of dc hopping conduction,
    J. Non-cryst. Sols. 11:1 (1972).
16. G.E. Pike and C.H. Seager, Percolation and conductivity, Phys. Rev. B
    10:14 (1974).
17. S. Summerfield and P.N. Butcher, A unified equivalent-circuit approach
    to the theory of ac and dc hopping conductivity in disordered
    systems, J. Phys. C 15:7003 (1982).
18. B. Movaghar and W. Schirmacher, On the theory of hopping conductivity
    in disordered systems, J. Phys. C 14:859 (1981).
19. B. Movaghar, B. Pohlmann and G.W. Sauer, Theory of ac and dc
    conductivity in disordered hopping systems, Phys. Stat. Sol. b
    94:533 (1980).
20. S. Summerfield, Universal low frequency behaviour in the ac hopping
    conductivity of disordered systems, Phil. Mag. B, to be published
    (1985).
21. S. Summerfield and P.N. Butcher, Low-frequency hopping conductivity
    in amorphous semiconductors, Phil. Mag. B 49:L65 (1984).
22. A.M. Patterson, A.J. Craven, J.N. Chapman and A.R. Long, Elemental
    analysis of sputtered hydrogenated amorphous silicon films,
    in: "Electron Microscopy and Analysis", P. Doig, ed., Institute
    of Physics, London and Bristol (1984).

23. A.M. Patterson, A.R. Long, A.J. Craven and J.N. Chapman, Structural studies of a-Si:H films in a high resolution analytical STEM, J. Non-cryst. Sols. 59 & 60:225 (1983).

24. A. Bienenstock, Structural characterisation of amorphous semiconductors, in: "Amorphous and Liquid Semiconductors", J. Stuke and W. Brenig, eds., Taylor and Francis, London (1974).

25. N.J. Shevchik and W. Paul, Voids in amorphous semiconductors, J. Non-cryst. Sols. 16:55 (1974).

26. P. Thomas, A. Barna, P.B. Barna and G. Radnóczi, The transport properties of an inhomogeneous model of a-Si and a-Ge films, Phys. Stat. Sol(a). 30:637 (1975).

27. W.E. Spear, Localised states in amorphous semiconductors, in: "Amorphous and Liquid Semiconductors", J. Stuke and W. Brenig, eds., Taylor and Francis, London (1974).

28. P. D'Antonio and J.H. Konnert, Small-angle-scattering evidence of voids in hydrogenated amorphous silicon, Phys. Rev. Lett. 43:1161 (1979).

29. J. Knights, Growth morphology and defects in plasma-deposited a-Si:H films, J. Non-cryst. Sols. 35 & 36:159 (1980).

30. J. Stuke, Esr in amorphous germanium and silicon, in: "Amorphous and Liquid Semiconductors", W.E. Spear, ed., Centre for Industrial Consultancy and Liaison, University of Edinburgh, Edinburgh (1977).

31. N.F. Mott, E.A. Davis and R.A. Street, States in the gap and recombination in amorphous semiconductors, Phil. Mag. 32:961 (1975).

32. M. Stutzmann and J. Stuke, Paramagnetic states in doped amorphous silicon and germanium, Sol. State Comm. 47:635 (1983).

33. M.L. Knotek, M. Pollak, T.M. Donovan and H. Kurtzman, Thickness dependence of hopping transport in a-Ge films, Phys. Rev. Lett. 30:853 (1973).

34. N. Apsley and H. Hughes, Temperature and field-dependence of hopping conduction in disordered systems II, Phil. Mag. 31:1327 (1975).

35. M. Pollak and I. Riess, A percolation treatment of high-field hopping transport, J. Phys. C, 9:2339 (1976).

36. M. Van der Meer, R. Keiper and R. Schuchardt, Non-ohmic hopping conduction-a treatment by directed percolation theory, J. de Physique(Paris) 42:C4-175 (1981).

37. W. Beyer and J. Stuke, Influence of evaporation parameters on electrical properties of a-Ge and a-Si, Phys. Stat. Sol.(a) 30:511 (1975).

38. A.J. Lewis, Use of hydrogenation in the study of the transport properties of a-Ge, Phys. Rev. B 14:658 (1976).

39. A.R. Long, N. Balkan, W.R. Hogg and R.P. Ferrier, Ac loss in sputtered hydrogenated a-Ge: measurements at around liquid nitrogen temperatures, Phil. Mag. B 45:497 (1982).

40. A.R. Long, W.R. Hogg, M.C. Holland, N. Balkan and R.P. Ferrier, Frequency dependent loss in sputtered a-Ge films: measurements at low temperatures, Phil. Mag. B 51:39 (1985).

41. P.N. Butcher and K.J. Hayden, Ac hopping conductivity in degenerate systems in: "Amorphous and Liquid Semiconductors", W.E. Spear, ed., Centre for Industrial Consultancy and Liaison, University of Edinburgh, Edinburgh (1977).

42. A.R. Long and N. Balkan, Ac loss in amorphous germanium, Phil. Mag.B 41:287 (1980).

43. N. Balkan, P.N. Butcher, W.R. Hogg, A.R. Long and S. Summerfield, Analysis of frequency dependent loss data in a-Si and A-Ge, Phil. Mag. B 51:L7 (1985).

44. A.R. Long, W.R. Hogg, N. Balkan and R.P. Ferrier, Ac loss in a-Ge at low temperatures, J. de Physique(Paris) 42:C4-107 (1981).

45. B. Movaghar, Paramagnetism and diamagnetism of amorphous semiconductors,. in: "Magnetism in Solids: Some current topics", A.P. Cracknell and R.A. Vaughan, eds., SUSSP Publications, Edinburgh (1981).

46. R. Bachus, B. Movaghar, L. Schweitzer and U. Voget-Grote, The influence of the exchange interaction on the esr linewidth in a-Si, Phil. Mag. B 39:27 (1979).

47. S.J. Hudgens, Low temperature magnetic properties of a-Ge and a-Si, Phys. Rev. B 14:1547 (1976).

48. B. Movaghar and L. Schweitzer, Esr and conductivity in a-Ge and a-Si, Phys. Stat. Sol.(b) 80:491 (1977).

49. B. Movaghar, L. Schweitzer and H. Overhof, Esr in a-Ge and a-Si, Phil. Mag. B 37:683 (1978).

50. M. Grünewald, B. Pohlmann, D. Würtz and B. Movaghar, Hopping transport of correlated electrons, J. Phys. C 16:3739 (1983).

51. M. Pollak, On the frequency dependence of conductivity in amorphous solids, Phil. Mag 23:519 (1971).

52. J.A. Chroboczek, L. Eaves, P.S.S. Guimares, P.C. Main, I.P. Roche, H. Mitter, J.C. Portal, M. Ketkar and S. Summerfield, Hopping conduction in n-GaAs at high frequencies and high magnetic fields, in: Proceedings of 17th Int. Conf. on the Physics of Semiconductors, to be published (1985).

53. A.R. Long, W.R. Hogg and N. Balkan, The effect of hydrogen on ac and dc conduction in sputtered a-Ge films – A temperature shift, Phil. Mag. B 48:L55 (1983).

54. M.L. Knotek, Temperature and thickness dependence of low temperature transport in a-Si thin films: a comparison to a-Ge, Sol. State Comm. 17:1431 (1975).

55. M.H. Gilbert and C.J. Adkins, Ac conductivity of a-Ge by time-domain spectroscopy, Phil. Mag. 34:143 (1976).

56. N. Apsley, E.A. Davis, A.P. Troup and A.D. Yoffe, Some effects of ion bombardment in elemental disordered films, in: "Amorphous and Liquid Semiconductors", W.E. Spear, Ed., Centre for Industrial Consultancy and Liaison, University of Edinburgh, Edinburgh (1977).

TIME RESOLVED OPTICAL MODULATION SPECTROSCOPY OF

AMORPHOUS SEMICONDUCTORS

J. Tauc

Division of Engineering and Department of Physics
Brown University
Providence, Rhode Island 02912

Recent results of studies of amorphous Si:H using time resolved photomodulation spectroscopy in the time domain from a fraction of a picosecond to milliseconds are reviewed. Information has been obtained on energy dissipation rates of hot carriers, transport processes associated with carrier trapping in shallow and deep states, recombination at high tempertures involving multiple trapping and at low temperatures by tunneling.

1. INTRODUCTION

Illumination of a semiconductor produces changes in the electron distribution and consequently in the optical spectrum. Recent studies of these changes in the steady state (CW "optical modulation (OM) spectroscopy") in amorphous semiconductors have been reviewed by Tauc and Vardeny[1]. In this paper, we summarize some recent studies on time dependent photoinduced spectra in amorphous semiconductors and the information obtained from them on the dynamics of photogenerated carriers involving processes such as hot carrier thermalization, trapping and recombination.

The method consists in illuminating the sample with a pulse ("pump") which is short compared to the relaxation time of the studied processes, and then measuring the time dependant changes in transmission or reflection using a CW "probe" beam with a sufficiently fast detection system. For relaxation times shorter than 1 nsec the detectors are less and less sensitive and eventually not available for response times in the picosecond range. In this case, a powerful correlation method is applicable in which relatively slow detectors can be used. It is usually referred to as "pump and probe technique". The probe is a pulse deduced from the pump pulse but of much smaller intensity and delayed with respect to the pump pulse by a longer optical path produced by a translation stage. The probe measures transmission or reflection as a function of the time delay. From this function information about the decay of the excitation can be deduced. This technique works even for extremely short relaxation times (in the femtosecond range) but in the nanosecond range it becomes impractical because the translation stage has to be too long.

As discussed below, the electronic effects that we are interested in in this paper are associated with the changes of the electron state occupation produced by light leading to changes in optical transitions. However, the pump pulse always produces some additional effects. Among them the most important are heat and stress generation which produces changes in the electronic structure leading to changes in the transmission and reflection of the probe. Recently, these effects have been observed and used for studying phonon propagation and heat dissipation in thin films[2]. In the cases discussed in this paper, the heat and stress effects were neglected because the conditions of the experiment were such that they were either small or their time constants were much longer than the relaxation times of the studied processes.

In the time interval between the pump absorption and the probe absorption a relaxation process occurs which is the subject of our study. If the time interval is short and the pump and probe are produced by the same laser a coherent effect may occur during the overlap of the subpicosecond pump and probe pulse. This effect may be used to obtain information about some very fast relaxation processes in the material but in the context of this paper it is considered as an experimental nuisance (a "coherent artifact"). In highly disordered systems (such as amorphous semiconductors) one may often get rid of it by using cross-polarized pump and probe beams. The effects about which we report in this paper are incoherent. One may clearly understand the distinction between the coherent and incoherent effects by referring to light emission for which a well known coherent effect is Raman scattering while incoherent effects are referred to as luminescence.

In our experiments, we measure the time-dependent change in the optical absorption coefficient $\Delta\alpha(t)$. Since we only consider effects that redistribute the electrons, it follows from the sum rules[3] that at every time t

$$\int_0^\infty \Delta\alpha(\omega, t)d\omega = 0 \qquad (1)$$

Therefore, there must be spectral regions where $\Delta\alpha > 0$ (photoinduced absorption PA) and other regions where $\Delta\alpha < 0$ (photoinduced bleaching PB); of course, these regions change with time.

In this paper we discuss the results of time resolved optical studies of a-Si:H and related tetrahedral semiconductors.

A. <u>PICOSECOND STUDIES</u>

2. PICOSECOND RELAXATION PROCESSES

In the range from 0.5 psec to 2 nsec we used the pump and probe technique. The laser was a cavity dumped passively mode-locked dye laser. The photon energy was 2 eV for both the pump and the probe, pulse duration was 0.5 to 0.8 psec, pulse energy about 2 nJ and the repetition rate 10^5 to 10^6 sec^{-1}. The pump and probe were focused onto the same spot of diameter less than 50μm. The same photon energy for the pump and probe and the lack of tunability were a limitation which will be removed in our new experimental set-up.

The processes that occur after the pump produces an electron-hole pair are schematically shown in Figure 1. The generation of an electron and a hole reduces the density of initial states and of available final states, and therefore the absorption coefficient seen by the probe is diminished. This process is characterized by a "bleaching" cross-section σ_0. The photogenerated electron and hole can absorb the probe photons and be excited into higher states. In Figure 1 only processes for the electron are shown. They include absorption by hot electrons characterized by absorption cross-section σ_1; absorption by electrons thermalized to the bottom of the conduction band (σ_2); and absorption by electrons in deep states in the gap (σ_3). The pump and probe technique works if these absorption cross-sections σ_i change as the electron dribbles into lower and lower states. Experience has shown that σ_i in the studied amorphous semiconductors is smaller if the initial state is deeper; we interpret it as related to the fact that the lower the initial state energy of the electron the smaller the density of final states. If for a particular transition $\sigma_i > \sigma_0$ photoinduced absorption (PA) is observed; if $\sigma_i < \sigma_0$ one observes photoinduced bleaching (PB).

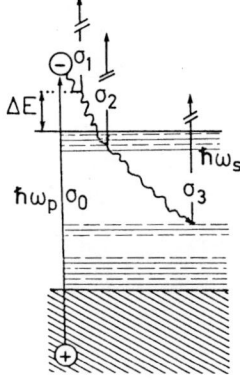

Fig. 1. Relaxation processes of an electron generated by the pump ($\hbar\omega_p$) as sampled by the probe ($\hbar\omega_s$).

3. HOT CARRIER THERMALIZATION

Effects occurring in the few picoseconds following the excitation have been studied in a-Si and a-Si:H[4-6], and a-GaAs[6]. One observes an onset of induced absorption which appears as instantaneous followed by a decay that often has a fast and a slower component. The fast effect has been interpreted[5] as due to photoinduced absorption by hot carriers whose absorption cross-section σ_1 decreases with decreasing energy ΔE (Fig. 1). The assignment of the effect to hot electrons is in agreement with the data on reflectivity; hot carriers reduce the index of refraction and consequently the reflectivity, as observed[6]. Using this interpretation, one obtains information about the hot carrier thermalization times t_0 and the excess energy dissipation rate R_d. It has been found that $R_d = 0.5 \, \text{eVpsec}^{-1}$ in a-Si and $R_d = 0.1 \, \text{eVpsec}^{-1}$ in a-Si:H[5,6]. Using the reflectivity data of Kuhl et al.[6] we estimate R_d in a-GaAs to be $(\hbar\omega - E_g)/2t_0 = (2 - 0.6)/0.6 \times 2 = 1.1 \, \text{eV/psec}$. R_d has a similar value for electron excess energy dissipation in c-GaAs ($R_d = (2 - 1.4)/0.6 = 1 \, \text{eV/psec}$; the factor $1/2$ is omitted since the electron mass is much smaller than the hole mass and most of the excess energy goes to the electron. Our conclusion that the dissipation rate is about the same in a-GaAs and c-GaAs differs from the conclusion of

Kuhl et al.[6] that the dissipation rate is substantially higher in c-GaAs. It was proposed that the dissipation mechanism in a-Si and a-Si:H is carrier scattering by optical deformation potential[7] while in GaAs the higher dissipation rate may be explained by the Frohlich interaction[6].

4. TRAPPING PROCESSES

The fast processes occuring within a few picoseconds (associated with hot carrier thermalization) are followed by considerably slower processes that are apparently associated with the decay of carriers from states close to the mobility edge into states deeper in the gap. In high quality a-Si:H this decay is very slow; in nonhydrogenated a-Si the decays are much faster[8]. It holds in general that if the defect density is higher the decay is faster. One may interpret these observations by assuming that in high defect density samples there are states close to the band edges that trap the carriers quickly. Once the carriers are trapped they do not move easily. If the defect density is lower the carriers go into states deeper in the gap where their absorption cross-section is smaller; the transfer takes a longer time.

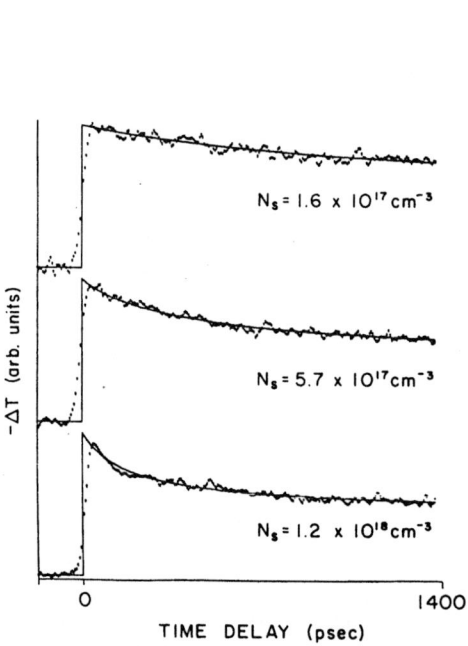

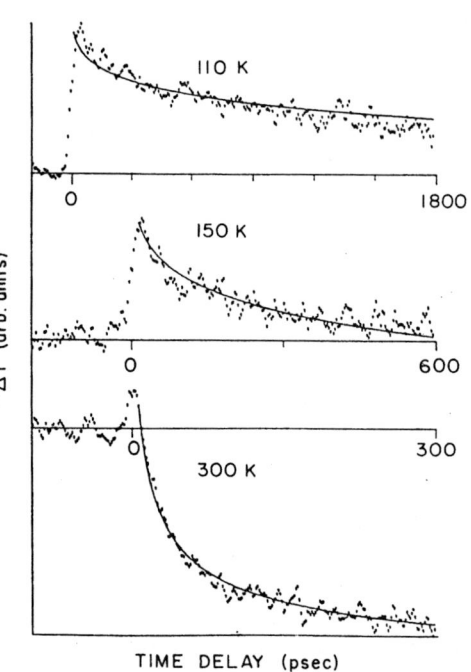

Fig. 2. Photoinduced absorption in an electron irradiated sample of a-Si:H (GD) at 300K. The lowest curve was measured before annealing, the upper two curves after annealing. N_p is the spin density (From Strait's Thesis[10]). The solid lines are the fits (see text).

Fig. 3. Photoinduced absorption in P-doped sample of a-Si:H measured at different temperatures after exposure to the pump pulse over an hour. The solid lines are the fits (see text) (From Strait's Thesis[10]).

The processes involved in these decays can be much better identified in samples in which the defects were deliberately introduced by irradiation or doping[9,10]. Strait has studied some of these cases and we will discuss two of his results[10].

In Figure 2 the decay of photoinduced absorption (observed as the decay of photoinduced transmission $\Delta T(<0)$) are shown in a sample that was irradiated by high energy electrons and subsequently annealed. As determined by ESR, the spin concentration (which is a measure of the defect density) was $1.2 \times 10^{18} cm^{-3}$ in the irradiated sample and decreased to $5.7 \times 10^{17} cm^{-3}$ and $1.6 \times 10^{17} cm^{-3}$ after annealing. The response is photoinduced absorption in the whole range; it is faster when the spin density is higher.

Figure 3 shows decays observed in a phosphorus doped sample on a spot that was exposed to the pump light for an hour. Without this exposure, the response is photoinduced absorption ($\Delta T < 0$) similarly as in the irradiated sample. However, after the exposure, the high temperature response ($T = 300$ K in the figure) corresponds to an onset of PA followed by photoinduced bleaching; the time range was too short to see the return of the bleaching response to zero.

The responses in Figures 2 and 3 have been identified[9] as due to carrier transfer from shallow traps (states close to the mobility edge) to deep traps (states deep in the gap). If the density of carriers in shallow traps is $n(t)$ and the density of carriers in deep traps is $n(0) - n(t)$ the change in the photoinduced absorption coefficient $\Delta\alpha(t)$ is

$$\Delta\alpha(t) = (\Delta\alpha_0 - \Delta\alpha_s)n(t)/n(0) + \alpha_s \qquad (2)$$

$$\Delta\alpha(t)/\Delta\alpha_s = \Delta T(t)/\Delta T_s = R_\sigma n(t)/n(0) + 1. \qquad (3)$$

$\Delta\alpha_0$, $\Delta\alpha_s$ are the values of $\Delta\alpha(t)$ when all the carriers are in the shallow or deep traps respectively, and $R_\sigma = \Delta\alpha_0/\Delta\alpha_s - 1$. From the measured $\Delta T(t)/\Delta T_s$ one can determine $R_\sigma (= \Delta T(0)/\Delta T_s - 1)$ and the decay of carriers in shallow traps $n(t)$. It is not exponential but could be fitted by

$$\frac{n(t)}{n(0)} = \frac{1}{1 + \left[\dfrac{t}{\tau}\right]^\beta} \qquad (4)$$

which is of the form one expects for a trapping (or recombination) process involving dispersive transport. From its long time limit $n(t) \sim t^{-\beta}$ one sees that β is the dispersion parameter; τ is the trapping time. Equation (4) was deduced by Orenstein and Kastner[12] for the case that the recombination occurs by deep trapping involving dispersive transport by multiple trapping in an exponential tail. Dispersion parameter β is related to the width of the tail kT_0:

$$\beta = T/T_0 \qquad (5)$$

and the trapping time is

$$\tau = \nu^{-1}(b_t N_t/b_p N_p)^{1/\beta} \qquad (6)$$

where b_t, b_p are the trapping coefficients, N_t, N_p are the densities of shallow and deep traps, and ν is an attempt frequency. Equations (5) and (6) describe remarkably well the data obtained at temperatures higher than 100K. For example, the fits in Figure 2 were obtained with $R_\sigma = 1.2$, $T_0 = 320$K and $\tau = 3.6 \times 10^5 \times (10^{15}/N_s)^{1/\beta}$ (psec) where N_s is the spin density (cm^{-3}) in the sample which was assumed to be proportional to the density of deep traps N_p. For the fits in Figure 3 $R_\sigma = -1.8$, $\nu = 9 \times 10^{12} sec^{-1}$, $b_t N_t / b_p N_p = 26$ and $T_0 = 430K$. In the phosphorus doped samples, there are strong reasons to believe that the carriers whose dispersive transport parameters we have determined are holes[9]. We also note that the deep trap density N_p in P- doped, as well as in B-doped a-Si:H samples increases with exposure to light. This is a manifestation of photostructural changes (a kind of Staebler-Wronski effect).

We have shown the usefulness of the optical modulation technique for studying transport effects at extremely short time. We will complete this section by showing how one can understand the difference between the decay in the irradiated samples (Figure 2) where $\Delta\alpha$ is always positive, and P-doped samples where $\Delta\alpha$ can be negative (Figure 3) at high temperatures, provided the exposure of the sample to light was long enough. The observed responses can be explained if one assumes that the relevant defect is the dangling bond.

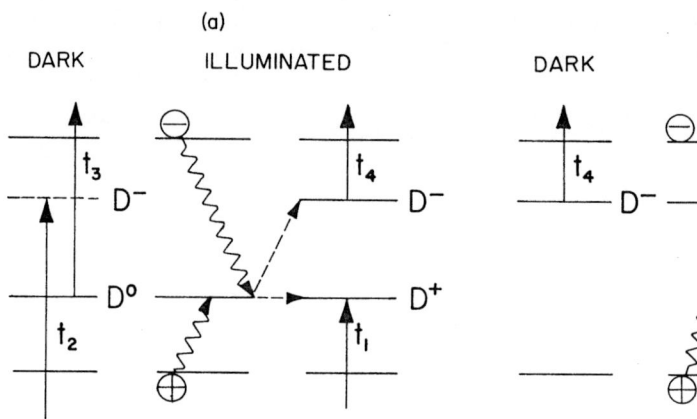 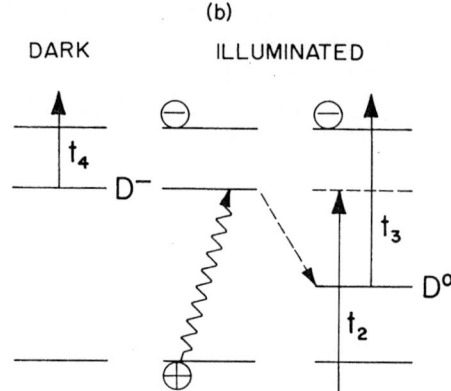

Fig. 4a. Optical transitions in irradiated a-Si:H in the dark (t_2, t_3) and after photocarrier trapping (t_1, t_4).

Fig. 4b. Optical transition in P-doped a-Si:H in the dark (t_4) and after hole trapping (t_2, t_3).

In the electron irradiated sample, the transitions associated with the dangling bond are seen in the dark as a broadening of the absorption edge. The dangling bond is in the D^0 state ($\equiv T_3^0$ in Adler's notation) and the transitions are from D^0 into the conduction band, and from the valence band into D^0 which, however, end up in state D^- because of the correlation energy. They are shown as transitions t_2 and t_3 in Figure 4a. Pump absorption produces electron hole pairs which transform some of the D^0 states into D^+ and D^- states. As shown in Figure 4a this produces transitions t_1 and t_4 (PA) and reduces transitions t_2 and t_3 (PB). With the pump and probe technique we follow the transfer of carriers from the initial states (carriers at the band extremes) to the final states (carriers at D^+ and D^-). The observed response is a superposition of the effects produced by electrons and holes. The recombination of electrons and holes trapped in D^- and D^+ states (discussed in Part B) occurs on a much longer scale that our range of 2 nsec for the pump and probe technique.

In P-doped samples (Figure 4b), the ground state is D$^-$ associated with transition t_4. The photogenerated hole is quickly trapped by this negatively charged defect transforming it into D^0. It is this hole trapping process that we follow with the pump and probe technique. The photogenerated electron is not trapped by the D$^-$ states and its trapping by the D^0 state (recombination) occurs on a time scale exceeding our range. Picosecond photoinduced response is photoinduced absorption (transitions t_2 and t_3) and photoinduced bleaching (transition t_4).

The above identification of the final states indicates why PB may prevail over PA in P-doped samples but does not in undoped samples. Apparently in undoped samples the absorption produced by the t_1 and t_4 transitions, which are from states close to the band edges, prevails over the bleaching (t_2 and t_3) which is associated with states far from the band edges. In P-doped samples, the states associated with absorption and bleaching are reversed and bleaching may prevail. A paper containing a comprehensive analysis of this problem is under preparation[11].

B. DECAY OF THE MIDGAP PA BAND

5. PHOTOINDUCED ABSORPTION BAND IN a-Si:H

Illumination of a-Si:H and similar semiconductors in the main absorption band produces a photoinduced absorption band below the absorption edge[13]. In undoped Si:H this band has been associated with the transitions of photogenerated holes trapped in the valence band tail into the valence band[14,15]. If the density of states in the gap is exponential the density of initial states sharply peaks at the quasi-Fermi level in the steady state or at the demarcation energy level in the transient case. The energy difference between the quasi-Fermi level (or the demarcation energy) and the band edge determines the onset of absorption. At high temperatures, the time and temperature dependence of this onset studied by Ray et al.[16] can be interpreted using the ideas underlying the multiple trapping model[17]. In doped or irradiated a-Si:H the PA spectrum differs from the spectrum of undoped a-Si:H and some exploratory work[18] has been done on transient responses; however, detailed studies of doped materials have not been completed. While the experimental results and their interpretation discussed in this section refer to undoped a-Si:H, the discussion of a description based on recombination time distributions has a general applicability.

In this paper we are interested in the time evolution of the total oscillator strength of the PA band defined as $\int \Delta \alpha d\omega$ over the whole band. For this measurement, the conditions of the experiment have to be chosen so that the optical and detection systems measure the induced absorption of the whole band, even when the spectral distribution changes with time. In our experiments on undoped a-Si:H this condition was satisfied; in this case the decay of the total PA band gives the rate at which the holes disappear from the tail. The data are consistent with the interpretation that this occurs by recombination whose rate at high temperatures is determined by multiple trapping processes of electrons at the conduction band mobility edge. At low temperatures, hopping or tunneling processes prevail over multiple trapping processes.

6. HIGH TEMPERATURE DECAYS

At high temperatures, experimental evidence supports the assumption that the recombination rate is determined by the rate at which the electrons are able to reach the slower carriers, the holes. Therefore the dispersive transport of electrons plays an important role. We will consider the simplest case in which the electron energy distribution in the tail is exponential so that the dispersion parameter β is given by Eq.(5). The decay is given by Eq.(4) in which τ now is the recombination time given by Eq.(6), b_p is the recombination coefficient and N_p is the concentration of recombination centers. At long times, $n(t) \sim t^{-\beta}$ as actually observed over a few decades of time. The experimental data also agree with Eq.(5) from which the width of the conduction band tail can be determined. As an example, the results obtained on a high quality a-Si:H sample[19] are shown in Figure 5. The slope of the straight lines in the log-log plot of the PA signal vs. t determines the dispersion parameter β which is a linear function of T (Figure 6). The value of β obtained from the PA measurements agrees in the studied temperature range with the dispersion parameter of electrons obtained from the time-of-flight measurements on the same sample.

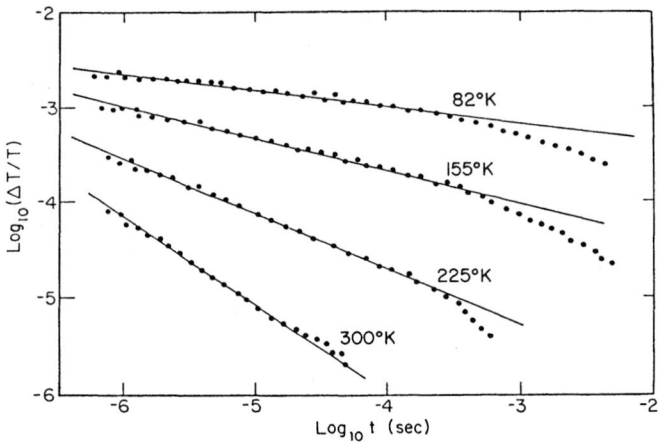

Fig. 5. PA decay in high quality sputtered a-Si:H at high temperatures (after Kirby et al.[19]).

When the tail is not exponential, as is the case in samples with a higher defect concentration, one may still represent the data as $n(t) \sim t^{-\beta}$ but β is much less temperature dependent and is a slowly varying function of time (the power law is a reasonable approximation only over a limited time range).

7. LOW TEMPERATURE RECOMBINATION

At low temperatures, we will assume that tunneling is the process that determines the decay rates. There may be some hopping processes occurring before the tunneling process and a more complete picture should include them.

In undoped a-Si:H the process may be tunneling of an electron in a c.b. tail to a hole in the v.b. tail. A more likely process will include defects, such as the dangling bonds. The electron may tunnel from the c.b. tail to D^{o} to form D^{-} and from this state to the hole in the v.b. tail. The process may be reversed; the hole may tunnel to D^{o} to form D^{+} to which an electron will tunnel from the c.b. tail. One of these processes is likely to determine the recombination rate. We note that in P-doped a-Si:H according to the results on picosecond decays discussed above (Figures 3 and 4b) the hole trapping on D^{-} states is the fast process and the recombination is completed through electron tunneling into the D^{o} states.

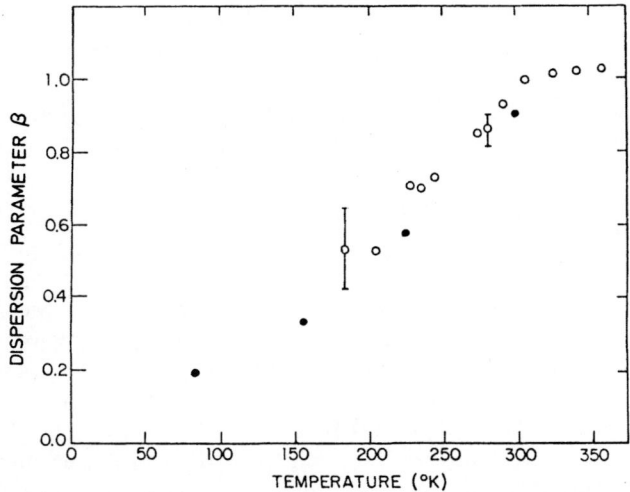

Fig. 6. Temperature dependence of the dispersion parameter β obtained from PA decays shown in Figure 5 (filled circles) and time-of-flight measurements on electrons (open circles) (after Kirby et al.[19]).

A convenient way of describing recombination in highly disordered materials is the use of the recombination time distribution function $g(\tau)$[20,21]. In undoped a-Si:H, at the time when the trapping of holes in the v.b. tail is completed, their total density has a certain value n_{o}. The decay of the hole density $n(t)$ is determined by the equation:

$$n(t)/n_{o} = \int_{0}^{\infty} e^{-t/\tau} g(\tau) d\tau \qquad (7)$$

If we know $n(t)$ we can determine $g(\tau)$ from Eq.(7) as the inverse Laplace transform. In the following, we will use an approximate approach. Since the exponential term is small for $\tau < t$ and approaches one for $\tau \gg t$ one has approximately

$$n(t)/n_{o} = \int_{t}^{\infty} g(\tau) d\tau \qquad (8)$$

For the derivative

$$R(t) = \frac{-1}{n_{o}} dn(t)/dt \qquad (9)$$

one obtains

$$R(t) = g(t) \qquad (10)$$

Eq.(10) indicates that one can determine the recombination time distribution from measuring the time evolution of R.

Following Pollak[22], the power law decays that are characteristic of dispersive transport are obtained from Eqs. (8) through (10) if we write $\nu\tau = e^X$ where X is a random variable with an exponential distribution $G(X) = X_0^{-1}e^{-X/X_0}$ where X_0 is a constant; ν is an attempt frequency. In this case, $g(\tau) = G(X)dX/d\tau = \tau^{-1}e^{-\ln\nu\tau/X_0} = \nu(\nu\tau)^{-1-1/X_0}$ which, according to Eq.(10) gives $R(t) = \nu(\nu t)^{-\beta-1}$ with $\beta = 1/X_0$. In the case of multiple trapping $X = E/kT$ (E = energy of electron state in the tail), $X_0 = T_0/T$ and $\beta = T/T_0$. For distributions of X other than exponential β is a time dependent quantity; at a certain time, β defined from the decay of n(t) differs from β defined from the decay of $dn(t)/dt$[23].

Measuring R(t) instead of n(t) has important advantages. This function directly gives the recombination time distribution $g(\tau)$. It is a quantity that can be directly compared with luminescence intensity which is proportional to the rate of decay of carrier densities (from which $g(\tau)$ of radiative recombination can be determined). Finally, direct measurement of $dn(t)/dt$ overcomes a serious experimental problem with measuring n(t) directly; at high temperatures PA is small and it is extremely difficult to determine the base line to which to refer n(t). This uncertainty is strongly enhanced in the log-log plots of the responses. The position of the base line does not influence R(t). A disadvantage of directly measuring R(t) instead of n(t) is a smaller signal whose detection requires longer integration times.

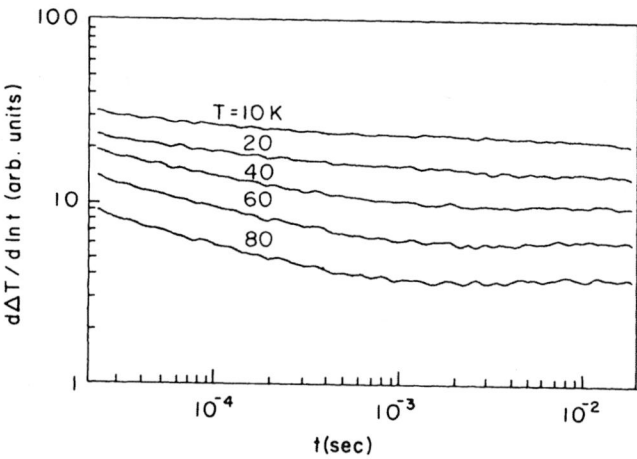

Fig. 7. Decay of tR(t) in a-Si:H at low temperatures. The curve spacing is arbitrary.

In a simple tunneling model the recombination time is related to the distance r between the two states by the relation

$$\tau = \nu^{-1}e^{2r/a} \qquad (11)$$

where a is the extent of the wave functions, ν is the attempt frequency. Since $r = (a/2)\ln\nu\tau$ the function $g(\ln\tau)$ is proportional to the distribution of distances $G(r)$. It follows from $g(\tau) = g(\ln\tau)d\ln\tau/d\tau = \tau^{-1}g(\ln\tau)$ and $g(t) = R(t)$ that $tR(t) = g(\ln t) \sim G(r)$. A plot of $tR(t) = n_0^{-1}dn/d\ln t$ vs. ln t has the same shape as the distribution function of r. Let's note that the negative slope of log $tR(t)$ vs. log t gives the dispersion parameter β which in general is time–dependent[23].

As an example, a plot of log $tR(t)$ vs. log t measured on a-Si:H is shown in Fig. 6. The shape corresponds to log $G(r)$ vs. r. It has the general features that one expects for this distribution, a cut-off at short r's and a decrease towards large r's. We may compare it with the distribution of nearest neighbor distances of randomly distributed electrons and holes with density n:

$$G(r) = 4\pi n r^2 e^{-\frac{4}{3}\pi n r^3} \tag{12}$$

The distribution calculated from Eq.(12) decays much faster at large distances r than the measured distribution (Fig. 7). A possible reason for the discrepancy may be the dependence of the distribution of distances on time as analyzed by Dunstan[24] ("distant pair model"). Recombination process quickly removes carriers at short distances so that as time evolves the r distribution shifts to longer r's. Calculations have shown that although the Dunstan's model works in the right direction, it does not provide a complete agreement with the data. At this time, the origin of the discrepancy is not understood.

ACKNOWLEDGEMENT

I benefited very much from discussions with M. Pollak, H. A. Stoddart, J. Strait and Z. Vardeny. The work was partly supported by the National Science Foundation grant DMR82-09148.

REFERENCES

[1]   Z. Vardeny and J. Tauc, Phil. Mag. B, 1985, Sir Nevill Mott's Festschrift (to be published).

[2]   C. Thomsen, J. Strait, Z. Vardeny, H. J. Maris, J. Tauc and J. J. Hauser, Phys. Rev. Lett. 53, 989 (1984).

[3]   M. Altarelli, D. L. Dexter, H. M. Nussenzweig and D. Y. Smith, Phys. Rev. B 6, 4502 (1972).

[4]   D. E. Ackley, J. Tauc and W. Paul, Phys. Rev., Lett. 43, 715 (1979).

[5]   Z. Vardeny and J. Tauc, Phys. Rev. Lett. 46, 1223 and 47, 700 (1981).

[6]   J. Kuhl, E. O. Gobel, Th. Pfeiffer and A. Jonietz, Appl. Physics A34, 105 (1984).

[7]   J. Tauc in Semiconductors and Semimetals, (Acad. Press) 21 Part B (editor J. Pankove), 299 (1984).

[8]   Z. Vardeny, J. Strait and J. Tauc, Appl. Phys. Lett. 42, 580 (1983).

[9]   Z. Vardeny, J. Strait, D. Pfost, J. Tauc and B. Abeles, Phys. Rev. Lett. 48, 1132 (1982).

[10]   J. Strait, Ph.D. Thesis in Physics, Brown University, 1984.

[11]   J. Strait and J. Tauc, to be published.

[12]   J. Orenstein and M. A. Kastner, Sol. State Commun. <u>40</u>, 85 (1981)

[13]   P. O'Connor and J. Tauc, Phys. Rev. Lett. <u>43</u>, 311 (1979); Phys. Rev. B <u>25</u>, 2748 (1982).

[14]   P. O'Connor and J. Tauc, Sol. State Commun. <u>36</u>, 947 (1980).

[15]   D. Pfost, Hsiang-na Liu, Z. Vardeny and J. Tauc, Phys. Rev. B <u>30</u>, 1083 (1984).

[16]   S. Ray, Z. Vardeny and J. Tauc, J. de Physique, <u>42</u>, C-4, 555 (1981).

[17]   T. Tiedje and A. Rose, Sol. State Commun. <u>37</u>, 49 (1981); J. Orenstein and M. A. Kastner, Phys. Rev. Lett. <u>43</u>, 161 (1981).

[18]   D. Pfost and J. Tauc, Sol. State Commun. <u>48</u>, 195 (1983).

[19]   P. B. Kirby, W. Paul, S. Ray and J.Tauc, Sol. state Commun. <u>42</u>, 533 (1982).

[20]   C. Tsang and R. A. Street, Phys. Rev. B <u>19</u>, 3027 (1979).

[21]   G. S. Higashi and M. A. Kastner, Phil. Mag. B <u>47</u>, 83 (1983).

[22]   M. Pollak, private communication and Ref. 23.

[23]   H. A. Stoddart, M. Pollak and J. Tauc, Proc. 17th Int. Conf. on Physics of Semiconductors, San Francisco 1984 (to be published).

[24]   D. J. Dunstan, Phil. Mag. B <u>46</u>, 579 (1982); Phil. Mag. B 1985, to be published.

COPLANAR TRANSIENT PHOTOCURRENTS AND THE

DENSITY OF STATES IN a-Si:H

E. A. Schiff

Department of Physics
Syracuse University
Syracuse, NY 13210 USA

INTRODUCTION

One of the most enduring concepts in amorphous semiconductors is that of a mobility gap (Mott, 1966). One-electron excitations of the semiconductor having energies in this gap are localized, and hence contribute substantially less to transport than extended excitations lying beyond well-defined mobility edges. The discovery by Spear and LeComber (1975) that the electronic properties of plasma-deposited amorphous silicon (a-Si:H) were sensitive to dopants, and the fabrication only one year later of an a-Si:H solar cell (Carlson and Wronski, 1976) precipitated an enormous worldwide research effort on this material whose themes are largely suggested by the concepts of mobility edges and gaps. In this paper I shall discuss the interrelationships of two of these themes in a-Si:H: the transport of photocarriers, and the energy distribution g(E) of states within the mobility gap.

The most influential measurements of the density of states in a-Si:H are "junction" measurements such as the field effect and deep level transient spectroscopy (DLTS); the history of this class of measurements has been thoroughly reviewed by Cohen (1984). Of course the explication of photocarrier dynamics and recombination is perhaps the principal application of g(E) measurements in a-Si:H. It is therefore very disturbing that the most refined junction measurements - capacitance DLTS - yield a density of states which is very different from the expectations from photoconductivity experiments. In Figure 1 we illustrate the densities of states proposed from DLTS and from transient photocurrent measurements. Transient photocurrents are the electrical response of a specimen to a photoillumination impulse, and are measured in two configurations. Time-of-flight (TOF) transient photocurrents are measured transverse to the surface of a thin-film specimen, and coplanar (CTP) photocurrents flow parallel to the surface. The reader is referred to the standard monograph (Mott and Davis, 1979) for a more complete discussion of these techniques and their applications in non-crystalline materials. The TOF g(E) of Fig. 1 is based on Tiedje (1984); the coplanar g(E) will be discussed further later in this paper. It is immediately apparent that the g(E) estimates based on coplanar photocurrent measurements do not agree with the DLTS results.

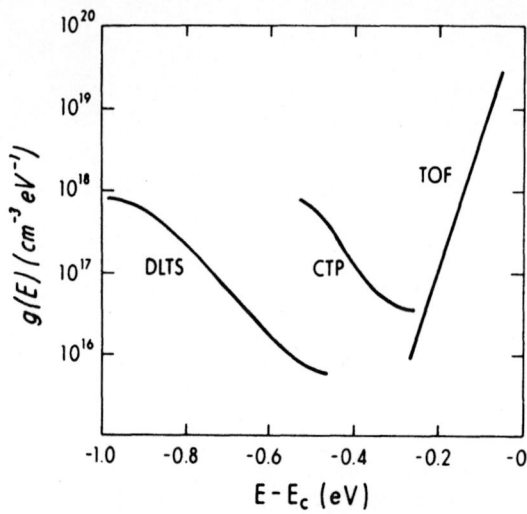

Fig. 1: Three estimates of the density of states in a-Si:H. DLTS - after Cohen (1984) for a phosphorus doped specimen. TOF - after Tiedje (1984) for an undoped specimen. CTP - for an undoped specimen (dangling bond spin density 7 x 10^{16} cm^{-3}).

One response to this discrepancy between coplanar based estimates of g(E) and DLTS is to prefer the DLTS density-of-states. There are several arguments in favor of this procedure, of which the most persuasive are:

(1) A fully satisfactory general procedure for inverting photocurrent data to obtain g(E) does not exist. The only widely accepted inversion requires that photocurrent transients exhibit a simple power-law decay. By way of contrast, coplanar photocurrent transients often have very complex structures. An example will be given later in this paper.

(2) Coplanar measurements are susceptible to confusion of surface and bulk densities of states. Capacitance DLTS is rather insensitive to surface states, and hence should be preferred. A similar argument pertains to preferring capacitance DLTS to field-effect and current mode DLTS measurements.

(3) Several important auxiliary experiments can be interpreted with some success in terms of the DLTS g(E). Such experiments include high-temperature luminescence (Street, Biegelsen, and Weisfield, 1984), time-of-flight charge collection experiments (Street, 1982), and high sensitivity optical absorption experiments (Jackson, 1982).

In this paper the present status of these three arguments against coplanar photocurrent determinations of g(E) will be explored. In particular I shall endeavor to demonstrate that none of the three arguments given above is compelling. First, inversion procedures for photocurrent transients will be discussed from a new viewpoint which substantially clarifies their utility and limitations. I shall then review the available information about the electronic properties of a-Si:H interfaces and surfaces from the point of view of determining whether a photocurrent transient is bulk or surface dominated; in particular it appears that a very simple experiment is sufficient to resolve this issue. Finally, I shall discuss the influence of doping on photocarrier studies in a-Si:H, and suggest that the discrepancy between the DLTS and coplanar photocurrent measurements of g(E) is more likely a consequence of the use of doped specimens for DLTS and undoped for coplanar photocurrents than of any artefacts in the coplanar measurements.

OBTAINING g(E) FROM PHOTOCURRENT TRANSIENTS

Two distinct processes are involved in transient photocurrent response: the time-evolution of the dominant photocarrier's dynamics, and the eventual loss of the photocarrier either to sweepout at an interface (in the time-of-flight geometry) or to recomination (in the coplanar contact geometry). It is always possible in principle to separate a measured transient into two domains: a short-time domain in which photocarrier dynamics are clearly observed, and a long-time domain in which the transient response shows features due both to dynamics and to sweepout/recombination.

In practice for amorphous semiconductors only one class of transient response has a widely known and accepted interpretation: a dispersive, power-law decay $t^{-(1-\alpha)}$ of the photocurrent in the short-time domain, followed by a steeper, long-time power law characteristic of sweepout or recombination (Scher and Montroll, 1975). The observation of a linear temperature dependence of the dispersion parameter $\alpha = T/T_c$ is furthermore generally accepted to be a consequence of a simple multiple-trapping model. In this model, electrical transport occurs only when a photocarrier occupies an energy level at or above a mobility-edge E_c. The observation of a power law decay is then viewed as evidence for an exponentially rising density of localized states below E_c. A photocarrier above E_c may be "trapped" by such states and later thermally re-excited above E_c.

In the present paper, I shall restrict the discussion to situations in which the multiple-trapping model transport applies, and will neglect other transport models (eg. the remarkable recent work on hopping by Monroe {1985}). However, even with this restriction the analysis of transient photocurrents can present formidable difficulties if the transient does not have the simple form described above. Fig. 2 shows such a transient obtained at Syracuse for an undoped a-Si:H specimen. The transient was obtained using a weak, uniformly absorbed laser illumination impulse; the photocurrent response was linear in the impulse strength. The relationship between the measured photocurrent i(t) and the plotted "drift mobility" is:

$$\mu_d(t) = i(t) \, (1^2/eNV) \tag{1}$$

where 1 is the spacing between the specimen's contacts, e is the electronic charge, N is the number of absorbed photons from the impulse, and V is the bias voltage. The specimen was intentionally deposited at a temperature (160 C) below the range considered optimal for a-Si:H, and is both less photoconductive and possesses a higher dangling bond spin density than typical for "device-grade" a-Si:H. The specimen was 2 μm thick. Similar transients have been published elsewhere for higher quality specimens (Huang, Guha, and Hudgens, 1983; Pandya, Schiff, and Conrad, 1984; Conrad, Pandya, and Schiff, 1985).

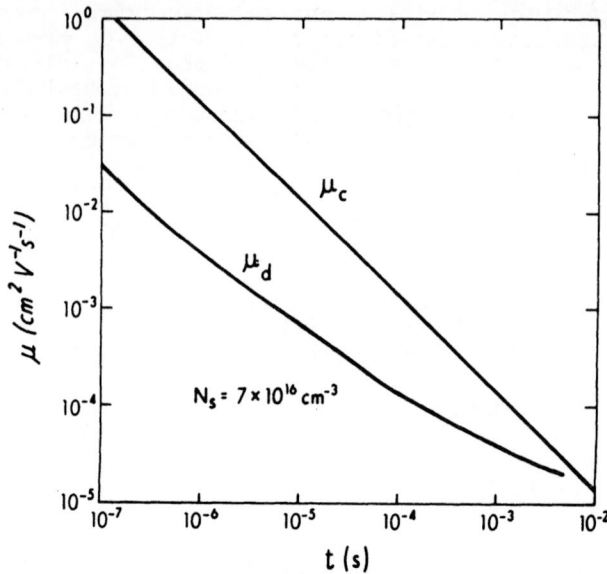

Fig. 2: μ_d : Experimental coplanar drift mobility at 300 K for an undoped a-Si:H specimen. The specimen was not optically biased. The relation between the measured photocurrent i(t) and $\mu_d(t)$ is given in the text.
μ_c : Calculated critical drift mobility defined in the text.

The first issue in the interpretation of transient photocurrents is the determination of the photocarrier - electron or hole - which dominates transport. This determination cannot be done using the information from coplanar transients. Time-of-flight transient measurements have explored both electron and hole dynamics at short times in a-Si:H, and can be used to make a short time judgment for coplanar measurements. The photo-thermopower technique yields the carrier which dominates the average photocurrent. I assume that observed coplanar transients represent transport of only one carrier type, and in particular for undoped a-Si:H I assume the dominant photocarrier to be electrons.

Recombination Effects on Transient Photocurrents

The interpretation of a complex transient next requires that the onset of recombination effects in the transient be recognized. In our view, the failure to deal adequately with this issue is a principal shortcoming of earlier analyses. There is in fact a simple procedure to recognize the onset of recombination for the case of linear transient photocurrents (measured using photoexcitation impulses sufficiently weak to yield a photocurrent response which is linear in the photoexcitation impulse strength); for such transients the recombination rate for a photocarrier excited above E_c is time independent. The procedure involves measuring both the direct photocurrent and (using an integrating capacitor) the time-integral of the transient. In particular, the time T_R at which the integrated photocurrent reaches half its asymtotic (infinite time) value estimates the time after which recombination (or more generally carrier loss) effects qualitatively modify the transient. This procedure is similar to (and in fact inspired by) charge collection measurements in the time-of-flight geometry (Street, 1982).

Surprisingly, analysis of the transient following T_R is quite straightforward. This ease of analysis results because the transient following T_R is "emission limited" (cf. Tiedje, 1984): the observed photocurrent following T_R is simply proportional to the rate of emission from traps. A simple approximate expression obtains for slowly varying $g(E)$ using the time-dependent demarcation energy approach ("TROK") of Tiedje, Rose, Orenstein and Kastner (Tiedje, 1984; Orenstein, Kastner, and Vaninov, 1982):

$$(kT\ g(E_d(t)))\ e^{-(E_c-E_d)/kT}\ \propto i(t)/T_R \tag{2}$$

$$E_c-E_d(t) = kT\ \ln(\nu t) \tag{3}$$

and hence:

$$g(E_d)\ \propto i(t)t \tag{4}$$

It is a simple matter to work out the constants of proportionality; since these relations will not be exploited here, their detailed forms are not given. If $g(E)$ is rapidly varying (as evidenced by photocurrent decays faster than t^{-1}) the analysis must be modified (Marshall and Main, 1983). An example of such a modification will be given shortly.

In actual practice our integrated coplanar photocurrent transient data on undoped a-Si:H show that T_R occurs very late in these transients (Conrad, Pandya, and Schiff, to be published). This result confirms the assumption made in earlier analyses of CTP data that a very late cutoff in the transient was due to the onset of recombination (Pandya, Schiff, and Conrad, 1984; Huang, Guha, and Hudgens, 1983). The confirmation of this assumption has important implications for recombination models.

An Inversion Procedure to Obtain g(E)

Given that transient data such as Fig. 2 must be interpreted as due to photocarrier dynamics alone, the problem of inversion of such data to obtain $g(E)$ remains. A schematic version of coplanar photocurrent transients in a-Si:H at 300 K is given in the upper portion of Fig. 3. Region I corresponds to the non-dispersive electron transport observed in TOF (Tiedje, 1984). The sharply rising conduction band tail at the right of the lower portion of the figure has essentially no effect on photocarrier transport at 300 K. Region II is transitional and corresponds to the transfer of electrons from energies near the mobility edge to traps lying below the conduction band tail (Street, 1982); the detailed form of the trap distribution is inconsequential in this region. Finally, Region III will be viewed as a consequence of the multiple-trapping dynamics associated with the distribution of shallow traps (curves A and B in Fig. 3).

Huang, et al (1983) ascribed region III of the transient to an exponentially rising $g(E)$ shown as curve A in Fig. 3. This form accounts for dispersive transport in region III, but it is not the only form which does. As I shall now show, a density of states which has a peak below E_c also accounts for the photocurrent; this possibility is represented by curve B of Fig. 3. The transient photocurrent associated with curve A is obtained in two parts. First, I review the analysis when the demarcation energy $E_d(t)$ lies below the peak (and in the familiar exponentially rising region).

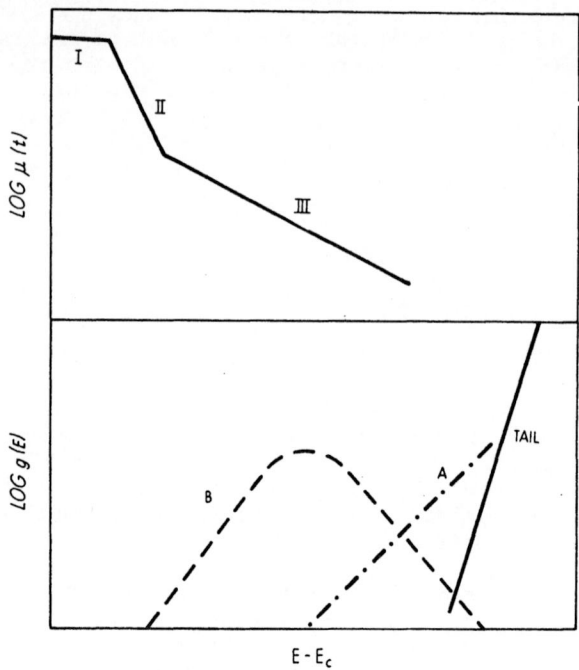

Fig. 3: The upper portion of the figure is a schematic representation of observed photocurrent transients in a-Si:H broken into three regions. The lower portion of the figure shows the possible forms for the density of states g(E) which account for the photocurrent data.

Taking for the density of states:

$$g(E) = (N_t/kT_c) \, e^{(E-E_t)/kT_c} \qquad E < E_t$$

where N_t is the total density of defects in the band and E_t is the energy of the band's maximum density, an adequate approximation to $\mu_d(t)$ is:

$$\mu_d(t) = \mu_0 \frac{N_c}{kT_c \, g(E_d(t))} \, (\nu t)^{-1} = \frac{\mu_0}{N_t b_t t} \, \frac{N_t}{kT_c \, g(E_d)} \qquad (5)$$

$$= (N_c/N_t) \, e^{-E_t/kT_c} \, (\nu t)^{-(1-\alpha)} \qquad (6)$$

where μ_0 is the microscopic mobility for electrons, N_c is the effective density of states at the mobility edge, b_t is the trap capture coefficient, and ν is the "attempt to escape" frequency. The detailed balance relation $b_t N_c = \nu$ has been used to simplify the expressions. Since $kT_c \, g(E_d) < N_t$, expression (5) is only valid if:

$$\mu_d(t) > \mu_c(t) = \frac{\mu_0}{N_t \, b_t \, t}$$

Two features of these equations should be noted. First, (5) provides a simple approximate procedure for obtaining a trial g(E) from the drift

362

mobility. One simply substitutes any reasonable constant for kT_c. An iterative procedure may be employed to refine the estimate if necessary. Second, the expression only aplies if the drift mobility is larger than the critical drift mobility μ_c; otherwise $E_d(t)$ lies above the peak of the distribution at E_t.

The drift mobility for $E_d(t) > E_t$ is remarkable. This region corresponds to a density of states falling in energy. Assuming:

$$g(E) = (N_t/kT_c) \ e^{(E_t-E)/kT_c} \qquad\qquad E > E_t \qquad\qquad (7)$$

An expression for the drift mobility is obtained by arguments similar to those given previously for emission-limited transients following recombination:

$$\mu_d(t) = (\mu_0 N_c/N_t^2) \ kT_c g(E_d) \ (\nu t)^{-1} = \mu_c(t) \ (kT_c g(E_d)/N_t) \qquad (8)$$

$$= (\mu_0 N_c/N_t) \ (\nu t)^{-(1-\alpha)} \ e^{-(E_c-E_t)/kT_c}$$

These expressions show that the inversion of coplanar drift mobilities to obtain g(E) first of all requires that the critical drift mobility μ_c be determined. It is not possible even to obtain the slope of g(E) without this knowledge. Segments of the drift mobility curve which lie below the critical curve correspond to the passage of the demarcation energy through a density of states which falls with increasing energy; segments above the critical curve correspond to a rising g(E). Remarkably, the drift mobility itself needs to change very little as the demarcation energy passes over the peak of g(E), and in fact for the model solved above there is no structure at this time.

In Fig. 2, the critical drift mobility curve was calculated using $b_T/\mu_0 = 10^{-10}$ cm-V and $N_t = 7 \times 10^{16}$ cm^{-3} (the specimen's dangling bond spin density). The g(E) curve labelled CPT in Fig. 1 is based on the inversion of Fig. 2 using these values and the expression (8) for the drift mobility. The estimate of b_T/μ_0 is based on optical bias effects on the transient photocurrent (Pandya and Schiff, to be published). I have assumed that the filled shallow traps are doubly occupied dangling bonds, as suggested by time-of-flight charge collection experiments (Street, 1982). The energy scale is determined directly by our choice of $\nu = 10^{11}$ s^{-1}. The energy scale is also determined indirectly by the value of b_T/μ_0, because for the inversion procedure given above the integral of g(E) is constrained to be N_t. A reasonable estimate of the error in the energy scale is 0.1 eV.

THE ROLE OF SURFACE STATES

The second major difficulty in the interpretation of coplanar transient photocurrents is distinguishing between the role of surface and bulk states in determining the measured g(E). It is particularly attractive to attribute the g(E) from coplanar transients (cf. Fig. 1) to surface effects because of its similarities with current mode DLTS results (Fuhs, 1984). Such measurements are much more sensitive to surface states than the capacitance methods normally used to obtain g(E). However, there is rather compelling evidence from two experiments on a-Si:H that photoconductivity is dominated by the bulk density of states.

The most important type of experiment is spectral-dependence of photoconductivity. Such experiments can determine the relative photosensitivity of the specimen to surface absorbed or uniformly absorbed photons. Because the photoconductivity in a-Si:H is a nonlinear function of the absorbed photon flux, spectral dependence experiments should either be performed in the linear, small-intensity limit or should be conducted at constant photoresponse.

The three possible outcomes of such an experiment are readily interpreted. If the response of the specimen is stronger for surface absorbed than for bulk absorbed light, then clearly surface processes are dominating the photoconductivity. Furthermore, such an outcome is only possible for specimens thicker than the characteristic depths of their surface regions. This depth represents the combined effects of the space-charge region near the surface and of photocarrier diffusion into the space-charge. In thin specimens all photons will effectively be surface absorbed. If surface and bulk absorbed photons give the same response, it is not possible to resolve whether surface states are important. Finally, if surface absorbed photons yield less photoresponse than bulk absorbed light, the specimen must again be in the thick film regime and furthermore its photoconductivity is bulk dominated.

The work of Persans (reported by Fritzsche, 1984) on the spectral dependence of photoconductivity in undoped a-Si:H clearly shows the features expected from bulk-dominated photoconductivity for uniformly absorbed light. For thicker specimens (d > 1 µm) a substantially reduced photosensitivity was found for surface absorbed light relative to uniform absorption. For thin specimens the photoconductivity for uniformly absorbed light was greatly diminished relative to thicker specimens, and the spectral dependence of the photoconductivity was also greatly reduced. Both of these observations are consistent with the hypothesis that thicker specimens were essentially uninfluenced by the state of the substrate interface or the surface. Furthermore, by ambient treatments Persans was able to substantially modify the free surface of his specimens, affecting both the dark current and also the photoresponse to surface absorbed light. The photoresponse to uniformly absorbed light was only slightly affected, again in agreement with the bulk photoconductivity model. Extensive results on the thickness-dependence of photoconductivity in undoped a-Si:H reported by Hasegawa and Imai (1982) also support Persan's findings.

These experiments demonstrate that bulk dominated photoconductivity is possible and may even be prevalent in a-Si:H. However, the fascinating recent experiments on the photo-field effect (Jackson, Street, and Thompson, 1983; Kagawa and Matsumoto, 1983; Vaid and Fritzsche, 1984) show that surface-dominated results are also possible under certain conditions. In these experiments a substantial enhancement of photoconductivity accompanied the creation of a surface space-charge layer by a nearby field electrode. Unfortunately, extensive spectral dependence studies of this effect have not been undertaken, and thus these photo-field results cannot be used to confirm the classification proposed above.

In conclusion, spectral-dependence studies with a favorable outcome (reduced photosensitivity for surface absorbed light) should be a sufficient test to rule out qualitative surface effects on photocurrent measurements. Such a favorable outcome is only anticipated in specimens thicker than their surface and interfacial regions. The existing data on photoconductivity for specimens without externally generated space-charge are not consistent with the attribution of substantial surface influence on photocurrents measured with uniformly absorbed light.

CONCLUSIONS

How can the discrepancy between coplanar determinations of g(E) and DLTS determinations be resolved? Although there has been some controversy regarding the interpretation of DLTS measurements (Lang, Cohen, Harbison, Chen, and Sergent, 1984), there is sufficient unanimity among its practitioners to give the DLTS g(E) substantial credibility. Furthermore, the DLTS g(E) (in conjunction with refinement of spin-dependent DLTS {Cohen, 1984}) has substantial predictive value for luminescence and electron spin-resonance results in phosphorus doped, n-type specimens which are comparable to those used for DLTS (Street, et al, 1984).

The DLTS g(E) is not as successful for undoped specimens. For example, based on judicious extrapolations of results for n-type a-Si:H, Street, et al (1984) attribute the defect-related luminescence band near 0.8 eV to electron capture by a filled dangling bond. Wilson, Sergent, Wecht, Williams, Kerwin, Taylor and Harbison (1984) maintain that this transition is in fact associated with electron capture by an unfilled dangling bond. Similar difficulties arise in accounting for photoconductivity and electron spin resonance measurements in undoped a-Si:H; a thorough discussion of these difficulties enters into the area of recombination, and will be presented elsewhere.

The availability of a new determination of the density of mobility gap states presents an opportunity for exploring it's predictive power for other experiments. This exploration is only beginning for the coplanar g(E) presented here, and thus this density of states has not yet been fuly tested. Nonetheless this g(E) appears sufficiently credible that the possibility of doping effects as the origin of the discrepancy between it and the DLTS g(E) deserves deeper consideration.

ACKNOWLEDGMENTS

I wish to thank Kevin Conrad for the data presented in Fig. 2, and Kevin Conrad, Ranjana Pandya, and Michael Parker for many illuminating conversations. This work was supported by the National Science Foundation through grant DMR 83-06083.

REFERENCES

Carlson, D. E. and Wronski, C. R. (1976). Appl. Phys. Lett. 28, 11.

Cohen, J. D. (1984). Semiconductor and Semimetals, Volume 21C (edited by J. I. Pankove, Academic Press), 9.

Fritzsche, H. (1984). Semiconductor and Semimetals, Volume 21C (edited by J. I. Pankove, Academic Press), 309.

Fuhs, W. (1984). Festkörperprobleme XXIV (edited by P. Grosse, Vieweg), 133.

Hasegawa, S. and Imai, Y. (1982). Phil. Mag. B 46, 329.

Huang, C.-Y., Guha, S., and Hudgens, S. J. (1983). Phys. Rev. B27, 7460.

Jackson, W. B. (1982). Solid State Comm. 44, 477.

Jackson, W. B., Street, R. A., and Thompson, M. J. (1983). J. Non-cryst. Solids 59 & 60, 497.

Kagawa, T. and Matsumoto, N. (1983). J. Non-cryst. Solids 59&60, 465.

Lang, D. V., Cohen, J. D., Harbison, J. P., Chen, M. C., and Sergent, A. M. (1984). J. Non-Cryst. Solids 66, 217.

Orenstein, J., Kastner, M., and Vaninov, V. (1982). Phil. Mag. B 46, 23.

Marshall, J. M., and Main, C. (1983). Phil. Mag. B 47, 471.

Monroe, D. (1985). Phys. Rev. Lett. 54, 146.

Mott, N. F. (1966). Phil. Mag. 13, 989.

Mott, N. F. and Davis, E. A. (1979). Electronic Properties of Non-Crystalline Materials, Second Edition (Oxford University Press).

Pandya, R., Schiff, E. A., and Conrad, K. A. (1984). J. Non-cryst. Solids 66, 193.

Scher, H. and Montroll, E. W. (1975). Phys. Rev. B 12, 2455.

Spear, W. E. and LeComber, P. G. (1975). Solid State Comm. 17, 1193.

Street, R. A. (1982). Appl. Phys. Lett. 41, 1061.

Street, R. A., Biegelsen, D., and Weisfield, R. L. (1984). Phys. Rev. B30, 5861.

Tiedje, T. (1984). Semiconductor and Semimetals, Volume 21C (edited by J. I. Pankove, Academic Press), 207.

Vaid, J. and Fritzsche, J. (1984). J. Appl. Phys. 55, 440.

Wilson, B. A., Sergent, A. M., Wecht, K. W., Williams, A. J., Kerwin, T. P., Taylor, C. M., and Harbison, J. P. (1984). Phys. Rev. B30, 3320.

TRAPPING OF ELECTRONS AND HOLES IN HYDROGENATED AMORPHOUS SILICON

R. Carius, W. Fuhs, and A. Schrimpf

Universität Marburg
Renthof 5
D-3550 Marburg, FRG

SUMMARY

We report on a study of non-equilibrium carrier distributions in a-Si:H using IR-excitation as a probe. In a dual beam experiment we observe quenching and enhancement of the photoconductivity and quenching of the photoluminescence. At low temperature after excitation with band gap light non-equilibrium distributions of trapped electrons and holes persist with long lifetimes. The recombination of these metastable carrier distributions is studied by light induced electron spin resonance and by the transients of photocurrent and photoluminescence which are created, when the trapped carriers are excited by IR-light.

INTRODUCTION

Studies of photoluminescence (PL), photoconductivity (PC) and electron spin resonance (ESR) have lead to a widely accepted model for recombination in a-Si:H. At low temperature photoexcited carriers are trapped in band tail states from where they either recombine radiatively with the carrier in the other band tail or non-radiatively by tunneling to defect states[1]. For radiative recombination two models have been suggested which lead to different kinetics. In the geminate model[1] it is assumed that the photo-excited electrons and holes do not diffuse apart to large separations such that the recombination occurs between trapped geminate pairs. In the distant pair model[2] it is supposed that the carriers are able to diffuse and are trapped at random in band tail states. Common to both models is that the distribution of intra-pair separations leads to a broad distribution of lifetimes. In addition, it is an inherent feature of both models that at low temperatures a metastable distribution of trapped electrons and holes builds up during illumination. These pairs have large intra-pair separations and correspondingly long lifetimes[3,4]. The non-equilibrium distributions persist after illumination and the long time decay of the residual light induced-ESR signal has been used to study their recombination[5-8]. At higher temperatures (T>50K) electron-hole pairs are broken by thermal excitation which leads to thermal quenching of PL when the carriers are able to diffuse to defects. The mobile carriers contribute to photoconduction which is determined by multiple trapping processes. Detailed studies of spin dependent photoconductivity and photoluminescence have lead to the conclusion that the Si-dangling bonds are the most important centers

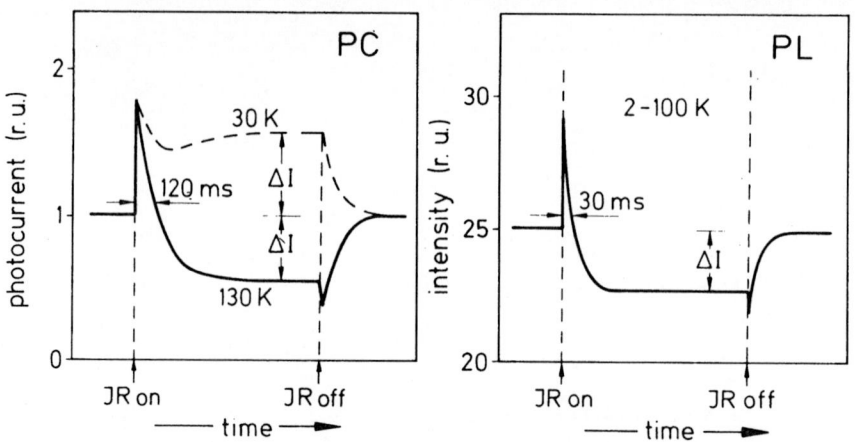

Fig.1: Transients of photoconductivity (PC) and photoluminescence (PL) in a dual beam measurement showing enhancement and quenching effects.

for non-radiative recombination[1,9]. It has been suggested that hopping of trapped carriers among band tail states is a rate limiting process for non-radiative recombination at least in samples of low defect density[9]. The non-equilibrium carriers can be detected by photoinduced absorption[10], LESR[5-8] and can be excited by IR-light which in a dual beam experiment gives rise to optical quenching and enhancement of photoconductivity and photoluminescence. In addition, at low temperature optical excitation from the metastable distribution of trapped carriers causes transients of photoconductivity and photoluminescence[6,7,11] which can be used to study the trap occupation and recombination.

In this report we summarize our recent investigations of non-equilibrium carriers in a-Si:H using IR-excitation as a probe. a-Si:H-films were studied which had been prepared by glow discharge decomposition in a capacitively coupled glow discharge system using the following parameters: 5% SiH_4 diluted in He, rf-power 4W, frequency 13.56 MHz, total pressure 0.5 mbar, flow rate 80 sccm, substrate temperature 280°C, deposition rate 2.5 Å/s. The Fermi level positions as determined from the dark conductivities amounted to $E_C-E_F = 0.87$ eV (undoped), $E_C-E_F = 0.59$ eV (33 vppm PH_3), $E_F-E_V = 0.72$ eV (100 vppm B_2H_6) and $E_F-E_V = 0.57$ eV (500 vppm B_2H_6).

DUAL BEAM EXPERIMENTS

In dual beam arrangement the photoconductivity of undoped a-Si:H (n-type) of high quality is effectively quenched by IR-light[12-14]. This effect has been explained by the enhancement of the recombination due to optical excitation of minority carriers from deep traps. In undoped films the relevant hole traps are considered to be band tail states and the recombination centers have been identified as Si-dangling bonds[9]. It seems, however, that the general situation is more complex: IR-excitation gives rise to quenching and enhancement effects of photoconductivity and photoluminescence in various temperature ranges and these effects are not restricted to undoped films. Fig. 1 shows examples for the transient response of the photocurrent (PC) and photoluminescence (PL) of an undoped film, when additional IR-light is turned on and off. At high temperature (T=130K) a short enhancement of PC is followed by quenching in the steady state. At low temperature the pulse shape of the same sample is quite different with

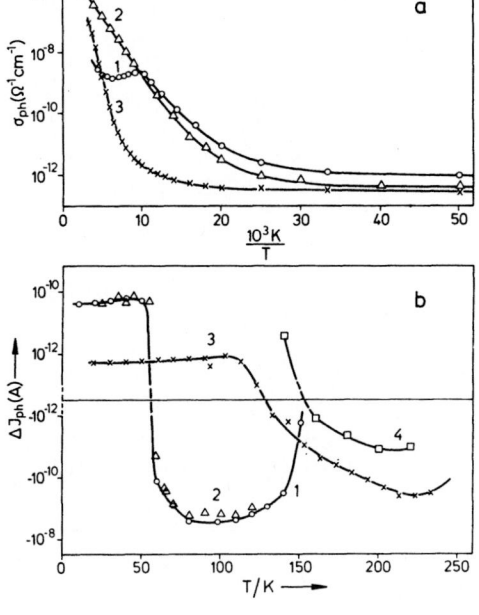

Fig.2: Photoconductivity (a) and change
of the photocurrent, ΔI_{ph}, by
additional IR-illumination (b)
as a function of temperature.(o)
(undoped), (Δ) 33 vppm PH_3, (x)
100 vppm B_2H_6, ($\square$) 500 vppm B_2H_6

a pronounced enhancement of PC also in the steady state. The pulse shape of
the PL-intensity is very similar to the PC-pulse at high temperature, the
short enhancement being always followed by quenching in the steady state.
Such behaviour is observed in the temperature range 5-100K. In Fig. 2 the
change of the steady state photocurrent, ΔJ_{ph}, induced by additional IR-
illumination is plotted as a function of temperature for the undoped and
doped samples. For comparison the temperature dependences of the photoconductivity are given. In these experiments the films were excited simultaneously by two unmodulated light beams of photon energies $h\nu_1 = 1.92$ eV
and $h\nu_2 < 0.7$ eV. The IR-light was obtained by passing the light of a tungsten
lamp through a Ge-filter. The n-type films (undoped and phosphorus doped)
exhibit IR-quenching above 60K and the effect vanishes above 150K when the
relevant hole traps are emptied by thermal excitation. This thermal quenching effect leads to a decrease of the photoconductivity with rising temperature which is nicely seen in the $\sigma_{ph}(T)$-curve of the undoped film. Most
surprisingly no comparable structure is observed in the doped samples. In
the p-type film (boron-doped), which has clearly p-type photoconduction,
quenching is observed above 130K and is still observable at room temperature.
PC-quenching effects are generally associated with the emptying of minority
carrier traps. It is then surprising to find that the thermal detrapping of
electrons in p-type material occurs at considerably higher temperatures than
the detrapping of holes in the n-type films. It is generally accepted that
the trap depth in the band tails is less by a factor of two for electrons
than for holes. It is tempting to assume that the electrons are excited from
negatively charged dangling bonds. Such a process can lead to PC-quenching
in p-type films. The contribution of such transitions to photoinduced absorption recently has been suggested from spin-dependent studies[15]. The

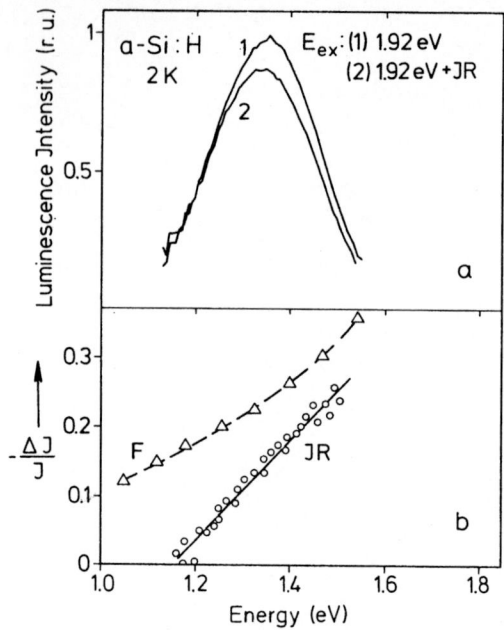

Fig.3: Influence of IR-illumination on
the PL-spectrum (a) and spectral
dependence of the quenching effect $\Delta I/I$ (b) in a dual beam measurement. $G_{ex}=0.5\text{mW/cm}^2=10^{-2}\cdot G_{IR}$
Curve F: field quenching at
10^5 V/cm.

spectral dependence of the PC-quenching effect does not give more insight.
The spectrum of these samples is a featureless one band structure with a
low energy cut off near 0.6 eV for both n- and p-type samples.

At low temperature (T<60K) the PC-quenching vanishes in all kinds of
a-Si:H films and the photocurrent is enhanced by IR-light ($\Delta I_{ph}>0$). In this
temperature range the quasi Fermi levels are very near to the mobility edges.
Then the photoconductivity is determined by the time for localization of the
excited carriers near the mobility edges and the $\eta\mu\tau$-product amounts to
nearly 10^{-11}cm^2/V irrespective of the defect density in the film[16]. This
leads to a temperature independent photoconductivity for all samples at low
temperature (Fig. 2). This interpretation is in accordance with measurements
of the response time, which is observed to be extremely short in the low
temperature range where multiple trapping processes are unimportant[16]. The
onset of PC-enhancement, as the temperature is decreased, is closely related
to the transition from a multiple trapping determined mechanism of photoconduction to a temperature independent behaviour of $\sigma_{ph}(T)$. We therefore
attribute the enhancement effect to electrons and holes which are excited by
IR-light from the deep trapping levels to the conducting states where they
contribute to conduction until they are localized near the mobility edges
or recombine at defect states.

The optical excitation from the non-equilibrium distribution of trapped
carriers results in addition in quenching of the photoluminescence[7,17]. The
quenching rate, defined by $\Delta I/I$ (Fig. 1), depends sensitively on the inten-

sities of the two light beams. When the IR-intensity, G_{IR}, is kept constant, $\Delta I/I$ increases with decreasing intensity of the excitation beam, G_{ex}, and tends to saturate at low excitation levels. At high levels of G_{ex}, e.g. 0.5 mW/cm² , $\Delta I/I$ is roughly proportional to $G_{IR}^{1/2}$, at low excitation intensities this dependence is considerably weaker.

Fig. 3 shows the influence of the IR-excitation on the spectral dependence of the photoluminescence. In the dual beam measurement with unmodulated light beams the spectrum becomes unsymmetrical and shifts slightly to lower energy. The relative change of the intensity by IR-light, $\Delta I/I$, decreases with decreasing photon energy and there is much less quenching on the low energy side of the PL-band. This trend is found on all kinds of samples although there may be differences in the detailed shape of the $\Delta I/I$-curve. The important result thus is that IR-photons preferably quench the luminescence from electron-hole pairs of high energy. It is remarkable that the quenching of the PL in a high electric field has a similar spectral dependence. This is shown by curve F in Fig. 3 which was measured on the same sample at 77K at an electric field of 10^5V/cm across coplanar electrodes. This result is in agreement with the observations by Street et.al.[18] but differs from similar measurements on sandwich structures[19].

It is interesting to note that raising the temperature, application of a high electric field and IR-excitation in a dual beam arrangement have a similar influence on the photoluminescence: The quantum efficiency decreases and the spectral distribution shifts to lower energy. Thermal quenching of photoluminescence in a-Si:H is described by thermal activation of electrons from the conduction band tail and subsequent nonradiative recombination at defects[1]. The shift of the PL-spectrum then is a result of the shift of the demarcation level with increasing temperature to lower energy. In this model field quenching occurs preferably by breaking pairs of high energy, because those most easily can be ionized. The question remains open, by what mechanism IR-excitation leads to quenching of photoluminescence from high energy pairs predominantly.

At present these data do not allow to decide between the conflicting models for recombination. In the geminate pair- and in the distant pair-model IR-quenching can arise when electrons and or holes are excited from band tail states and diffuse to defects where they recombine non-radiatively just like in case of thermal quenching and field quenching. In the geminate pair model one should expect that optical excitation of pairs does not necessarily enhance non-radiative recombination. If the excitation occurs well inside the Onsager radius, diffusion may even enhance radiative recombination by transforming pairs of large separation to those of short separation which can recombine radiatively within the duration of the experiment. This may be the origin of the transient enhancement effect in Fig. 1. Thus the PL-quenching should originate from the pairs of high energy which have a large intra-pair separation. In the geminate pair model, where each electron recombines with its geminate hole, the distribution of the intra-pair separations does not depend on the exciting light intensity, G_{ex}. Therefore, the fraction of pairs which can contribute to the IR-quenching should be independent of G_{ex} in accordance with the observation that $\Delta I/I$ attains a constant value at low excitation levels[7]. In the distant pair model where the carrier are trapped at random in band tail states the lifetime depends on the density of trapped carriers and like in case of the geminate pair model the transient enhancement of the luminescence intensity, when the IR-light is turned on (Fig. 1), can be caused by the redistribution of the trapped carriers prefering short radiative lifetimes. This enhancement effect is related to the PL-transients induced by IR-light, which will be described in the next section.

In these experiments the traps were populated at low temperature (5-100K) by excitation with light of 1.92 eV photon energy. After switching off the illumination the decay of the residual light induced electron spin resonance signal (LESR) was recorded with the magnetic field set at the peak of the absorption derivative of the electron-dangling bond line. The LESR-signal consisted as usual of a superposition of the contribution of trapped majority carriers and neutral dangling bonds in doped samples and of both band tail carriers in undoped films. No change in line shape was observed during the decay i.e. the concentrations of trapped majority carriers and neutral dangling bonds decayed at the same rate. Fig. 4a displays such a transient recorded with an integration time of 0.1s. The general form consists of an initial rapid decay followed by an extremely slow long time decay, which suggest the existence of a metastable trap population. After a dark period t_D=10s the LESR indicates that in undoped films the concentration of carriers trapped in the band tails with lifetimes in excess of 10s is near $5 \cdot 10^{16}$cm^{-3} at T=15K. When the sample is exposed to IR-light ($h\nu$<0.7eV) the residual LESR-signal is effectively quenched (Fig. 4a). Connected with this removal of the trapped carriers are transients of photoluminescence (PL, Fig. 4b) and photoconductivity (PC, Fig. 4c). Such signals are found in undoped and doped a-Si:H films of widely different doping level and film quality. The detection system in these measurements had rise times of below 30 ms for the PC and below 10 ms for the PL-transients. The total PL-intensity was detected by a S1-photomultiplier, hence the light emission is above 1.1 eV, i.e. in the intrinsic luminescence band.

The observation that IR-excitation quenches the LESR signal and thereby generates transients of PL and PC strongly indicates, that the IR-induced effects, reported here, originate from the same non-equilibrium carrier distribution. By IR-illumination trapped carriers are excited into states above the mobility edges where they contribute to conduction until they are immobilized in states near the mobility edges or recombine via defect states. At low temperatures (T<5oK) thermal detrapping is negligible and the response time, τ_{resp}<<10^{-6}s[16]. If the IR-excitation is removed, the PC-transient therefore decays almost instantly on the time scale of this experiment and when it is turned on again the transient recommences at the current level to which it is expected to decay during the interruption. From this behaviour it is evident that the response time of the photocurrent is much smaller than the halfwidth of the signal. In addition, if one calculates the number of carriers which are contained in such a current transient using the value of $\mu\tau \approx 10^{-11}$cm^2/V, which is typically found in the low temperature range[16], one obtains concentrations in excess of $5 \cdot 10^{18}$cm^{-3} in accordance with recent results of Boulitrop[11]. This is much higher than what is indicated by the residual LESR. The form of these signals therefore has to be assigned to the time decay of the generation rate of free carriers, which is proportional to the number of trapped electrons and holes. Retrapping and multiple excitation then are supposed to lead to the large halfwidths of these signals.

The PL-intensity, on the other hand, is determined by the number of recombination processes per sec and therefore is related to the first time derivative of the IR-induced generation rate. Obviously at least part of the excitation processes result in electron-hole pairs of shorter separation, when the carriers are retrapped, which enables radiative recombination on the time scale of experiment. From an integration of the PL-transient therefore a lower limit for the density of trapped carriers can be obtained. For the undoped sample this leads to $2 \cdot 10^{16}$cm^{-3} at t_D=10s and T=15K. In view of the inherent experimental uncertainties this value compares favorably with that one of the LESR-spin density. It is not possible to decide whether these radiatively recombining pairs are geminate. If they were, they would not contribute to photocondution. On the other hand, the PC-transient can

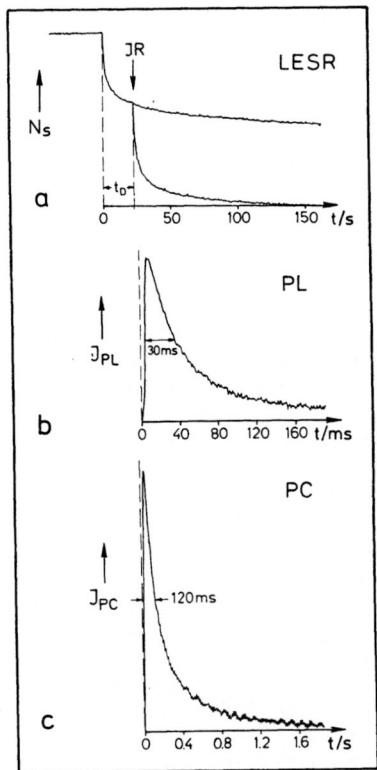

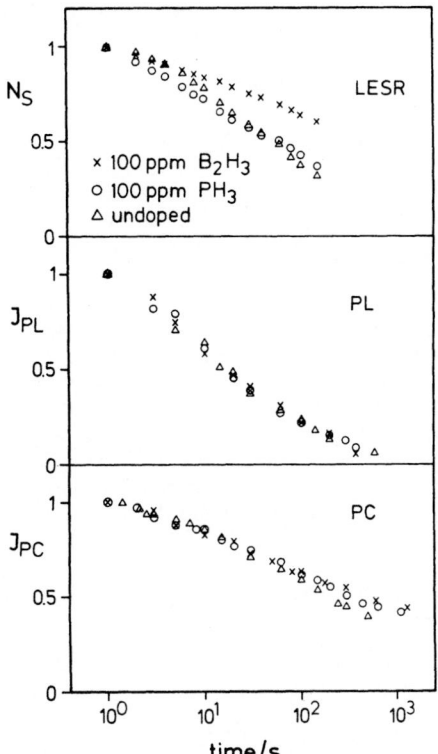

Fig.4: Transients of LESR (a),
PL (b) and PC (c) in-
duced by IR-light.
T=15K

Fig.5: Decay of the LESR, PL and
PC-peak heights at 30K
normalized to their value
after t_D=1s.

originate only from those electrons and/or holes which finally recombine radiatively or non-radiatively following distant pair kinetics. It seems thus possible that although both transients are due to excitation from the same metastable distribution they represent only part of the excited carriers. For instance, the PL-transient may arise from those electrons and holes, which before excitation already had a small separation and which after being excited may easily diffuse towards each other.

If recombination proceeds by tunneling, the lifetimes of pairs with intra-pair separation r is given by[1]

$$\tau = \tau_0 \exp\left(2\frac{r}{r_0}\right) \tag{1}$$

In this expression r_0 denotes the extension of the wavefunction of the more extended state which is supposed to be near 10Å for band tail electrons. The prefactor is assumed to amount to $\tau_0=10^{-12}$s for non-radiative tunneling and to $\tau=10^{-8}$s for radiative recombination. Excitation with 1.92 eV photons leads to a random distribution of intra-pair separations and according to relation (1) to a broad distribution of lifetimes. After a dark period of t_D=10s all carriers with τ<10s have recombined and following relation (1) the remaining pairs have large separations, r>100Å for radiative and r>150Å for non-radiative recombination. This limits the concentration of frozen in e-h-pairs to $10^{17}cm^{-3}$ and $4\cdot10^{16}cm^{-3}$, respectively, which is in accepteable accordance with the experimental observations. We thus conclude that in a-Si:H at low

373

temperatures a metastable distribution of trapped carriers exists with a concentration of 10^{16}-10^{17}cm^{-3}.

The growth of the metastable distribution can be examined by varying the number of absorbed photons, N_{abs}, by changing either the intensity G_{ex} of the exciting light (1.92 eV) or the exposure time t_{ex}. After each exposure most of the electron-hole pairs have recombined at $t=t_D$ and the pairs with longer lifetimes are not yet saturated if N_{abs} is small. In this range it is possible to obtain from the PL-data an estimate for the production efficiency of pairs with long lifetimes by comparing the number of photons, which are emitted during the excitation time t_{ex}, with the number of photons contained in the IR-induced PL-transient. Since not all of the IR-excited pairs may recombine radiatively, this procedure leads to a lower limit for the production efficiency. Assuming for the steady state an efficiency of 0.3 one finds that at low N_{abs} about 6% of the created e-h-pairs have lifetimes in excess of 10s. This estimate is in reasonable agreement with the conclusions of Street et al.[3] from a detailed comparison of PL and LESR. At higher values of N_{abs} saturation is attained i.e. an increase in G_{ex} and t_{ex} does not cause any further increase of the density of metastable pairs[6]. At t_D=10s, the number of absorbed photons where this occurs is roughly 10^{19}cm^{-3}.

The recombination of the metastable trap population can be examined by varying the delay time t_D. In these measurements the number of absorbed photons at each set of t_D was chosen such that the trap polpulation had been saturated i.e. N_{abs}>10^{19} cm^{-3}. Fig.5 shows the decay of the residual LESR and of the peak amplitude of the PL and PC transients of differently doped samples. The peak values, I_{PL} and I_{PC}, are determined by the IR-induced excitation rates and therefore can be taken as a measure for the concentration of trapped carriers instead of the more complicated integrals. Indeed, with increasing intensity of the IR-light, I_{PL} and I_{PC} increase following a power law with an exponent near 1. On the other hand, with decreasing I_{PL} there is also some increase in the half width of the transients, which indicates that the integral of the transients might be a more reliable measure for the concentration of non-equilibrium carriers. However, these differences are relatively small. In case of the PL-transient the time decay of the integral is nearer to a linear behaviour in a lin-log-plot than that of I_{PL} in Fig. 5. The LESR-decays are quite similar to those reported in the literature[5], the decay being considerably faster for n-type and undoped than for p-type films. In the doped films the LESR-spectrum consists of the signals of the trapped majority carriers and of the neutral dangling bonds. It is thus reasonable to assume that the recombination proceeds by tunneling of band tail carriers to the neutral dangling bonds, i.e. to the minority carriers which are trapped at the defect states. The differences between the decays of n- and p-type samples then indicate that the band tail holes are more strongly localized than the band tail electrons. It is surprising that in the decays of the PL- and PC- signal heights these differences do not reveal, n- and p-type films behave quite similarly.

In order to enable a comparison of the decays of these different quantities it is convenient to plot instead of the peak heights, I, the logarithmic derivative

$$\frac{1}{I} \cdot \frac{dI}{dt_d} = \tau_d^{-1} \tag{2}$$

This quantity does not depend on the normalization of the curve and has the meaning of a reciprocal differential lifetime. Fig. 6a then demonstrates that the LESR, PL and PC decay in much the same way. In the LESR-decay the

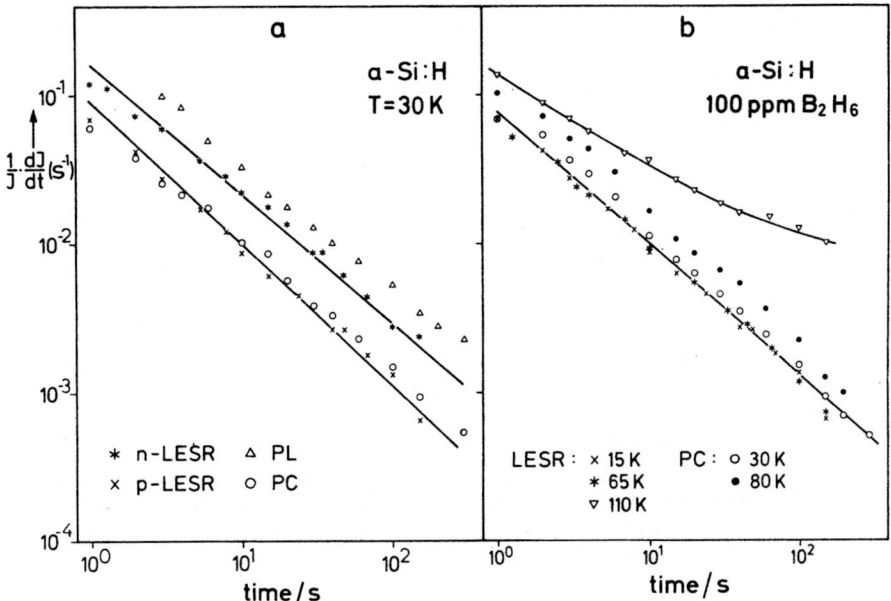

Fig.6: (a) Plot of the relaxation rate 1/I dI/dt versus delay time
t_D. The full curves are calculated from relation (3), details see text. (b) Relaxation rates of LESR and PC of a
p-type films versus delay time at various temperatures.

relaxation rate is smaller for the p-type than for the n-type and undoped
films. It is interesting to note, that the rates for the PC-decay are the
same for all kinds of samples and coincide with the lower values of the
LESR in the p-type films. On the other hand, the PL-data of all films lie
near to the LESR-relaxation rates of n-type material.

The form of these decays is typical of a broad spectrum of time constants which originates in a natural way from a distribution of intra-
pair separations according to relation (1). Assuming a simple rectangular
distribution of intra-pair separations the relaxation rate can easily be
calculated. If after a time t, all pairs with $r<r_t=r_o/2 \ln t/\tau_o$ have recombined the rate is given by

$$\frac{1}{N_p} \frac{dN_p}{dt} = \frac{r_o}{r_m} \frac{1}{t(1 - r_o/2r_m \ln t/\tau_o)} \qquad (3)$$

where r_m is the maximum intra-pair distance in the distribution. N_p denotes
the concentration of pairs which may be trapped band tail electrons and
holes in undoped samples or trapped majority carriers and neutral dangling
bonds in doped films. This simple model gives a surprisingly good description of the experimental results. The full curve through the data of n-type
films in Fig. 6a has been calculated using $r_o/r_m=1/20$ and $\tau_o=10^{-12}$ s. It coincides fairly well with the LESR-decay. The decay of the p-type films can
be fitted using a value of r_o which is smaller by 20%. If it is assumed that
r_m is the same in both types of samples and amounts to $r_m=200$Å, the radius
of the band tail electrons is 10Å and that of the band tail hole 8Å. Thus
the localization is stronger in the valence band tail. These numbers agree
satisfactorily with other estimates in the literature[5].

The long-time decay of the metastable trap population is essentially independent of temperature below T=80K. Fig. 6b shows LESR- and PC-decays of the p-type sample at various temperatures, n-type films behave quite similarly. Up to 80K the curves coincide and only at 110K the relaxation rate is considerably enhanced. This result supports the assumption that at low temperature recombination occurs by tunneling processes between localized states. Above about 80K thermal detrapping of band tail carriers or hopping processes of carriers among the band tail states may enhance the recombination when the carriers diffuse towards defect states.

CONCLUSION

At low temperatures in a-Si:H non-equilibrium carrier distributions persist after illumination with long lifetimes. Optical excitation from the metastable distributions by IR-light leads to quenching of the residual LESR-signal and to transients of photocurrent and photoluminescence. It is found that the LESR, PC- and PL-transients decrease at the same rate with the delay time between the end of the excitation with band gap light and the onset of the excitation with the probing IR-light. From LESR and from the PL-transient it is concluded that the concentration of metastable carriers with lifetimes above 10s is near $5 \cdot 10^{16} \text{cm}^{-3}$. The concentration derived from the PC-transient is considerably larger due to retrapping and multiple excitation. The recombination of these non-equilibrium carriers proceeds by tunneling processes. The results indicate stronger localization for valence band tail states and the data are consistent with localization radii of 8Å and 10Å for band tail holes and electrons, respectively. In dual beam experiments excitation by IR-Light leads to quenching of the photoluminescence in the temperature range 5-100K. The quenching effect occurs predominantly on the high energy side of the PL-spectrum and at low excitation intensities.

ACKNOWLEDGEMENT

The authors gratefully acknowledge financial support by the Bundesministerium für Forschung und Technologie (BMFT).

REFERENCES

1   R.A. Street, in Semiconductors and Semimetals 21B,
            J.I. Pankove, ed., Academic Press, New York (1984)
            chapter 7.
2   D.J. Dunstan and F. Boulitrop, Journ.Phys. 42, C4-331 (1981)
3   R.A. Street and D.K. Biegelsen, Solid State Commun. 44, 501 (1982)
4   D.J. Dunstan, Solid State Commun. 49, 395 (1984)
5   D.K. Biegelsen, R.A. Street and W.B. Jackson, Physica 117B/118B,
            899 (1983)
6   R. Carius and W. Fuhs, AIP Conf. Proceed. 120, 125 (1984)
7   R. Carius, W. Fuhs and M. Hoheisel, J.Non-Cryst.Solids 66, 151 (1984)
8   F. Boulitrop, J. Dijon, D.J. Dunstan and A.Herve, Int.Conf.Physics
            of Semicond., San Francisco (1984)
9   H. Dersch, L. Schweitzer and J. Stuke, Phys. Rev. B28, 4678 (1983)
10 J. Tauc, in Festkörperprobleme (Advances in Solid State Physics)
            XXII, P. Grosse, ed., Vieweg, Braunschweig 1982, p.1
11 F. Boulitrop, AIP Conf. Proceed. 120, 178 (1984)
12 D.E. Vanier and R.W. Griffith, J. Appl. Phys. 53, 3098 (1982)
13 P.D. Persans, Phil. Mag. B46, 435 (1982)

14   W. Fuhs, H.M. Welsch and D.C. Booth, phys.stat.sol. (b)120, 197 (1983)
15   L. Hirabayaschi and K. Morigaki, J.Non-Cryst.Solids 59/60, 133 (1983)
16   M. Hoheisel, R. Carius and W. Fuhs, J.Non-Cryst.Solids 63, 313 (1984)
17   C. Varmazis, M.D. Hirsch and P.E. Vanier, AIP Conf. Proceed. 120,
              133 (1984)
18   R.A. Street, C. Tsang and J.C. Knights, Inst.Phys.Conf.Series
              43, 1139 (1979)
19   T.S. Nashabishi, J.G. Austin and T.M. Searle, Int.Conf. on Amorphous
              and Liquid Semiconductors, W.E. Spear, ed., CICL,
              University Edinburgh (1977) p. 392

HOLE TRANSPORT IN GLOW DISCHARGE a-SiGe$_x$:H,(F) ALLOYS

S. Oda, S. Takagi, S. Ishihara and I. Shimizu

Imaging Science and Engineering Laboratory
Tokyo Institute of Technology
Nagatsuda, Midori-ku, Yokohama 227, Japan

INTRODUCTION

Recently amorphous silicon germanium alloys (a-SiGe$_x$) with low spin densities and high photoconductivity have been prepared by glow-discharge of gaseous mixture of fluorides and hydrogen. Measurements of transient photocurrent have clarified that transport of electrons is non-dispersive which is similar to that of a-Si:H and that transport of holes, of particular interest, looks like non-dispersive. In this paper we discuss the electronic states in a-SiGe$_x$ alloys on the basis of time-of-flight measurements in conjunction with the experimental facts of infrared spectra, x-ray photoemission and absorption spectra.

PREPARATION AND PROPERTIES OF a-SiGe$_x$

Conventionally a-SiGe$_x$ alloys have been prepared by rf glow-discharge of gaseous mixture of SiH$_4$ and GeH$_4$ as starting materials. However, obtained films contained a large number of spin densities presumably due to weak Ge-H bonds. In order to reduce densities of dangling bonds and to obtain highly photoconductive a-SiGe$_x$ films, various combination of gases were examined; viz., SiH$_4$, Si$_2$H$_6$, SiF$_4$, GeH$_4$ and GeF$_4$. Among them a combination of SiF$_4$-GeF$_4$-H$_2$, where the concentration of hydrogen in the gases is typically 30%, has turned out to provide good results.[1] In SiF$_4$-GeF$_4$-H$_2$ system, Ge atoms are incorporated in the network quite effectively. Amorphous Si$_{0.7}$Ge$_{0.3}$, for example, was prepared from a gaseous ratio of GeF$_4$/SiF$_4$-0.004, which was surprisingly different from the chemical constituent of the film determined by x-ray photoemission measurements. Band gap of the film was 1.4eV. ESR density was less than $7 \times 10^{15} cm^{-3}$. Photoconductivity measured under AM1 light was $9 \times 10^{-5} S/cm$, which did not deteriorate during 5-hour irradiation. The mobility-lifetime product measured at 780nm was between 10^{-5}-$10^{-6} cm2/V$. Photosensitivity of a-SiGe$_x$ alloys has been extended into longer wavelength lights compared with a-Si:H. This means that a-SiGe$_x$ alloys are promising materials for photoreceptor of laser printers as well as amorphous solar cells.[2]

MEASUREMENTS OF TRANSIENT PHOTOCURRENT

Carrier transport properties of a-SiGe$_x$ alloys are investigated by time-of-flight measurements. Electron transport is non-dispersive with carrier drift mobility of 0.2-$0.3 cm^2/Vs$. These characteristics are similar to a-Si:H. Transient photocurrent for holes, on the other hand, as shown in Fig. 1 is different from that of a-Si:H. The structure of the device for hole current measurement is, as shown in the inset, Al/a-SiN$_x$(300A)/a-Si$_{0.7}$Ge$_{0.3}$:H,(F)(1.5μm)/p$^+$-a-Si:H(800A)/Pd-Cr/glass. Pulsed laser light, duration

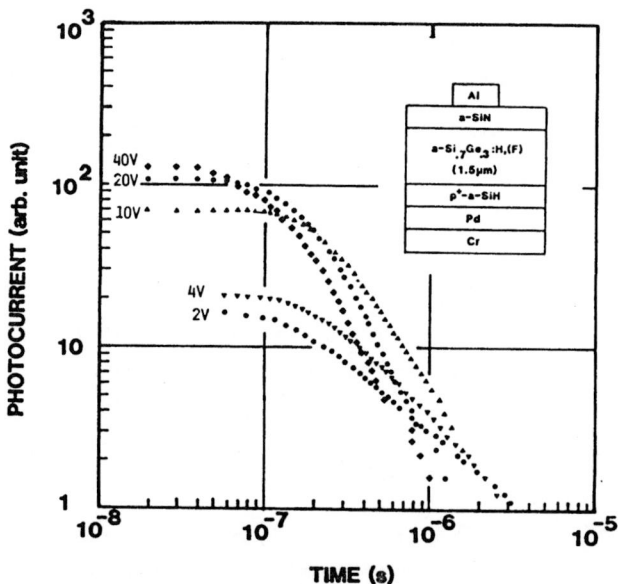

Fig. 1   Transient photocurrent of holes for a-Si$_{0.7}$Ge$_{0.3}$:H,(F) measured at various applied bias. The structure of the device for the measurement is shown in the inset. Pulsed laser light of 490nm impinges on semitransparent Al electrode, which is biased positively with respect to Pd-Cr electrode. The hole drift mobility is about $10^{-2} cm^2/Vs$.

of 150ps, 490nm in wavelength, impinged on semitransparent Al electrode, which is biased positively with respect to Pt-Cr electrode. a-SiN$_x$ and p$^+$-a-Si:H layers are blocking layers for holes and electrons, respectively. It should be noted that such a blocking configuration is essential in the measurements of primary photocurrent particularly for lower mobility type of carriers; viz. holes in a-Si:H and a-SiGe$_x$. Small amount of boron is doped in a-SiGe$_x$ layer (B$_2$H$_6$/(SiF$_4$+GeF$_4$)=10ppm) so as to increase the carrier range of holes. As shown in the figure, transient photocurrent for holes is consisted of two processes; the fast non-dispersive path and emission from deep defect states. At low applied bias, photocurrent is determined by deep trapping lifetime. When the applied bias exceeds 10V, the transit time decreases with increasing applied bias, keeping the collected charge

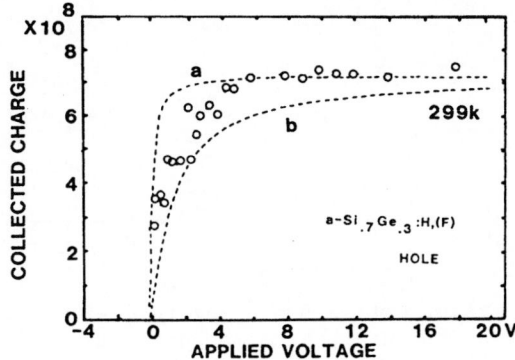

Fig. 2   The collected charge deduced from the integral of the hole current transient taken as a function of the applied bias. Dotted lines denote for theoretical curve of simple Hecht analysis assuming uniform field across the sample with the hole range, i.e., the mobility-lifetime product, a; 1x10^{-7}cm^2/V and b; 1x10^{-8}cm^2/V.

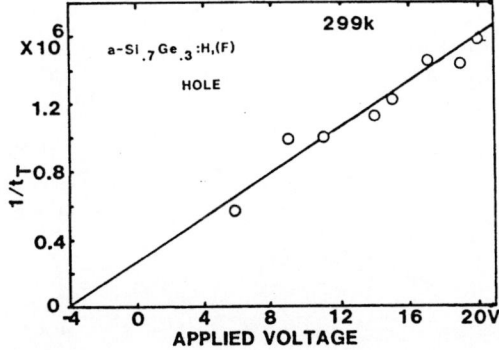

Fig. 3   The field dependence of the inverse of the transit time deduced from the peak of the derivative of the hole current transient of a-SiGe$_x$. The straight line gives the hole drift mobility of 1.5x10^{-3}cm^2/Vs.

381

constant. The hole drift mobility is about $10^{-2} cm^2/Vs$ at room temperature. The collected charge and the inverse of the transit time are plotted as a function of the applied bias in Figs. 2 and 3, respectively, for another sample of a-Si$_{0.7}$Ge$_{0.3}$:H,(F) with lower hole drift mobility of $1.5 \times 10^{-3} cm^2/Vs$. The initial photocurrent increases with increasing light intensity (Fig. 4) as well as with increasing applied bias (Fig. 5). Although the straight line in Fig. 5 may suggest that the fields are almost uniform across the sample,

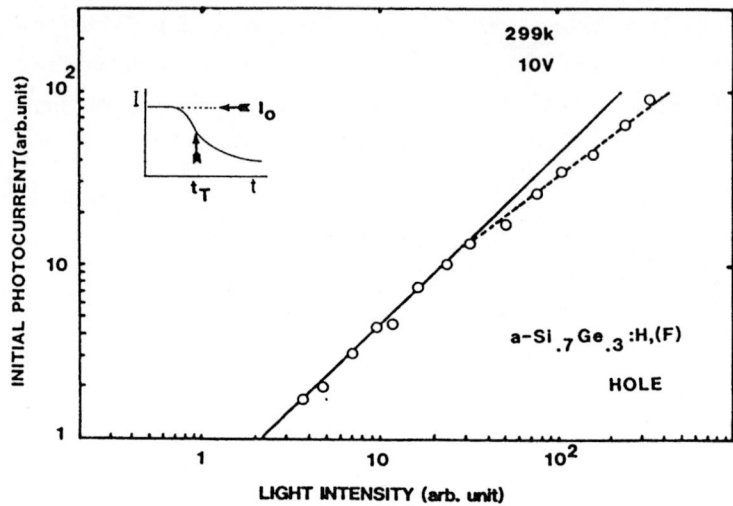

Fig. 4   The initial photocurrent versus light intensity. The inset shows a typical current transient with the definition of the initial photocurrent I_0 and the transit time t_T. The measurements presented in the other figures were performed at the lowest light level.

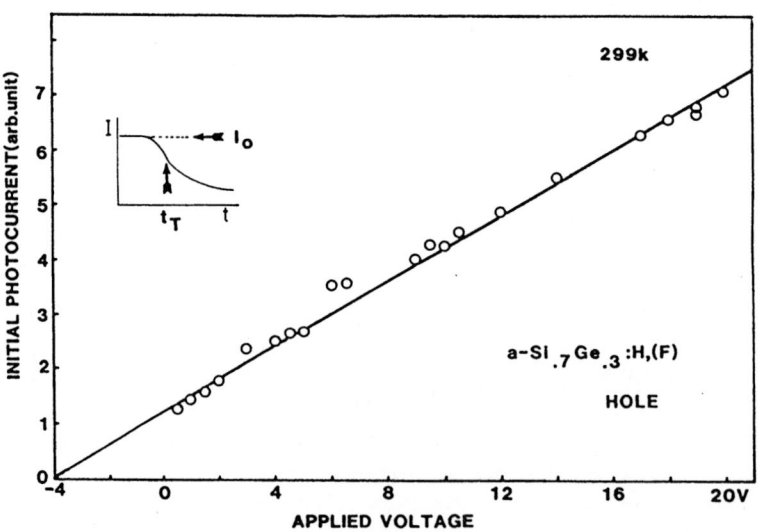

Fig. 5   The initial photocurrent versus applied voltage.

382

the plots in Fig. 2 do not fit with a simple Hecht-type analysis which assumes the uniform distribution of field. The extrapolation of the plots of the voltage dependence (Figs. 3 & 5) yields the "built-in voltage" of 4V. It seems unrealistic that such a high voltage for the built-in voltage. However, Street and Thompson[3] also obtained nearly the same flat-band voltage in the similar device structure of a-Si:H/a-SiN interface. As shown in Fig. 6, the hole drift mobility increases as temperature is increased with the activation energy of between 0.09 and 0.14eV.

It is very surprising that hole transport of a-SiGe$_x$ looks like non-dispersive. One may claim that this behavior might be due to some improper experimental conditions; such as space charge perturbation or trap saturation. However, these possibilities can be excluded. The collected charge per single pulse is about $10^{-10}C/cm^2$. That is three orders of magnitude smaller than space charge limitation determined by the sample capacitance and applied voltage of 10V. Photon densities for a single pulse is about $10^{14}/cm^3$ which is almost negligible compared to $10^{18}/cm^3$ at that light level Hvam and Brodsky[4] observed the trap saturation. It should be noted that in spite of non-dispersive nature of the hole transport, microscopic mobility in the valence band should be very small, about $10^{-1}cm^2/Vs$, which might reflect the fluctuation of the valence band due to alloying of Si and Ge.

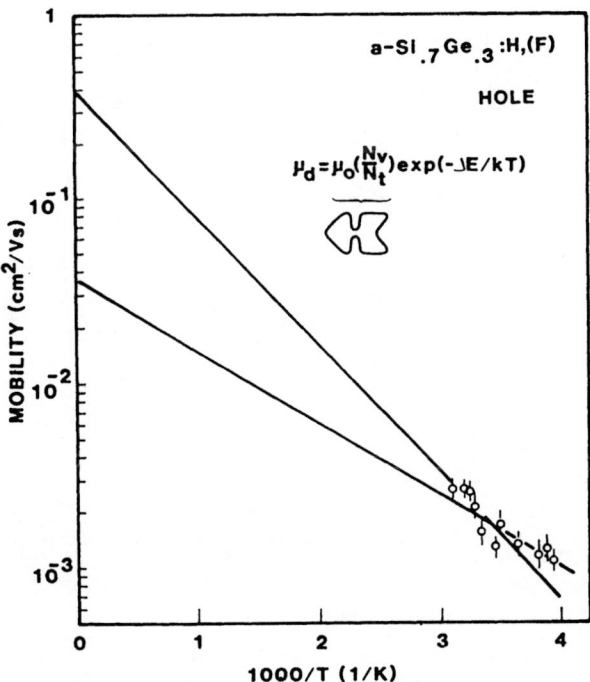

Fig. 6 The hole drift mobility plotted as a function of the reciprocal temperature.

The measurements of ir spectra and x-ray photoemission spectroscopy suggest[2] that (1) hydrogen atoms attach preferentially to Si, (2) the concentration of fluorine in the film is less than 1%, and (3) Ge dangling bonds may be terminated by F. Figure 7 shows ir spectra of $a\text{-}Si_{1-x}Ge_x$:H,(F), where x=0, 0.2 and 0.5. An ir spectrum of $a\text{-}Si_{0.7}Ge_{0.3}$:D,(F) made from the glow-discharge of gaseous mixture of $SiF_4\text{-}GeF_4\text{-}D_2$ is also shown in order to distinguish hydrogen related bands from others. The dominant bands which appear at $630cm^{-1}$ and $2000cm^{-1}$ are of Si-H bonds. It should be noted that the Ge-H band to appear at $1850cm^{-1}$ is very small even when Ge concentration in the film exceeds 50%. Therefore, hydrogen atoms attach preferentially to Si rather than to Ge. In the ir spectrum of $a\text{-}SiGe_x$:D,(F), now Si-H band of $630cm^{-1}$ shifts to Si-D band of $510cm^{-1}$, another band appears at $630cm^{-1}$, which might be due to Ge-F_x bonds.

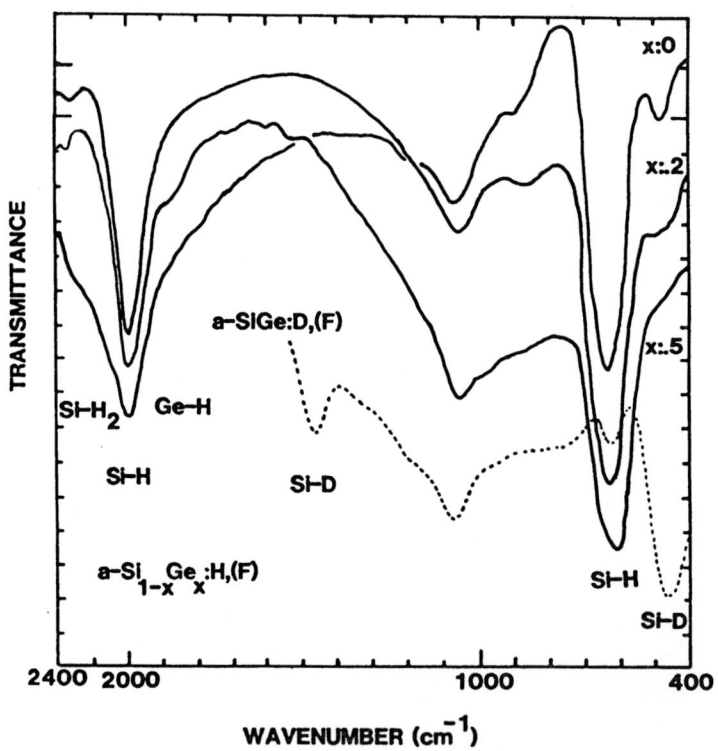

Fig. 7 Infrared absorption spectra of $a\text{-}Si_{1-x}Ge_x$:H,(F) films for x=0, 0.2 and 0.5. An ir spectrum of an $a\text{-}SiGe_x$:D,(F) film is also shown.

Figure 8 shows the absorption spectra of a-Si:H and a-SiGe$_x$:H,(F) taken by various methods of measurements. Absorption spectra in low absorption region obtained by photoacoustic spectroscopy show the evidence that there are deep defect states in a-SiGe$_x$.

A MODEL FOR ELECTRONIC STATES IN a-SiGe$_x$:H,(F)

By taking into account above experimental facts we propose a schematic picture which describes a model for electronic states in a-SiGex:H,(F) as shown in Fig. 9. Tail states of the conduction band might be exponential with characteristic temperature of less than 300K since electron transport is non-dispersive at room temperature, which is, more or less, the same as in a-Si:H. The mid gap density-of-states with spins are less than 10^{16}/cm^3, which is also the same as the best a-Si:H. The states near the valence band, however, are not the same as in the case of a-Si:H. They consist of two types of states; a tail state whose characteristic temperature might be lower than 300K, which is responsible for the observed non-dispersive nature of hole-transport, and deep defect states which manifest itself in the absorption spectra.

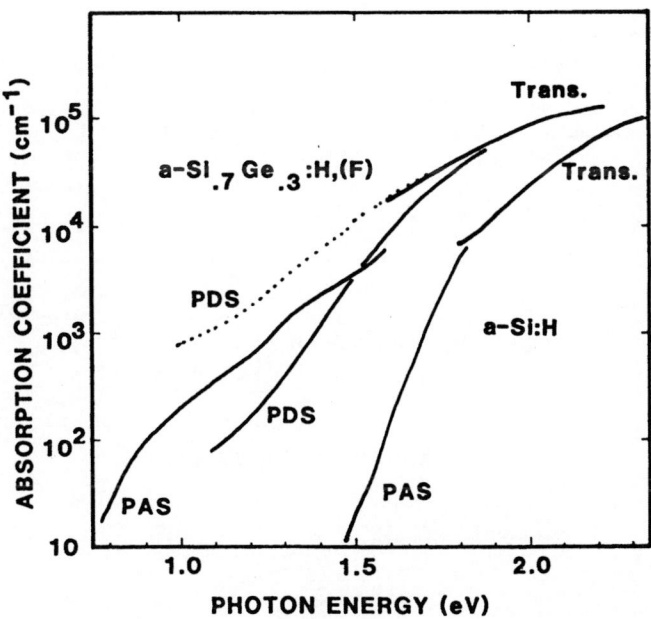

Fig. 8 Absorption spectra of a-Si:H and a-SiGe$_x$:H,(F) measured by various methods, i.e., transmittance, photothermal deflection spectroscopy and photoacoustic spectroscopy.

Now, let us discuss about the nature of the deep states. Hudgens and co-workers[5] have found new defect states in a-SiGe$_x$, for which they attributed to Ge dangling bonds. However, in our samples spin densities are as small as the best a-Si:H. In view of the transport properties these deep states should have small capture rates for holes. Two-fold coordinated Ge atoms may be the candidates for the relevant defect states. Adler[6] suggested that these types of states are likely in tetrahedrally bonded amorphous solids. Divalent Ge, as can be seen in GeF$_2$, GeS or GeO, is more stable than its Si counterpart. Divalent Ge has lone-pairs which will not contribute to ESR signal. Electrons in these states should have very small interaction to lattice electrons, which might be responsible for the small capture rate for holes.

It should be noted that we could interpret the nature of the deep states in a-SiGe$_x$:H,(F) in another way; i.e., in terms of inhomogeneity. It was reported[5] that microstructures were observed in a-SiGe$_x$ alloys. It is also generally believed that microcrystalline Si can be made easily by the glow-discharge of gaseous mixture containing fluorides. However, it is very unlikely in our case. First of all, we havenot found any sign of microcrystallization in our samples by Raman scattering or conductivity measurements. We prepare a-SiGe$_x$ alloys at even smaller RF power densities than a-Si:H, while high RF power favors microcrystalline Si:F.

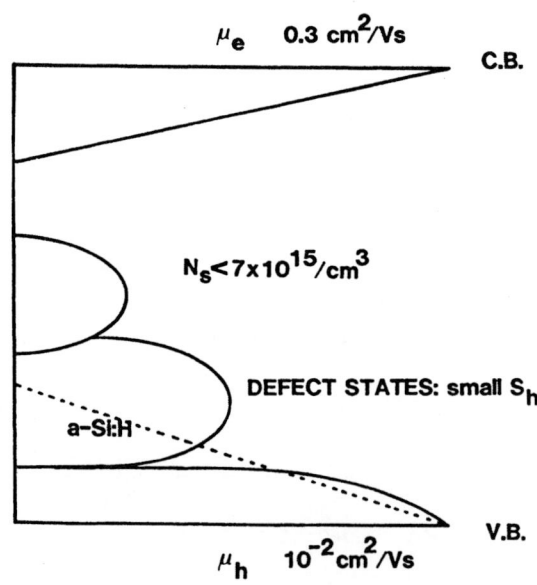

Fig. 9 A schematic model of the electronic states in a-SiGe$_x$:H,(F). The dotted line denotes the valence band tail states of a-Si:H.

THE ROLE OF FLUORINE IN THE PREPARATION OF a-Si:Ge$_x$:H,(F)

In view of the experimental results of IR spectra and x-ray photo-emission, the role of fluorine in a-SiGe$_x$:H,(F) alloys is not as dangling bond terminator for Si, although may be partly acting as Ge-F or Ge-F$_2$ bonds. This fact may suggest that the key issue in preparing tetrahedral amorphous alloys is the use of multiple terminators for dangling bonds of different types of atoms. As shown in the IR spectra hydrogen atoms attach preferentially to Si while fluorine atoms make bonds only with Ge. Recent-ly we have prepared a-SiC$_x$ alloys by glow-discharge of gaseous mixture of SiF$_4$-C$_2$F$_6$-H$_2$. Infrared spectra of the film reveal that hydrogen preferen-tially attach to Si and fluorine to C.

Cohen and co-workers[7] have predicted that the width of the tail of the valence band edge is determined by dihedral-angle disorder, which can be reduced by incorporating atoms of an element with a large electronegativity difference from Si(or Ge). Actually we have observed the non-dispersive transport of holes, which may be due to the narrower tails of the valence band edge. However, the concentration of fluorine in our samples is less than 1%, which is too small to be accounted as the fixer of dihedral-angle variation.

The deposition rate of a-SiGe$_x$ alloys depends strongly upon plasma conditions such as RF power, pressure or the concentration of hydrogen. It seems that the deposition rate is determined by the balance between deposi-tion and etching. The use of fluoride gas may be effective in preparing defect-free amorphous alloys via etching out weaker bonds leaving only pro-per bonds.

However, the most important role of F lies in the growth kinetics. The chemical reaction between fluorides and hydrogen creates HF. The chemical potential obtained by the reaction makes it possible to prepare amorphous films by the selected chemical species which is effective in making the network properly.

SUMMARY AND CONCLUSIONS

(1)   Amorphous SiGe$_x$ films with low spin densities and high photo-conductivity were prepared by glow-discharge of the gaseous mixture of SiF$_4$-GeF$_4$-H$_2$.

(2)   Transport properties of a-SiGe$_x$:H,(F) were investigated by time-of-flight method. It was found that electron transport was non-dispersive with drift mobility of 0.2-0.3cm^2/Vs. Contrary to the case of a-Si:H, transient hole current was consisted of fast non-dispersive process and

slow emission from deep states.

(3) Infrared spectra and x-ray photoemission spectra clarified that fluorine concentration in the film was less than 1%, that hydrogen atoms preferentially attached to Si atoms and that Ge might be terminated by F atoms.

(4) Divalent germanium atoms could be the possible candidate for the origin of deep defect states in a-SiGe$_x$:H,(F).

(5) The role of fluorine in the growth of a-SiGe$_x$:H,(F) was discussed. Only selected chemical species, which would be useful to form a proper network, might be produced in the plasma.

ACKNOWLEDGEMENT

The authors are indebted to S.R. Ovshinsky, G. Lucovsky, M. Paesler, S.J. Hudgens, S. Guha, Y. Yamaguchi and S. Yamasaki for their helpful discussions. This work was supported in part by the Grant-in-Aid for Scientific Reasearch form the Ministry of Education, Science and Culture of Japan.

REFERENCES

1. K. Nozawa, Y. Yamaguchi, J. Hanna and I. Shimizu, J. Non-Cryst. Solids 59/60, 533 (1983).
2. S. Oda, Y. Hamaguchi, J. Hanna, S. Ishihara, R. Fujiwara, S. Kawate and I. Shimizu, Technical Digest of the International PVSEC-1, Kobe, 429 (1984).
3. R.A. Street and M.J. Thompson, to be published.
4. J.M. Hvam and M.H. Brodsky, J. Phys. (Paris) 42, 551 (1981).
5. C.-Y. Huang, S. Guha and S.J. Hudgens, J. Non-Cryst. Solids 66, 187 (1984).
6. D. Adler, Phys. Rev. Lett. 41, 1755 (1978).
7. M.H. Cohen, H. Fritzsche, J. Singh and F. Yonezawa, Proc. 15th Int. Conf. Physics of Semiconductors, Kyoto, 1980; J. Phys. Soc. Japan 49, Suppl. A, 1175 (1980).

THE EFFECT OF DIFFUSION LIMITED KINETICS ON CURRENT FLOW IN AMORPHOUS MATERIALS

M. Silver and V. Cannella*

Department of Physics and Astronomy
University of North Carolina
Chapel Hill, North Carolina 27514

*Ovonic Display Systems
1896 Barrett Street
Troy, Michigan 48084

ABSTRACT

A short electronic mean-free-path results in diffusion limited recombination kinetics. The consequences regarding current-voltage characteristics of p-i-n structures is discussed. In particular, comparison between single injection (n-i-n) and double injection (p-i-n) currents are made. With trapping in a distribution of tail states and recombination only through extended states, the double injection currents are shown to be significantly larger than the single injection space charge limited currents. Comparison with experimental data is made.

Many widely held ideas regarding recombination and trapping came from concepts originating from classical semiconductor physics where the extended state band mobility was large and selection rules governed transitions. In general, this led to the notions that trapping and recombination rate constants could be described by σv_o (where σ was the cross section and v_o the random velocity) and that often band to band transition rates were small. In amorphous materials, however, the view is generally held[1] that the mobility of free carriers is small and band to band transition rates are large. In this case the rate constants would be controlled by diffusion processes rather than the actual transition rates. It is therefore instructive to examine these processes in some detail. We will concentrate our focus on the physical concepts rather than attempt rigorous calculations since these latter generally require complex computer simulations, particularly when one is considering a complicated density of localized gap states.

We can describe the recombination in a simplified way as

$$\frac{1}{e} \ | \nabla \cdot j | = |b_r n \ N_R| \tag{1}$$

where j is the current density, n is the density of free carriers, N_R

the density of recombination centers and b_r is the recombination coefficient. If we consider only a volume Ω around one recombination center, so that $\int N_R \, d\Omega = 1$, then

$$\int (\nabla \cdot j) \, d\Omega = I_{in} - I_{out} \simeq b_r ne \qquad (2)$$

The mathematics is further simplified without serious loss of content if we further assume that $I_{out} = 0$. This implies that the recombination lifetime at the recombination center is low and the center acts as a current sink. When we do this, we obtain

$$b_r = I_{in}/ne \qquad (3)$$

Now to evaluate b_r it is only necessary to calculate I_{in}. We can demonstrate the utility and ease of employing this treatment by considering two cases: (1) the ballistic case for which $|j| = |env_0|$ where v_0 is the random velocity and (2) the diffusion limited case $|j| = |eD \, dn/dr|$, where $D = \mu kT/e$ is the diffusion constant of the free carriers. In both cases we will assume spherical symmetry. Case (1) applies when the mean free path is large compared with $(1/N_R)^{1/3}$ (N_R is the density of recombination centers) so ballistic non-scattering motion is involved. On the other hand, case (2) applies when the mean free path is small compared with $(1/N_R)^{1/3}$ so that diffusive motion is involved. For case 1, the current that one is interested in is at the recombination center whose radius is R, consequently

$$I_{in} = 4\pi R^2 j = 4\pi R^2 nev_0 = b_r en \qquad (4)$$

and one obtains the usual expression

$$b_r = 4\pi R^2 v_0 = \sigma v_0. \qquad (5)$$

Likewise, we can use our simplified scheme to calculate I_{in} for case (2),

$$I_{in} = -4\pi r^2 De \, dn/dr \qquad (6)$$

$$\frac{4\pi \, Dne}{I} = \frac{1}{r} \Big|_{(N_r)^{-1/3}}^{R} \simeq \frac{1}{R} \qquad (7)$$

because the radius of the center R is generally very small compared with the average distance between centers. Thus, we obtain for the diffusion limited case,

$$b_r = 4\pi DR = 4\pi \mu kTR/e \qquad (8)$$

It is important to point out that it is the extended state mobility that is applicable here and not the drift mobility because it is only the free carriers that contribute to the flux into the recombination center. If there is shallow trapping as well as recombination then, as shown by Rose[2], the increase in recombination lifetime τ_r enters through the ratio of free to trapped carriers.

$$\tau_r \simeq (1/bN_R)(n_t/n_f) \qquad (9)$$

where n_t is the density of shallow trapped carriers and n_f the density of mobile carriers.

Before exploring the consequences of diffusive motion, it is interesting to compare the $\mu\tau$ products for cases (1) and (2). In the ballistic limit, case (1),

$$\mu\tau = \mu/N_r \sigma v_0 \qquad (10)$$

and depends upon the transport parameters as well as the density and character of the recombination center; in the diffusive limit, however,

$$\mu\tau = 4\pi e/N_R kTR \qquad (11)$$

what becomes apparent is that for the diffusive limit, the $\mu\tau$ product does not depend upon the transport parameters but only upon the density and character of the recombination centers and, of course, temperature.

Another interesting aspect of the diffusion limited case is that only the radius of the recombination center enters and not the cross section (radius squared). This means that the effect of charged centers compared with neutral ones is not as great in the diffusion limited as it is in the ballistic limited regime.

For the diffusion limited case, from equation 8, $b_n/b_c = R_n/R_c$ where R_n is the radius of neutral center, R_c is the radius of the charged center ($R_c = e^2/4\pi\epsilon\epsilon_0 kT$), b_n is the rate of capture by the neutral center and b_c is the rate of capture by the charged center.

In order to make our comparison between single carrier injection space charge limited currents (n-i-n devices) and two carrier injection recombination limited currents (p-i-n devices) we have to assume an energy dependance for the density of localized states. Several have been postulated for a-Si:H alloys. Tiedje et al.[3] assumed a single exponential behavior while Spear[1] assumed a modified exponential. On the other hand, Stutzmann et al.[4] assumed an exponential distribution of tail states plus an added density of defect states closer to mid-gap than the tail states. We will consider the Spear[1] density of states because this makes it easier to reach general conclusions regarding the processes. We will, however, also comment briefly on the effect of deep defect levels as in the Stutzmann model[4].

Further, for simplicity, we will assume that the density of the hole states is the same as for electrons which allow us to preserve the simple closed analytic forms. A schematic diagram of the density of localized states is shown in Figure 1. Without loss of generality we can take $g(\epsilon)$ = const. for $\epsilon > \epsilon_1$.

We first consider a unipolar steady state space charge limited current. If the intrinsic thermal current is extremely small so that without injection the sample can be considered to be a perfect insulator, then the two basic equations are:

$$j = e n_f \mu E \qquad (12)$$

and

$$\epsilon \, dE/dx = e(n_f + n_t) = en_f(1 + 1/\theta) \qquad (13)$$

where θ is defined as n_f/n_t. (Notice it is all the trapped charge n_t that determines θ and consequently j, and not just the ones in shallow states).

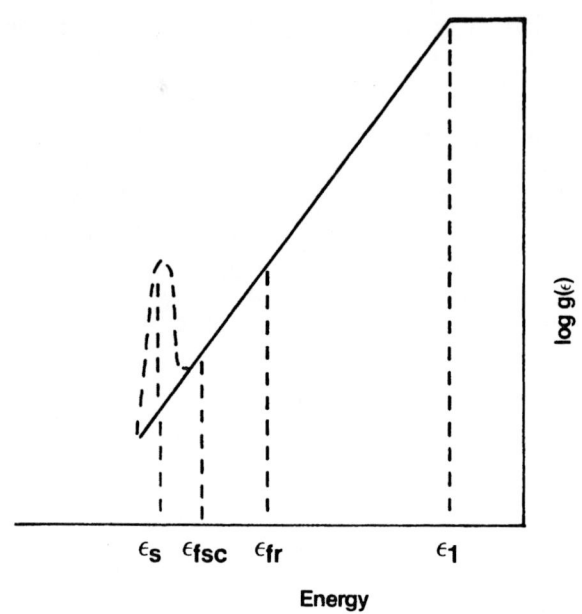

Figure 1. Schematic Density of States Diagram

Above energy ϵ_1, $g(\epsilon)$ = const. = g_c and below
ϵ_1, $g(\epsilon) \propto \exp[(\epsilon - \epsilon_1)/kT_c]$.
ϵ_{fsc} indicates the quasifermi level for single injection;
ϵ_{fr} indicates the quasifermi level for double injection
and the dotted density of states represents an almost single
level at an energy ϵ_s.

Integrating equations 12 and 13 gives the well known result when θ is independent of x

$$j = (9/8)\theta\epsilon\mu \, V^2/L^3 \qquad (14)$$

and it is easily shown that the total amount of charge in the sample $Q = (3/2)CV$. For a wide energy distribution of traps Rose[2] showed that if it is the density of trapped charge n_t which is treated as independent of x rather than θ then $Q \simeq 2CV$ which is not substantially different. Consequently, considering the density of states shown in Figure 1 with the approximation that $g(\epsilon) = g_c$ above ϵ_1 we will employ the constant n_t case. It is now only necessary to calculate n_t which we do by

392

considering that there is a quasifermi level at ε_{fsc} as shown in Figure 1.

$$n_t \simeq \int_{-\infty}^{\varepsilon_c} f(\varepsilon)g(\varepsilon)d\varepsilon \simeq \frac{g_c kT}{(1-\alpha)} \exp[-(\varepsilon_1 - \varepsilon_{fsc})/kT] \simeq 2CV/eL \qquad (15)$$

where $f(\varepsilon)$ is the probability of occupation of a state at energy (ε), $\alpha = T/T_c$ and T_c characterizes the assumed exponential dependence in energy of the density of localized states below ε_1 (see Figure 1). Equation 15 holds providing that[2] $\alpha < 0.8$. For $\alpha > 0.8$ a more complex relationship than that given by Eq. 15 applies. The quasifermi level ε_{fsc} is easily obtained from Eq. 15 and

$$\varepsilon_{fsc} = \varepsilon_1 - kT_0 \ln \left[\frac{g_c kTeL}{2CV(1-\alpha)}\right] \qquad (16)$$

where it is important to note that ε_1 and ε_{fsc} are negative because we take $\varepsilon_c = 0$. Recalling that $\theta = n_c/n_t$ and $n_c = g_c kT \exp(\varepsilon_{fsc}/kT)$ then the space charge limited current for unipolar injection is

$$j_{sc} = \frac{e\mu}{L} V g_c kT \exp(\varepsilon_1/kT) \left[-\frac{2CV \alpha (1-\alpha)}{g_c kTeL}\right]^{1/\alpha} \qquad (17)$$

Equation 17 follows from the specific density of states and the absence of recombination. Consequently, for all states below ε_{fsc}, $f = 1$.

Double injection differs from single injection insofar as the currents are limited by recombination as well as or rather than by space charge effects and therefore, below the quasifermi level $f < 1$. (This will be discussed later). Further, more than CV of each sign of charge may exist in the sample provided of course $Q^+ - Q^- < CV$, that is $Q^+ \simeq Q^-$. The net effect is to produce a quasifermi level e_{fr} higher than e_{fsc} because Q may be larger than CV and $f < 1$ below ε_{fr}. Therefore, the expected current in a p-i-n will be larger than in an n-i-n. To show this effect, we rely on the basic treatment of Lampert and Mark[6]. We will assume that the injecting contacts can supply all of the current necessary which means that all the voltage drop is across the i (intrinsic) layer.

To insure space charge neutrality and recombination limited currents, the theory of Lampert and Mark[6] requires

$$Q = |Q^+| = |Q^-| = \frac{CV\tau}{t_n + t_p} \simeq \frac{CV\tau}{2t_n} \qquad (18)$$

where τ is the average common lifetime of the carriers, t_n and t_p are the free carrier transit times and because of the assumed symmetry in the density of states $t_n = t_p$. According to Rose[2], all the trapped charge below ε_{fr} acts as recombination centers. Because[2] approximately $(1 - \alpha)Q$ lies below ε_{fr} we can calculate τ/t_n in the diffusion limited regime and find $Q \simeq CV(1/1-\alpha)^{1/2}$ because the recombination centers are presumably charged when occupied. We can also use simple arguments to obtain f. Since the rate of population of a given recombination level is zero in steady state, then it follows that

$$nb_n(1-f) = pb_c f \qquad (19)$$

where as before b_c is the capture rate of holes to a negatively charged center while b_n is the capture rate of electrons to neutral unpopulated recombination centers. For the diffusion limited regime the $f \simeq R_n/R_c$ because of the symmetry in n and p densities of states. Even though $f<1$ we can calculate n_t and therefore ε_{fr} in a manner similar to that given by equations 15, 16 and 17. In this case

$$j_r = -\frac{e\mu V}{L} \ g_c \ kT \ \exp(\varepsilon_1/kT) \ [-\frac{CV\alpha(1-\alpha)}{fg_c kTe}]^{1/2} \, ^{1/\alpha} \qquad (20)$$

and the ratio of

$$j_r/j_{sc} = [-\frac{1}{2f\sqrt{(1-\alpha)}}]^{1/\alpha} = [-\frac{R_c/R_n}{2\sqrt{(1-\alpha)}}]^{1/\alpha} \qquad (21)$$

Here, again, we must remember that μ is the band mobility and that $\alpha < 0.8$.

Figure 2 shows experimental I vs V characteristics for p-i-n and n-i-n devices. As one can see, the ratio of experimental values j_r/j_{sc} is quite large, having values of order 10^3.

From equations 17 and 20, the slope of log j vs log V is $(1+1/\alpha)$. Using the value of α from the experimental slope and estimates of R_n and R_c along with equation 21, we can calculate an expected ratio j_r/j_{sc}. Since we find experimental values of α close to unity and R_c/R_n is only about 20, the expected value of j_r/j_{sc} is $< 10^2$ rather than the observed 10^3. Several possible explanations for this difference come to mind. If α was smaller than that given by the slope in Figure 2, and was, in fact, less than 0.5 then from Eq. 21, we could explain the results. This would be possible if there was a residual field at the injecting contact. Then the current would be reduced and we would have $j \propto (V_B-V_o)^{1+1/\alpha}$ where V_B is the bulk i layer voltage and V_o characterizes the residual field at the injecting contact. Another possible explanation would be if $p \gg n$ because of assymetries in the density of states and the extended state mobility[5] then f would be reduced from the value given in Eq. 21 and the ratio would be increased. Finally, if the recombination kinetics were ballistic rather than diffusion limited then $f \propto (R_n/R_c)^2$ and again the ratio of the currents would be predicted to be larger. At this time, it is not clear which of these possibilities is the appropriate explanation. Here we should add that the change in the functional form of Eq. 21 which is necessary when α approaches 1 from 0.8 does not substantially alter the expected value of j_r/j_{sc}.

One last question can be addressed concerning the effect on these predictions of an added density of almost single level deep states. If one uses a typical set of deep states[4], for example, $N_s = g_s(\varepsilon)\Delta\varepsilon > 10^{16} cm^{-3}$ then $2CV/eL < N_s$ and these states would not normally be filled by single injection in a 1 micron thick device with one volt applied. Hence the quasifermi level would lie at the deep level ε_s. On the other hand, for a double injection device with $Q \simeq 2CV$ but $f \ll 1$ the quasifermi level might lie above E_s. Unfortunately the value for the ratio $fN_s eL/CV$ is so close to unity that we cannot predict the ratio j_r/j_{sc}.

Experimental measurements of j_r/j_{sc} vs N_s would show the effect of the deep states, particularly for large N_s. To date these experiments have not been performed and so we cannot say for sure from these arguments that these deep states act as the primary recombination centers.

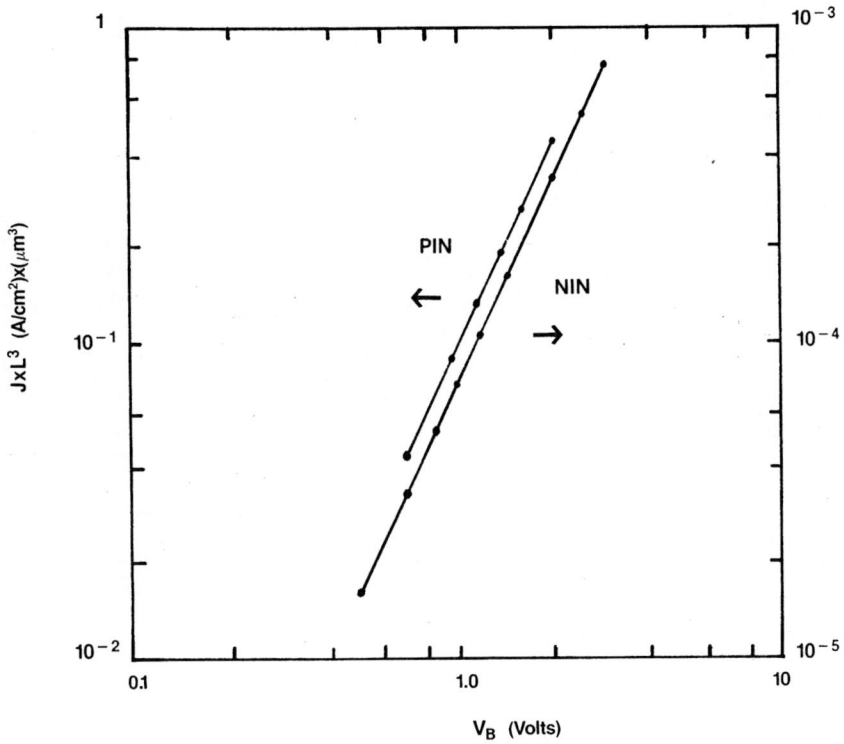

Figure 2.   Experimental values for jL^3 vs V_B
for p-i-n (left scale) and n-i-n (right scale) devices.

The L^3 factor normalizes j for different thicknesses when
$\alpha \simeq 1$. V_B is the voltage across the bulk material.
For p-i-n devices the junction voltages must be subtracted
from the applied voltage to obtain V_B.

In summary, we have analyzed diffusion limited recombination currents and compared them with space charge limited currents. If the deep tail states are the recombination centers, one predicts that the recombination limited currents are very large compared with the space charge limited currents.

ACKNOWLEDGEMENT

The authors wish to thank S. R. Ovshinsky for his continued support and encouragement and to acknowledge very fruitful discussions with Mike Hack.

REFERENCES

1. W.E. Spear, J. Non. Cryst. Sol. 59/60 1, (1983).
2. A. Rose "Concepts in Photoconductivity and Allied Problems",
   Interscience, New York (1963).
3. T. Tiedje, B. Abeles, D.L. Morel, T.D. Moustakas and C.R. Wronski
   Appl. Phys. Lett. 36, 695 (1980).
4. M. Stutzmann, W.B. Jackson and C.C. Tsai, A.I.P. Conference
   Proceedings, #120 "Optical Effects in Amorphous Semiconductors," 1984.
5. See for example: M. Hack, S. Guha and M. Shur, Phys. Rev. B 30, 6991
   (1984).
6. M.A. Lampert and P. Mark. "Current Injection in Solids" Academic
   Press, New York (1970).

GEMINATE, NONGEMINATE AND EXCITON RECOMBINATION IN a-Si:H

B. A. Wilson

AT&T Bell Laboratories

Murray Hill, N.J. 07974

ABSTRACT

This paper examines the current models of carrier recombination in a-Si:H in the context of recent cw and time-resolved photoluminescence experiments. Exciton recombination is found to provide the most consistent description of the submicrosecond dynamics, with distant-pair nongeminate processes contributing at later times. The most commonly cited geminate-pair model is found to be least able to account for the experimental observations.

INTRODUCTION

The issue of geminate versus nongeminate recombination in a-Si:H is a controversial point in the interpretation of optical and transport measurements. It emerged as a central concern in the 1981 Conference on Tetrahedrally Bonded Amorphous Semiconductors,[1] and although a great deal of related literature has appeared in the intervening few years,[2-12] it remains an important unresolved problem in our understanding of carrier dynamics in these materials. At the heart of this controversy lies the question of exciton formation and dissociation. In this paper I review some recent photoluminescence measurements which bear on this question. The implications of these results are examined for various models of geminate and nongeminate recombination.

The low-temperature cw photoluminescence (PL) in plasma-deposited a-Si:H consists of a band at ~ 1.4 eV and weaker bands generally attributed to coordination defects.[13,14] The main band at ~ 1.4 eV has been studied extensively using time-resolved spectroscopy, and the results prior to 1981 have been reviewed in detail by Street.[15] The model that emerged from this work is one of geminate recombination by radiative tunnelling of electrons and holes independently trapped in localized band-tail states. The shift of the spectrum with delay has been attributed to carrier thermalization in the band tails for delays less than ~ 1 μs and to energy-dependent recombination rates for delays greater than ~ 1 μs. If the dynamics of the carriers are dominated by microscopic potential fluctuations and local electric fields rather than by the effects of intrapair Coulomb interactions, then excitons will not form,[16] and the carriers cascade independently through their respective band tails. This is the model proposed by Street, referred to here as the geminate-pair model. In this picture pairs diffuse to an average separation of ~ 50 Å during thermalization, and recombine as loosely-correlated (i.e. nonexciton) geminate pairs. If, on the other hand, the Coulomb interaction predominates, then carriers are not free to diffuse independently, and thermalization and recombination occur within localized exciton states. Excitonic recombination has been proposed by a number of authors.[7,10,17] The main argument used to support geminate-pair over exciton recombination is the ease with which the pair model explains the wide range of lifetimes in the PL decay.[15] Radiative tunnelling rates depend exponentially on pair separation so that factors of 5 in separation correspond to more than 2 decades in the recombination rate. Recently, however, theoretical work on excitons in disordered systems[18] has predicted that the lifetime of trapped excitons depends on the fifth power of the radius of the more localized carrier. Thus a physically realistic range in the degree of localization can also account for the observed decay statistics, and the two geminate models may not be so easily distinguished on this basis.

It has recently been pointed out[8,19] that the observed independence of the PL decay statistics on excitation density does not preclude certain types of nongeminate recombination processes, as previously assumed. A nongeminate distant-pair model has been proposed that appears to be equally capable of explaining the observed PL characteristics. Low temperature photoconductivity data also appear to support this model.[20] In contrast to the geminate picture, carriers are thought to diffuse away from each other so that they lose all spatial correlation and are independently and randomly trapped. Due to the tremendous range of carrier lifetimes, a metastable population of long-lived carriers builds up as samples are exposed to light at low temperatures. Eventually the population saturates, and thereafter provides a constant background density of thermalized carriers. As new pairs are photoexcited they too diffuse away from each other, and are trapped at random positions with respect to this thermalized population. Recombination of newly created carriers with those of the metastable background should exhibit a simple, linear dependence on excitation intensity indistinguishable from that of geminate processes. The presence of a metastable background of long-lived carriers in low-temperature optical experiment is not disputed, only the absolute density of the population, and thus the extent of its contribution to the PL dynamics.[21,22]

To escape from its geminate partner, a carrier must diffuse a distance r_0 which is comparable to or larger than the Coulomb capture radius, $r_c(T) = e^2/4\pi\epsilon kT$. At room temperature, theoretical[6] and experimental[23] estimates place $r_0 \sim 40\text{-}55$ Å, roughly equal to $r_c(300K) = 46$ Å. While r_c rises as $1/T$, r_0 appears to decrease as the temperature is lowered.[23] Thus at the low temperatures of most PL experiments, photoexcited pairs are expected to remain well within r_c during thermalization, supporting a model of geminate recombination. These estimates, however, are based on the Onsager model of thermalization[24,25] which was derived for the case where the Coulomb energy alone determines the energy difference between states. The appropriateness of such an approach in a-Si:H has been questioned,[19] as it is known that thermalization continues within the localized band-tail states that have a disorder-induced spread in electronic energies comparable to the magnitude of Coulomb interaction energies. Another unresolved question concerns the extended-state mobility. The magnitude sets the distance scale for carrier thermalization on the ps time scale prior to trapping into localized states. Although most workers in the field assume a modest value $\sim 10^2$ cm^2/V-s, some experiments suggest a much larger value,[26] which would tend to support a nongeminate model of recombination. Thus an unambiguous theoretical determination of the carrier dynamics is not currently possible, and one must turn to experimental evidence for further insight. In this paper I review the information provided by four recent PL experiments, and examine the ability of the exciton, geminate-pair and nongeminate distant-pair models to account for these results.

DISCUSSION OF EXPERIMENTS AND MODELS

The first experimental results to be considered involve the effects of sub-bandgap excitation. The cw spectrum is independent of excitation energy above the optical gap, but shifts monotonically to lower energy when the excitation energy is lowered below the gap.[2,3,5] Based on the cw data, various conflicting models of the dynamics were proposed.[2,3,5,27] Time-resolved spectra[4] have greatly clarified this point. As shown in Fig. 1a, the shift of the cw spectrum as the excitation energy is lowered is due to the reduction and eventual loss of the initial thermalization shift. For

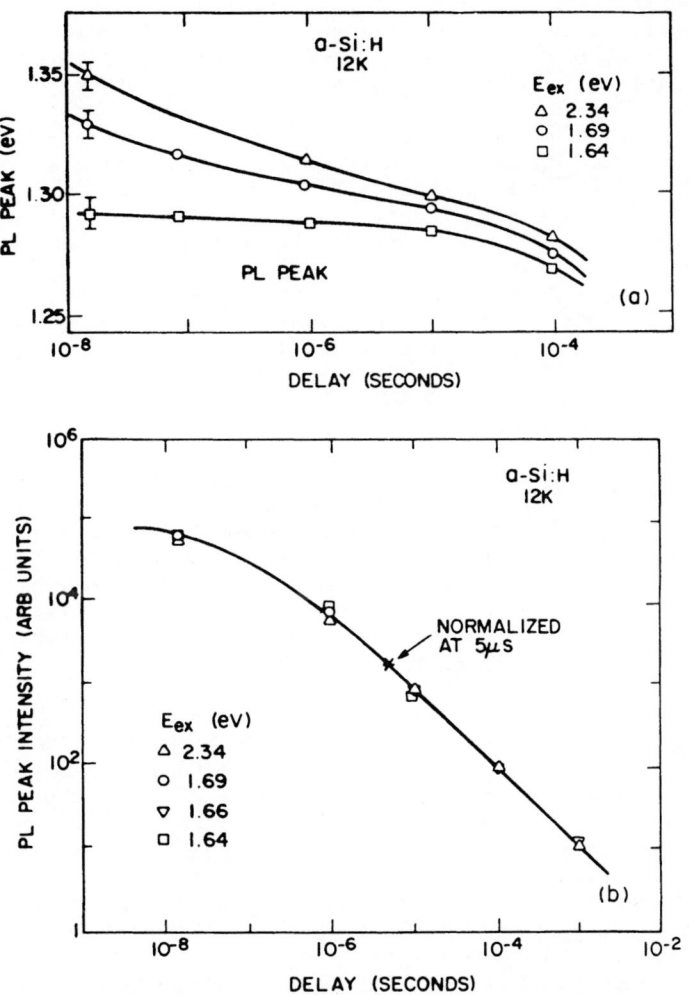

Fig. 1. PL peak position (a) and intensity (b) versus delay as a function of excitation energy, E_{ex} (Ref. 4). As E_{ex} is lowered the early shift of the spectrum due to thermalization is reduced, but there is no significant change in the decay rates.

excitation well below the optical gap, the spectrum starts at lower energy, and shifts little until beyond 10 μs. Apparently sub-bandgap light directly populates localized states deep in the gap, where subsequent thermalization is negligible. Significantly, there is no noticeable change in the decay rate of the PL, as evident from Fig. 1(b).

The absence of thermalization processes at early times places severe limitations on the distance to which carriers can diffuse. The pairs have no opportunity to move apart rapidly, as they did when excited well above the bandgap, and most carriers must remain close to their geminate partner at early times. The lack of a concomitant change in the decay rates means that the recombination rates are uncorrelated with the distance carriers diffuse apart. This result is manifestly at odds with geminate-pair recombination, in which the rate depends exponentially on pair separation. In the absence of rapid early thermalization it also appears unrealistic to advocate the continued existence of the extensive diffusion required by the distant-pair model. Thus these results provide strong arguments against the applicability of both the geminate-pair and distant-pair models. Only exciton recombination can offer a consistent picture of these observations. When an exciton is formed in a crystal, the memory of the original electron-hole separation is lost so that it does not affect the subsequent recombination. The observed decoupling of the recombination dynamics and the thermalization dynamics observed in these sub-bandgap experiments suggests that this is also the case in a-Si:H. This conclusion further refines the concept of the trapped exciton in a-Si:H. It must be pictured as a specific state into which electron-hole pairs coalesce, independent of their original separation. Only the efficiency of exciton formation is then affected by the original distribution of pair separations, not the subsequent recombination. A specific model with these characteristics has been suggested by Kivelson and Gelatt.[18] They picture an exciton composed of a highly localized, trapped hole and an electron in a "hydrogenic" state, bound together by their mutual Coulomb interaction.

As the excitation energy is lowered, a changeover to a sublinear dependence on excitation density occurs, apparently simultaneously with the onset of the spectral changes.[5] This puzzling behavior is not inherent to any of the models discussed here. It has been suggested[5] that interference from a lower energy PL band with different nonradiative dynamics may be responsible. Since the density of available states falls exponentially with excitation energy, another possible source of sublinear behavior is that the excited population is approaching saturation.

Another aspect of these sub-bandgap results is difficult to explain by either the geminate- or distant-pair models. In both nonexcitonic models there is a Coulomb interaction term in the excited state energy. This should result in a shift to higher energy with delay as nearby pairs with larger Coulomb interactions decay faster. It has been argued that the Coulomb interactions may not play a significant role in a-Si:H,[19] but the shift is, in fact, observed in alloys of a-Si:H with O, C and N, where the dielectric constants are smaller, and thus the Coulomb energies larger.[28,29] The smaller shift expected in a-Si:H had previously been assumed[28] to be masked by the larger, competing thermalization shift. In the sub-bandgap experiments, however, the thermalization shift is eliminated, and the shift due to Coulomb energies would have been observable in the time-resolved measurements[4] had it been present. Its absence implies that there is no asymmetry of Coulomb energies in the ground and excited states, i.e. that the recombination is excitonic or that it occurs at a charged defect. As no other evidence for a high density of charged defects has emerged, this result strongly supports the role of excitons in the early PL decay.

This conclusion calls for a reevaluation of the alloy data. The radiative rate of the trapped exciton is predicted[18] to increase rapidly with the reduced dielectric constant of the alloys, and, indeed, a faster PL decay has been observed.[29] Thus the exciton states may recombine rapidly, leaving the separated pairs to dominate the spectrum at later times. This is consistent with the observations[30] that a gap opens up in the distribution of decay rates: there are very fast rates and very slow rates with few states radiating on intermediate time scales. The Coulomb interaction in the excited states of the separated pairs may then be responsible for the upward motion of the spectrum observed in the 100 ns to 1 μs time regime.

The picture that emerges from these experiments using sub-bandgap excitation is that the PL is dominated by exciton recombination at early times, and by radiative tunnelling of more distant nongeminate pairs later. The cw PL then contains contributions from both processes, with the latter being responsible for the nongeminate properties observed in frequency resolved lifetime (FRL) measurements.[8]

A very recent detailed study[12] of the optically determined thermalization rate as a function of temperature is the next experiment to be discussed in the context of these models. Carrier thermalization in the band tails is thought to underlie the dispersive transport observed in a-Si:H and other amorphous materials.[21,31,32] Room-temperature conductivity (PC) and photoinduced absorption (PA) data[6,33] have been successfully modelled in the multiple-trapping (MT) picture in which carriers trapped in localized band-tail states are thermally excited to transport states, and subsequently retrapped. It has been recognized for some time that the thermalization shift observed in the PL at early times is too rapid to be consistent with the MT model. Similar discrepancies have recently been reported in low temperature PA and PC data[34,35,36] as well, and a model based on direct hopping between band-tail states as the dominant mode of thermalization[32] appears to provide a unified description of all the low-temperature results.

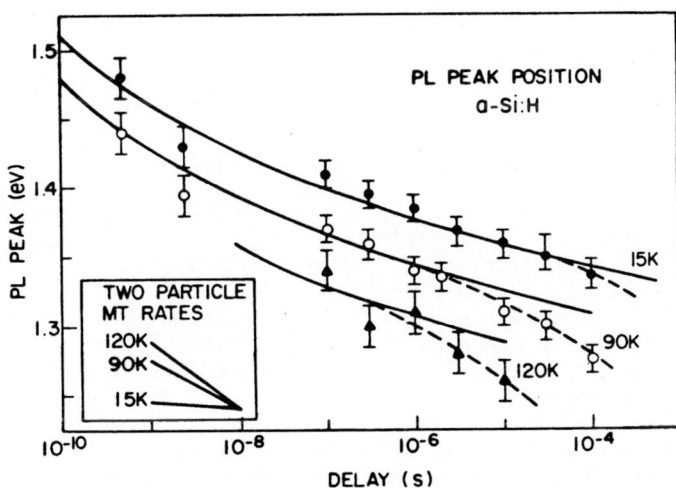

Fig. 2. PL peak positions versus delay in a-Si:H films as a function of temperature (Ref. 12). The solid lines are fits to a model of thermalization based on direct hopping between localized states. The inset depicts calculated shift rates per decade for both carriers thermalizing simultaneously by a multiple-trapping process.

Fits of this model to the optical data in the early-time regime are shown in Fig. 2.

Since the direct-hopping rate decreases rapidly with delay, the dominant thermalization mechanism should gradually cross over from direct-hopping to MT processes. Thus at later times the net thermalization rate should asymptotically approach the MT rate:

$$\Delta E \text{ per decade} = \begin{cases} - kT \, \ell n 10 & \text{single-particle thermalization} \\ -2kT \, \ell n 10 & \text{both carriers thermalizing} \\ & \text{independently.} \end{cases}$$

Inspection of the 90K data in Fig. 2 reveals that at intermediate times the thermalization rate per decade is actually slower than $2kT \, \ell n 10$. This means that the MT process for the electrons and/or holes is inhibited in some way, or that thermalization involves a single particle. Within the geminate-pair model, electrons and holes thermalize independently, and in order to explain this result, one of the carriers must be deeply trapped within 500 ps of photoexcitation. Possible sources of rapid relaxation include lattice relaxation around the excited carrier, and self-trapping. Both mechanisms have been suggested,[15] specifically for holes in a-Si:H, to account for energy scale discrepancies between absorption and emission data. Consider, however, the energy scale required for this rapid trapping process to explain the data of Fig. 2. The solid-line fit to the 90 K data indicates that a net energy loss of ~ 0.3 eV occurs by 1 µs, which in this picture is to be assigned to thermalization by the electron. For hole thermalization to remain inhibited at 1 µs implies that it is trapped even more deeply. Thus the 1.4 eV PL observed at this delay should be more than 0.6 eV below the band edge. This is inconsistent with the measured 90 K optical gap of 1.80 eV in this sample.[12]

Single-particle thermalization is, however, an inherent feature of both the exciton and distant-pair models. For the distant-pair model, recombination at early times occurs between the thermalizing carrier and a fully-thermalized nongeminate partner from the metastable background population. Thus each recombination event involves only one thermalizing carrier, as observed. As with the geminate-pair model, however, there is an energy-scale problem. The fully-thermalized carriers would need to be > 0.3 eV below the bandgap inconsistent with the measured value. An additional complication arises in that the conduction and valence band tails are not identical. This inequivalence means that electrons and holes thermalize at different rates and to different depths, causing the predicted PL spectrum to consist of 2 peaks, with a time-dependent separation. To explain the absence of a second peak, one must argue that only one of the thermalizing carriers can undergo radiative recombination, a suggestion for which there is no additional support. Thus the exciton model is the only one which has no difficulty in accounting for these data. Thermalization by definition involves a single particle in this case, with hopping occurring within the band tail of the exciton density of states.

For the third set of relevant experiments I turn to sub-ns results from various laboratories.[9,10,11] Experimenters using detectors sensitive only to the higher energy portion of the spectrum report that the distribution of lifetimes in the PL extends back to ps time scales.[10,11] These fast rates are not observed, however, in spectrally-integrated measurements.[9] Instead, the shortest decay times are limited to about 8 ns, indicating that the faster rates are associated with spectral shifts which move the spectrum rapidly to lower energy, rather than an actual decay of the luminescing population. This interpretation is verified by sub-ns time-resolved spectra, which reveal rapid thermalization shifts occurring on these short time

scales.[9,10] The absence of a decay (or rise) in the luminescence intensity
during this rapid thermalization implies that the population of states radiating on ∿ 8 ns time scale is not affected by the thermalization mechanism.
This is not possible within the framework of the geminate-pair model in which
the independent thermalization of electrons and holes through their respective band tails would naturally alter the initial pair distribution. Only
if the distribution were already random, and the Coulomb effects negligible,
could <u>independent</u> hopping leave the pair distribution unaltered. This, of
course, is the description offered by the distant-pair model. The other
possibility is that electrons and holes are constrained to thermalize together, in which case an exciton picture is being invoked. Thus both the
exciton and distant pair models are consistent with these sub-ns data, the
geminate pair model is not.

The last experiments to be discussed in the context of these models
involve quantum efficiency (QE) measurements of the main PL band and the
0.7 eV band in annealed a-Si:H films.[7,14] Annealing drives off hydrogen
creating dangling bond (DB) defects whose density, N_S, can be measured by
ESR. Although the N_S dependence of the QE in each of the individual bands
is rather complex, the relative intensity in the 0.7 eV band to that in the
main band scales linearly with N_S as shown in Fig. 3. This is true both for
the initial intensity after pulsed excitation as well as for the cw intensities. These results can be well accounted for with a simple branching
ratio model[7] based on the competition for initial capture of photoexcited
electrons and holes at band-tail and DB sites. The recombination of electrons and holes both initially trapped in the band tails is the source of
the main PL band. The 0.7 eV band is attributed to recombination between
an electron trapped in a band-tail state and a hole at a DB site. This
interpretation requires the <u>independent</u> trapping of electrons and holes,
which can occur only if the photoexcited pairs become spatially uncorrelated

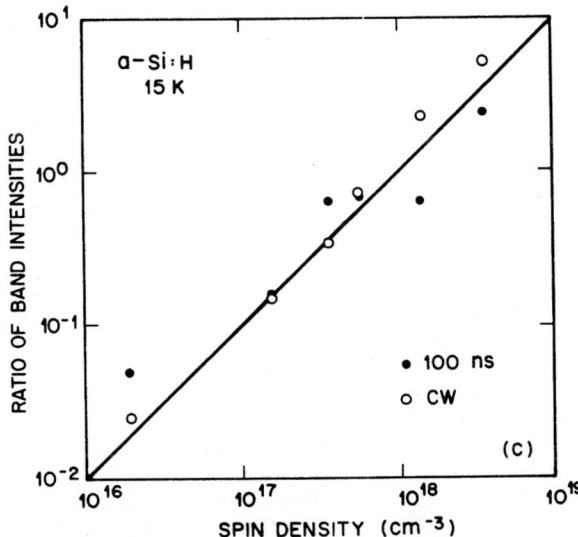

Fig. 3. Ratio of the intensity in the 0.7 eV
        PL band to that in the main PL band
        as a function of the spin density
        in annealed a-Si:H films (Ref. 7).
        Closed circles represent the initial
        intensities after pulsed excitation,
        open circles the cw intensities.

during thermalization to the band edges. This means that the two carriers must individually diffuse far enough to randomize their relative spacing prior to the first capture events, which PC and PA measurements[38,39] place in the ps range. Strictly speaking, this random behavior is only consistent with the distant-pair model. Note, however, that diffusion covering 50 Å corresponds to a random sampling of $\sim 10^3$ sites, which might be sufficient to account for reasonable fits to this simple model without precluding geminate recombination. A study of the branching ratios during sub-bandgap excitation where thermalization is restricted would be useful to clarify this point.

CONCLUSIONS

In this paper 3 models of recombination have been examined in the context of a number of recent optical experiments. Of the three models, the geminate-pair model appears to be least able to account for the new observations. The distant-pair model fares better, but requires additional assumptions in some cases. In general the exciton model provides the most straightforward description of the early-time PL, although a consistent interpretation of the branching ratio data remains unclear. While these experiments do not definitively determine the recombination dynamics, the most consistent picture of low temperature PL that emerges from this analysis is the following. Photoexcitation creates electron hole pairs that thermalize rapidly and relatively independently through high mobility states to the band edges where they become trapped in localized states. At low temperatures most localized pairs remain relatively close and coalesce into a trapped exciton state, thereby losing all memory of their original separation. Subsequent thermalization occurs by direct hopping between exciton band-tail states. The early luminescence is dominated by exciton recombination with a distribution of rates arising from a distribution in the degree of localization. At very long delays after pulsed excitation, and in cw PL, a significant contribution may arise from nongeminate pairs. This nongeminate recombination occurs between recently excited carriers which escaped exciton formation, and thermalized carriers in the metastable background population created during the previous history of light exposure.

The author thanks D. Monroe for a careful reading of the manuscript and is grateful to M. Silver, M. A. Kastner, D. Adler, J. Orenstein and D. Monroe for valuable discussions.

REFERENCES

1. "Tetrahedrally Bonded Amorphous Semiconductors", (Carefree, Arizona), Topical Conference on Tetrahedrally Bonded Amorphous Semiconductors, edited by R. A. Street, D. K. Biegelsen and J. C. Knights (AIP, New York, 1981).
2. W.-C. Chen, B. J. Feldman, J. Bajaj, F.-M. Tong and G. K. Wong, Solid State Commun. 38, 357 (1981).
3. J. Shah, A. Pinczuk, F. B. Alexander and B. G. Bagley, Solid State Commun. 42, 717 (1982).
4. B. A. Wilson and T. P. Kerwin, Phys. Rev. B 25, 5276 (1982).
5. P. K. Bhat, T. M. Searle and I. G. Austin, Solid State Commun. 45, 481 (1983).
6. J. Tauc in "Festkörperprobleme" Vol. XXII of Advances in Solid State Physics, edited by J. Treusch (Vieweg, Braunschweig, 1982), p. 85 and references therein.
7. B. A. Wilson, A. M. Sergent and J. P. Harbison, Phys. Rev. B 30, 2282 (1984); B. A. Wilson, A. M. Sergent, K. W. Wecht, A. J. Williams and T. P. Kerwin, Phys. Rev. B 30, 3320 (1984).

8. D. J. Dunstan, S. P. Depinna and B. C. Cavanett, J. Phys. C $\underline{15}$, L425 (1982).

9. B. A. Wilson, P. Hu, J. P. Harbison and T. M. Jedju, Phys. Rev. Lett. $\underline{50}$, 1490 (1983); B. A. Wilson, P. Hu, T. M. Jedju and J. P. Harbison, Phys. Rev. B $\underline{28}$, 5901 (1983).

10. D. G. Stearns, Phys. Rev. B $\underline{30}$, 6000 (1984).

11. T. E. Orlowski and H. Scher, Phys. Rev. Lett. $\underline{54}$, 220 (1985).

12. B. A. Wilson, T. P. Kerwin and J. P. Harbison, to be published.

13. R. A. Street, D. Biegelsen and J. Stuke, Philos. Mag. B $\underline{40}$, 459 (1979).

14. U. Voget-Grote, W. Kümmerle, R. Fischer and J. Stuke, Philos. Mag. B $\underline{41}$, 127 (1980).

15. R. A. Street, Adv. in Physics $\underline{30}$, 593 (1981) and references therein.

16. N. F. Mott, J. Phys. C $\underline{13}$, 5433 (1980).

17. D. Engemann and R. Fischer, Phys. Status Solidi, B $\underline{79}$, 195 (1977).

18. S. Kivelson and C. D. Gelatt, Jr., Phys. Rev. B $\underline{26}$, 4646 (1982).

19. D. J. Dunstan and F. Boulitrop, Phys. Rev. B $\underline{30}$, 5945 (1984) and references therein.

20. M. Hoheisel, R. Carius and W. Fuhs, J. Noncryst. Solids, $\underline{59}$ & $\underline{60}$, 457 (1983); Ibid J. Noncryst. Solids, $\underline{63}$, 313 (1984).

21. R. A. Street and D. K. Biegelsen, Solid State Commun. $\underline{44}$, 501 (1982).

22. D. J. Dunstan, Solid State Commun. $\underline{49}$, 395 (1984).

23. J. Mort, J. de Physique $\underline{C4}$, 433 (1981).

24. L. Onsager, Phys. Rev. $\underline{54}$, 554 (1938).

25. Kwok-Leung Yip, Leonard S. Li and I. Chen, J. Chem. Phys. $\underline{74}$, 751 (1981).

26. M. Silver, E. Snow and David Adler, Solid State Commun. $\underline{51}$, 581 (1984).

27. F. Boulitrop and D. J. Dunstan, Phys. Rev. B $\underline{28}$, 5923 (1983).

28. R. A. Street, Solid State Commun. $\underline{34}$, 157 (1980).

29. R. Carius, K. Jahn, W. Siebert and W. Fuhs, J. Luminescence $\underline{31}$ & $\underline{32}$, 354 (1984).

30. B. A. Wilson and J. P. Harbison, unpublished, and W. Siebert, private communication.

31. T. Tiedje and A. Rose, Solid State Commun. $\underline{37}$, 49 (1980).

32. J. Orenstein and M. A. Kastner, Solid State Commun. $\underline{40}$, 85 (1981).

33. J. M. Hvam and M. H. Brodsky, Phys. Rev. Lett. $\underline{46}$, 371 (1981).

34. D. Monroe, J. Orenstein and M. A. Kastner, J. de Physique $\underline{C4}$, 559 (1981).

35. I. K. Kristensen and J. M. Hvam, Solid State Commun. $\underline{50}$, 845 (1984).

36. T. Tiedje in "Semiconductors and Semimetals" Vol. 21, edited by J. I. Pankove, (Academic Press, NY) 1984, Part C.

37. Don Monroe, Phys. Rev. Lett. $\underline{54}$, 146 (1985).

38. A. M. Johnson, D. H. Auston, P. R. Smith, J. C. Bean, J. P. Harbison and A. C. Adams, Phys. Rev. B $\underline{23}$, 6816 (1981).

39. Z. Vardeny and J. Tauc, Phys. Rev. Lett. $\underline{46}$, 1223 (1981).

BAND-TAIL DIFFUSION AND PHOTOLUMINESCENCE

IN a-Si:H

H. Scher

The Standard Oil Company (Ohio)
4440 Warrensville Center Road
Cleveland, Ohio 44128

INTRODUCTION

Time resolved recombination luminescence (PL) between suitably prepared impurity state populations in crystalline semiconductors have been extensively studied. For example an important study[1] of low temperature PL in GaP presents a clear picture of randomly situated localized electrons and holes recombining through a radiative tunneling mechanism. This has become a model for a system exhibiting a large distribution of (PL) lifetimes. The ability to carefully control the type and concentration of the impurity states, as well as the initial population, has enabled one to analyze and separate effects due to recombination kinetics, diffusion and Coulomb interaction. Application of these ideas have been made to interpret PL measurements of amorphous semiconductors. Here the picture is considerably complicated by both the presence of a quasi-continuum of states near the band edges -- band tail states, and large concentrations of deeper defect states. Besides the large distribution of radiative rates attributed to the tunneling recombination between separately localized electrons and holes,[2,3] there has been discussion of electron-phonon interactions (Stokes shift)[2], thermalization in the band tail states,[4] competing non-radiative processes, diffusion, Coulomb effects, geminate pairs[2], and excitonic states.[5] In a-Si:H, Street et. al. and others[2] have given a comprehensive account of the 1.4 eV emission band. Nonetheless, due to the possible presence of a number of mechanisms (above), the uncertainty of initial conditions, the relatively featureless emission spectra, and the variation in sample quantum efficiency, there is still controversy over most of the details of the proposed models.[6] As the time resolution in PL experiments improves there is increasing opportunity to obtain more detailed information concerning the initial distribution of pair separations created by optical excitation (and the subsequent time evolution of this distribution) as well as the maximum or limiting recombination and relaxation rates. PL studies in the picosecond regime[7,8] can elucidate the nature of the electronic states in the energy region between the exponential band tail and the mobility edge as well as excitonic effects involving these states.

Our analysis of the earliest time PL data is close in spirit to that of Dunstan and Boulitrop[6] -- add complications only as necessary. As described below our focus on the highest PL energies emitted permits us to detail the initial thermalization process. This process is consistent with separately localized electrons and holes relaxing via downward transitions in the band tail states. An alternate explanation[5] in terms of an excitonic state, where the electron and hole relax[9] "together", is based on evidence for an exponential initial decay with a lifetime of ~8 nsec. One can advance a simple argument to reconcile these pictures. We maintain that electron and hole diffusion compete with radiative tunneling and the possible existence of a "cutoff" at small radiative lifetimes can be due to a reaction-rate limitation. That is, at sufficiently close encounter the electron and hole decay is limited by the intrinsic radiative process W^{-1} (such as the fluorescence of an excited molecular state) and not the wavefunction overlap. One can associate an exciton-like state with this closest approach, however, it is one degenerate in energy with states consisting of separately diffusing electrons and holes. The radiative decay is limited by diffusion and/or pair overlap when the time for this process exceeds W^{-1}. The transition from reaction-rate limitation to diffusion controlled limitation has been used to model exciton annihilation in organic solids.[10]

As an historical aside, the relation between impurities in crystalline semiconductors and amorphous semiconductors was first promoted by Sir Nevill Mott. Since he first suggested the hopping process[11] for low concentration impurity conduction in 1956 one can trace these ideas as an important precursor of his interest in amorphous semiconductors. On a personal note, I recall a stimulating and enlightening discussion I had with Prof. Mott in 1969 when I outlined to him the preliminary ideas of a new approach to the dynamics of impurity conduction. The stochastic nature of the hopping conduction was modelled with a continuous time random walk formalism (CTRW)[12]. For each step of the transport, the probability distribution for the hop time and displacement was approximated by the time-dependent spectral distribution function for donor-acceptor pair PL in semiconductors.[1] The quasi-algebraic time dependence of the total PL intensity, $I(t)$[1], became the basis of all the key features of dispersive transport.[13]

The present paper describes an admixture of both hopping transport (diffusion) and PL. I am pleased to add this contribution to this volume honoring Sir Nevill Mott on the occasion of his 80th birthday. Many of us in this field have felt directly the influence of the stimulation of his ideas/insights on our work.

SUMMARY

In this paper we review the results of recent picosecond PL measurements[7,8] in a-Si:H. An excitation of energy of 2.33 eV is used, well above the optical gap (1.8 eV at 300K) resulting in the creation of mobile carriers. Because of the time (10 psec - 4 nsec) and spectral (2.1-1.5 eV) regimes associated with the streak camera detection scheme used, our measurements select the closest pairs emitting at the highest PL energies. We find rapid PL decay in this time/spectral regime with power law behavior. A comprehensive model of nonradiative relaxation is developed in which band-tail thermalization procedes initially via efficient electron transfer (hopping) among states within an energy bandwidth determined by $\hbar\omega$, an effective phonon frequency, and kT. The model correlates the thermalization process with radiative decay and T-dependent local diffusion and provides results in excellent agreement with experiment. Recently subnanosecond (>0.25 nsec) PL measurements[5] at lower PL energies (<1.6 eV) were reported in a-Si:H. We compare those findings with the present results. A brief comparison of our model with their exciton interpretation has already been discussed (above). This comparison will be expanded in another paper.[14]

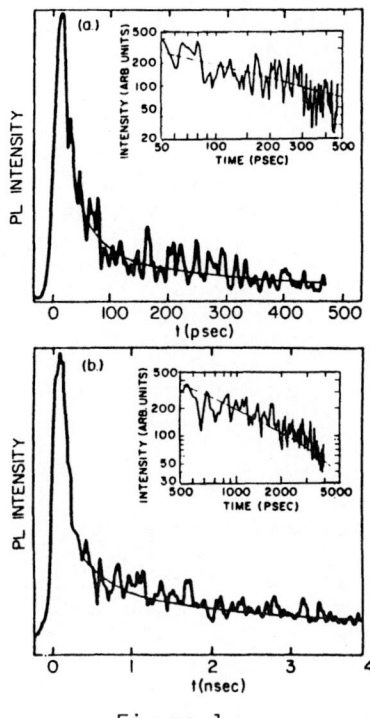

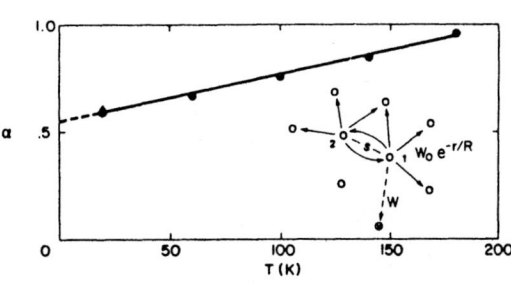

Figure 2.

Figure 1.

Fig. 1.   Measured buildup and decay of PL (2.1 - 1.5 eV) in a-Si:H at 20K. Also shown are least squares fits assuming a power law, $t^{-\alpha}$, decay. The inserts show log I vs. log t plots of the data and fits obtained using the model described in the text.

Fig. 2.   Temperature dependence (filled circles) of the power law exponent, α, for the initial PL decay in a-Si:H and the fit (solid line) using the model described in the text. The insert illustrates the mechanism for rapid PL decay involving competition between electron band-tail hopping (among a cluster of sites, O) at a rate $W_o e^{-r/R}$ with radiative recombination at a maximum rate, W. The hole is taken to be immobile at site ⊗.

RESULTS

Figure 1 displays the PL intensity I(t) measured at 20K and averaged over 20 laser shots for fast (500 psec) and slow (4 nsec) streak camera total sweep times. In all of our measurements the PL buildup time is less than the system response time which places an upper limit of ~10 psec on the time for photoexcited carriers to relax into states emitting within our spectral observation window (2.1 - 1.5 eV). Since the exponential optical absorption edge (Urbach edge) in a-Si:H extends to ~1.85 eV at 20K[3,15], and the fastest radiative lifetimes are ~8 nsec[5], these experiments probe the closest pairs emitting at the highest PL energies as they thermalize within the manifold of band-tail states. This thermalization process, which manifests itself as a PL spectral shift,[3,4] represents relaxation of carriers to lower energy states and not recombination. Recent experiments[5] have shown that at

409

15K the PL spectrum at 500 psec following excitation is quite broad with appreciable intensities at energies >1.6 eV. Using a high energy cutoff filter at 1.8 eV we examined the PL buildup and decay characteristics and, within experimental accuracy, found no variation in these times, although the peak PL intensity decreased about a factor of three with the reduced spectral bandwidth. <u>This indicates that the PL spectrum extends to energies >1.8 eV for $t < 100$ psec.</u> In this range any exciton binding energies are small compared to potential energy fluctuations.

The PL decay data at 20K is best fit by a power law, $t^{-\alpha}$, for 50 psec $< t <$ 3 nsec. At earlier times the decay is convoluted with the system response. From the least squares fits in Fig. 1 we find for the fast sweep $\alpha = 0.59 \pm 0.03$ and for the slow sweep $\alpha = 0.61 \pm 0.03$. Thus, within experimental uncertainty the decay at 20K is characterized from 50 psec to 3 nsec by $\alpha = 0.60$. Attempts at fitting the data to an exponential decay over the same time range gave poorer fits in all cases.

Measurements of the temperature dependence of the initial PL decay were performed over the range 20 to 180K. We find that the power law exponent, α, is temperature dependent taking the form $\alpha(T) = \alpha_0 + \beta T$. Shown in Fig. 2 is a plot of α vs. T where we find from the intercept a value of $\alpha_0 = 0.55 \pm .03$ and from the slope a value of $\beta = .0022 \pm .0003$ K^{-1}. No change in the initial (maximum) PL intensity is found up to 100K where it begins to decrease gradually, down a factor of two at 140K. At 180K the initial PL intensity is down a factor of four from that at 20K. Although the power law decay behavior seen in these experiments differs from the exponential decay reported in Ref. 5 for the spectral regime below 1.6 eV, the temperature dependence of the initial intensity does agree.

MODEL

As emphasized earlier, our measurements use PL as a probe of thermalization processes which are much faster than the fastest radiative recombination rates. We assume, along with a number of authors[2,6], that the higher energy luminescence is due to a band-tail electron recombining with a band-tail hole. A typical configuration for this interaction is shown in the insert of Fig. 2. The arrows indicate the possible electron transfer steps {among a random distribution (energy, separation) of sites, o } and the hole is taken to be immobile (at site ⊗). As discussed above, these experiments select the closest pairs emitting at the highest PL energies. The pair can be geminal (although we neglect any specific Coulomb interaction), a localized exciton (in the sense discussed above as a radiative reaction-rate limitation), or randomly separated as a result of relaxation of an electron with as much as 0.4 eV energy above the optical gap.

The single (fastest) radiative recombination rate is designated as W. Competing with W are the electron transfers away from the hole offset by the back transfers. One can readily assume that the back transfer occurs preferentially from a small number of nearby sites. This local diffusion involving interaction among a cluster of sites imbedded in a fixed random background of additional sites is solved exactly. The electron can hop back and forth an arbitrary number of times to the sites within the cluster and from each of these, hop once to the fixed random background of sites. In the absence of Coulomb effects, the time

410

to return to the cluster after leaving is on the order of a large
sampling time of all the other sites. The configuration average (c.a.)
over these latter site positions/energies is then carried out exactly,
retaining all the correlations between the cluster geometry and the
background. Referring again to Fig. 2, we are interested in determining
the (averaged) probability $\langle P(1,t)\rangle$ for the electron to be found at site
1 at time t if it started there at t=0. The PL intensity is then $I(t) =$
$W\langle P(1,t)\rangle$. Since we are considering one (i.e. the maximum) radiative
rate, $I(t)$ is proportional to the probability and not the rate of change
of the probability. The latter case applies when all the decay channels
are radiative.[1] Hence $I(t) \sim t^{-\alpha}$ (as we will show) and not $t^{-1-\alpha}$. We[14]
specialize the cluster to two sites {1,2} separated by s and obtain

$$W\langle P(1,t)\rangle = W\exp\{-t(W_{21} + W)\}\Phi_1 +$$

$$W(W_{12}W_{21})^{1/2} \int_0^t d\tau \exp\{-(t-\tau)W_{12}-\tau(W_{21}+W)\}\tau^{1/2}(t-\tau)^{-1/2}I_1(\sigma)\Phi_2(t-\tau,\tau), \quad (1)$$

$$\Phi_1(t) = \exp\{-\int d^3r \, d\epsilon \, p(r,\epsilon)[1-\exp(-tW(r-r_1))]\}, \quad (2)$$

$$\Phi_2(t-\tau,\tau) = \exp\{-\int d^3r \, d\epsilon \, p(r,\epsilon)[1-\exp(-(t-\tau)W(r-r_2)-\tau W(r-r_1))]\}. \quad (3)$$

Here, $p(r,\epsilon)$ is the probability of finding a site at r with energy
difference ϵ and $W(r)$ is the transition rate to this site (the ϵ-
dependence is implicit). $I_1(\sigma)$ is the modified Bessel function of order
unity and $\sigma \equiv \{4W_{12}W_{21}(t-\tau)\tau\}^{1/2}$, where W_{12}, W_{21} are the transition
rates between sites {1,2}. The function Φ_2 is a generalization (to two
sites) of the more familiar[1,12] Φ_1, the c.a. probability to remain on a
single site in a random distribution of sites.

In order to calculate $I(t)$ from Eqns. (1)-(3) we must specify $W(r)$
and $p(r,\epsilon)$. To initially simplify the calculation of Φ_1 and Φ_2 we
consider the transition rate to be non-zero and a function of r only in
an energy width $\hbar\omega$, an effective phonon energy, for $\epsilon < 0$ (hopping down)
and kT for $\epsilon > 0$ (hopping up). The basic parameters can now be defined:

$$W(r) = W_0\exp(-r/R) \quad (4)$$

$$p(r,\epsilon) = g(\epsilon_1-\epsilon)\equiv g_0\exp\{-(\epsilon_1-\epsilon)/kT_0\}. \quad (5)$$

$$n = \int_{\hbar\omega}^{kT} d\epsilon \, p(r,\epsilon) = kT_0 g(\epsilon_1)\{\exp(T/T_0) - \exp(-\hbar\omega/kT_0)\}. \quad (6)$$

Here ϵ_1 is the energy, measured from the conduction band (CB) edge, of
the electron at 1 and kT_0 is the width of the exponential density of
band-tail states (~30 meV in a-Si:H)[16]. The estimates for the
parameters are: $R\simeq 7\text{Å}$ for $\epsilon_1 < 0.1$ eV, $W_0 \simeq 10^{13}$ sec^{-1} (order of
magnitude estimate obtained from donor state calculations[12] in
crystalline Si), and $g_0 \sim 4 \times 10^{20}$ eV^{-1} cm^{-3}.

Now, using the large argument asymptotic form for
$I_1(\sigma) \sim \exp(\sigma)(2\pi\sigma)^{-1/2}$ and the Laplace method one can evaluate the
expression in Eq. (1) analytically as a function of $\eta \equiv 4\pi nR^3$ and W_0t
and compare with the PL $I(t)$ in Fig. 1. One obtains a power law decay
$I(t) \sim t^{-\alpha}$ for nearly two decades of W_0t with a departure toward
increasing α at large W_0t. An excellent fit of the theory to the PL
$I(t)$ over the entire experimental range (50 psec $\leq t \leq$ 4 nsec), is
obtained using $\eta=5 \times 10^{-3}$ and $W_0 = 3 \times 10^{13}$ sec^{-1} as shown in the log I
vs. log t plots in the inserts to Figs. 1(a) and 1(b).

The theoretical values for α show a linear dependence on η over the range of α values shown in Fig. 2. Using Eq. 6 one can obtain the <u>relative</u> change in η as a function of T for various values of $\hbar\omega$, and therefore α(T). The linear relation (solid line) shown in Fig. 2 corresponds to $\hbar\omega$ = 20 meV. Thus 20 meV is an <u>effective</u> measure of the energy width for relaxation involving single hops in the band-tail. Since there is a peak in the acoustic phonon density of states[17] in a-Si at 20 meV, it is tempting to ascribe the relaxation we probe to interaction with acoustic phonons.

THEORETICAL RESULTS

One can now determine the electron position (energy) in the band tail by again using Eq. 6 with the value of $\hbar\omega$ obtained from the α(T) data and the above value of η at T=20K. We obtain ϵ_1=50meV. The nearest neighbor hopping time at ϵ_1 (with width $\hbar\omega$) is 60 psec. This time is consistent with the notion that the electron relaxes, by hopping down, to a level where the hopping time is on the order of the observation time of the experiment ($\sim$100 psec). At ϵ_1 the release time to the CB edge at T=20K is 10 sec. Only at T=180K is the release time comparable to the observation time. Therefore, multiple trapping is unlikely to play any role in the nonradiative transitions (via local diffusion) in the temperature range of Fig. 2. <u>With the model presented here, one can obtain a linear α(T) with a nonzero T=0 value simply by considering relaxation via electron hopping.</u>

If the initial fast relaxation of the electrons in the band-tail produces an occupation proportional to $g(\epsilon)$ then it can be presumed that the distribution is peaked at ϵ_1 (with a width T_0 + T) where ϵ_1 is relatively independent of T ($\Delta\epsilon_1$ < kT_0) for T < 180K. An important additional measurement would be the time resolved PL build-up at lower energy.

CONCLUSION

In this paper we have attributed the observed power law nature of the initial PL decay in a-Si:H to nonradiative relaxation processes which bring electrons to lower energy states within the band-tail manifold and have developed a comprehensive model of nonradiative relaxation which correlates the thermalization process with radiative decay and T-dependent local diffusion. The model incorporates both the site energy and position in determining the electron transfer efficiency. By considering hopping down to states within an effective phonon energy, $\hbar\omega$, and hopping up to states within kT, the model can explain the power law nature of the PL decay and the linear temperature dependence of the power law exponent as well as its T=0 intercept. Applying this new model to all of our PL data, we conclude the following about the initial PL decay dynamics in a-Si:H: 1) within 50 psec, electrons have moved down the band-tail 50 meV from the conduction band edge; 2) the pre-factor for the hopping rate, W_0, is 3 x 10^{13} sec^{-1}, and 3) the effective energy width for relaxation involving single hops in the band-tail is $\sim$20 meV, which is close to the peak in the acoustic phonon density of states[17] for a-Si. The rate W can be seen to be the radiative reaction-rate limitation. The effective radiative rate becomes diffusion-limited when these times exceed W_b^{-1}. A single step distant pair recombination can also be considered "diffusion-limited" (cf. Ref. (14)). Thus W^{-1} can be interpreted as the "initial decay time" in the total PL intensity discussed in Ref. (5).

412

REFERENCES

1. D.G. Thomas, J.J. Hopfield and W.M. Augustyniak, Phys. Rev. A140, 202 (1965).

2. R.A. Street, Adv. Phys. 25, 397 (1976); ibid., 30, 593 (1981), and references therein.

3. C. Tsang and R.A. Street, Phys. Rev. B 19, 3027 (1979).

4. B.A. Wilson and T.P. Kerwin, Phys. Rev. B25, 5276 (1982).

5. B.A. Wilson, P. Hu, J.P. Harbison and T.M. Jedju, Phys. Rev. Lett. 50, 1490 (1983); Phys. Rev. B28, 5901 (1983).

6. D.J. Dunstan and F. Boulitrop, Phys. Rev. B30, 5945 (1984), and references therein.

7. T.E. Orlowski, Bull. Am. Phys. Soc. 28, 239 (1983); T.E. Orlowski, B.A. Weinstein and H. Scher, Proceedings of the Conference on Optical Effects in Amorphous Semiconductors, Snowbird, 1984. The experimental set-up is discussed in T.E. Orlowski and B.A. Weinstein, Phil. Mag. B (to be published).

8. T.E. Orlowski and H. Scher, Phys. Rev. Lett. 54, 220 (1985).

9. Cf. Sec. III in D.G. Stearns, Phys. Rev. B30, 6000 (1984) for a discussion of this interpretation (in Ref. (5)).

10. M. Pope and C.E. Swenberg, Electronic Processes in Organic Crystals (Oxford, New York, 1982); T.E. Orlowski and H. Scher, Phys. Rev. B27, 7691 (1983).

11. N.F. Mott, Can. J. Phys. 34, 1356 (1956).

12. H. Scher and M. Lax, Phys. Rev. B7, 4502 (1973).

13. H. Scher and E.W. Montroll, Phys. Rev. B12, 2455 (1975).

14. H. Scher and T.E. Orlowski, to be published.

15. G.D. Cody, T. Tiedje, B. Abeles, B. Brooks and Y. Goldstein, Phys. Rev. Lett. 47, 1480 (1981).

16. T. Tiedje, J.M. Cebulka, D.L. Morel and B. Abeles, Phys. Rev. Lett. 46, 1425 (1981).

17. M.H. Brodsky and M. Cardona, J. Non-Crystalline Solids 31, 81 (1978).

ON THE DETERMINATION OF THE GAP DENSITY OF STATES IN AMORPHOUS SEMICONDUCTORS FROM INVESTIGATIONS ON DOPING SUPERSTRUCTURES

Gottfried H. Dohler

Hewlett-Packard Laboratories
1501 Page Mill Road, Palo Alto, Ca. 94 304, USA

On leave of absence from:
Max-Planck-Institut fuer Festkoerperforschung
Heisenbergstrasse 1, 7 000 Stuttgart 80, FRG

A new approach for determining the gap density of states in amorphous semiconductors is discussed. It is expected that the results of this method are neither influenced by any surface or interface effect. Also the method has the advantage that measurements are made in thermal equilibrium.

Some time ago the present author suggested the experimental investigation of amorphous semiconductors with periodic doping structures /1,2/. He claimed, that one of the most attractive properties of such systems would be their potential to provide information about the *bulk* properties of the amorphous host material. This is in contrast to the crystalline doping superlattices, where the bulk properties are well understood. Therefore, the main interest on the latter structures has been concentrated onto the observation of exotic properties and new phenomena expected from the theory, such as "tunability" of conductivity, absorption coefficient, luminescence spectra and two-dimensional subband structure. In fact, our experimental investigation of GaAs doping superlattices (also called "n-i-p-i crystals", because of the periodic sequence of n- and p- doped layers, possibly interspersed with intrinsic layers) /3,4/ has provided quantitative agreement with our former theoretical predictions /5,6/.

Very recently, amorphous silicon with doping superstructure has been grown and investigated by Hundhausen, Ley and Carius /7,8/ and by Kakalios and Fritzsche /9,10/. These experimental studies confirmed, that peculiarities which distinguish crystalline doping superlattices from the uniform host material such as dramatically increased carrier recombination lifetimes and extremely high photo sensitivity, will "survive" if the host material is a (highly disordered) *amorphous* semiconductor.

In Refs. 1 and 2 I had briefly outlined as an example of obtaining information about properties of the amorphous bulk the determination of the gap density of states. The purpose of this paper is to give a detailed discussion of this new approach.

The gap density of states is one of the most significant quantities of amorphous semiconductors. Unfortunately, all previously used methods have some intrinsic problems which make data obtained from such studies doubtful, in particular with respect to energy regions where the density of states is low. The results obtained from measurements of the field effect conductivity /a/ and of capacitance-voltage (C-V) curves /b/ relate to the material close to the interface to an oxide layer, which differs in its electronic properties from the amorphous bulk material /13/. This is also true for the results obtained from deep level transient spectroscopy (DLTS) /14/. Moreover, the field effect conductivity measurements may be obscured by interface gap states /11/. C-V experiments on amorphous junctions, which would reflect the bulk properties suffer from the following dilemma: Since one is dealing with a non-equilibrium state, the Fermi levels for the electrons and holes will be split like in a crystalline junction. The *trap quasi Fermi levels*, which describe the population of the *gap states* are only known to lie somewhere in between those for the electrons and holes. The actual distribution of these Fermi levels, which (in contrast to most crystalline junctions!) determines the local space charge density and, hence, the band bending in the depletion layer, depends in a most complicated way on generation recombination processes in the mobility gap.

This applies to forward and reverse biased junctions as well. Therefore, it is not possible, to relate the results of these capacitance voltage measurements directly to the gap density of states.

In principle, there are still a number of other possibilities for a determination of the density of state distribution, such as relating the position of the Fermi level (determined from the activation energy of the bulk conductivity) as a function of doping concentration to the changes in the density of gap states filled or depleted by the additional electrons coming from the donors and holes, respectively. This method, however, fails to provide reliable information about the gap density of states, as the doping efficiency is less than unity and the density of state distribution may (and apparently does) change as a function of doping level.

A major advantage of of our new method of deducing the density of states distribution from measurements on doping superstructures is that one is analyzing the band bending in the *bulk*, far away from interfaces to other materials, under *thermal equilibrium* conditions. That means, one is facing none of the previous problems.

There are many possibilities in designing periodic doping superstructures depending on the doping levels and the thicknesses of the individual layers. For our present purpose of determining the density of state distribution function $N(\epsilon)$ it is most appropriate to investigate structures consisting only of two components, each one characterized by a uniform doping level. The shape of $N(\epsilon)$ may be quite different for these two components as indicated in Fig.1 for an n-type and an intrinsic material. However, we assume, that the topological structure, the bond-length and bond-angle distributions will be essentially the same in the two materials. In this case it is a very plausible assumption to neglect the interfaces between the layers completely (We assume, that the samples are grown continuosly or quasi-continuously, as was the case in the experimental investigations reported so far /7-10/).

It is, of course, always possible to choose the thicknesses of the layers, d_n and d_i in our case, large enough, that the configuration represents a periodic sequence of familiar junctions with undepleted material in between. The most suitable quantity for obtaining information about bend bending in the space charge regions is the conductivity parallel to the layers of the

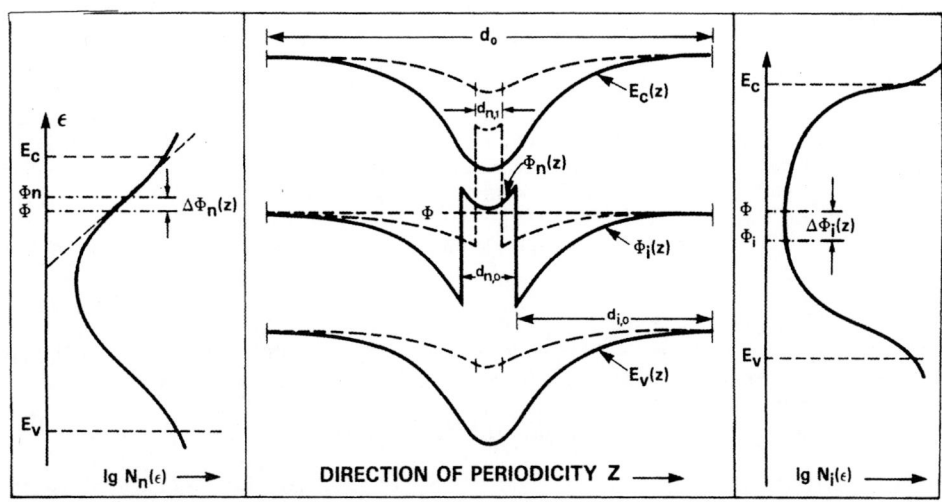

Fig.1 Band profiles of one period of a doping superlattice, consisting of
      intrinsic and n-doped amorphous layers. Full lines depict the
      situation of sufficiently thick layers, such that the spatial
      modulation of the bands is close to maximum. The dashed lines
      correspond to a configuration with significantly thinner n-layers. The
      depletion of normally occupied gap states in the n-layers and the
      filling of normally empty gap states in the i-layers is also shown for
      two specific points on the density of state curves.

superstructure. It is clear, that the conductivity is approximately

$$\sigma \simeq \sigma_{n,o} \, d_n^0/d \qquad\qquad (1)$$

in the present case, where d_n^0 is the thickness of the undepleted part of
the n-layers and d the period of the superstructure. Eq. (1) has to be
replaced by the more accurate expression

$$\sigma = (2/d) \left[\int_0^{d_n/2} \sigma_n(z) \, dz + \int_{d_n/2}^{d/2} \sigma_i(z) \, dz \right] \qquad\qquad (1')$$

if the undepleted part d_n^0 of the n-layer becomes small or the space charge
regions of two neighboring junctions even begin to overlap. The local
conductivities $\sigma_n(z)$ and $\sigma_i(z)$ are given by

$$\sigma_n(z) = \sigma_{n,o} \, \exp[-(E_c(z)-\phi)/kT \qquad\qquad (2)$$

and

$$\sigma_i(z) = \sigma_{i,o} \, \exp[-(E_c(z)-\phi)/kT \qquad\qquad (3)$$

The extensions to be made in order to include hole contributions if the
structure contains p-layers are obvious.

The distance between the mobility edge of the conduction band $E_c(z)$ and the
Fermi level ϕ is everywhere increased on the n-side of the space charge
region compared with its value E_c-ϕ_n in the neutral n-type bulk. This
shift of the Fermi level to lower energies (see left hand part of Fig.1)
results from the depletion of normally occupied gap states due to electron
transfer from the n-type material into the intrinsic one. Correspondingly,
the (large) distance between conduction band mobility edge and Fermi level

417

of the neutral intrinsic bulk is lowered on the i-side of the space charge region (see right hand part of Fig.1).

The full lines in the central part of Fig.1 depict the band profiles of a "n-i-n-i" structure in which the depletion layers of neighboring junctions just begin to overlap in both, the n- and the p-layers. Experimentally, this configuration can be achieved in the following way. Starting with large values of d_n and d_i one grows a set of samples which are identical, except for decreasing values of d_n. The conductivity per superstructure period depends linearly on d_n and extrapolates to zero for a finite value $d_{n,o}$ (roughly corresponding the $d_n^o = 0$ in Eq. (1)). Subsequently, one may create a set of samples with $d_n = d_{n,o}$ and decreasing thicknesses d_i. The thickness $d_{i,o}$ is reached, when the conductivity starts to *increase* upon further reduction of d_i. (If the left and the right space charge regions in the i-layers start to overlap, less electrons from the n-layers will be needed in order to establish thermal equilibrium in the i-layers. This results in a non-depleted region of finite width in the n-layers).

So far we have not yet related the conductivity to the density of state distributions $N_n(\epsilon)$ and $N_i(\epsilon)$. On a first glance this may look like a tremendous problem, requiring the analysis of a two-dimensional set of (d_n, d_i) values, in order to determine the *two* density of state distributions. We will see, however, that for actual cases it will be sufficient to investigate a one dimensional set of samples and that the analysis is nearly as simple as in the case of C-V measurements /12,15/.

First, we note, that $N(\epsilon)$ can be approximated by its value in the neutral bulk, i.e. by $N_n(\phi_n)$ and $N_i(\phi_i)$ in our present example, for sufficiently small shifts $\Delta\phi(z)$ of the local Fermi level. In actual amorphous semiconductors this may be applicable for a rather wide energy range around ϕ_i, whereas it will represent an acceptable approximation for strongly doped material only within a narrow range around ϕ_n.

In most cases we also can use zero-temperature statistics, i.e. approximate the Fermi distribution by the unit-step function. A reasonable criterion for the validity of this simplification is, whether a straight line of slope kT (or -kT), intersecting the curve $\ln N(\epsilon)$ at any energy ϕ, corresponding to a local Fermi level position somewhere in the sample has no other intersections with $N(\epsilon)$. This condition is probably always fulfilled in amorphous silicon at room temperature.

From the two preceeding paragraphs it follows, that we can approximate the space charge density by

$$\rho_n(z) = e\, N_n(\phi_n)\, [\phi_n(z) - \phi] \qquad (4)$$

and

$$\rho_i(z) = -e\, N_i(\phi_i)\, [\phi - \phi_i(z)] \qquad (5)$$

respectively, as long as $\Delta\phi_n(z)$ and $\Delta\phi_i(z)$ (see Fig. 1) are small enough. (e = elementary charge). In this case an analytical expression for $\Delta\phi(z)$ is obtained as solution of Poisson's equation

$$d^2\Delta\phi_n(z)/dz^2 = (4\pi e^2 N_n(\phi_n/\kappa_0)\, \Delta\phi_n(z)] , \quad \text{(n-layers)} \qquad (6)$$

and

$$d^2\Delta\phi_i(z)/dz^2 = (4\pi e^2 N_i(\phi_i/\kappa_0)\, \Delta\phi_i(z)] , \quad \text{(i-layers)} \qquad (7)$$

(κ_0 = static dielectric constant of the amorphous semiconductor, assumed to

418

be independent of doping). One finds

$$\phi_n(z) = \phi + \Delta\phi_{n,o} \cosh[(z-md)/z_{n,o}] \qquad (8)$$

and

$$\phi_i(z) = \phi + \Delta\phi_{i,o} \cosh[(z+d/2-md)/z_{i,o}] \qquad (9)$$

with

$$z_{n,o}^{-2} = 4\pi e^2 N_n(\phi_n)/\kappa_0 \qquad (10)$$

and

$$z_{i,o}^{-2} = 4\pi e^2 N_i(\phi_i)/\kappa_0 \qquad (11)$$

The values of $\Delta\phi_{n,o}$ (>0) and $\Delta\phi_{i,o}$ (<0) depend on the doping layer thickness and will be calculated later.

It is possible to calculate the d_n-dependence of $\sigma^{(2)}$ for the limiting case of thick n-layers, if the assumption of constant density of states $N_n(\phi_n)$ around ϕ_n applies over an energy range considerably larger than kT (with d_i assumed constant and equal or larger than $d_{i,o}$).

With

$$E_C(z) = \begin{cases} (E_C- \phi_n)_0 + \phi_n(z), & \text{(n-layers)} \\ (E_C- \phi_i)_0 + \phi_i(z) & \text{(i-layers)} \end{cases} \qquad (12)$$

inserted into Eq.s (2) and (3) we obtain for Eq. (1')

$$\sigma \simeq (2/d) \{ \sigma_{n,o} \exp[-(E_C-\phi_n)_0/kT] \int_0^{d_n/2}\exp[-(\phi_n(z)-\phi)/kT]\, dz +$$
$$\sigma_{i,o} \exp[-(E_C-\phi_i)_0/kT] \int_{d_n/2}^{d/2}\exp[(\phi-\phi_i(z))/kT]\, dz \} \qquad (13)$$

The prefactors of the integrals are the n- and i- bulk-conductivities. $(E_C-\phi_n)_0$ and $(E_C-\phi_i)_0$ are the "activation energies" of the conductivities /dd/ in the neutral bulk components of the superstructure. The value of the first integral in Eq. (13) is dominated by the contribution from the z-region where the exponent differs not by much more than unity from its minimum in the n-layers. In this region Eq. (8) represents a good approximation. We define d_{kT} as the distance from the interface at which $\phi_n(z)-\phi = kT$ for the case of thick n-layers with nearly non-overlapping depletion regions. From the properties of the hyperbolic functions it follows directly, that $\phi_n(\pm d_n/2)$ and the slope at these points (determined by the total space charge in the i-layer !) are nearly independent of d_n, as long as d_{kT} and d_n are significantly larger than $z_{n,o}$. In other words, $\phi_n(z)$ can be represented as the superposition of increasing and decreasing exponential functions through the points $z= \pm(d_n/2 - d_{kT})$ and kT with the increment $z_{n,o}$.

Within this approximation the n-layer integral in Eq. (13) can be evaluated numerically as a function of d_n. In the limiting cases we obtain

$$\int_0^{d_n/2} \exp\{-2\exp[-(d_n/2-d_{kT})/z_{n,o}] \cosh(z/z_{n,o})\}\, dz \simeq \begin{cases} d_n/2-d_{kT}, \\ \text{for} \quad (d_n/2-d_{kT}) \gg z_{n,o} \\ (\pi/2a)^{1/2}\, z_{n,o}\, e^{-a}, \\ \text{for} \quad d_n/2 < d_{kT}, \end{cases} \qquad (14)$$

with $a = 2 \exp[(d_n/2-d_{kT})/z_{n,o}]$.

419

The contribution of the i-layers to Eq.(13) is much smaller, unless $d_{i,0}$ is unreasonably large. Moreover, this contribution is independent of d_n, as long as our present assumptions hold. Therefore, it is now possible to determine d_{kT} and $z_{n,0}$ from measurements of the conductivity per superstructure period for samples differing by their n-layer thickness. The value of $z_{n,0}$ provides directly the density of states at the Fermi level in the n- bulk material, $N_n(\phi_n$, because of the relation (10).

In order to proceed further we will use in the following a very useful property of $N_n(\epsilon)$, which applies to (sufficiently strongly) doped amorphous semiconductors. Near ϕ_n the density of states $N_n(\epsilon)$ changes so fast as a function of energy (although, presumably, slowly compared with a room temperature Boltzmannfactor, as mentioned before), that the space charge density, determined by

$$\rho_n(\phi) = e \int_{\phi}^{\phi_n} N(\epsilon) \, d\epsilon \qquad (15)$$

becomes nearly independent of ϕ at large down-shifts of the Fermi level. (Note, that in Fig. 1 the *logarithm* of $N(\epsilon)$ is displayed). We will call this saturation value

$$\rho(\phi << \phi_n) = \tilde{n}_D \qquad (16)$$

in analogy to the situation in doped crystalline semiconductors. In the case of an exponential conduction-band tail with

$$N_n(\epsilon) = N_n(\phi_n) \exp[-(\phi_n - \epsilon)/\epsilon_0] \qquad (17)$$

this charge density becomes

$$\tilde{n}_D \simeq e \, N_n(\phi_n) \, \epsilon_0 \qquad (18)$$

if ϕ lies substantially below $\phi_n - \epsilon_0$. The space charge potential on the n-side becomes parabolic, like in a n-type crystalline semiconductor

$$E_c(z) = E_c(z=0) + (2\pi e^2 \tilde{n}_D / \kappa_0) \, (z - md)^2 \qquad (19)$$

whenever our argument for $\rho_n(\phi) \simeq \tilde{n}_D$ applies.

In a n-i-n-i structure with sufficiently thin n- and sufficiently thick i-layers the conditions for parabolic and cosh-shaped band-bending in the respective layers will be fulfilled. Such a situation is depicted in Fig.1 by the dashed lines. The term "sufficiently thick" means for the i-layers, that their central region should be close to neutral i-bulk, or in mathematical terms, $\Delta\phi_{i,0}$ should be less than kT. The term "sufficiently thin" means for the n-layers, that the band edge modulation, caused by the positive space charge in the n-layers, $e\tilde{n}_D d_n$, should not exceed the values for which the cosh-approximation is applicable in the i-layers.

Experimentally, one will start with an i-layer thickness larger than twice the the maximum value of $z_{i,0}$ to be expected. (Assuming $N_i(\phi_i) = 10^{15} \text{cm}^{-3} \text{eV}^{-1}$ we obtain $z_{i,0} \simeq 300$ nm). The minimum reasonable value of d_n will be determined by the n-layer thickness, $d_{i,1}$ at which the conductivity of the superstructure material is significantly increased (say by a factor of about 2) compared with the uniform i-bulk.

The corresponding mathematics for calculating this increase of the conductivity is rather simple. First we need to know the minimum distance of the mobility edge at the center of the n-layers. Using Eqs. (9) and (19) we find

$$E_c(z=0)-\phi = (E_c-\phi_i)_0 - (2\pi e^2 \tilde{n}_D/\kappa_0)(d_n/2)^2 - |\Delta\phi_{i,0}| \cosh(d_{i,0}/2z_{i,0}) \quad (20)$$

The continuity requirement for the electric field at the boundaries between the n- and the i-layers allows us to eliminate $\Delta\phi_{i,0}$ from Eq. (20),

$$E_c(z=0)-\phi = (E_c-\phi_i)_0 - (2\pi e^2 \tilde{n}_D/\kappa_0)(d_n/2)^2 [1 + 4(z_{i,0}/d_n)$$

$$ctgh(d_{i,0}/2z_{i,0})] \quad (20')$$

From Eq. (20') we see, that $(E_c-\phi)_0$ becomes nearly constant if the intrinsic layer thickness is significantly larger than $2z_{i,0}$. Using Eqs. (9) and (19) in (13) we obtain for the conductivity

$$\sigma = (2/d) \{ \sigma_{n,0} \exp[-(E_c(z=0)-\phi)/kT] \int_{d/2}^{d_n/2} \exp[-(z/\tilde{z}_0)^2] \, dz +$$

$$\sigma_{i,0} \exp[-(E_c-\phi_i)_0/kT] \int_{d_n/2}^{0} \exp[((4z_{i,0}d_n/\tilde{z}_0^2 \sinh(d_n/2z_{i,0}))$$

$$\cosh(z/2z_{i,0})] \, dz \} \quad (21)$$

with $(E_c(z=0)-\phi)$ given by (20') and

$$\tilde{z}_0^{-2} = (2\pi e^2 \tilde{n}_D/\kappa_0 kT) \quad (22)$$

Although Eq. (21) has to be evaluated numerically in general, it can be approximated by the first term, integrated to $z \to \infty$, if this term becomes dominant under sufficiently large spatial modulation of the mobility edge. For compatibility with our former assumptions it is necessary, that the gap density of states in the intrinsic material can be assumed as constant over an energy range of several kT. In this case Eq. (21) reduces to

$$\sigma = (\pi^{1/2}d/\tilde{z}_0) \sigma_{n,0} \exp[(E_c(z=0)-\phi)/kT] \quad (23)$$

Under these circumstances the experimental determination of the two unknown quantities $\tilde{n}_D$ and $N_i(\phi_i)$ becomes particularly simple. The conductivities of two samples with the same n-layer thickness, $d_{n,1}$, but with different intrinsic layer thicknesses will only differ by their activation energies, given by Eq. (20'). The dotted line in Fig. 2 illustrates the lowering of the activation energy which is associated with a reduction of the intrinsic layer thickness from $d_{i,0}$ to $d_{i,1}$. Note, that this method for determining $\tilde{n}_D$ and $N_i(\phi_i)$ is expected to be sensitive enough to variations of d_i for obtaining reasonably accurate results even at moderate reproducibility of the growth conditions of the amorphous material.

After having determined the *values* of n_D and $N_i(\phi_i)$ one can proceed quite easily to determine the *function* $N_i(\epsilon)$ in the range $\phi_i < \epsilon < \phi_n$. by investigating a set of samples with constant i-layer thickness and increasingly thick n-layers. If we assure that $\Delta\phi_{i,0} \ll kT$ for all the samples we can replace the cosh by a simple exponential function. This implies, that the solution of Poisson's equation on the i-side of the n-i interface has a fixed *shape*, determined by $N_i(\epsilon)$ and the boundary condition $\phi_i(z\to\infty) = \phi$.

With increasing n-layer thickness and, therefore, increasing positive space charge in the n-layers, $e\tilde{n}_D d_n$, the whole curve shifts to the right with respect to the n-i interface. This is analoguous to the shift discussed in

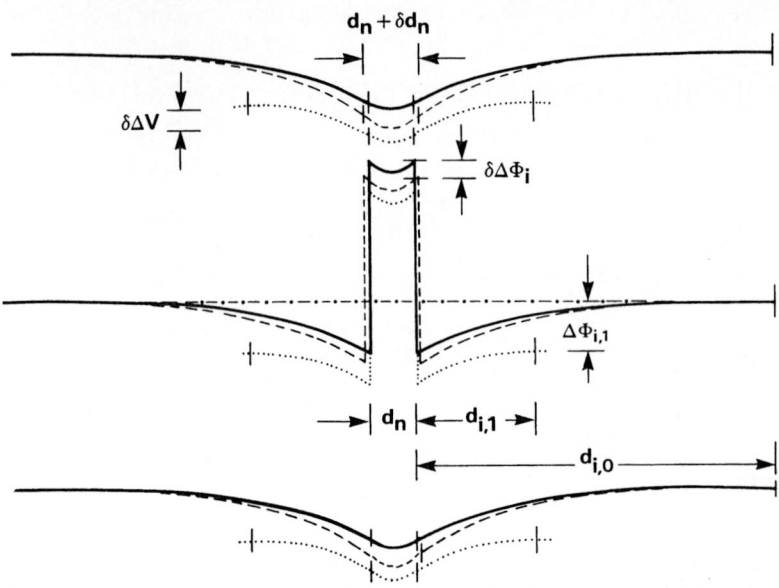

Fig.2 Band profiles of amorphous n-i-n-i structures with thin n-layers. Full lines: thick i-layers. Dashed lines: slightly increased n-layer thickness. The total band-bending is increased by the amount $\delta\Delta V$, the band banding on the i-side by $\delta\psi\phi_i$. Dotted line: Same n- layer thickness as for full lines, however, with reduced i-layer thickness.

Refs. /12,15/ in connection with the C-V measurements. A translation by the amount of dz to the right implies an increase of band-bending on the i-side by

$$d\phi_i = eFdz = (4\pi eQ/\kappa_0)\ dz \qquad (24)$$

and an increase dQ in space charge per unit area by

$$dQ = \rho dz \qquad (25)$$

The experimental determination of $N_i(\epsilon)$ is particularly simple if our condition for constant effective doping concentration $\tilde{n}_D$ holds. From $\ln \sigma$ versus n-layer thickness we obtain directly $E_C(z=d/2)-E_C(z=0)$ as a function of d_n. Subtracting the n-layer contribution to the spatial modulation of the mobility-edge which follows from Eq. (14) and using

$$Q = e\tilde{n}_D d_n \qquad (26)$$

we can express the down-bending $\Delta\phi_i(d_n/2)$ in the i-layer as a function of the space charge Q. Using Eqs. (24) and (25) we find

$$\rho = (4\pi eQ/\kappa_0)\ dQ/d\phi \qquad (27)$$

and with

$$N(\epsilon) = d(\rho/e)/d\epsilon \qquad (28)$$

finally

$$N(\phi) = (4\pi/\kappa_0)\, d/d\phi\, (Q\, dQ/d\phi). \qquad\qquad (29)$$

We have described in detail a procedure for determining the density of state distribution in i-material for energies above the bulk Fermi level ϕ_i. It is clear, that we can obtain $N_i(\phi)$ for $\phi < \phi_i$ in an analogous way from studies of p-i-p-i superstructures.

So far we have limited our analysis to the case of i-layers which were thick enough, so that the hole-depletion layers did not overlap. In order to cross-check these results and also to obtain more accurate details about certain energy ranges it will be interesting to complement these studies by investigations of superstructures with thinner i-layers and, possibly thicker n-layers. From such structures interesting information may also be obtained from measurements of the conductivity *perpendicular* to the layers. This conductivity and its temperature dependence is determined by the distance between mobility-edge *maxima* and Fermi level. Note, that the transport is still determined by one kind of carriers in n-i-n-i and p-i-p-i structures. Therefore one has not to take into account the problems due to different quasi Fermi levels, as discussed as a serious problem for the determination of $N(\epsilon)$ from C-V measurements in amorphous p-n junctions.

Finally I want to make some critical remarks. By attributing uniform bulk properties to the components within each layer I have tacitly assumed that $N(\epsilon)$ does not change as a function of the Fermi level variation. Also I have not considered correlation effects which may cause deviations from the simple picture. Also the carrier mobility can change as a function of Fermi level position. Most of these effects do not relate to our special system only, but they may also influence field-effect measurements, for instance. In our system they may, however, not be obscured by competing stronger effects, like inhomogeneities of the amorphous material close to interfaces to other materials.

In summary I have shown how the density of state distribution in the mobility gap of amorphous semiconductors can be obtained from conductivity studies in amorphous films with periodic doping superstructure. This method probes the properties of quasi-uniform bulk material under thermal equilibrium conditions. Therefore, one can expect, that this new approach will prove as more reliable and accurate than the methods used so far.

I am grateful for stimulating discussions with M. Hundhausen and L. Ley, with J. Kakalios, and with R.A. Street and his colleagues.

References

/1/ G.H. Dohler, Verhandl. DPG (VI) 17, 745 (1982)
/2/ G,H, Dohler, Proceedings of the 17th Int. Conf. on the Physics of Semiconductors, to be published
/3/ G.H. Dohler, in: *Festkoerperprobleme (Advances in Solid State Physics)*, Vol. XXIII, 207, P. Grosse (ed.) Vieweg Braunschweig 1983
/4/ For a review see: K.Ploog and G.H. Dohler, Advances in Physics 32, 285 (1983).
/5/ G.H. Dohler, phys. stat. sol. (b) 52, 79 and 533 (1972)
/6/ G.H. Dohler, J. Vac. Sci. Technol. 16, 851 (1979) and B1, 278 (1983)
/7/ M. Hundhausen, L. Ley, and R. Carius, Proceedings of the 17th Intl. Conf. on the Physics of Semiconductors, to be published
/8/ M. Hundhausen, L. Ley, and R. Carius, Phys. Rev. Lett.53, 1598 (1984)
/9/ J. Kakalios and H. Fritzsche, Phys. Rev. Lett. 53, 1602 (1984)
/10/ J. Kakalios and H. Fritzsche, Proceedings of the 17th Intl. Conf. on the Physics of Semiconductors, to be published

/11/ A. Madan, P.G. LeComber, and W.E. Spear, J. Non-Cryst. Solids 20, 239 (1976)

/12/ G.H. Dohler and M. Hirose, in: *Amorphous and Liquid Semiconductors* W.E. Spear, Ed. (Edinburgh, 1977) p. 372

/13/ D.G. Ast and M.H. Brodsky, Phil. Mag. B 41, 273 (1980)

/14/ J.D. Cohen, D.V. Lang, and J.P. Harbison, Phys. Rev Lett. 45, 197 (1980)

/15/ M.J. Powell and G.H. Dohler, J. Appl. Phys. 52, 517 (1981)

/16/ G.H. Dohler, Phys. Rev. B 19, 2124 (1979) 60

EFFECTIVE MEDIUM EXPRESSION FOR THE OPTICAL

PROPERTIES OF PERIODIC MULTILAYER FILMS

H. Ugur, R. Johanson, and H. Fritzsche

James Franck Institute and Department of Physics
The University of Chicago
Chicago, Illinois 60637

ABSTRACT

We derive an expression for the normal incidence transmittance of light through a multilayer film consisting of an arbitrary number of sequential layers of two materials with different optical constants. For wavelengths large compared to the thicknesses of the two materials, d_a and d_b, the optical properties of the multilayer film are determined by a complex effective medium dielectric constant $\varepsilon_e = \varepsilon_a d_a/(d_a+d_b) + \varepsilon_b d_b/(d_a+d_b)$ where ε_a and ε_b are the complex dielectric constants of the two materials. We verify the correctness of the effective medium expression by comparing the optical spectra of multilayer films predicted by this formula with numerical calculations obtained by solving the boundary value problem for all layers. After establishing the validity of the effective medium expression, we then use it to analyze optical transmittance data of multilayer films consisting of hydrogenated amorphous silicon (a-Si:H) and amorphous silicon nitride (a-SiN$_x$:H). The optical gap E_0 of the a-Si:H layers is found to increase with decreasing a-Si:H layer thickness whereas the slope B of the Tauc plot remains constant.

INTRODUCTION

The growing interest in amorphous semiconductor multilayers [1-12] created a need for calculating the optical properties of the individual layers from the spectral dependence of the normal incidence optical transmittance. The multilayer films consist of alternating layers typically of two substances such as amorphous hydrogenated silicon interlayered with amorphous silicon nitride, silicon oxide, or amorphous hydrogenated germanium. The number of layer periods varies and can be several hundred when the individual layers are thin.

Even though the method of calculating optical properties of materials, that are layered or that have spatially varying dielectric properties, has been treated in standard textbooks [13], we have been unable to find the simple effective medium expression for the complex dielectric constant which is derived in this paper. Its simplicity stems from the fact that it is limited to normal incidence of light onto periodic multilayer films whose layer thicknesses are small compared to the wavelength. Much more complicated expressions hold in composites and randomly inhomogeneous materials [14,15] and for arbitrary angles of incidence.

425

We first set up the general boundary value problem for calculating the normal incidence transmittance and reflectance of multilayer films containing two materials. Specializing to periodic multilayer films consisting of a-Si:H and a-SiN$_X$:H we then calculate the transmittance spectra for various layer thicknesses using the known bulk values of the optical constants of these two materials. We show that these spectra are the same as those of homogeneous films having the effective medium value of the complex dielectric constant. Conversely, the effective medium expression allows one to calculate the optical constants of one layer material from the transmission spectra of the multilayer film if the optical constants of the other layer material and the layer thicknesses are known.

Assuming that the optical constants of a-SiN$_X$:H are independent of the silicon layer thickness, we use the effective medium expression to obtain the optical constants of the silicon layers as a function of their thickness from the transmittance spectra of a-Si:H/a-SiN$_X$:H multilayer films.

THEORETICAL ANALYSIS

We calculate the normal incidence reflectance and transmittance of a multilayer film that consists of N layers having indices of refraction n_j with j=1----N. We assume the surrounding medium is air with $n_0=n_{N+1}=1$. For normal incidence of light, the electric and magnetic field vectors E and H lie in the layer planes so that we can write

$$E = [A\exp(ikz) + B\exp(-ikz)]\exp(-i\omega t) \tag{1}$$

$$H = n[A\exp(ikz) - B\exp(-ikz)]\exp(-i\omega t) . \tag{2}$$

Applying the boundary conditions for E and H and using Eqs. (1) and (2) for each layer m yields

$$A_m\exp(ik_m z_m)+B_m\exp(-ik_m z_m)=A_{m+1}\exp(ik_{m+1}z_m)+B_{m+1}\exp(-ik_{m+1}z_m) \tag{3}$$

$$n_m[A_m\exp(ik_m z_m)-B_m\exp(-ik_m z_m)]=n_{m+1}[A_{m+1}\exp(ik_{m+1}z_m)-B_{m+1}\exp(-ik_{m+1}z_m)]. \tag{4}$$

The two equations can be rewritten in matrix notation as

$$\begin{bmatrix} A_m \\ B_m \end{bmatrix} = M_m \begin{bmatrix} A_{m+1} \\ B_{m+1} \end{bmatrix} \tag{5}$$

where

$$M_m = \begin{bmatrix} \dfrac{n_m+n_{m+1}}{2n_m}\exp[-i(k_m-k_{m+1})z_m] & \dfrac{n_m-n_{m+1}}{2n_m}\exp[-i(k_m+k_{m+1})z_m] \\[3mm] \dfrac{n_m-n_{m+1}}{2n_m}\exp[i(k_m+k_{m+1})z_m] & \dfrac{n_m+n_{m+1}}{2n_m}\exp[i(k_m-k_{m+1})z_m] \end{bmatrix} \tag{6}$$

The coefficients of the zeroth layer (front surface) are expressed in terms of those in the $(N+1)^{th}$ layer (back surface) by

$$\begin{bmatrix} A_0 \\ B_0 \end{bmatrix} = M \begin{bmatrix} A_{N+1} \\ B_{N+1} \end{bmatrix} \tag{7}$$

where

$$M = \begin{bmatrix} M_{11} & M_{12} \\ M_{21} & M_{22} \end{bmatrix} = \prod_{m=0}^{N} M_m \tag{8}$$

When the light is incident on the front surface, one has $B_{N+1}=0$ because there is no reflected wave behind the back surface. Equation (7) relates then the coefficient A_{N+1} of the transmitted wave with A_O and B_O of the incident and reflected wave, respectively, as $A_O=M_{11}A_{N+1}$ and $B_O=M_{21}A_{N+1}$. The reflectance and transmittance coefficients R_A and T_A are

$$R_A = \left|\frac{B_O}{A_O}\right|^2 = \left|\frac{M_{21}}{M_{11}}\right|^2 \tag{9}$$

$$T_A = \left|\frac{A_{N+1}}{A_O}\right|^2 = \frac{1}{\left|M_{11}\right|^2} \tag{10}$$

These expressions were evaluated numerically for multilayer films consisting of alternating a-Si:H and a-SiN$_x$:H layers. We label the two kinds of materials as (a) and (b), respectively. We used for the real part of the refractive indices $n_a=3.4$ and $n_b=1.95$. We assumed that the absorption coefficient α of the amorphous silicon (a) has the spectral dependence typical for band to band transitions, given by the Tauc expression

$$\alpha = \frac{1}{h\nu} B^2(h\nu - E_O)^2 \qquad \text{for } h\nu > E_O \quad . \tag{11}$$

The optical gap was taken to be $E_O=1.75$ eV and for B we chose 900 (eV cm)$^{-\frac{1}{2}}$. Absorption in the silicon nitride layers (b) is negligible in the photon energy range considered here. The nitride layer thickness was kept constant at $d_b\dot{=}24$ Å while the silicon layer thickness was varied between 10 Å and 5000 Å.

Non-Absorbing Region

At long wavelengths absorption is negligible and the calculated transmittance shows the characteristic spectral dependence of interference fringes, oscillating as a function of $h\nu$ between a minimum value T_{min} and a maximum value $T_{max}=1$. In Fig.1 we have plotted $-\log_{10}(T_{min}/T_{max})$, which is equal to the optical density at T_{min}, as a function of the silicon layer thickness d_a. The calculations were carried out for three different total silicon thicknesses $D_a=N_a d_a$, as noted in the figure, in order to test whether T_{min} was, as expected, independent of the total thickness.

We found that the spectral dependence of the transmittance as well as the reflectance of the multilayer films are identical to those of a homogeneous film having a total thickness $D=D_a+D_b$, where $D_b=N_b d_b$, and having a complex effective medium dielectric constant

$$\varepsilon_e = \varepsilon_a d_a/(d_a+d_b) + \varepsilon_b d_b/(d_a+d_b) \tag{12}$$

where ε_a and ε_b are the dielectric constants of the materials a and b. The values of T_{min} calculated from Eq.(12) are represented by the full curve in Fig.1.

427

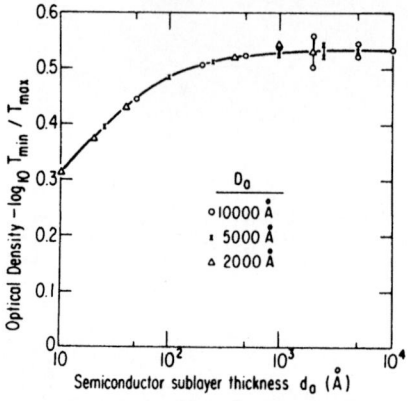

Figure 1. Optical density at interference minima in the transmittance of multilayer films as a function of silicon layer thickness d_a. Solid curve is predicted by Eq.(12) of text.

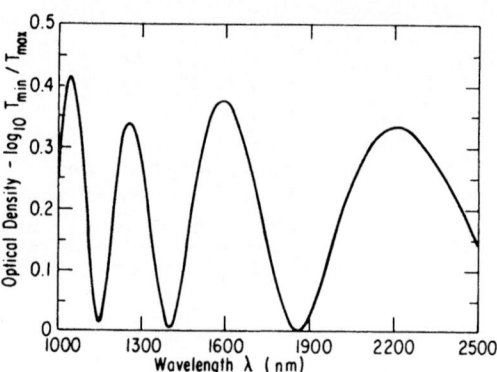

Figure 2. Transmittance spectrum of two 4000 Å thick a-Si:H layers separated by 60 Å of a-SiN$_x$:H.

We do not expect that a homogeneous effective medium can properly simulate the optical properties of a multilayer film when one or both layer thicknesses become comparable to or larger than the wavelength of light in the respective medium. In that case, interference effects within the layers are superimposed onto those of the total film. This is illustrated in Fig.2 which shows $-\log_{10}(T_{min}/T_{max})$ of experimental data for two a-Si:H layers of d_a=4000 Å separated by d_b=60 Å of a-SiN$_x$:H in the wavelength range $1000 \leqslant \lambda \leqslant$ 2500 nm. The variations in transmission minima due to the additional interference is shown as vertical bars in Fig.1. These effects become noticeable at silicon layer thicknesses $d_a \geqslant 1000$ Å. The three points at d_a=2000, 5000, and 10000 Å which correspond to D_a of the same value represent homogeneous unlayered a-Si:H films.

Absorbing Region

The calculations leading to Eq.(12) are valid even in the presence of absorption if one uses complex values of the dielectric constants. In the photon energy range in which only one material is absorbing, material (a) for instance, then Eq.(12) yields for the absorption coefficient α_e of the effective medium:

$$\alpha_e(h\nu) = \alpha_a(h\nu) \frac{n_a}{n_e} \frac{d_a}{(d_a+d_b)} \tag{13}$$

When α_a is described by Eq.(11) then the optical gap E_0 remains unchanged. The constant B, on the other hand, acquires the effective medium value

$$B_e = B_a \left[\frac{n_a}{n_e} \frac{d_a}{(d_a+d_b)} \right]^{1/2} \tag{14}$$

Before the onset of interband transition there is an absorption region that is attributed to transitions between localized bandtail states and extended states. Experimentally one finds in homogeneous, unlayered amorphous semiconductors an exponential absorption regime

$$\alpha = C \exp(h\nu/kT_0) \tag{15}$$

which is known as the Urbach absorption tail. Since α_e and α_a differ only by a multiplicative factor that is nearly independent of $h\nu$ (see Eq.(13)), we expect that the slope $1/kT_0$ of the Urbach tail remains unaffected if the layering does not change the optical constants of the materials used.

EXPERIMENTAL RESULTS

Using Eq. (12) we obtained the optical constants of the silicon layers from the transmittance data of silicon/silicon nitride multilayer films. The optical constants of the a-SiN$_x$:H layers were obtained from 1µm thick single layer films of this material and were assumed to apply to the 15-60 Å thick a-SiN$_x$:H layers in the multilayer films. From the amplitude of the interference fringes in the nonabsorbing region between λ=1000-2500 nm the effective medium index of refraction n_e was obtained. This value together with the spacings of the interference peaks yield the total sample thickness and hence d_a+d_b since we know the number of layers. The refractive index n_a of the amorphous silicon layers calculated from (12) is shown as a function of the silicon layer thickness d_a in Fig. 3.

The effective medium absorption coefficient $\alpha_e(h\nu)$ is computed from the transmittance data in the absorbing region. This yields with Eq.(11) the optical gap E_0 and the effective medium value B_e of the slope of the Tauc plot. According to Eq.(13) the values of E_0 are the same in the effective medium and in the a-Si:H layers since the nitride layers are non-absorbing. They are plotted in Fig. 3. On the other hand, B_e and B_a differ according to Eq.(14). These two values are shown as a function of silicon layer thickness in Fig. 4.

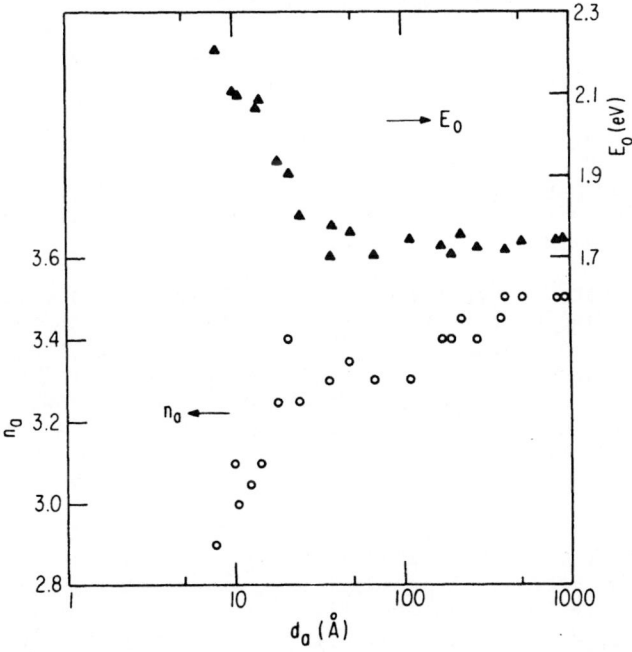

Figure 3. Optical gap E_0 and refractive index n_a of a-Si:H interlayered with a-SiN$_x$:H as a function of a-Si:H layer thickness.

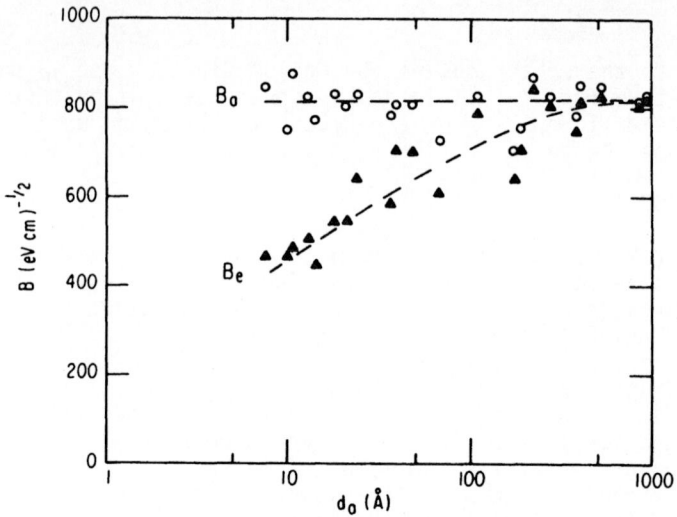

Figure 4. Slope of the Tauc plot of absorption of a-Si:H/a-SiN$_x$:H multilayer films: B$_a$ of a-Si:H and B$_e$ of the effective medium as a function of the a-Si:H layer thickness.

DISCUSSION

The effective medium expression for the complex dielectric constant for normal incidence of light onto periodic multilayers allows us to identify changes in the optical constants of the layers which may result from a number of reasons. These may include changes in the nature and energies of electronic states as the layers become thin compared to the extent of the wavefunctions, interface effects associated with strains, defects, and alloying, as well as electric fields due to charge transfer between layer materials having different work functions.

In a-Si:H/a-SiN$_x$:H multilayers the index of refraction n$_a$ decreases and the optical gap $_0$E$_0$ increases as the silicon layer thickness d$_a$ is decreased below 50 Å [1-3,8]. This effect has been associated with quantum confinement of the extended valence and conduction band states in the quasi-two dimensional a-Si:H layers [1,5,8].

However, contrary to earlier claims [3,8], we find no evidence that B$_a$, the slope of the Tauc plot is increasing in this regime of small d$_a$. The Urbach slope 1/kT$_0$ was found [1] to decrease with decreasing silicon layer thickness d$_a$. As argued above, this does not follow from Eqs.(12) and (13). It therefore suggests an increased subband gap absorption which probably comes from defect-rich interface regions whose relative importance increases with the number of interfaces.

CONCLUSION

The optical properties of periodic multilayers of two different substances can be described by an effective medium complex dielectric constant which is the algebraic thickness average of the complex dielectric constants of the two media. This relationship applies to normal incidence of light whose wavelength in the media is long compared to the sublayer thicknesses. We have applied this analysis to multilayers consisting of an amorphous semiconductor and an insulator. Significant changes occur in the optical constants of the semiconductor as its thickness is decreased below 50 Å.

ACKNOWLEDGEMENTS

We wish to take this opportunity to express our respect and thanks to Sir Nevill Mott whose stimulating interest and novel ideas greatly enhanced our joy of working in this field. This work was supported by the National Science Foundation Grant No. DMR8009225.

REFERENCES

1. B. Abeles and T. Tiedje, Phys. Rev. Lett. 51:2003 (1983).
2. H. Munekata, M. Mizuta, and H. Kukimoto, J. Non-Cryst. Solids 59-60:1167 (1983).
3. J. Kakalios, H. Fritzsche, N. Ibaraki, and S. R. Ovshinsky, J. Non-Cryst. Solids 66:339 (1984).
4. B. Abeles, T. Tiedje, K. S. Liang, H. W. Deckman, H. C. Stasiewski, J. C. Scanlon, and P. M. Eisenberger, J. Non-Cryst. Solids 66:351 (1984).
5. T. Tiedje, B. Abeles, P. D. Persans, B. G. Brooks, and G. D. Cody, J. Non-Cryst. Solids 66:345 (1984).
6. M. Hirose and S. Miyazaki, J. Non-Cryst. Solids 66:327 (1984).
7. H. W. Deckman, J. H. Dunsmuir, and B. Abeles, Appl. Phys. Lett. 46:171 (1985).
8. N. Ibaraki and H. Fritzsche, Phys. Rev. B 30:5791 (1984).
9. H. Ugur and H. Fritzsche, Solid State Commun. 52:649 (1984).
10. Rujuang Cheng, Shulin Wen, Jingwei Feng, and H. Fritzsche, Appl. Phys. Lett. 46:592 (1985).
11. C. B. Roxlo, B. Abeles, and T. Tiedje, Phys. Rev. Lett. 52:1994 (1984).
12. M. Hirose, S. Miyazaki, and N. Murayama, this volume.
13. M. Born and E. Wolf, "Principles of Optics", Pergamon Press 6th Edition, (1983).
14. D. Stroud, Phys. Rev. B 12:3368 (1975).
15. D. E. Aspnes and J. B. Theeten, J. Appl. Phys. 50:4928 (1979).

EFFECTS OF MEAN FREE PATH ON THE

QUANTUM WELL STRUCTURES OF AMORPHOUS MATERIALS [+]

Raphael Tsu

Instituto de Física e Química de São Carlos
Universidade de São Paulo, 13560, São Carlos, Brasil
and
Energy Conversion Devices Inc. Troy, MI, U.S.A.

ABSTRACT

Because of the presence of phonons, impurities and other defects such as structural disorder, a mean free path L must be introduced in dealing with quantum well structures and superlattices particularly important when amorphous materials are involved. The major effects involved the broadening and level shift of the mini-band states. For crystalline materials, level shift is negligible in comparison to level broadening, however, for amorphous materials, the opposite is true due to the extremely small L.

INTRODUCTION

To date, mean free path has not been taken into account for the size quantization of quantum well structures and mini-bands in man-made super-lattices. Whenver damping is involved, there is present line broadening and level shift of the energy eigenstates involved. The mean free path in crystalline bulk materials is such that, the product of the warevector k and L, $kL \gg 1$, so that the level shift may be safely ignored. Nevertheless, line broadening should be important for man-made quantum solids having mini-bands of at most ~0.1 e V. This is because L is usualy no more than several periods of a man-made superlattice[1]. The situation is very different when amorphous materials are used to construct quantum wells and periodic layer structures[2-5]. It is generally believed that the mean free path in amorphous hydrogenated silicon is of the order of 5 Å[6]. Qualitatively one would not have expected that quantum size effects may be observable whenever L < W with W being the well width or the period of the multilayer structure. This is based on the definition of L being a length characterized by the loss of any coherence effects due to phase randomization. Obviously, quantitatively there cannot be a sharp division between: with and without effects on the line broading and level shift. Therefore this work presents a quantitative description of the energy states. With finite L, damping does not allow the presence of an eigenstate which is independent on space and time. The situation is exactly analogous with the vibration of a string. If loss is small and thus negligible, eigen states exist. However, we may still deal with the spectrum of the response having resonances characterized by linewidth and level shift even for a highly damped system. The conventional method is to examine the Green function where the real and imaginary parts of the self-energy[8] give the level shift and line broadening respectively. Alternatively, as in optics and microwaves, we calculate the mean free path L based on randomizing scattering, a quality facter Q is obtained which gives line broadening. It is this method that has been used and fully discussed in this work. Together with the use of a quantum mechanical counterpart of the Langevin equation[9], the level shift may be obtained.

LINE WIDTH, Q, AND LEVEL SHIFT FOR UNBOUNDED SPACE

The Green function for a one-dimension wave equation having a velocity dependent damping term characterized by the operator $H_1 \equiv 2qd/dx$ is

$$G_{\pm}(k) = \frac{1}{(k_o^2 - k^2) \pm i2kq} \quad , \tag{1}$$

in which (+) sign for $x > 0$ and (−) sign for $x < 0$, and $k_o^2 \equiv 2mE/\hbar^2$, q, a spatial damping constant and k, the wave vector. Figure 1 shows a typical sketch of $\mathrm{Im}\,G_+(k)$ which is the spectral density. Note that $G_{\pm}(k)$ have poles at $k = \pm (k_o^2 - q^2)^{1/2} \pm iq$, $\mathrm{Im}\,G_+(k)$ has a maximum at

$k^2 = k_o^2 - q^2(1 - q^2/k_o^2)$. Since the $\mathrm{Im}\,G_+(k)$ has a maximum at $k^2 \approx k_o^2 - q^2$, i.e. $k_o^2 \approx k^2 + q^2$, the enegy, k_o^2, is shifted up. The width of $\mathrm{Im}\,G_+(k)$, $k \sim 2q$ for $k_o \gg q$. From Eq. (1), it is trivial to show that the wave function ψ_+ and ψ_- may be written as

$$\psi_{\pm} = e^{\pm ikx} \quad , \tag{2}$$

having $k = k_r + iq$, in which $k_r = (k_o^2 - q^2)^{1/2}$. The mean distance, L_o, for density $\psi^*\psi$ is defined by $2qL_o = 1$, giving $L_o = 1/2q$

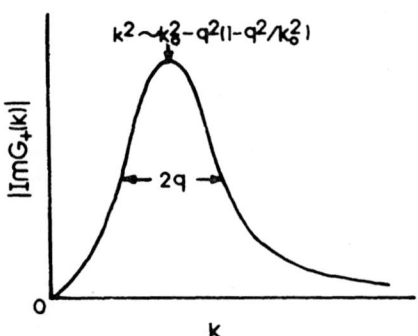

Fig. 1. $\mathrm{Im}\,G_+(k)$ versus k showing a maximum $k = k_o^2 - q^2(1 - q^2/k_o^2)$
and a linewidth $\Delta k \sim 2q$.

The definition for the quality factor Q is either $Q \equiv 2\pi(L_o/\lambda)$, or $Q \equiv 2\pi \times$ particle stored/particle lost in a cycle of oscillation. Using the former definition,

$$Q = k_r L, \qquad\qquad (3a)$$

and with the latter definition,

$$Q = 2\pi \langle \overset{*}{\psi}\psi \rangle_0 / \langle \overset{*}{\psi}\psi \rangle_0 (1 - e^{-2q\lambda}) \sim 2\pi/2q\lambda = k_r L_0. \qquad (3b)$$

We thus see that for $\lambda << L_0$, the two definitions for Q agree. In what follows, we shall use the first definition for Q being 2π times the number of wavelength in a given mean free path L.

The energy line width ΔE may be obtained from $\Delta k \sim 2q$. With $E \propto k^2$, we obtain $E/\Delta E = Q/2$, or $\omega\tau = Q/2$ giving $L = v\tau$ with $v = \hbar^{-1} dE/dk$. There is a difference between Eq. (1) and that of Thouless[10]. Equation (1) gives both the real and imaginary parts of the self energy while Thouless's has no level shift. In order to emphasize, let us summerize our results from $\text{Im } G_+ (k)$. The energy line width ΔE is given by

$$\Delta E \simeq 2E_0 / k_r L_0 = 2E_0 / Q \qquad\qquad (4a)$$

and the level shift ∂E is given by

$$\delta E \simeq E_0 / 4Q^2. \qquad\qquad (4b)$$

The ratio of the level shift to broadening, $\delta E / \Delta E = 1/8Q$. Thus for $Q >> 1$, the level shift may be neglected. Typically, for crystalline materials, $k \sim 10^7 \text{ cm}^{-1}$, $L_0 \sim 1000 \text{ Å}$, $\Delta E \sim 200 \text{ meV}$, and $\delta E \sim 25 \text{ meV}$, showing that the level is significant. Since the level shift is simpler to determine for the amorphous case, we shall use the level shift to determine the mean free path. In the next section we shall treat the case of an isolated well and the case of a double barrier which may serve to approximate the case of a multilayer superlattice.

LINEWIDTH, Q, AND LEVEL SHIFT FOR BOUNDED REGION

In this section we shall take ψ_+ and ψ_- of last section to find linewidth, Q, and level shift for a quantum well or a double barrier structure. The method is rather parallel to the problem of a lossy resonator. To find Q, we let an election bounce between two barriers having a magnitude of reflection denoted by $|r|$. It is important to use only the magnitude to prevent any phase coherence effects. After n-bounce, the density of the electron $\psi^* \psi$ is given by $|r|^{2n} e^{-2nqW}$ with W being the width of the well.

We define a mean free path $L = nW$ such that

$$L^{-1} = L_0^{-1} + L_B^{-1}, \qquad\qquad (5)$$

Where $L_0 = 1/2q$, and $L_B = W /2 \ln |1/r|$. Note that L_0 is the mean free path in absence of the confinement and L_B is the mean free path when damping is zero so that it only takes into account the tunneling out of the barriers.

In order to find shift δE for the double barrier, we use the method of phase. Since we want to find the resonance, we need to equate the total phase change after the wave has experinced two bounces from the two walls to

$$2n\pi, \text{ or } 2k_r W + 2\Phi = 2n\pi, \text{ thus}$$

$$k_r = (n\pi - \Phi) /W \qquad\qquad (6)$$

in which Φ is the phase change upon reflection from the barrier.

It can be shown that for wave ψ_+ incident from the left, having a reflected part $r \psi_-$ and a transmitted part $t \psi_+$, of a single barrier specified by a width B and height V,

$$|r|^2 = \frac{(k_r^2 - q^2 + \alpha^2)^2 + 4 k_r^2 q^2}{(k_r^2 - q^2 - \alpha^2 - 2\alpha q F)^2 + 4 k_r^2 (q + \alpha F)^2} \tag{7a}$$

and

$$\tan \Phi = \frac{2k_r}{\alpha^2 - k_o^2} \frac{F(k_o^2 + \alpha^2) + 2q\alpha}{(k_o^2 + \alpha^2) + 2q\alpha F} \tag{7b}$$

where

$$F = (1 + e^{-2\alpha B}) / (1 - e^{-2\alpha B}), \tag{7c}$$

in which $k_r^2 = k_o^2 - q^2$, $\alpha^2 = \alpha_o^2 - k_o^2$, and $\alpha_o^2 = 2mV / \hbar^2$.

Let us examine eq. (6) and eq. (7) for a known case of a single well with $q = 0$, $F = 1$, we know from Schiff[11], that

$$k \tan kW / 2 = \alpha$$

with

$$\tan \Phi = 2 \tan(\Phi/2) / 1 - \tan^2 \Phi / 2,$$

and using eq. (6), we have

$$\tan \Phi = \frac{2k}{\alpha^2 - k^2}$$

which is eq. (7b) for $q = 0$ and $F = 1$. Therefore, in case of $q = 0$, the method of phase is identical to solving the eigenvalue problem. However, for $q \neq 0$, our method should be identical to finding the peak in energy corresponding to the maximum in the density of states Im $G_+(k)$.

As an added example of our approach, we shall compare the method of calculating L from $|r|$ with a rather physical approach. Since $Q = 2\pi \times$ particle stored/particle lost in a period, we may write $Q = 2\pi / JT$ for J and T being the current and period respectively given by

$$J = |A_o|^2 \, \hbar k / m, \text{ and } T = 2\pi \hbar / E.$$

If we take a wavefunction in the well to be $\psi_w = (2 / W)^{1/2} \sin kx$, together with results from Schiff[11] (p.95),

$$|A_o|^2 = \frac{2}{W} e^{-2\alpha B} \frac{4\alpha^2 k^2}{\alpha_o^2},$$

we arrive at

$$Q_B = \frac{kW \alpha_o^4}{16 \alpha^2 k^2} e^{2\alpha B} \tag{8}$$

Using eq. (5) for L_B, with $q = 0$ in eq. (7a), we obtain the same expression as in eq. (8) for Q_B, the quality factor due to tunneling out of the double barrier. From that the linewidth ΔE_B is obtained. This expression may be compared with straight forward computer calculation[12] of the linewidth of the transmission $|t|^2$ in resonant tunneling.

To summerize, using a damped electron wave function ψ_+ (x > 0) and ψ_- (x < 0), we find the amplitude and phase of the reflection coefficient $|r|$ and Φ of a wave incident on a barrier. With the method of constructive phase, we arrive at the energy corresponding to the resonant tunneling which is identical to the maximum in the density of states. With the method of incoherent multiple reflection, we arrive at the Q, which is identical to the usual definition of allowing a wave to leak out of a storage system. All we need to do is to apply eqs.(7a) and (7b) with eqs.(5) and (6) together with $k_r^2 = k_o^2 - q^2$, we can find Q and thus ΔE or the scattering time τ, and similarly the self energy shift δE and again the mean free path L or the scattering time τ. Which is more useful depends on L. For kL >> 1, as we have pointed out before that δE should be used and for kL ~ 1, since the linewidth is so broad, it is more convenient experimentally to find δE. Therefore δE should be used for most amorphous cases.

RESULTS AND DISCUSSIONS

For kL >> 1, in principle we can experimentally determine the linewidth as in resonant tunneling in a double barrier[12,13]. The part due to the barrier and damping effects, Q_B and Q_L may be seperated similar to eq.(5). Therefore Q_L can be obtained from experiment involving optical or resonant tunneling techniques. This in turn gives us τ and therefore the mobility. Since it is expected that alternating epitaxial layers introduce interface states contributing to the lowering of mobility, we can study this using our solution and experimentally measured linewidth. Similarly for amorphous multi-well systems, kL ~ 1, the linewidth is so broad that generally it is difficult to be experimentally determined. However, as shown by Abeles and Tiedje[2], and Hirose and Miyazaki[5], one may obtain the position from $(\alpha h \nu)^{1/2}$ vs hν[14]. Thus the determination of E from fitting the experiment with our result allowing the calculation of the mean free path L. Figure 2 gives the result of our calculation of δE in eV versus L in Å. Note that for L >> W, the quantum states are those of the quantum well. However for L < W, new features appear. For example the curve for W = ∞ crosses that of W = 40 Å at L ~ 4.5 Å. That is to say for L ~ 4.5 Å, one would not expect any difference whether the well width is 40 Å or ∞. On the other hand there is an increase between W = 20 Å and 10 Å indicating that for L ~ 4.5 Å, the quantum size effect is basically operative. The insert of Fig. 2 shows the result of fitting our calculation to data given by Abeles and Tiedje[2]. In order to fit we need to select a different L for each W. Our result indicates that as the well width are made smaller, it is required to choose a smaller L as in the plot of L vs W. This clarly indicate that mean free path is reduced as the width W is reduced. Thus, it is reasonable to conclude that interface states are being introduced as W is reduced.

For kL ~ 1, the real physical situation may involve other than plane waves. Our results based on one-dimensional damped waves may not quantitatively describe well. Nevertheless the physics should be qualitatively correct, that is, there should be a substantial energy shift for kL ~ 1. Therefore one needs to pay particular attention to the effect of small mean free path dealing with amorphous multilayer work.

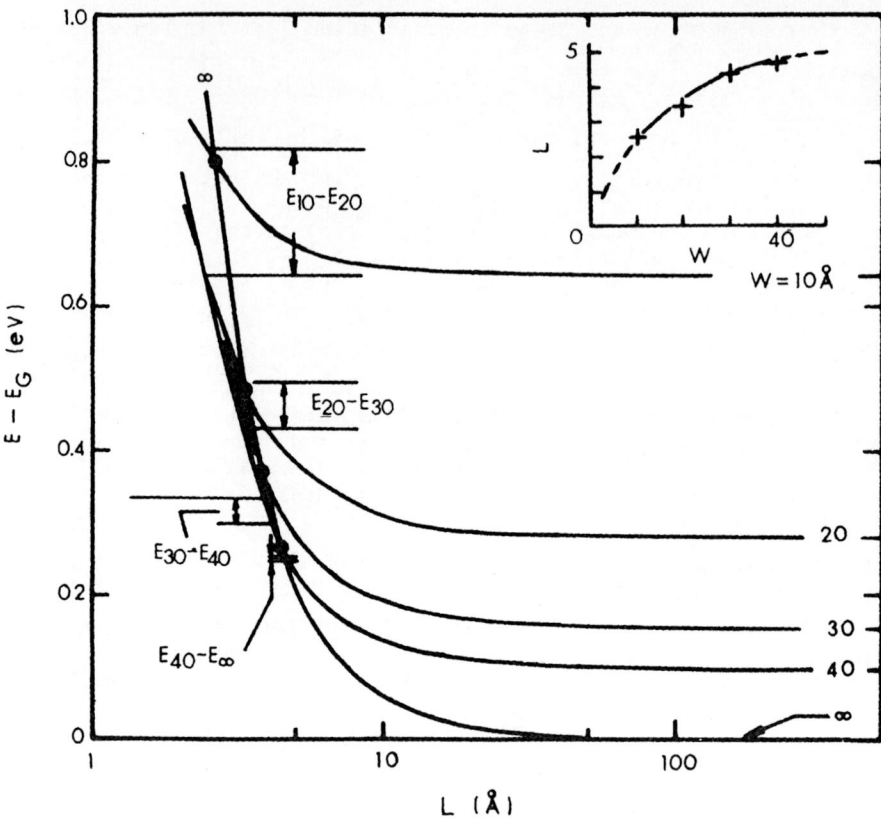

Fig. 2   Energy versus mean free path for a double barrier structure
(a – Si : H / a – Si N$_x$: H) having dimensions (B, W, B) =
= (35 Å, W, 35 Å) with W = ∞ , 40, 30, 20, and 10 Å.   The
parameters used were taken from Ref. 2: V_c = 1 eV, V_v= 0.6 eV,
m_c^* = 0.2 m$_e$, m_v^* = m$_e$. Chossing a particular value for q, E and
L are calculated. Insert shows the fit of data from Ref. 2 to
obtain L vs W. Note that our analysis gives a maximum L~ 5 Å.

Although this work gives a better description for amorphous superlattices and quantum wells, there is remained the question of increased luminescent efficiency observed by Hirose and Myazaki[5]. This author[15] has
observed effects of interface on the crystallization temperature of amorphous materials. And our conjecture is that a slight change of structure is
involved whenever a given amorphous material is constrained by two closely
spaced interfaces. For example, whenever an epi layer is constrained within
two closely spaced layers having lattice mismatch as the case of strained
layer superlattice[16], the epilayer may be grown without dislocations as
pointed out by Matthews and Blakeslee[17]. And therefore a slight change of
structure is involved. In this case, only a slight change of lattice constant
along the plane of the interface is involved. However, it is conceivable

that a - Si : H can grow into a structure resembling the barrier materials, resulting in a widening of the optical gap and presumably different luminescent bahaviors.

REFERENCES

1. L. Esaki and R. Tsu, IBM J. Res.Dev. $\underline{14}$, 61 (1970)
2. B. Abels and T. Tiedje, Phys. Rev. Lett. $\underline{51}$, 2003 (1983).
3. H. Munekata, H. Kukimoto, Jpn. J. Appl Phys. $\underline{22}$, L544, (1983).
4. J. Kakalinos, H. Fritzsche, N. Ibaraki, and S.R. Ovshisky, J.Non-Cryst. Solids $\underline{66}$, 339 (1984).
5. M.Hirose and S.Miyazaki, J. Non-Cryst. Solids $\underline{66}$, 327, (1984).
6. D. Adler, private communication.
7. N. F. Mott and E.A. Davis, Electronic Processes in Non-Cryst.Materials, p. 13, Clarendon Press, Oxford (1979).
8. For example, T.D.Shultz, Quantum Field Theory and Many-Body Problem, Gordon and Breach, N.Y. 1964.
9. See for example. A. Isihara, Statistical Physics, Acad.press, N.Y.1971.
10. D.J. Thouless, Phil. Mag. $\underline{32}$, 877, (1975).
11. L.I. Schiff, Quantum Mechanics, Mcgraw-Hill Inc. N.Y. 1955.
12. R. Tsu and L. Esaki, Appl. Phys. Lett. $\underline{22}$, 562 (1973).
13. L.L.Chang, L. Esaki and R. Tsu, Appl. Phys. Lett. $\underline{24}$, 593 (1974).
14. Strictly speaking $(\alpha h \nu)^{1/2}$ vs h ν plot is only applicable whenever the wave vector conservation is totally relaxed. Since for L ~ W, there is coherent size effects, one would expect some restoration of wave vector conservation at least along the multilayer direction together with polarization dependence.
15. To be publised.
16. G.C. Osbourn, R. M. Biefeld and P.L. Gourley, Appl. Lett. $\underline{41}$, 172 (1982).
17. For example: J.W. Matthews and Blakeslee, J. Cryst. Growth $\underline{32}$, 265 (1976). and references earlier.

+ This work was performed at IFQCS, Brazil, under the partial support of the ONR Grant No. N00014-83-G-0140.

LUMINESCENCE OF AMORPHOUS SILICON SUPERLATTICES

Masataka Hirose, Seiichi Miyazaki and Naoki Murayama

Department of Electrical Engineering, Hiroshima University
Higashihiroshima 724, Japan

INTRODUCTION

Remarkable progress in semiconductor technology has enabled us to control thin film growth in an atomic scale and to obtain the thin-film layered structures such as superlattices[1], high-electron-mobility transistors[2] and multiple quantum well lasers[3]. In these ultra-thin film structures, tunneling phenomena and quantum size effects become important when the layer thickness decreases down to the order of a carrier de Broglie wavelength. Since quantum mechanical phenomena involve novel properties, extensive studies on new functional electron devices have been carried out in theoretical and experimental fields. Recently ultra-thin layered structures consisting of amorphous semiconductor heterojunctions or amorphous semiconductor/insulator systems have been studied as a new class of semiconductor superlattices[4~12]. The existence of quantum size effects in these structures has been demonstrated for a potential well layer thickness below 50 A;i.e, T. Ogino et al.[4] have observed the blue-shift of optical absorption edge of ultra thin $As_{40}Se_{60}/Ge_{25}Se_{75}$ multiple layered structures and also unique change in the vibrational spectrum reflecting the presence of artificially induced long-range-order. Quantum size effects in ultra-thin layered structures containing a-Si:H layers sandwitched with a-Ge:H[7,8], semiconducting $a-Si_{1-x}N_x$:H[11,12], insulating $a-Si_{1-x}N_x$:H[6~11], insulating $a-Si_{1-x}C_x$:H[5,7,8] or insulating $a-Si_{1-x}O_x$:H[9] have been observed from the blue-shift of the optical band gap, increase in conductivity activation energy and corresponding increase of in-plane resistivity. Such quantum size effects in amorphous semiconductor superlattices will open up a new field of physics as well as possibility to design novel device structures.

In this paper, we describe the optical properties of two kinds of amorphous superlattice systems with a-Si:H well layers sandwitched by semiconducting a-Si$_{0.8}$N$_{0.2}$:H[13] barrier layers or by insulating a-Si$_{0.43}$N$_{0.57}$:H (stoichiometric a-Si$_3$N$_4$:H) barriers. We have studied on quantum size effects in a-Si:H/a-Si$_{1-x}$N$_x$:H superlattices by luminescence under external electric field applied perpendicularly to the superlattice. In the a-Si$_{0.8}$N$_{0.2}$:H barrier system, the photoluminescence comes only from the a-Si:H well layer and the luminescence quenching by electric field can be explained in terms of the field-assisted tunneling of holes confined in the well layer. For insulating a-Si$_3$N$_4$:H barrier system, the existence of quantum size effects is more clearly observed not only in the optical absorption spectra but also in the electric field quenching of luminescence.

EXPERIMENTAL

The a-Si:H/a-Si$_{1-x}$N$_x$:H superlattices were deposited on quartz or c-Si substrates by the rf glow discharge technique, in which the material gas for a-Si:H well layer was SiH$_4$ diluted to 10.3% in H$_2$, that for semiconducting a-Si$_{0.8}$N$_{0.2}$:H was a SiH$_4$ (10.3% in H$_2$) + NH$_3$ (1.63% in H$_2$) gas mixture corresponding to a molar fraction of N_{NH_3}/N_{SiH_4} = 0.24, and that for insulating a-Si$_{0.43}$N$_{0.57}$ (stoichiometric a-Si$_3$N$_4$:H) was SiH$_4$ (10.3% in H$_2$) + pure NH$_3$ (N_{NH_3}/N_{SiH_4} =10). In order to fabricate the superlattices having the designed compositions in each layer and the sufficiently abrupt interfaces in the heterojunctions, we controlled growth rate of the individual layers ranging from 0.2 to 0.4 A/sec. Since the residence time of the gas in the reactor is about 3 sec, during which only submonolayer growth of a-Si$_{1-x}$N$_x$:H can proceed, the sufficiently sharp interface in the layered structure may be achieved even under the continuous rf discharge. However, once the ammonia gas was introduced into the system, the decomposed species must contaminate the internal surfaces of the reactor and a minute amount of a residual NH$_3$ in the nozzle can not decay rapidly. Therefore, in the present experiment, after a-Si$_{1-x}$N$_x$:H deposition the system was pumped down to $\sim$10^{-3} Torr without discharge and purged by a SiH$_4$ gas and then the next a-Si:H layer was deposited. The a-Si:H/semiconducting a-Si$_{0.8}$N$_{0.2}$:H heterojunction superlattices have equal film thicknesses for both well layers and barriers, which were varied from 50 to 200 A. The total thicknesses of the superlattices were kept almost constant. In the case of a-Si:H/insulating a-Si$_{0.43}$N$_{0.57}$:H superlattice, the thickness of stoichiometric silicon nitride layer was fixed at 100 A and a-Si:H thickness was varied between 8 and 400 A. The total thickness of the multiple-layers was held approximately constant for all specimens. The optical gaps of a-Si:H and a-Si$_{1-x}$N$_x$:H were determined by the optical absorption spectra of the respective thick films. The photoluminescence measurement was carried out using the 514.5 nm (2.41 eV, 120 mW) light from an Ar$^+$ laser. The electric field quenching of luminescence was measured for the superlattices grown on n$^+$c-Si substrate. A semitransparent Au top electrode with a thickness of 200 A was evaporated in order to apply the bias across the layered structure as shown in Fig. 1. In this case, a laser power of 250 mW was used by taking into account the attenuation depth of the laser light.

The periodic potentials in a-Si:H/a-Si$_{1-x}$N$_x$:H superlattices can be estimated if the electron affinities for the both materials are known. In the a-Si:H/semiconducting a-Si$_{0.8}$N$_{0.2}$:H system, the potential well depths in the conduction and valence bands of a-Si:H were estimated to be 0.03 eV and 0.21 eV, respectively, by using the measured electron affinities for a-Si:H (3.93 eV)[14] and a-Si$_{0.8}$N$_{0.2}$:H (3.90±0.10 eV)[12] and by using the optical bandgaps of a-Si:H (1.72 eV) and a-Si$_{0.8}$N$_{0.2}$:H (1.96 eV) as shown in Fig. 2. Even considering that the carriers in both well and barrier layers must move to equalize the Fermi level throughout the semiconductor heterojunctions, the band bending in the respective layer was found negligibly small. The positions of measured E$_F$ in bulk a-Si:H (E$_C$-0.61

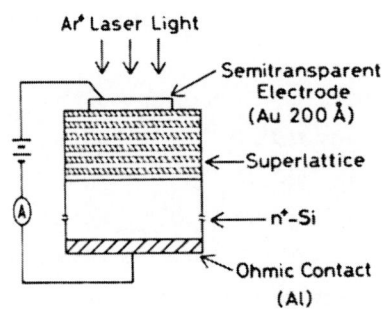

Fig. 1. Sample structure for luminescence measurement under applied bias.

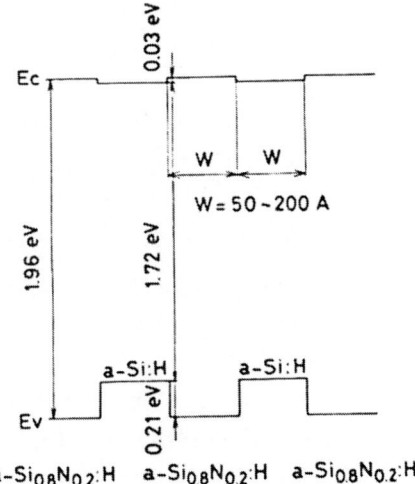

Fig. 2. Energy-band diagram for a-Si:H/semiconducting a-Si$_{0.8}$N$_{0.2}$:H superlattice.

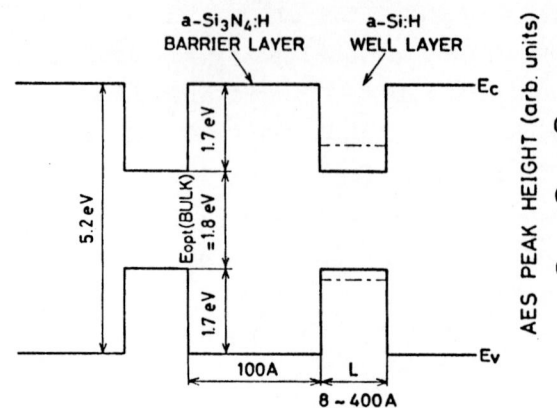

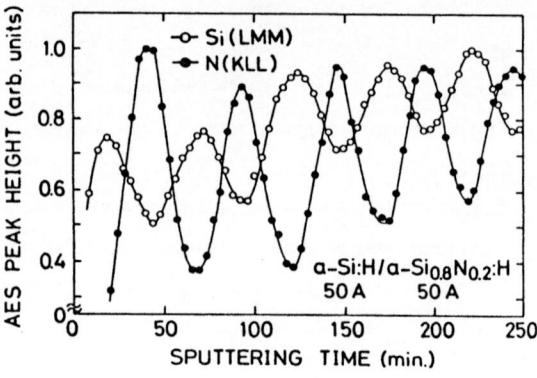

Fig. 3. Energy-band diagram for a-Si:H/insulating a-Si$_3$N$_4$:H superlattice.

Fig. 4. Auger in-depth profile of an a-Si:H (50 A)/semiconducting a-Si$_{0.8}$N$_{0.2}$:H (50 A) superlattice.

eV) and in bulk a-Si$_{1-x}$N$_x$:H (E$_C$–0.64 eV) are consistent with the result of Fig. 2. Nevertheless, it is not clear whether the potential discontinuity is correctly determined in Fig. 2, because there exist some ambiguities of the measured electron affinities for the both materials and because the band discontinuity is not a simple function of the electron affinity and the Fermi level in the superlattice structure. In the a-Si:H/insulating a-Si$_{0.43}$N$_{0.57}$:H (stoichiometric a-Si$_3$N$_4$:H) superlattices the well depth of the a-Si:H conduction band was obtained to be 1.7 eV, and the optical band gaps for a-Si:H (1.80 eV) and a-Si$_3$N$_4$:H (5.2 eV) yielded the 1.7 eV well depth also for the a-Si:H valence band as shown in Fig. 3.

Existence of multiple layered structure was examined Auger electron spectroscopy (AES) and X-ray photoelectron spectroscopy (XPS). Figure 4 shows an example of the in-depth profile of the compositions for the a-Si:H (50 A)/semiconducting a-Si$_{0.8}$N$_{0.2}$:H (50 A) superlattice structure obtained by high-spatial-resolution AES, where the signal intensities of Si (LMM) at 92 eV and N (KLL) at 380 eV are normalized by each maximum signal intensity. From this figure it is shown that the a-Si:H/ a-Si$_{1-x}$N$_x$:H superlattice has the well-defined periodic structure. The layer spacing estimated by the sputtering rate for a-Si:H and a-Si$_{0.8}$N$_{0.2}$:H agree well with the thickness determined by the deposition rate of the corresponding layers. The a-Si:H/a-Si$_{1-x}$N$_x$:H interface appears to be abrupt on an atomic scale when the profile is compared with Auger in-depth profile of the thermally grown SiO$_2$/Si system, which has the structural transition layer of at most a few monolayers[15].

RESULTS

In the superlattice structure, when the thickness of the potential well is reduced to the order of a carrier de Broglie wavelength and when the inelastic diffusion length exceeds the well width, it is expected to observe the blue shift of the optical absorption edge, namely, the increase of the effective optical band gap of the well layer arising from the quantization of the extended states of the well layer. Figure 5 shows the optical band gap E_{opt} of the superlattice as a function of the well layer thickness. We determined the optical band gap for a–Si:H (8∿400 A)/a–Si$_3$N$_4$:H (100 A) by a Tauc plot of $(\alpha h\nu)^{1/2}=A(h\nu-Eopt)$, as is normally done with amorphous semiconductors. In calculating optical absorption coefficient of the a–Si:H well layer from optical absorption spectra, we take into account the effective refractive index in the layered structure instead of the refractive index in bulk a–Si:H. Increase in E_{opt} begins to be observed with decreasing a–Si:H well layer thickness below 50 A. The increase in E_{opt} indicates the quantization of the extended states as expected. It is unlikely that the undesirable incorporation of nitrogen into the a–Si:H well layer causes an increase of E_{opt} because the slope of the Tauc plot, A, increases from 800 to 1100 $(eV\ cm)^{-1/2}$ when the a–Si:H well layer thickness decreases. The number of A associated with the steepness of the band edge also increases with decreasing well width. If the a–Si:H layer is contaminated with nitrogen, the number of A must decrease. Additionaly, in Fig. 5, the experimental data are good agreement with the calculated optical band gap shown by solid curve, where the first quantization state for both electrons and holes is calculated by assuming the electron mass $m_e^*=0.6m_0$ and hole mass $m_h^*=m_0$ in a single potential well with a depth of 1.7 eV (Fig. 3).

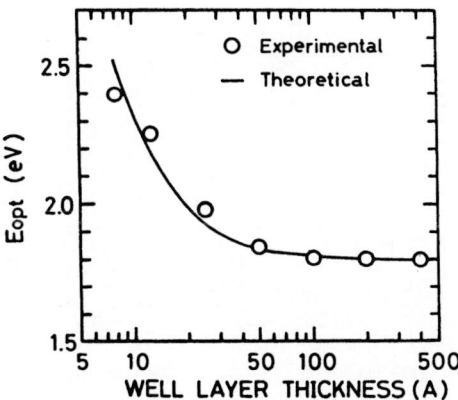

Fig. 5. The optical gap E_{opt} as a function of well layer thickness. Solid line indicates the calculated E_{opt} by assuming the mass for electrons $m_e^*=0.6m_0$ and for holes $m_h^*=m_0$.

In order to examine the influence of the quantization of the extended states on the carrier recombination through band tail or gap states, we measured photoluminescence (PL) spectra for a-Si:H/Si$_{1-x}$N$_x$:H superlattices. In the a-Si:H (50∿200 A)/a-Si$_{0.8}$N$_{0.2}$:H (50∿200 A) superlattices, the photoluminescence (PL) spectrum taken at 80 K exhibits a single peak at ∿1.37 eV arising from the radiative recombination through tail states in the a-Si:H well layers (E_{opt} =1.72 eV), and there is no significant yield from the a-Si$_{0.8}$N$_{0.2}$:H barrier layers, where the luminescence peak at ∿1.48 eV is expected as a characteristic band for a-Si$_{0.8}$N$_{0.2}$:H (E_{opt} =1.96 eV). In Fig. 6 the PL spectrum for the superlattice composed of the twenty periods of a-Si:H (50 A) and a-Si$_{0.8}$N$_{0.2}$:H (50 A) alternatively deposited on a crystalline n$^+$-Si(100) substrate (sample A) is compared with that of an a-Si$_{0.8}$N$_{0.2}$:H (1000 A)/ a-Si:H (1000 A) double layerd structure (sample B), in which the respective thicknesses are equal to the total thicknesses of the barrier layers and the well layers in sample A. The PL spectrum for a single layer of a-Si$_{0.8}$N$_{0.2}$:H (1000 A) (sample C) and that of a-Si:H (1000 A) (sample D) are also shown in the figure as references. If it is assumed that the PL spectrum is composed of a few Gaussian-shape emission bands, the PL spectrum from the superlattice (sample A) consists of a strong emission band peaking at 1.37 eV and a very weak emission at 1.19 eV, both of

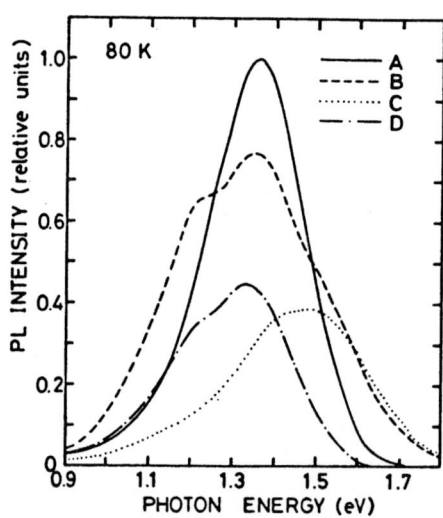

Fig. 6. Photoluminescence spectra for the superlattice consisting of twenty layers of a-Si:H (50 A) and of a-Si$_{0.8}$N$_{0.2}$:H (50 A) on n$^+$ c-Si substrate (sample A), a-Si$_{0.8}$N$_{0.2}$:H (1000 A)/a-Si:H (1000 A)/n$^+$ c-Si double layered structure (sample B), a-Si$_{0.8}$N$_{0.2}$:H (1000 A)/n$^+$ c-Si single layer structure (sample C) and for a-Si:H (1000 A)/n$^+$ c-Si single layer structure (sample D).

446

which have the same full width at half maximum of 0.24 eV. The PL spectum
from the superlattice is essentially similar to that from the single layer
a-Si:H (sample D) except for the difference in their intensities.
Furthermore, in the superlattice, there is no emission band at 1.48 eV
which originates in the $a-Si_{1-x}N_x$:H barrier layer as can be observed in
the spectra of sample B and C. This result clearly indicates that the
luminescence from the superlattice system originates in the radiative
recombination in the a-Si:H well layer. The luminescence intensity of the
superlattice (sample A) is most intense among the four samples as a result
of significant enhancement of the emission band at 1.37 eV, probably
because the photocarriers generated in the barrier layers flow into the
a-Si:H well layers. Note that the PL intensity of the emission band at
1.19 eV in the superlattice is weak compared with that in the single layer
a-Si:H. This implies that the capture rate of electrons and holes by the
deep lying gap states is also lowered in the superlattice structure
presumably due to quantization of the extended states reduces the
effective transition rate of carriers to the deep gap states, the
radiative recombination through gap states providing the 1.19 eV emission
is effectively quenched.

Photoluminescence spectra for the a-Si:H (8∿400 A)/stoichiometric a-
Si_3N_4:H (100 A) superlattices consist of an emission band originated only
from a-Si:H well layers. The spectrum for the superlattice having a well
layer thickness of more than 100 A is basically similar to that for bulk
a-Si:H, while the spectrum for the thinner well width than 50 A is
different from the bulk one. The luminescence peak energy and the full
width at half maximum (FWHM) of the emission band plotted as a function of

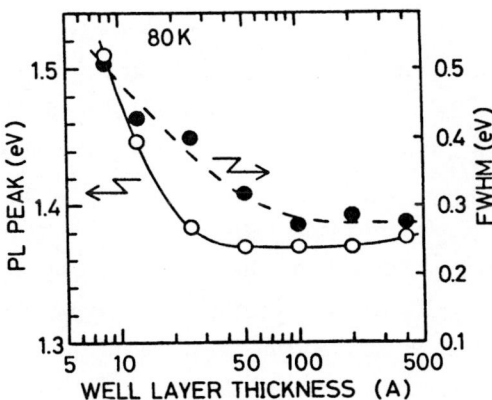

Fig. 7. The peak energy and the FWHM in photoluminescence spectra as a
function of well layer thickness.

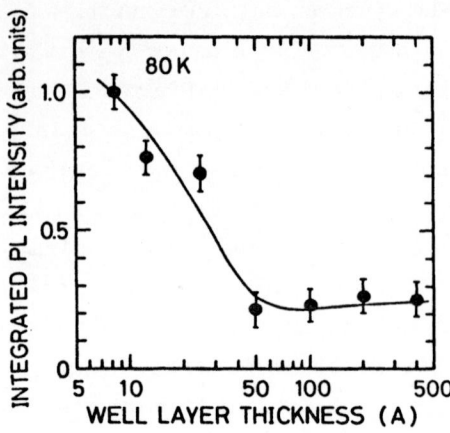

Fig. 8. Integrated PL intensity as a function of well layer thickness.

a-Si:H well layer thickness are shown in Fig. 7. When the well layer thickness decreases less than 50 A, the blue-shift of the peak energy and broadening of FWHM are observed. This suggests that the quantization of carriers in the extended states (see Fig. 5) enhances the carrier capture rate by radiative recombination centers existing close to the band edge. Namely, quantization of carriers in the extended states newly creates the emission band at higher energy side and hence the FWHM of the emission spectrum becomes broad. Additionally, the integrated PL intensity clearly increases with decreasing the well layer thickness below 50 A as shown in Fig. 8, in which the differences in the total a-Si:H layers thickness and in the effective refractive index of each sample are carefully taken into account to determine the integrated PL intensity for the same amount of absorption of the excited laser light. The result indicates that the quantization reduces the nonradiative recombination through gap states because the capture rate of carriers by deep lying states is decreased due to the shift of the band edge toward higher energy.

The luminescence quenching by electric field applied perpendicularly to the superlattices is studied, in order to reveal the existence of the well-defined a-Si:H quantum well to confine electrons and holes in the a-Si:H/semiconducting $a-Si_{0.8}N_{0.2}$:H superlattices (Fig. 2) and also in the $a-Si:H/a-Si_3N_4$:H superlattices (Fig. 3). The absence of built-in electric field in the superlattices and the spatial extent of carrier wave functions in the well layer are also clarified by PL quenching experiments. Figure 9 shows the integrated PL intensity of the luminescence from the superlattice consisting of twenty layers of a-Si:H (50 A) and of $a-Si_{0.8}N_{0.2}$:H (50 A) formed n^+c-Si substrate (sample M) and from a single layer a-Si:H (2000 A) on n^+c-Si substrate (sample S) as a function of electric field applied through a semitransparent Au electrode. In the figure the integrated PL intensity is normalized by that without external applied field. The field quenching of the luminescence for a single layer a-Si:H (sample S) is influenced by the built-in electric field which arises from the Au/a-Si:H Schottky barrier. Therefore, when the positive bias is applied to the Au electrode, the internal electric field decreases with increasing in the external applied field up to $\sim 5 \times 10^4$ V/cm at which the external applied bias is ~ 1 volt[14], because of the forward bias of the Au/a-Si:H Schottoky barrier whose height is about 1.0 volt. Since the field separation of electron and holes becomes less effective, the luminescence intensity slightly increases as the applied field increases up to $\sim 5 \times 10^4$ V/cm. At applied field above 5×10^4 V/cm, the luminescence intensity starts to decrease as a consequence of an increase in the internal electric field. When the negative bias is applied to the Au electrode, the reverse bias of the system results in the efficient field separation of electrons and holes, and the luminescence intensity gradually decreases. In contrast to this, the luminescence intensity of the superlattice (sample M) is almost independent of the polarity of applied bias and hardly influenced by the bult-in field induced by the Schottky barrier, possibly because more strong perturbation comes from the periodic potentials. When the internal electric field is about 1.5×10^5 V/cm, the luminescence intensity rapidly decreases and almost complete quenching occurs at 3.5×10^5 V/cm. Since the

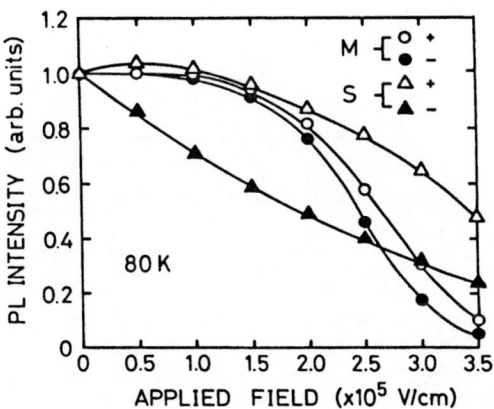

Fig. 9. Integrated PL intensity as a function of applied field for the superlattice sample denoted by M consisting of twenty layers of a-Si:H (50 A) and of a-Si$_{0.8}$N$_{0.2}$:H (50 A) and for a single layer a-Si:H sample S (2000 A thick).

electrons can move freely out of the shallow quantum well at low electric fields, the rapid quenching of the luminescence in the region $2 \times 10^5 \sim 3.5 \times 10^5$ V/cm could be attributed to the hole tunneling through the a-$Si_{0.8}N_{0.2}$:H barrier which leads to an efficient separation of holes and electrons in the superlattice. We will more quantitatively discuss on the quenching phenomenon associated with the field-asisted tunneling of hole in a later section. The change in the luminescence spectrum by varying the applied bias is very little in both the superlattice (sample M) and the single layer a-Si:H (sample S); The luminescence peak at 1.37 eV shifts by ~ 0.05 eV toward the low energy side when the applied field is increased to 3.0×10^5 V/cm. This implies that the built-in electric field in the superlattice (sample M) is basically quite small.

In the case of the a-Si:H/insulatinga-Si_3N_4:H superlattice the luminescence quenching due to field separation of carriers hardly occurs, since photocarriers can be confined in deep potential well (1.7 eV) separated by the 100 A barrier layers. The luminescence quenching in such a system, if exists, could be caused by the separation of photogenerated holes and electrons in the well layers. Figure 10 shows the luminescence intensity in the a-Si:H/insulating a-Si_3N_4:H superlattices as a function of applied electric field perpendicularly to the layered structures. In the range of the well layer thickness from 400 to 100 A, the magnitude of the luminescence quenching due to the applied field becomes more pronounced with decreasing the well layer thickness, indicating that photogenerated holes and electrons are effectively separated. For samples having a well layer thickness less than 50 A, the quenching effect becomes less pronounced as the well width decreases down

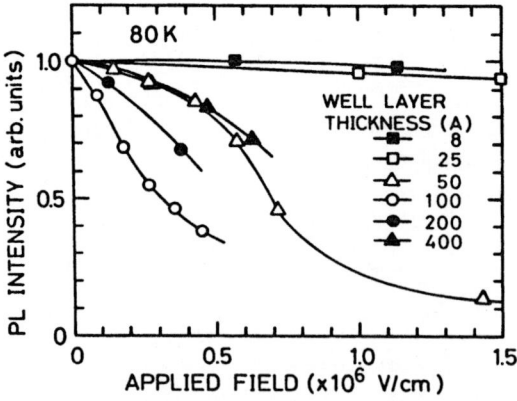

Fig. 10. Integrated PL intensity as a function of applied field for the
a-Si:H/insulating a-Si_3N_4:H (100 A) superlattices.

to 25 A. In particular below 25 A in the well layer thickness, the quenching effect is hardly observed because the photocarriers in the quantized levels can not be spatially separated even at an electric field strength of 10^6 V/cm. Therefore, it is likely that the spatial extent carrier wave function is larger than 25 A. Note that the clear observation of PL quenching by external electric field is a direct evidence on the absence of built-in field induced by defect in the interface.

DISCUSSION

In order to quantitatively discuss the electric field quenching phenomena of the photoluminescence of a-Si:H/a-Si$_{0.8}$N$_{0.2}$:H system, we calculate the internal electric field dependence of the spatial distribution of carrier wave function $\psi_i(z)$ which occupies the lowest eigen state in the one-dimentional a-Si:H quantum well. In this wave function calculation, it is assumed that (1) the depth of potential well at the conduction band and at the valence band are 0.03 eV and 0.21 eV as shown in Fig. 2, respectively, (2) the single quantum well with a width of 50 A thick is sandwiched by two thick a-Si$_{1-x}$N$_x$:H barrier layers, implying that the coupling of the energy eigen states with those of the adjacent wells in the superlattice is neglisible, and (3) the concentration of photogenerated carriers is insufficient to screen the electric field in

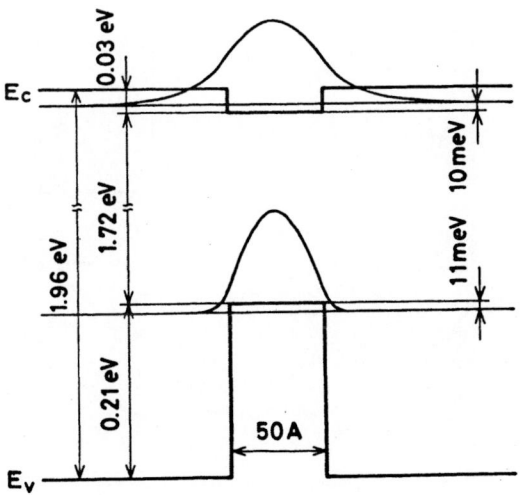

Fig. 11. Calculated energy eigen states and the wave functions in the quantum well at zero electric field. The electron mass $m_e^*=0.6m_0$ and hole one $m_h^*=m_0$ were assumed.

the well layers. From these approximations, the wave function $\psi_i(z)$ for holes and electrons has been obtained by solving the one-dimentional Schodinger equation. The parameters used in the numerical calculation are the carrier effective mass m^* and the internal electric field strength E. Figure 11 shows the calculated wave function and its energy eigen state in the quantum well without the external applied field, in which the electron mass $m_e^*=0.6m_0$ and hole one $m_h^*=m_0$ are assumed as in the case of Fig. 5. Figure 11 shows that the electron wave function significantly penetrates into the barrier layer and the confinement of electrons in the quantum well is not complete, while the hole wave function is effectively confined in the quantum well. In the case of the existence of the internal electric field, the spatial distribution of carriers wave function $\psi_i(z)$ is calculated as shown in Fig. 12 (a) and (b). It is clear from Fig. 12 (a) that electrons are no longer confined in the well layer when the electric field exceeds 1×10^4 V/cm, above which the pronounced electron tunneling through the a-$Si_{0.8}N_{0.2}$:H barrier takes place. Also, the thermal smearing effect (7 meV at 80 K) assists the broadening of the spatial distribution of electrons over the barriers. In contrast to this, the hole wave function is apparently confined in the well layer even at an electric field strength of 3×10^5 V/cm as shown in Fig. 12 (b). However, the spatial symmetry of the hole distribution is lost by the electric field and the penetration of the wave function into the barrier layer becomes significant. In fact the calculated tunneling probability at fields above 2×10^5 V/cm is considerably large as shown in Fig 13. This model calculation indicates that in the superlattice the hole tunneling through the barrier results in the leakage of confined holes from the well layer

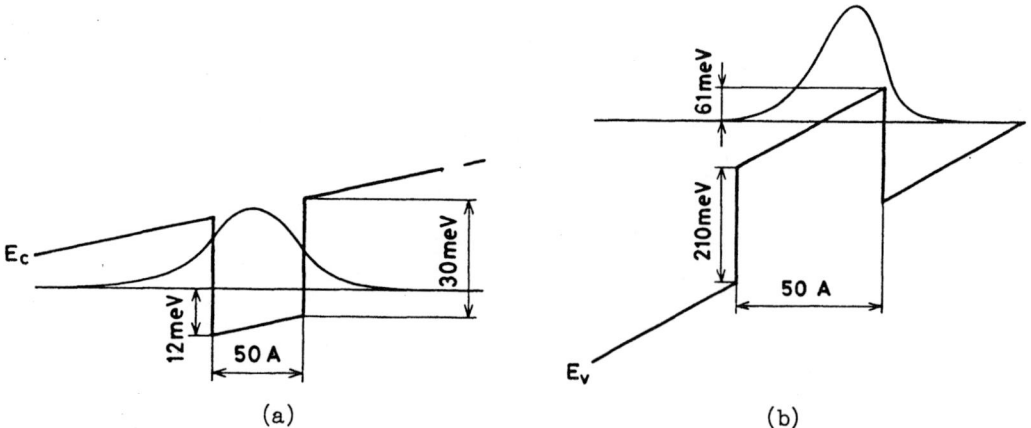

(a)   (b)

Fig. 12. Calculated energy eigen state and the wave function in the conduction band quantum well at a field of 1.0×10^4 V/cm (a), and those in the valence band quantum well at a field of 3.0×10^5 V/cm (b).

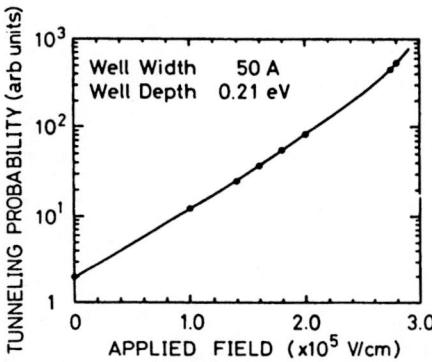

Fig. 13. Tunneling probability of hole as a function of applied field. The well width and the well depth are 50 A and 0.21 eV, respectively.

at field strengths above 2×10^5 V/cm. Therefore, the luminescence quenching by the electric field in the superlattice is primarily interpreted in terms of the field-assisted extraction of holes confined in the quantum wells, which becomes significant at electric fields above $\sim 2 \times 10^5$ V/cm as shown in Fig. 13.

In a-Si:H (8$\sim$400 A)/a-Si$_3$N$_4$:H (100 A) superlattices, the result shown in Fig. 10 can be explained as follows: when the well layer is sufficiently thick, there exists the weak electric field region in the layer due to the screening effect of photogenerated carriers against external electric field. Therefore, the PL quenching hardly occurs by applying electric field. From 400 to 100 A of the well layer thickness, along with decreasing the well layer thickness, photogenerated holes and electrons are effectively separated by internal electric field due to the decrease in the screening effect. In the case of further decrease in the well layer thickness, the PL quenching is getting weak because the spatial extent of carrier wave functions in the well layers becomes nearly equal to the well layer thickness. This result is consistent with the onset of quantum size effect in optical band gap (Fig. 5) and in photoluminescence spectrum (Figs. 7 and 8). Moreover, the very weak PL quenching below 25 A well width shows that the spatial extent of carrier wave function in the tail states near band edge must exceed 25 A and hence the spatial extent of extended states wave function is also larger than 25 A, in consistent with electron mass $m_e^* = 0.6 m_0$.

CONCLUSION

Two types of amorphous superlattices consisting of a-Si:H well layers sequentially alternating with semiconducting a-Si$_{0.8}$N$_{0.2}$:H barrier layers or insulating a-Si$_3$N$_4$:H barrier layers are fabricated by the rf glow discharge tecnique. In the a-Si:H/semiconducting a-Si$_{0.8}$N$_{0.2}$:H superlattice, the photoluminescence comes only from the a-Si:H well layer. The luminescence quenching by the electric field applied perpendicularly to the superlattices is compared with the result of the model calculation of the carrier wave functions in the well layer and explained in terms of the field-assisted tunneling of the confined holes in the well layer. As for the a-Si:H/a-Si$_3$N$_4$:H superlattices, the existence of quantum size effect is demonstrated by photoluminescence and optical absorption measurements. In particular, the photoluminescence intensity is enhanced with decreasing the well layer width as a consequence of quantization. The field quenching of the luminescence is clearly observed even for sample with a well width of 50 A, indicating the absence of the built-in field due to the interface defects. The spatial extent of carrer wave functions in extended states is found to be larger than 25 A as evidenced by PL quenching experiments.

ACKNOWLEDGEMENTS

The authors wish to thank Prof. M. Yamanishi for his assistance in the calculation of carrier wave functions in quantum well, Prof. Y. Osaka for his helpful discussions, and S. Ohkawa for his help in experiments.

REFERENCES

1. L. Esaki and L. L. Chang, New Transport Phenomenon in a Semiconductor "Superlattice", Phys. Rev. Lett., 33: 495 (1974).
2. T. Mimura, S. Hiyamizu, T. Fujii, and K. Nanbu, A New Field-Effect Transistor with Selectively Doped GaAs/n-Al$_x$Ga$_{1-x}$As Heterojunctions, Jpn. J. Appl. Phys., 19: L225(1980).
3. N. Holonyak, R. M. Kolbas, R. D. Dupuis, and P. D. Dapkus, Quantum-well Heterostructure Lasers, IEEE J. Quantum Electron., QE-16: 170 (1980).
4. T. Ogino, A. Takeda, and Y. Mizushima, Long-Range Interaction in Multi-Layered Amorphous Structure, in; Collected papers of 2nd Intern. Sympo. in Molecular Beam Epitaxy and Related Clean Surface Techniques, Tokyo, 65 (1982).
5. H. Munekata and H. Kukimoto, Optical Properties of a-Si:H Ultrathin Layers, Jpn. J. Appl. Phys., 22: L544 (1983).
6. B. Abeles and T. Tiedje, Amorphous Semiconductor Superlattices, Phys. Rev. Lett., 51: 2003 (1983).
7. T. Tiedje, B. Abeles, P. D. Persans, B. G. Brooks, and G. D. Cody, Bandgap and Resistivity of Amorphous Semiconductor Superlattices, J. Non-Cryst. Solids, 66: 345 (1984).

8. B. Abeles, T. Tiedje, K. S. Liang, H. W. Deckman, H. C. Stasiewski, J. C. Scanlon, and P. M. Eisenberger, Growth and Structure of Layered Amorphous Semiconductors, ibid., 66: 351 (1984).

9. J. Kakalios, H. Fritzsche, N. Ibaraki, and S. R. Ovshinsky, Properties of Amorphous Semiconducting Multilayer Films, ibid., 66: 339 (1984).

10. T. Tiedje, C. B. Roxlo, B. Abeles, and C. R. Wronski, Interface States in Amorphous Silicon/Amorphous Silicon Nitride Superlattice Structures, in: Extended Abstracts of the 5th Intern. Conf. on Solid State Devices and Materials, Kobe, 531 (1984).

11. S. Miyazaki, N. Murayama, M. Hirose, and M. Yamanishi, Novel Properties of a-Si:H/a-Si$_{1-x}$N$_x$:H Superlattices, in: Technical Digest of 1st Intern. Photovoltaic Science and Engineering Conf., Kobe, 425 (1984).

12. M. Hirose and S. Miyazaki, Luminescence and Transport in a-Si:H/a-Si$_{1-x}$N$_x$:H Quantum Well Structures, J. Non-Cryst. Solids, 66: 327 (1984).

13. H. Kurata, M. Hirose, and Y. Osaka, Wide Optical-Gap, Photoconductive a-Si$_x$N$_{1-x}$:H, Jpn. J. Appl. Phys., 20: L811 (1981).

14. T. Yamamoto, Y. Mishima, M. Hirose, and Y. Osaka, Internal Photoemission in a-Si:H Schottky-Barrier and MOS Structures, Jpn. J. Appl. Phys., Suppl. 20-2: 185 (1981).

15. C. R. Helms, W. E. Spicer, and N. M. Johnson, New Studies of the Si-SiO$_2$ Interface Using Auger Sputter Profiling, Solid State Commun., 25: 673 (1978).

455

AMORPHOUS SEMICONDUCTOR HETEROSTRUCTURES

F. Evangelisti

Dipartimento di Fisica
Universita' di Roma "La Sapienza"
P.le A.Moro 2, 00185 Roma, Italy

INTRODUCTION

The microscopic study of crystalline-semiconductor heterostructures by photoelectronic techniques has been the subject of much work in the last years.[1] This effort is justified by fundamental reasons and by the widespread applications of semiconductor heterojunctions in solid-state electronics. Recently, very promising heterojunction devices made by amorphous semiconductors have been realized in the field of superlattices[2] and photovoltaic solar cells.[3] Since the behavior of a heterojunction device is strongly dependent on band discontinuities, interface states and diffusion potential, there was an immediate interest in extending to amorphous semiconductors the microscopic investigation of these parameters.[4-6] In the present work we will review some of the results obtained in this field which is still at its first stages of development. It will be shown, moreover, that the technique offers the interesting possibility to investigate fundamental problems like the effect of disorder and/or hydrogenation on the electronic structure of the materials.

EXPERIMENTAL TECHNIQUE

The surface sensitivity of U.V. photoemission spectroscopy makes this technique an ideal probe for interface studies. The physical information carried out by the ejected electrons is inherent to the outermost region of the solid. The thickness of this region varies with the kinetic energy of the electron according to the escape depth curve. Therefore, by using a tunable photon source like synchrotron radiation, it is possible to modulate the surface sensitivity of the technique and to enhance surface effects.

The microscopic study of a heterojunction imply the determination of parameters like the band discontinuities, the position of the Fermi level at the interface and the magnitude of band-bending effects. Photoemission spectroscopy is the most direct way to measure these parameters. The experimental approach consists in growing in situ and step-by-step one semiconductor on top of the other and following the evolution of the valence band and core levels at each step.

The formation of the heterojunction between two semiconductors of

457

different energy gap will produce either a valence band or a conduction band discontinuity or both. Valence band discontinuities larger than few tenths of eV can be measured directly in the spectra by growing on one semiconductor a thin overlayer of the other semiconductor. If the escape depth of the electrons is larger than the thickness of the overlayer, both valence band edges are directly measurable. ΔE_v is obtained by taking the difference between the extrapolation to zero of the two leading edges. This situation is schematically shown in Fig.1.

When the valence band discontinuity is small (0-0.4 eV) the two valence edges are not resolved, however, and ΔE_v is to be derived from the evolution of the leading edge provided that band-bending changes during the interface formation are duly taken into account. This second case requires an analysis of both valence and core structures. First of all, we must remember that a photoemission measurement is always affected by the initial band bending since the probed region is, in general, much smaller than the extention of the band bending itself. When this undergoes a variation all the valence and core structures shift rigidly relative to the Fermi level. Therefore, the determination of ΔE_v proceeds as follows. The valence band maximum E_v and one or more core levels are measured on the clean substrate. Then the measurements are repeated at different overlayer thicknesses. The valence-band top usually shifts and so do the core levels. The shift of E_v is the sum of the band discontinuity plus the change in band bending caused by the variation of the local charge distribution. The band-bending changes at different coverages can be deduced from the shift of the substrate core-level peaks. Therefore, ΔE_v is determined as the difference between the total shift of E_v and the band bending change. It is worth mentioning, however, that often the situation is made more complicated by chemical shifts of the core levels due to the formation of interface chemical bonds. In this last occurrence the analysis is more complex and requires either the measurement of different core lines or a fitting procedure to unfold the different components. It is important to point out that the major source of error in the value of ΔE_v is due to the determination of the valence-band top from the experimental energy distribution curves (EDC). Usually this is done by a linear extrapolation to zero of the leading spectral edge and can be easily affected by 0.1-0.15 eV uncertainty.

For heterojunctions involving amorphous semiconductors it is necessary to clarify which electronic levels are compared by defining the valence band maximum E_v as the linear extrapolation to zero of the valence-band EDC's leading edge. Under usual experimental conditions this edge merges into the noise at a signal level 0.05-0.1 times the intensity of the highest in energy valence-band feature (3p or 4p states for Si or Ge, respectively), i.e. E_v corresponds to $\sim 10^{21}$ states cm^{-3} eV^{-1}. In the crystaline case, this definition does not cause any ambiguity because the top of the valence band is a well defined concept and the density of states $g(E)$ increase rapidly (f.e. in c-Si, $g(E)$ varies from 0 to 10^{21} in ~ 0.05 eV, i.e. over an energy distance smaller than the experimental uncertainty). Things are different in the amorphous case due to the continuum of localized states tailing toward the Fermi level. However, a state density in the 10^{20}-10^{21} cm^{-3} eV^{-1} range is a reasonable estimate[7] for the demarcation level between extended and localized states. Therefore, we can assume that E_v as defined above locates the valence-band mobility-edge, within 0.1-0.2 eV uncertainty.

Finally, let us mention that the conduction-band discontinuity too could be determined experimentally by studying the evolution of the onset of the core-to-conduction-band transitions as measured by partial yield spectroscopy. In practice, however, these transitions are affected by large excitonic effects[8] which make difficult to locate the bottom of conduction band.

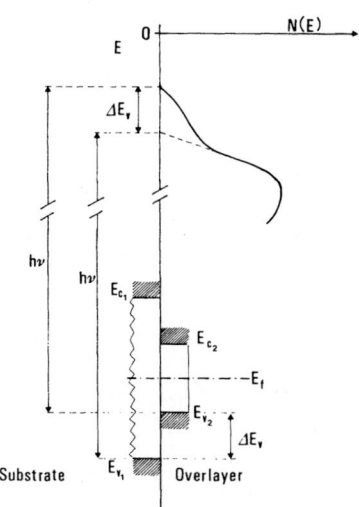

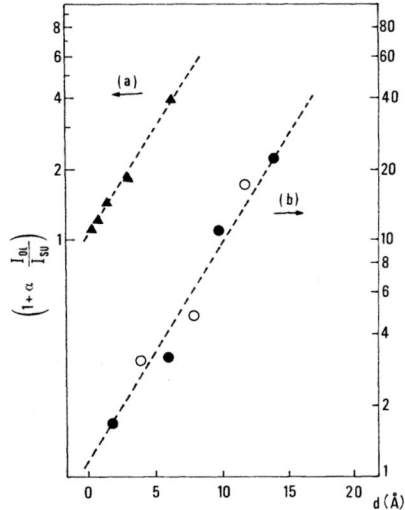

Fig. 1 Transition scheme showing the low binding-energy side of the valence-band photoemission spectra in the case of a thin overlayer. N(E) is the photoemission intensity.

Fig. 2 Ratio of the core-level intensity as a function of overlayer thickness. The expression in abscissa is discussed in the text. a) a-Si:H/a-Ge interface; b) a-Si$_x$C$_{1-x}$:H/a-Si (full dots) and a-Si$_x$C$_{1-x}$:B, H/a-Si (open circles) interfaces.

RESULTS AND DISCUSSION

In the following we review and discuss briefly the results obtained on a number of heterojunctions involving amorphous silicon and germanium as well as silicon-carbon alloys.

An important characterization of a heterojunction is the quality of the interface, i.e. its abruptness or the possible interdiffusion of the different species or the growth of the overlayer through island formation. This information can be obtained by the intensity dependence of the core lines as a function of the overlayer thickness. In particular, if the interface is abrupt and the overlayer is uniform we have:

$$(1 + \alpha\, I_{OL}/I_{SU}) = \exp(d/\lambda)$$

where I_{OL} and I_{SU} are the core intensities of the overlayer and substrate, respectively, α is a factor which accounts for the different photoemission cross-sections, λ and d are the overlayer escape depth and thickness. The experimental values found for the interfaces a-Si$_x$C$_{1-x}$:H/a-Si and a-Si:H/a-Ge are reported in Fig.2. The exponential behavior is strictly obeyed showing that the growth at room temperature of amorphous interfaces between tretravalent elements proceeds without interdiffusion. An equally abrupt junction has been obtained by Abeles et al.[6] by growing a-Si:H on a-SiN$_x$:H .

As mentioned previously, the heterojunctions between amorphous semiconductors studied until now exhibit small valence-band discontinuities, very close to the experimental limit of the technique (a notable exception

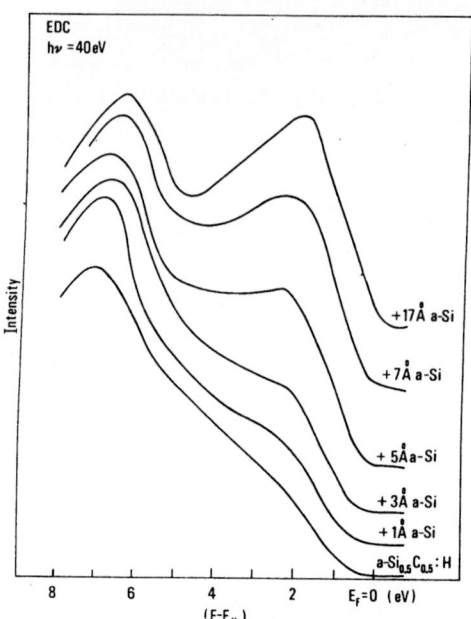

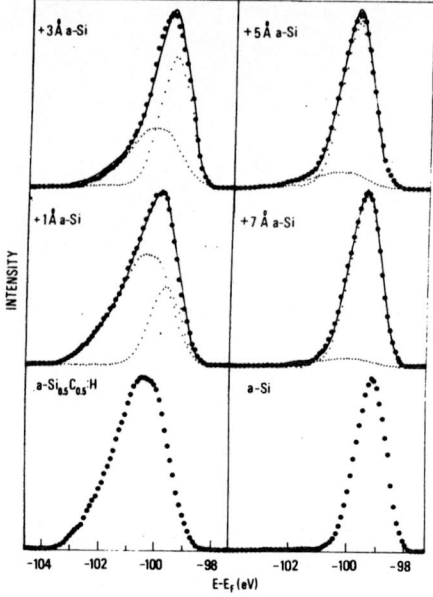

Fig. 3 Photoelectron energy
distribution curves (EDC's)
of a clean a-Si_xC_{1-x}:H substrate covered by a-Si
overlayer of increasing
thickness. Energy scale
is referred to the Fermi
level E_F.

Fig. 4 Si 2p spectra for clean
and a-Si covered a-Si_xC_{1-x}:H
taken at hν = 135 eV. The
solid lines correspond to
the results of the fitting
procedure explained in the
text.

is the system a-SiN_x:H/a-Si:H at large nitrogen concentration[6]). As a
consequence, at small overlayer-thicknesses the configuration sketched in
Fig.1 is not found experimentally and the band discontinuity cannot be inferred unambiguously from the inspection of the valence band edge. Quite
to the contrary, the edge evolves continuously from that of the substrate
to that of the overlayer. An example is shown in Fig.3 for the system
a-Si_xC_{1-x}:H/a-Si. The silicon-carbon alloy was grown by glow-discharge
in a mixture of 1:3 SiH_4 to CH_4. The atomic composition was determined by
Auger spectroscopy and a relative concentration x=0.4-0.5 was estimated.
It is not certain, however, whether this concentration reflects the bulk
composition or is the consequence of a preferential removal of one component by Ar sputtering used to clean the surface. Two features are evident
in Fig.3 for increasing overlayer thickness: the increase of the first
structure (extending in the range 0-5 eV) due to Si 3p bonding states and
the movement of the onset toward the Fermi level. However, from the behavior of the valence band top is not possible to establish whether a discontinuity is developing or a simple band bending is occurring. In order
to distinguish between these two possibilities it is convenient to analyze
the core level spectra. The Si 2p core level EDC are shown in Fig.4. The
photon energy (hν = 135 eV) was selected to obtain a photoelectron escape
depth similar to that of the valence bands of Fig.3. The two bottom
curves refer to the clean substrate and to a thick a-Si overlayer respectively. Notice the larger linewidth and the energy shift of the
a-Si_xC_{1-x}:H peak with respect to the a-Si peak. The lineshape of the
a-Si_xC_{1-x}:H core is due to the simultaneous presence of large amounts of C
and H which introduce a large compositional disorder and a variety of

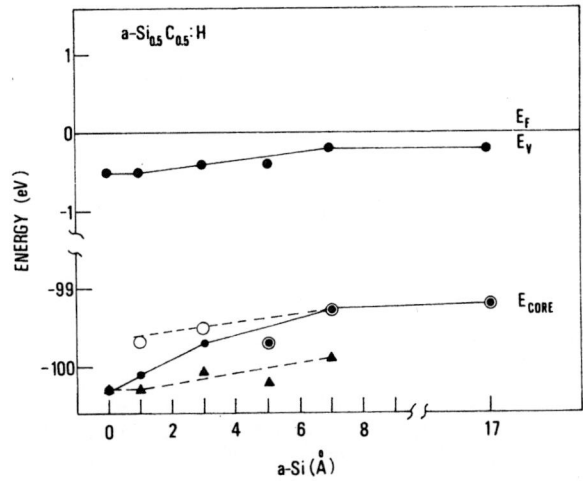

Fig. 5 Position of the valence-band maximum
and of Si 2p core level (dots and solid
lines) as a function of a-Si coverage.
Open circles and triangles refer to the
two components of the core level decon-
volved with the procedure explained in
the text.

bonding configurations for the Si atoms. The resulting lineshape is the
envelope of different Si 2p peaks affected by different chemical shifts.
The middle and top spectra of Fig.4 show the evolution of the linshape
from that of a-Si_xC_{1-x}:H to that of a-Si for progressive a-Si coverages.
These spectra were fitted by a linear superposition of the a-Si and
a-Si_xC_{1-x}:H lineshapes with fixed energy separation, using the peak intensities as fitting parameters. The solid lines show the results of the
fitting procedure and the dotted lines show the two components for each
spectrum.

The results of the analysis of valence-band and core level spectra
can be summarized as shown in Fig.5. The top part shows the shift of the
valence-band maximum as a function of a-Si coverage. The bottom part of
the figure shows the position in energy of the two components of the fit
(dashed lines) together with the overall shift of the Si-2p (solid line).
All energies are reported as a distance from the constant Fermi level.
Notice that the two components of the Si-2p band closely follow the shift
of E_V . This demonstrates that all these shifts are simply due to changes
in the band bending during the a-Si deposition and that there is no discontinuity in the valence band.

Identical results are obtained when the a-Si_xC_{1-x}:H is doped by adding
0.1% of B_2H_6 to the gas mixture. This doped material, with $E_F \sim 0.3$ eV
above the top of the valence band, is the one which is actually used as a
p-doped layer in the p-i-n photovoltaic devices.

By combining the photoemission information with the knowledge of the
optical gap and Fermi level position in the bulk, a complete characterization of the heterostructure during its formation is obtained, as sketched
in Fig.6. Notice the p-character of the clean surface, a result found
consistently in the hydrogenated samples investigated and probably due to

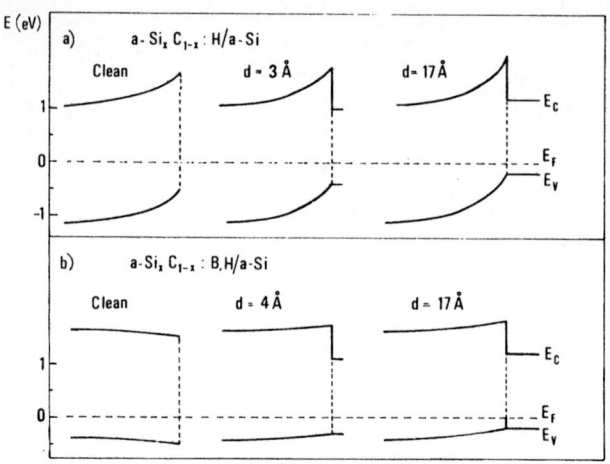

Fig. 6 Scheme of the heterojunction formation
showing the change of surface band
bending and the band discontinuities.

the sputtering procedure used for cleaning the samples which introduces
surface states that pin the surface Fermi level close to the valence band
maximum. The density of these surface states was estimated by Wagner et
al.[9] for a-Si:H and found to be 10^{13} cm^{-3} eV^{-1}. For the boron doped material, however, the distance $E_F - E_v$ at surface is larger than that in
the bulk causing a depletion layer for holes. With increasing overlayer
thickness the p-character is enchanced and results in a change from depletion to accumulation of holes.

As an example we have discussed in same detail the results obtained
for the heterostructure a-Si$_x$C$_{1-x}$:H/a-Si. However, the
a-Si$_x$C$_{1-x}$:H/a-Si:H system too has been studied[4,6] and a negligible
ΔE_v found. It seems, therefore that alloying the amorphous hydrogenated
silicon with a percentage of C not higher than 40-50% does not cause any
substantial recession of the valence-band maximum, although an appreciable
depletion of states in the Si 3p region is evident. The negligible magnitude of ΔE_v implies that the difference between the two pseudogaps is
entirely accomodated by the conduction band discontinuity, as sketched in
Fig.6.

It is worth emphasizing that the knowledge of the band discontinuities is fundamental to the understanding of the behavior of a heterostructure device. In particular, the use of a-Si$_x$C$_{1-x}$:H as a p-doped layer in
the p-i-n photovoltaic devices resulted in a substantial improvement of
the efficiency of amorphous silicon solar cells.[3] Besides the beneficial
window effect due to the larger optical gap, the band alignment we have
shown suggests the presence of a second advantage, i.e. the conduction
band discontinuity hinders the back-diffusion of the electrons while the
hole collection is not affected at all due to the negligible ΔE_v .

The interfaces between Si and Ge have been extensively investigated
in the crystalline case[10] and are one of the test systems[11] for the theories on the heterojunction formation. As a consequence, the study of
their amorphous counterpart, besides new information on these disorder
systems, can provide new insight on the driving force that determine the

462

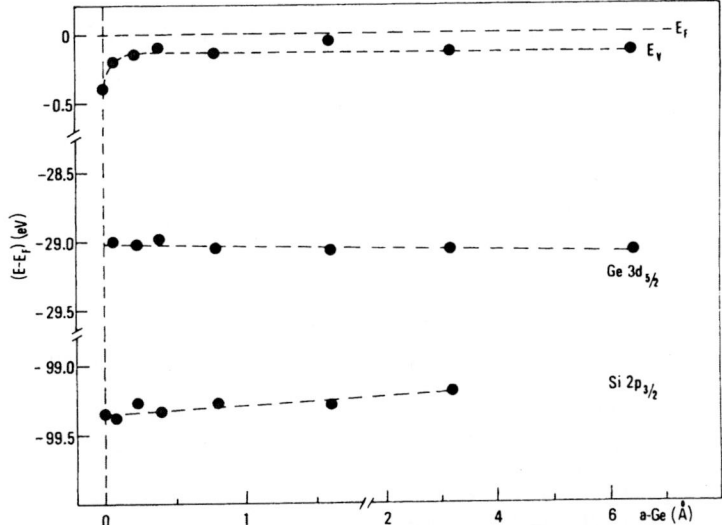

Fig. 7 Position of the valence-band maximum,
Si 2p and Ge 3d core levels as a
function of a-Ge coverage for the
a-Si:H/a-Ge heterojunction.

band alignment, diffusion potential, etc. Moreover, the study of the
amorphous Si/Ge interface has an immediate technological interest, since
amorphous hydrogenated silicon-germanium alloys are used in photovoltaic
devices to obtain an absorbing layer with optical gap smaller than that of
a-Si:H. The band discontinuity of the amorphous Si/Ge system gives an
upper limit to the discontinuities that can occur at the interface between
silicon and the alloys.

The results for the a-Si:H/a-Ge heterojunction[12] can be summarize as
shown in Fig.7, where the energy of the valence band maximum E_V, of the
Ge 3d and the Si 2p core levels are shown as a function of the overlayer
thickness. A shift of E_V by 0.2 eV occurs upon deposition of a fraction
of monolayer, while the Si 2p does not move, indicating that a valence
band discontinuity is setting in. By increasing the overlayer thickness,
a slight decrease of E_V -Si 2p distance is detectable, without any displacement of the valence-band maximum and of the Ge 3d core level. It is
a small effect, close to the sensitivity of the technique, nevertheless,
it points to a complex evolution in the interface formation. It is worth
remarking that the valence band discontinuity is practically established
at the very beginning of the overlayer growth. This surprising and somewhat mysterious behavior is consistently found in the study of crystalline
heterjunctions as well as of the Schottky barrier formation.[1]

Quite to the contrary, no valence band discontinuity is found for
a-Ge grown on a-Si.[12] This result must be compared with the crystalline
counterpart where ΔE_V = 0.17 eV is found.

One particularly interesting aspect of the research on amorphous heterojunctions is the possibility to investigate the effects of disorder
and/or hydrogenation on the valence band edge of the semiconducting material. This is obtained by studying an amorphous overlayer of a given material on the crystalline or hydrogenated amorphous counterpart. This investigation has been performed on c-Si/a-Si and a-Si:H/a-Si systems[5].
The results are summarized in Fig.8.

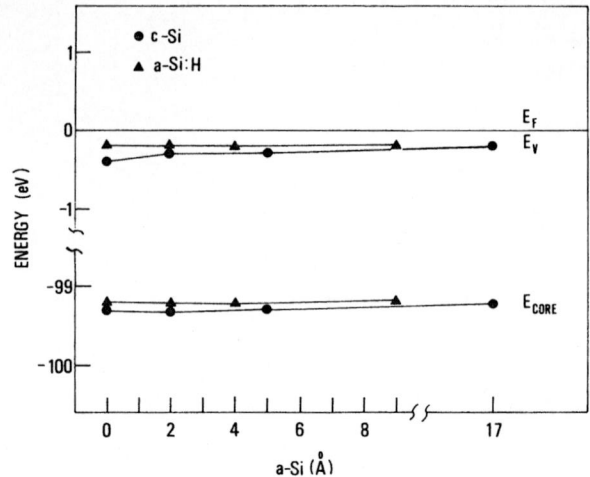

Fig. 8 Position of the valence-band
maximum and of Si 2p core level
for a-Si:H and c-Si as function
of a-Si coverage.

The data on c-Si/a-Si interface suggest a possible upward shift of 0.1 eV which is, however, comparable to the experimental uncertainty. This result clearly demonstrate that the valence-band mobility-edge coincides within 0.1 - 0.2 eV with the valence-band maximum of the corresponding crystal. Since it is commonly estimated[13] that the mobility gap in amorphous silicon is larger than the forbidden gap of the crystal, it follows that the conduction-band mobility-edge should undergoes the larger shift.

A similar situation results from the study of the a-Si:H/a-Si system. Fig.8 shows that E_V of amorphous hydrogenated silicon is aligned to that of a-Si. The data refer to samples with an hydrogen content of ~10 %. By comparing the distance of E_V from the Si 2p core level the same conclusion can be inferred from data published recently.[14,15] However, from ref.15, it is evident that at far larger hydrogen concentration (C_H >25%) the valence-band maximum moves closer to the core level, i.e. there is recession toward higher binding energy. On the other hand, at 10 % hydrogen content the optical gap of a-Si:H is already much larger than that of a-Si (~1.7-1.8 eV compared to ~1.2 eV). The recent photoemission data show, therefore, that this widening of the optical gap is primarily due to a shift in energy of the conduction band edge.

It is instructive to compare some of the experimental results reviewed above with the current theories on the heterojunctions. For crystalline substrates, the problem of the interface formation was the subject of many theories and experiments over the past 20 years. It cannot be regarded as completely solved, however. For amorphous substrates, the problem is even harder due to the presence of several kind of disorder in the matrix.

From the theoretical point of view, there are few microscopic band calculation trying to determine the local electronic structure of the interface in detail, by taking into account the local structure, coordination, etc.. None computes ΔE_V for the crystalline counterpart of the

464

systems we have discussed. On the other hand, there are several theories trying to calculate the band lineup from bulk crystal properties by suggesting different "driving forces" for the alignement. The first example of these theories is the well-known Anderson's "electron affinity rule",[16] which states that the conduction-band discontinuity is equal to the difference between the electron affinity of the constituent semiconductors. This rule, which proved hard to test experimentally due to the lack of reliable values for the electron affinity within an accuracy of a tenth of eV, was later criticized because it uses free-surface parameters to describe interface parameters. Presently, there are several different suggestions for the "driving force" determining ΔE_v.[17-20] It was recently shown by Katnani and Margaritondo[11] that all the above models predict values for ΔE_v in unsatisfactory agreement with the experimental findings. However, the Harrison's and Frensley-Kroemer's models give the best agreement and the correct chemical trend within 0.2 eV, i.e. very close to the experimental uncertainty itself. More recently a new and interesting suggestion for the "driving force" for band alignement was put forward by Tersoff.[21] These three models, which are not necessarily in contrast, will be discussed briefly in the following.

Harrison used a tight-binding approach to calculate the absolute position of the valence-band maximum in the bulk relative to the vacuum level and argued that in a heterojunction ΔE_v is simply the difference between E_v of the constituent semiconductors. Let us apply this criterion to amorphous interfaces. Moving from crystalline to amorphous semiconductors the valence-band top is blurred by localized states induced by disorder. The width of this tail (which is mainly the result of fluctuation of the dihedral angle) is expected to be of the order of 0.3 eV.[22] The mobility edge should move over a distance less than that or not move at all. As a consequence within few tenth of eV uncertainty, the Harrison criterion is in agreement with the experimental results for c-Si/a-Si and a-Si/a-Ge heterojunctions.

In the Frensley-Kroemer's criterion the valence-band maximum is calculated relative to the average interstitial potential and it is assumed that this average potential coincide in the two semiconductors when the heterostructure is formed. Since in the amorphous case the average local coordination, bond length and bond angle remain unchanged just as the average valence density of states does, we expect that the average interstitial potential too does not change. Therefore, the only difference between amorphous and crystalline case is again a slight shift of the mobility edge. The Frensley-Kroemer model also is in agreement with the experimental results.

Tersoff's criterion states that the interface dipole is the dominant factor determining band lineup and that the actual lineup is that which gives a zero interface dipole. In order to find this "canonical" lineup one has to find the demarcation energy E_B where the gap states shift from dominant valence (i.e. bonding) character to dominant conduction (i.e. antibonding) character. A zero-dipole lineup is obtained by aligning E of the respective semiconductors. This theory, which is remarcably successful in predicting the experimental ΔE_v for a series of crystalline heterojunctions, gives $\Delta E_v = 0.18$ eV for the c-Si/c-Ge system resulting from $E_B = 0.36$ eV above the valence-band maximum in c-Si and 0.18 eV in c-Ge, to be compared to $\Delta E_v = 0.17$ eV found experimentally. Therefore, the zero valence-band discontinuity found for the amorphous Si/Ge interface would mean that E_B is located at equal distance from E_v in the two amorphous semiconductors and that there is a displacement of the demarcation level between bonding and antibonding caracter in one or both semiconductors. On the other hand, the data on the c-Si/a-Si heterojunction are consistent with a displacement of 0.1 eV of E_B in a-Si bringing the expected ΔE_v for a-Si/a-Ge system to 0.08 eV, provided that E_B of Ge does not change. The study of the c-Ge/a-Ge system will be, therefore, an experimental test for the application of the theory to the amorphous semiconductors. It is worth noticing that the determination of E_B is of great importance because it directly impinges on the crucial problem of the extent of donor- and acceptor-like states into the pseudo-gap.

From this short discussion we have seen that the most successful theories on the heterojunction formation are consistent with the results obtained on the amorphous semiconductors. On the other hand, it would be even more interesting to use the above theoretical criteria to infer from the data new properties of the amorphous semiconductors, above all of the more complexes ones like a-Si:H and the hydrogenated alloys. However, to do so, we must be fully confident on the theory and, therefore, we have to wait until the problem of heterojunction formation can be considered completely solved, at least for the less difficult case, i.e. crystalline heterojunctions.

REFERENCES

1. G. Margaritondo, Solid State Electron. 26, 499 (1983), and references therein.
2. B. Abeles and T. Tiedje, Phys. Rev. Lett. 51, 2003 (1983).
3. Y. Tawada, H. Okamoto, and Y. Hamakawa, Appl. Phys. Lett. 39, 237 (1981); A. Catalano, R.V. D'Aiello, J. Dresner, B. Faughman, A. Firester, J. Kane, H. Schade, Z.E. Smith, G. Swartz and A. Triano, 16th IEEE Photovoltaic Specialists Conference, San Diego 1982.
4. F. Evangelisti, P. Fiorini, C. Giovannella, F. Patella, P. Perfetti, C. Quaresima, and M. Capozi, Appl. Phys. Lett. 44, 764 (1984).
5. F. Patella, F. Evangelisti, P. Fiorini, P. Perfetti, C. Quaresima, M.K. Kelly, R.A. Riedel, and G. Margaritondo, "Optical Effects in Amorphous Semiconductors", AIP Conf. Proc. 120, 402 (1984).
6. B. Abeles, I. Wagner, W. Eberhardt, J. Stohr, H. Stasiewski, and F. Sette, "Optical Effects in Amorphous Semiconductors", AIP Conf. Proc. 120, 394 (1984).
7. D. Adler and F.R. Shapiro, Physica 117B-118B, 932 (1983).
8. F. Evangelisti, F. Patella, R.A. Riedel, G. Margaritondo, P. Fiorini, P. Perfetti, and C. Quaresima, Phys. Rev. Lett. 53, 2504 (1984).
9. I. Wagner, H. Stasiewski, B. Abeles,, and W.A. Lanford, Phys. Rev. B28, 7080 (1983).
10. G. Margaritondo, N.G. Stoffel, A.D. Katnani, and F. Patella, Solid State Commun. 36, 215 (1980); P. Perfetti, N.G. Stoffel, A.D. Katnani, G. Margaritondo, C. Quaresima, F. Patella, A. Savoia, C.M. Bertoni, C. Calandra, and F. Manghi, Phys. Rev. B24, 6174 (1981).
11. A.D. Katnani and G. Margaritondo, Phys. Rev. B28, 1944 (1983).
12. F. Evangelisti, S. Modesti, F. Boscherini, P. Fiorini, P. Perfetti, and C. Quaresima, MRS 1985 Spring Meeting, San Francisco, 1985 and to be published.
13. M.H. Cohen, C.M. Soukoulis, and E.N. Economou, "Optical Effects in Amorphous Semiconductors", AIP Conf. Proc. 120, 371 (1984).
14. D. Wesner and W. Eberhardt, Phys. Rev. B28, 7087 (1983).
15. J. Reichardt, L. Ley, and R.L. Johnson, J. Non-Cryst. Solids 59-60, 329 (1983).
16. R.L. Anderson, Solid-State Electron. 5, 341 (1962).
17. W.R. Frensley and H. Kroemer, Phys. Rev. B16, 2642 (1977)
18. W. Harrison, J. Vac. Sci. Technol. 14, 1016 (1977).
19. M.J. Adams and A. Nussbaum, Solid-State Electron. 22, 783 (19749.
20. O. von Ross, Solid-State Electron. 23, 1069 (1980).
21. J. Tersoff, Phys. Rev. B30, 4874 (1984).

22. F. Yonezawa and M.H. Cohen, "Fundamental Physics of Amorphous Semiconductors", Ed. F. Yonezawa, Springer-Verlag, 1981, p.119.

INTERFACES BETWEEN CRYSTALLINE AND AMORPHOUS

TETRAHEDRALLY COORDINATED SEMICONDUCTORS

Frank Herman
IBM Research Laboratory
San Jose, California 95193

Philippe Lambin
Facultes Universitaires Notre-Dame de la Paix
5000 Namur, Belgium

ABSTRACT. By carrying out a series of geometrical operations, we transform a superlattice composed of alternating slabs of ordered and random close-packed spheres into a superlattice composed of alternating crystalline and amorphous tetrahedrally coordinated networks. In this way we generate an atomic-scale model of the interface between the [001] face of crystalline Si and amorphous Si. The construction is fully automatic, being determined by computer algorithms which take account of physical and chemical constraints such as tetrahedral coordination and most probable bond lengths and angles. The first stage of the construction leads to a preliminary structural model having interfacial transition zones extending over a few atomic layers. The density of dangling bonds increases rapidly as one moves across the interfacial regions from the crystalline toward the amorphous sides. Calculations of the local electronic density of states indicate the presence of states in the thermal gap at the interface and in the amorphous region. These gap states arise primarily from dangling bonds, and to a lesser extent from structural disorder, i.e., statistical variations in bond geometries. Because of the conventions used to enumerate bonds, the preliminary model appears to have an excessive number of dangling bonds. Most of these dangling bonds are eliminated in the second stage, as neighboring pairs of unsaturated atoms are linked together, additional atoms are introduced into accommodating voids, and the lattice is concurrently relaxed.

INTRODUCTION

Ball-and-stick models of disordered bulk solids[1,2] and of interfaces between ordered and disordered materials[3-5] have been constructed by hand by various investigators. We do not regard these models as trustworthy because they represent the outcome of a long series of subjective and largely arbitrary human judgments. Although hand-crafted structures can be improved by subsequent computer-assisted lattice relaxation, such refinements are usually not designed to modify the intrinsic topology or connectivity. Our principal concern is that a structure that satisfies one's own preconceived notions can usually be hand-fashioned. For example, a model of amorphous Si containing no dangling bonds can be built in the form of a continuous random network,[1,2] but is this what actually occurs in nature?

Again, starting out with the premise that the Si/SiO_2 interface is atomically abrupt and essentially free of dangling bonds, it is possible to construct a model by hand having these properties.[4] Actually, this is quite a feat, demanding a considerable amount of experimentation and ingenuity in arranging and rearranging the structure so as to eliminate all dangling bonds. The end result is an abrupt interface between the [001] face of crystalline Si and a continuous random SiO_2 network. Unfortunately, this interface model is inconsistent with experiment,[6] which indicates that the interface between Si and SiO_2 is several atomic layers wide, extending a few atomic layers into both the Si and the SiO_2. The detailed nature of the Si/SiO_2 interface remains a mystery.

We thought it would be instructive to develop a computational method for building interfaces between crystalline and amorphous semiconductors that would be free of subjective influences. Our approach is based on a set of computer algorithms that express plausible physical and chemical constraints, such as tetrahedral coordination and most probable bond geometries. Since some aspects of the construction are random, the structures we obtain are not unique, but form a statistical ensemble. Although the detailed atomic arrangements may vary considerably from structure to structure, some features, such as the range of the interface and the spatial distribution of dangling bonds, are common to all generated structures.

As a first step, we generate Lennard-Jones crystal/melt interfaces by Monte Carlo simulation. This leads to adjoining regions of ordered and disordered close-packed spheres, which are then transformed into adjoining ordered and disordered covalently bonded networks. The geometrical transformation[7] is similar to that used by earlier workers to generate bulk amorphous Si clusters from random close-packed configurations,[8,9] We will illustrate how this process leads to an atomic-scale model of the interface between c-Si and a-Si, though our structural results apply equally well to C, Ge, and grey Sn. The first stage of the construction was discussed briefly in an earlier publication,[10] and is discussed more fully in the present paper. The second stage, which is still under development at this writing, is outlined here, and some preliminary results are reported. A more complete discussion of the computer algorithms underlying the first-stage construction will be published elsewhere.[11]

ORDERED AND DISORDERED CLOSE-PACKED STRUCTURES

One of the key ideas underlying the construction is that it is considerably easier to generate a system of co-existing ordered and disordered close-packed spheres than it is to generate co-existing ordered and disordered tetrahedrally coordinated networks. Accordingly, we will start out with arrays of close-packed spheres and transform these into arrays of covalent networks. The second key idea is that all regions in our system — crystalline, amorphous, and interfacial — are characterized by tetrahedral coordination. Therefore, nearly all the atoms should be surrounded by atomic tetrahedra whose shape may be regular or irregular depending on their location.

In order to explain the rationale behind the geometrical transformation that lies at the heart of our treatment, let us momentarily work backwards, and begin with a hypothetical composite ordered/disordered covalently bonded structure that represents the ultimate goal of our work. Removing the atoms lying at the centers of (roughly) half the tetrahedra, and then enlarging the remaining atomic spheres so that they form a close-packed array, we are led to adjoining regions of ordered and disordered close-packed spheres. In the ordered regions, we have simply removed one of the two fcc lattices that together form the diamond structure. The treatment of the disordered regions is geometrically analogous to that of the ordered regions, at least in any local neighborhood.

In our actual construction, we begin by generating ordered and disordered arrays of close-packed spheres by computer simulation, using as our prototype the crystal/melt interface of a Lennard-Jones system,[12] for which solid/liquid co-existence data are available.[13] Having established this starting configuration by computer simulation, we then shrink the close-packed atomic spheres to a size appropriate to Si, say, and then put back the Si atoms that we had imagined removing earlier. Proceeding in this manner, we expect to obtain a final structure composed of co-existing ordered and disordered covalently bonded networks. The challenge, of course, is to find ways of putting back as many atoms as possible consistent with tetrahedral coordination and realistic bond lengths and angles. The adequacy of the solutions for the amorphous regions can be tested by comparing such quantities as the radial distribution function and atomic density with experiment. Once there is reasonable agreement between theory and experiment for the amorphous regions, we would expect our predictions for the interfacial regions to be reasonable as well.

Instead of dealing with an isolated interface between ordered and disordered regions, however, we consider a superlattice composed of alternating slabs of ordered and disordered close-packed spheres. These slabs may be regarded as co-existing frozen and melted regions of systems governed by a Lennard-Jones interatomic potential having the form $V(r) = 4\varepsilon[(\sigma/r)^{12} - (\sigma/r)^6]$, where ε and σ are the units of energy and length, respectively. The starting configuration is a fcc lattice composed of 27 [001] planes, each containing 50 atoms. We use a fcc lattice because we want the ordered slabs to have the diamond structure eventually. The choice of parameters is based on co-existence data[13] for the Lennard-Jones system at the temperature $kT=1.35\varepsilon$. In the ordered regions (planes 1 to 5 and 23 to 27), the density was set equal to $\rho_s=1.053\sigma^{-3}$, and in the disordered region to $\rho_{liq}=0.964\sigma^{-3}$, the density of the liquid phase. The corresponding interplanar spacings are 0.780σ and 0.852σ. The 400 atoms in planes 1 to 4 and 24 to 27 are held fixed at their fcc positions, while the positions of the 950 atoms in the remaining planes are rearranged according to Monte Carlo techniques, with periodic boundary conditions imposed in directions parallel to the interfaces.

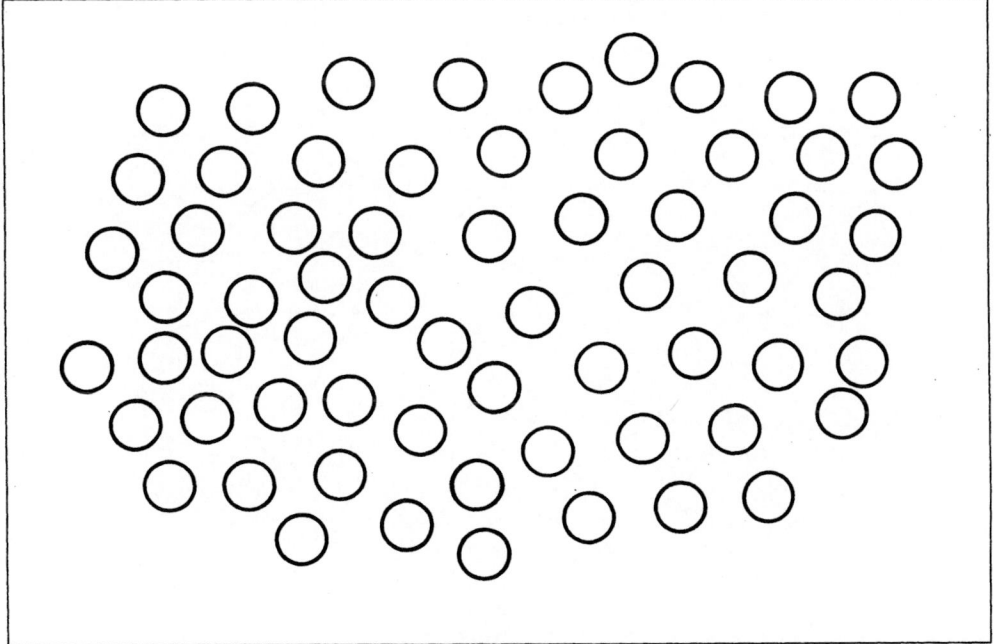

Fig. 1. Randomly arranged spheres in disordered region, shown in two dimensions.

A representative segment of the disordered region is shown schematically in Fig. 1. For simplicity, we will use two-dimensional drawings to illustrate the various steps of our construction. The actual construction is, of course, three-dimensional. In the actual model, the spheres are more closely packed than indicated in the drawing. The centers of these spheres define the vertex atom positions in subsequent drawings.

TETRAHEDRAL NETWORKS AND SIMPLICIAL GRAPHS

Once the crystal/melt/crystal Lennard-Jones system has been created, we suppress periodic boundary conditions in planes parallel to the interfaces for computational convenience. We then partition the three-dimensional system into space-filling tetrahedra by connecting the center of each sphere to the centers of neighboring spheres, making sure that each tetrahedron lies outside all other tetrahedra.[11] The lines and vertices forming the network of space-filling tetrahedra constitute the so-called Delauney or simplicial graph.[1,2,7,14] For our system, this graph contains about 6700 tetrahedra. The analogous simplicial graph in two dimensions is the network of triangles shown in Fig. 2.

Each of the tetrahedra in the simplicial graph is then examined, and those having highly irregular shapes are discarded so as to avoid unphysical atomic arrangements. (The way this is done is indicated below.) After eliminating the ill-shaped tetrahedra, we select a subset of the remaining tetrahedra which satisfy the condition that adjoining tetrahedra touch only at their common vertices. We will call this residual tetrahedral network the simplicial subgraph. Thus, tetrahedra belonging to the subgraph share vertices with one another, but not edges or faces. Many of the connecting lines that belong to the simplicial graph do not belong to the simplicial subgraph, as is illustrated in Fig. 3 for the two-dimensional case. In this figure, the discarded connecting lines are seen to be edges that are shared by the shaded triangles, which lie outside the area enclosed by the unshaded triangles constituting the simplicial subgraph. The reader will find it instructive to construct alternate two-dimensional subgraphs by starting off with a triangle that in Fig. 3 is shaded. How many different subgraphs can be generated?

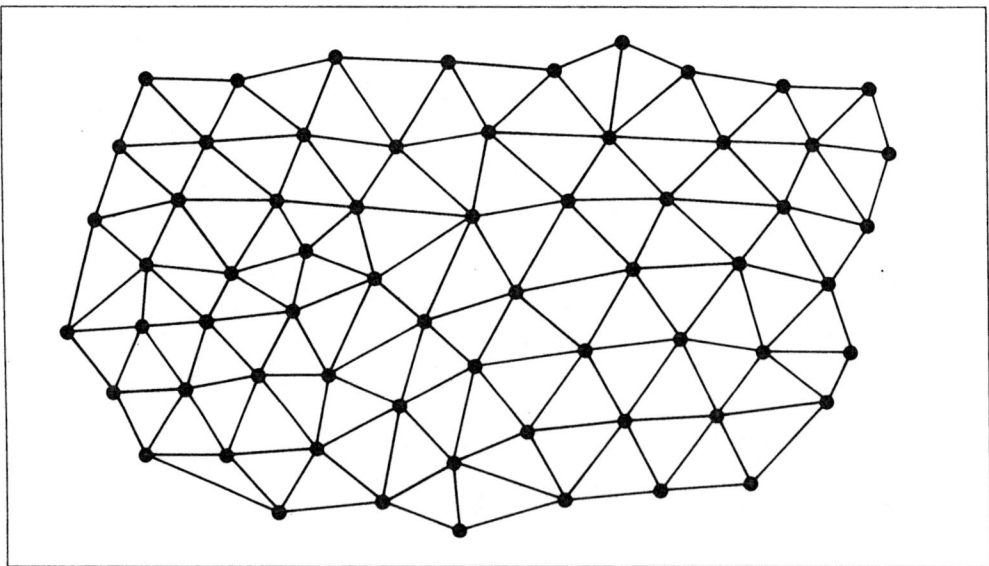

Fig. 2. Simplicial graph in two dimensions, obtained by connecting neighboring centers to form triangular network.

To illustrate some of these ideas in three dimensions more fully, let us construct the simplicial graph and subgraph for a fcc lattice. For this particular lattice, it is actually easier to construct the subgraph first and then the graph, rather than the other way around. Let us begin by considering a diamond lattice, which consists of two interpenetrating fcc lattices, which we will call A and B. We will construct the simplicial graph and subgraph for A, using B as a guide or template, and then we will discard B, since it has served its purpose.

As a first step, let us choose some particular atom on the B lattice — call this B_0 — and let us connect the four nearest neighbors of B_0 to one another. In this way we form a regular tetrahedron from atoms on the A lattice, with atom B_0 at its center. Repeating this process for all atoms on B, we generate a set of regular tetrahedra which touch one another only at their vertices, these always being points on A. This set of regular tetrahedra — the vertices and the connecting lines — is the simplicial subgraph of fcc lattice A. We can now remove fcc lattice B since this only served to identify the centers of the regular tetrahedra. Note that the construction of the simplicial subgraph of fcc lattice A is unique.

The regular tetrahedra belonging to the simplicial subgraph are separated from one another by regular octahedra. In a suitably chosen coordinate system, the six corners of one of these regular octahedra lie at $(-a/2,0,0)$, $(a/2,0,0)$, $(0,a/2,0)$, $(0,-a/2,0)$, $(0,0,a/2)$, and $(0,0,-a/2)$, where a is the unit cube edge. Each of these octahedra can be decomposed into four irregular tetrahedra by connecting a pair of opposite vertices, e.g., $(-a/2,0,0)$ and $(a/2,0,0)$. Note that this can be done three different ways because there are three different sets of opposite vertices. The simplicial graph consists of the simplicial subgraph previously constructed plus these additional connections between next nearest neighbors. These additional connections do not belong to the simplicial subgraph because they represent edges of (irregular) tetrahedra which share edges as well as faces with one another and also with the simplicial graph tetrahedra. Note that the construction of the simplicial graph is not unique because there are three different ways to decompose each regular octahedron into four irregular tetrahedra.

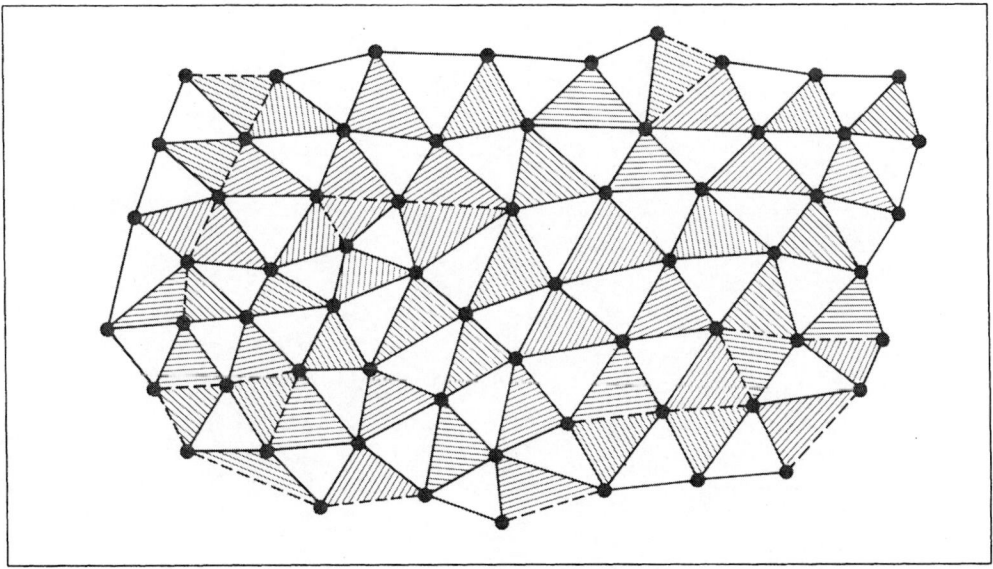

Fig. 3. Simplicial subgraph in two dimensions, composed of (unshaded) triangles which touch one another only at their vertices. The dashed lines are discarded connections that do not belong to the subgraph.

Returning to the more general case of adjoining ordered and disordered regions, we note that in the ordered regions the simplicial subgraph can be taken as either of the two fcc sublattices. As we move across the interfaces into the disordered regions, extending the subgraph by adding acceptable tetrahedra to the ordered subgraph, many different subgraphs can be generated because of the many ways that tetrahedra can be added. Our algorithms[11] do not lead to a unique subgraph for the composite structure, but rather to members of an ensemble. Some statistical features of this ensemble will clearly depend on physical factors such as the Lennard-Jones interaction parameters and the atomic density chosen for the disordered region. Other features will depend on the construction algorithms. Fortunately, statistical averages over the ensemble appear to be relatively insensitive to algorithmic details and to such design features as the number of atoms included in the supercell.

DECORATING THE SIMPLICIAL SUBGRAPH

Having constructed a particular simplicial subgraph for the composite system, we now decorate it by placing Si atoms at the vertices and centroids of all tetrahedra belonging to this subgraph. Holding the positions of the vertex atoms fixed, we then optimize the positions of the centroid atoms by minimizing the Keating elastic energy[15] of each five-atom cluster composed of the four vertex atoms of a given tetrahedron and its centroid atom. This is done for each tetrahedron independently, using elastic energy parameters appropriate to Si. The Keating expression involves bond stretching and angle bending terms, and so serves to optimize bond lengths and bond angles. The same procedure was used earlier to eliminate ill-shaped tetrahedra. There, tetrahedra were discarded if their minimum elastic energies were larger than some specified threshold energy.

Note that this procedure does not optimize the positions of the vertex atoms, which are determined by the frozen and melted Lennard-Jones systems from which we started. Moreover, the positions of the centroid atoms are optimized with respect to their nearest neighbors, but not with respect to more distant neighbors.

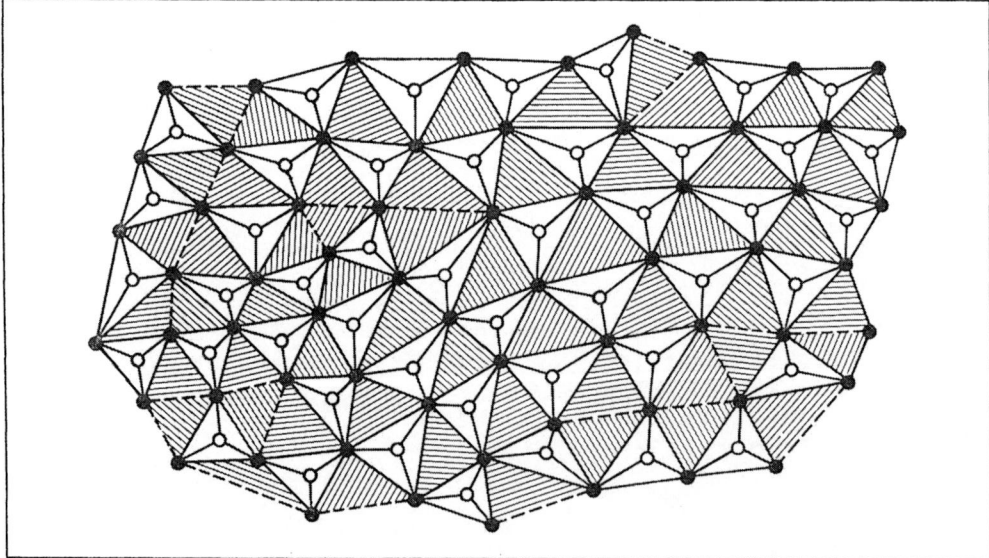

Fig. 4. Covalently bonded structure in two dimensions, with vertex atoms represented by closed circles and centroid atoms by open circles. The drawing represents the unrelaxed structure.

The decoration of the simplicial subgraph is completed by drawing nearest-neighbor bonds between vertex and centroid atoms. For our system, the simplicial subgraph contains about 800 tetrahedra having about 2,100 atoms at the vertices and centroids. In the ordered regions, this procedure generates a diamond lattice, with each centroid atom and each vertex atom lying at the center of a regular tetrahedron. Thus, atoms located in planes 1 to 4 and 24 to 27 of the initial fcc configuration are automatically arranged in the diamond structure. The choice of geometry clearly insures the creation of the [001] c-Si/a-Si interface.

In the disordered regions and at interfaces, this procedure generates a discontinuous random network, that is to say, the structure contains dangling bonds which at best are isolated from one another, and at worst are neighbors, forming voids of various sizes and shapes. In three dimensions, some of the vertex atoms are connected to less than four centroid atoms, while each centroid atom is necessarily connected to four vertex atoms (by construction). The analogous distinction between vertex and centroid atoms in two dimensions is clearly seen in Figs. 4 and 5, where some vertex atoms are connected to less than three centroid atoms, while all centroid atoms are connected to exactly three vertex atoms.

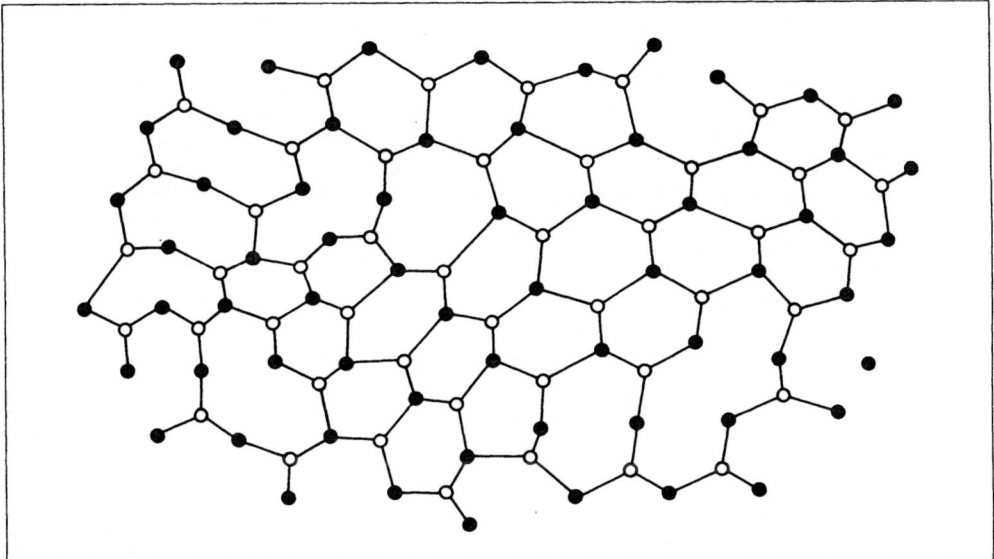

Fig. 5. Covalently bonded structure in two dimensions, with triangular construction lines and shading removed. The drawing represents the unrelaxed structure.

PRELIMINARY MODEL: STRUCTURAL RESULTS

The entire process just described was repeated five times in order to generate five independent final configurations. The mean interatomic distance in the amorphous phase was found to be 0.701σ, with an r.m.s. deviation of 0.068σ. The corresponding values for the bond-angle distribution were 110° and 12°. The Si-Si distance in the crystalline phase is 0.676σ, so the crystalline phase is slightly more dense than the amorphous phase. This result is due partly to the presence of voids and dangling bonds in the amorphous region, and partly to the initial choice of the interplanar spacings in the ordered and disordered regions. (We used slightly different interplanar spacings initially in order to insure melting of the disordered region.) In Fig. 6 we show connectivity (coordination number) and atomic density profiles obtained by averaging the results for the five final configurations.

The principal conclusions that can be drawn from Fig. 6, and from an examination of the structural model, are the following:

- The construction leads only to even-membered rings. Topologically, this is a consequence of the fact that the structure is composed of two interpenetrating sublattices,[7] namely, the vertex-atom and centroid-atom sublattices.
- The connectivity (average number of bonds per atom) is four in the ordered region, as it should be. In the amorphous regions, the connectivity is less than four because of the presence of dangling bonds and voids.
- The connectivity alternates from layer to layer in the disordered regions, particularly near the interfaces, because these layers are occupied preferentially by vertex and centroid atoms, which are introduced into the structure differently.
- The alternation in connectivity from layer to layer decreases as one moves into the interior of the disordered region because of the increasing importance of spatial averaging.
- In the interior of the disordered region, the average connectivity is three. Since the connectivity of the centroid atoms is everywhere four, the connectivity of the vertex atoms is only about two well inside the disordered region.
- Connectivities as small as two are more a reflection of the conventions (algorithms) used to define bonds than an indication of serious deficiencies in the model. Large numbers of neighboring pairs of dangling bonds will be eliminated during the second stage of construction as more sophisticated algorithms are used to identify such pairs and to relax the structure concurrently.
- As can be seen from both the connectivity and atomic density profiles, the interface extends over a few atomic layers. This range is determined by the nature of the original crystal/melt interface, and as such is physically reasonable.

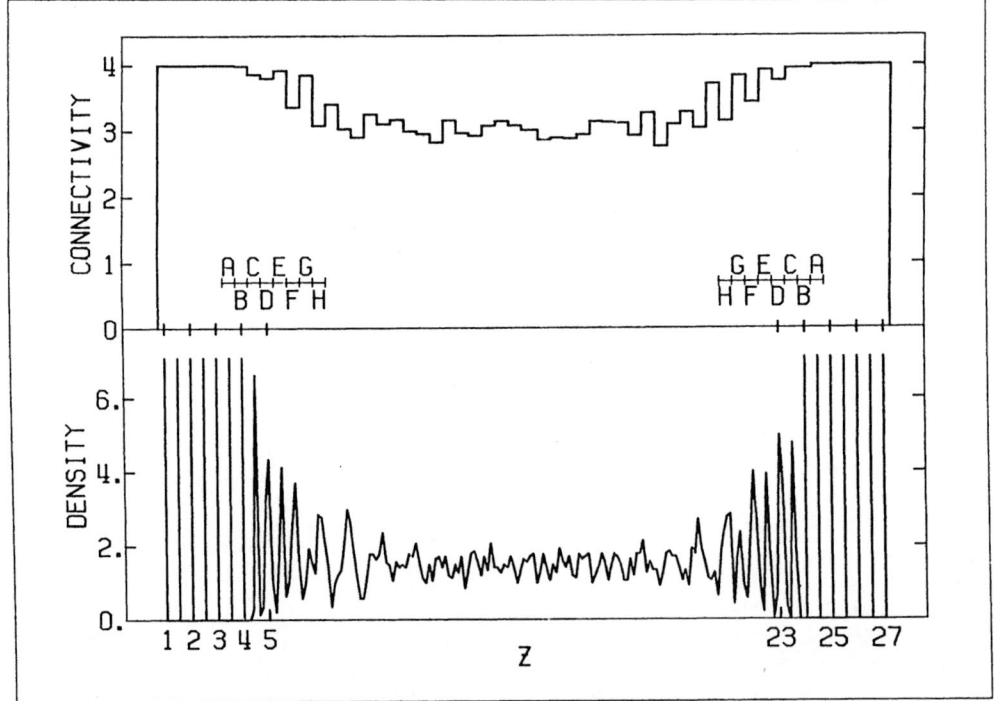

Fig. 6. Profiles of coordination number and atomic density (in units of atoms per σ^3) averaged over five configurations. Lateral boundary effects are avoided by ignoring atoms lying near the surface. Face-centered cubic planes in the initial configuration are labeled 1 through 27.

ELECTRONIC DENSITY OF STATES AT THE INTERFACE

The local density of states (LDOS) for various planes was calculated using a first-neighbor tight-binding Hamiltonian whose two-center integrals were obtained by fitting a recent LCAO band structure of c-Si.[16] Periodic boundary conditions were applied only in the direction perpendicular to the interfaces. In order to avoid boundary effects, the only atoms considered were those lying far away from the lateral surfaces. The moments of the LDOS were computed using the recursion method. All of this was done for each of the five final configurations, and the results were then averaged. The results are shown in Fig. 7.

Since the number of moments included was small, the energy resolution of the LDOS is relatively low. Nevertheless, the existence of states in the gap is clearly seen. The states in the gap are due primarily to dangling bonds, and to a lesser degree to lattice distortions. As one moves across the interface from the crystalline toward the amorphous region, the gap state density increases, the range of variation being a few atomic spacings. There is again a slight alternation from layer to layer due to the preferential occupancy of these layers by vertex and centroid atoms.

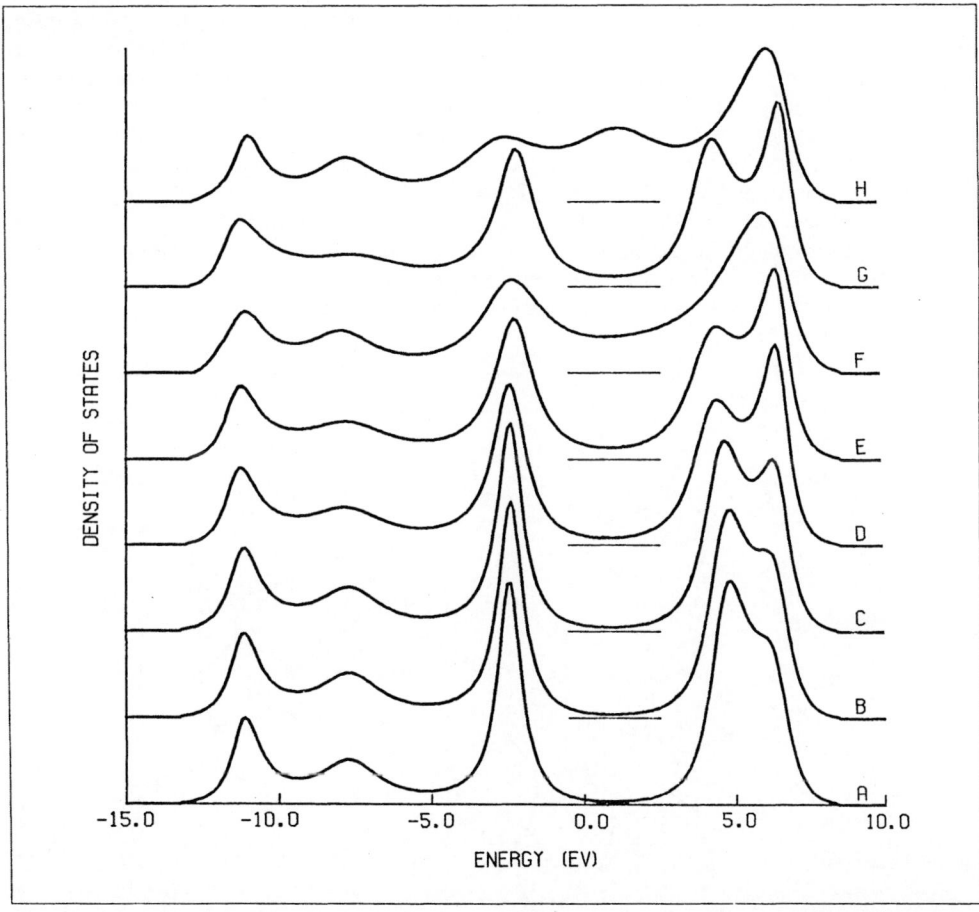

Fig. 7. Local electronic density of states, based on low-resolution LCAO/recursion method calculations, averaged over five configurations. Panels A through H correspond to the intervals identified in Fig. 6. The zero of energy is placed at the top of the valence band of c-Si ($\Gamma_{25'}$). The states in the gap are clearly seen above the zero of energy.

SECOND STAGE OF THE CONSTRUCTION

As already noted, the excessive number of dangling bonds in the preliminary model is only apparent, being a consequence of the conventions used to identify interatomic bonds. Most of these dangling bonds will disappear during the second stage as more sophisticated algorithms are used to identify interatomic bonds and otherwise improve the structure. Briefly, the second stage involves the following:

- Pairs of atoms that are reasonably close together and contain dangling bonds are connected to one another, eliminating the dangling bonds.
- Additional atoms are introduced into voids bounded by unsaturated atoms, thereby increasing the atomic density and reducing the number of dangling bonds.
- The structure is relaxed locally in order to minimize the elastic energy in each neighborhood.
- The average layer spacing in the disordered regions is varied in order to minimize the total elastic energy of the system.
- Periodic boundary conditions are used systematically throughout.

The most important improvement involves the elimination of neighboring pairs of dangling bonds, as is illustrated for the two-dimensional case in Fig. 8. The dangling bonds necessarily belong to unsaturated vertex atoms, since all centroid atoms are fully saturated, being 3-fold and 4-fold coordinated in two and three dimensions, respectively. In the unrelaxed structure, it is usually a simple matter to identify and eliminate closely spaced pairs of dangling bonds, provided these are well isolated from other dangling bonds. The local structure can then be relaxed, improving the bond geometries in that neighborhood. However, there are frequently many dangling bonds in a given neighborhood, so that there may be many different ways that neighboring pairs of unsaturated atoms can be joined to one another. Once a particular attachment is made, however, others are ruled out. As in a chess game, it is desirable to examine the consequences of several possible sequences of moves before deciding to pursue any one of them.

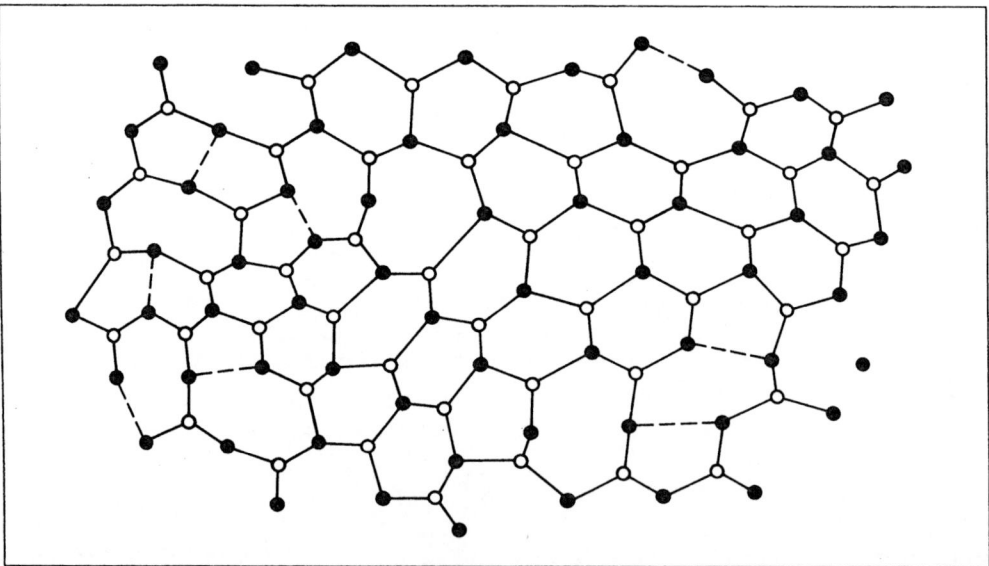

Fig. 8. Covalently bonded structure in two dimensions, indicating one of several ways of eliminating neighboring pairs of dangling bonds in the unrelaxed structure shown earlier (cf. Fig. 5). Dangling bonds that have been joined together are shown here by dashed lines.

Some attachments might eliminate a larger number of dangling bonds but inhibit subsequent lattice relaxation, while other attachments might eliminate fewer dangling bonds but pave the way for considerable lattice relaxation. It is clearly desirable to devise algorithms for eliminating dangling bonds and relaxing the lattice concurrently, taking possible energy tradeoffs into account. To do this properly, one must minimize the total energy, including not only the elastic energy, but also the electronic energy associated with the making and breaking of bonds. Minimizing only the elastic energy is not likely to induce topological changes.

Simplicial graphs and subgraphs clearly play a central role in the first stage of our construction. What role can these graphs and subgraphs play in the second stage? Returning to Figs. 3 and 4, we note that certain lines belonging to the simplicial graph are not included in the simplicial subgraph. The discarded lines are shown dashed in these figures. If we now examine the new connecting lines in Fig. 8, which are also dashed, we see at once that some of the discarded lines of Figs. 3 and 4 become the new connecting lines in Fig. 8. Since the first stage generates a complete catalogue of discarded lines (or bonds), we can begin the second stage by examining the suitability of using these discarded bonds to join neighboring pairs of unsaturated vertex atoms. Thus, many of the "virtual" bonds that were discarded during the first stage can be incorporated into the structure during the second stage.

In extreme cases, unsaturated atoms may surround a void which can accommodate one or more new atoms. It may be possible to reduce the net number of dangling bonds by introducing new atoms into the voids and bonding these to neighboring unsaturated atoms. This process of course leads to an increase in the atomic density, but is limited by chemical and steric constraints. Since we had to choose an atomic density for the disordered regions that was considerably lower than that for the ordered regions to insure melting, the filling of voids brings the atomic densities in the two regions closer together, as demanded by experiment.

Work in progress indicates that during the second stage of construction, the majority of the unsaturated vertex atoms become attached to one another, shifting the average connectivity in the interfacial and disordered regions much closer to four. In a subsequent paper[17] we hope to report results for the maximum average connectivity that can be achieved at the end of the first stage, using improved algorithms, and at the end of the second stage, taking into account the improvements outlined above.

INTERPENETRATING NETWORKS AND ODD-MEMBERED RINGS

We noted earlier that the preliminary structural model exhibits only even-membered rings. This is related to the fact that the overall structure can be decomposed into interpenetrating vertex-atom and centroid-atom sublattices. By construction, the vertex-atom network forms a simplicial subgraph everywhere in space. In the ordered regions, the centroid-atom network clearly forms a simplicial subgraph in its own right. This subgraph can be extended into the disordered regions, though not uniquely. In contrast to the vertex-atom simplicial subgraph, for which there is a centroid atom inside each tetrahedron, the centroid-atom simplicial subgraph tetrahedra may or may not have vertex atoms in their interior, at least for the unrelaxed structure. This apparent asymmetry between vertex-atom and centroid-atom simplicial subgraphs is greatly reduced after the lattice is relaxed, as we would expect, since the distinction between vertex and centroid atoms is merely a constructional fiction.

The reader might find it instructive to draw centroid-atom simplicial subgraphs for the unrelaxed two-dimensional case shown in Fig. 5, including as many vertex atoms as possible inside the triangles belonging to the subgraph. Some of the vertex atoms will be found to lie close to the edges of the triangles, but these will be brought into more central positions by lattice relaxation.

In the second stage of the construction, the elimination of dangling bonds and voids leads to the formation of odd-membered rings, to a blurring of the distinction between vertex and centroid atoms, and hence to a blurring of the distinction between the original interpenetrating networks. All of these factors are undoubtedly essential features of disordered covalently bonded networks.

A particularly simple way of introducing odd-membered rings into the diamond structure is that used recently by Wooten and Weaire to generate atomic models of bulk amorphous silicon.[18] For the purposes of discussion, it is sufficient to consider a single step in their process: Beginning with an ideal diamond structure, choose a neighboring pair of atoms at random, say atoms 2 and 5 in Fig. 9(a), and rebond these atoms to their neighbors as indicated in Fig. 9(b). After rebonding, all atoms are still four-fold coordinated. Many of the bond angles are severely distorted, but these distortions can be reduced by relaxing the lattice locally. The rebonding does not alter the total number of bonds but does alter the local topology. In particular, six-membered rings disappear and five- and seven-membered rings appear. Moreover, pairs of atoms that were originally next-nearest neighbors become nearest neighbors, blurring the identity of the interpenetrating fcc lattices locally. This is exactly the type of topological change that occurs during the second stage of our construction, when neighboring pairs of unsaturated vertex atoms are joined together: originally these atoms were destined to be second neighbors, but now they are first neighbors. In a subsequent paper we hope to discuss such atomic rearrangements in terms of simplicial graphs and subgraphs.

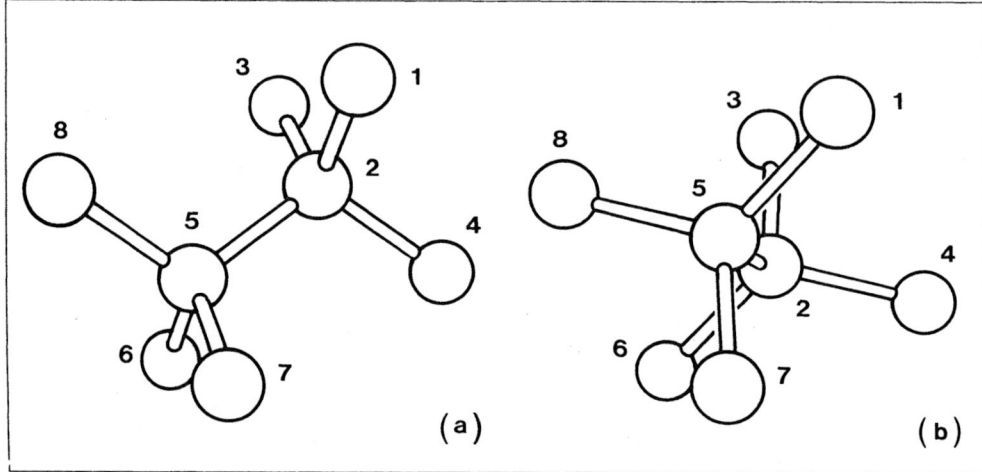

Fig. 9. Local rearrangement of bonds in the diamond lattice, showing the original structure in panel (a) and the rebonded structure in panel (b). Adapted from the work of Wooten and Weaire.[18]

EXTENSIONS TO OTHER SYSTEMS

Our general approach can be applied to a wide variety of systems by using different crystal structures for the ordered arrays of close-packed spheres, by using different elastic energy expressions to favor different bond geometries, and by decorating the simplicial subgraphs in different ways. Beginning with ordered fcc arrays of close-packed spheres, we can study sphalerite-type compounds such as cubic ZnS in addition to diamond-type crystals. By starting with ordered regions having the hcp rather than the fcc structure, we can extend the procedure to wurtzite-type compounds such as hexagonal ZnS. For binary compounds, vertex and centroid atoms correspond to anions and cations (or vice versa). The attachment of pairs of unsaturated vertex atoms to one another during the second stage would lead to nearest-neighbor anion or cation pairs. Such attachments might not be energetically favorable individually, but might be favorable in special combinations such as antisite defects.[19]

By modifying our procedures slightly, we can also generate interfaces between different materials, such as crystalline Si and amorphous SiO_2. We can then go a step further and simulate different growth conditions by introducing H, O, and OH groups in accordance with prescribed statistical rules and geometrical distributions, letting these groups saturate some or all of the dangling bonds during the construction or at the very end. We believe that such computer modeling of interface growth will provide new insights into the nature of the structural defects that arise during growth.[20,21]

CONCLUDING REMARKS

Although computer simulation studies of solid/liquid interfaces have already reached a high level of development for simple atomic systems,[22,23] comparable studies of covalently bonded systems based on realistic interactions are still in their infancy.[24] Such studies should lead to realistic structural models once they have been perfected. In the meanwhile, one can attempt to build models using considerably less computer time by simplifying the construction, as we are attempting to do. In the case of bulk amorphous Si, it has already been demonstrated by Wooten and Weaire[18] that a simplified construction based on random interchange of neighboring atoms followed by lattice relaxation leads to structural models that are in excellent agreement with experiment. Although our own work for composite systems is still in an early stage of development, we find the preliminary results encouraging.

Once the results of full-scale computer simulations become available, it will be instructive to examine their topological properties in detail to see whether we can learn something useful from such analyses. In particular, it would be interesting to construct the simplicial graphs and subgraphs, and see whether certain approaches to their construction are more satisfactory than others. Such topological analyses might provide new insights into the nature of disordered structures, complementing ring statistics, for example.

ACKNOWLEDGMENTS

The early stages of this work were partially supported by the Office of Naval Research under Contract Number N00014-79-C-0814, while one of us (P.L.) was an IBM World Trade Fellow at IBM San Jose Research Laboratory. We are grateful to the Office of Naval Research and IBM Belgium for their encouragement and partial support. We thank Frederick Wooten and Denis Weaire for sending us preprints of their computer simulation studies. Finally, we are pleased to acknowledge stimulating discussions with Farid F. Abraham, Bernie J. Alder, John A. Barker, Jean-Luc Bredas, Paul J. Flory, Douglas Henderson, and Robert K. Nesbet.

REFERENCES

1. For detailed references, see: J.M. Ziman, Models of Disorder (Cambridge University Press, 1979).
2. For additional references, see: R. Zallen, The Physics of Amorphous Solids (Wiley, New York, 1983).
3. F. Spaepen, Acta Metall. 26, 1167 (1978).
4. S.T. Pantelides and M. Long, in The Physics of SiO_2 and its Interfaces, edited by S.T. Pantelides (Pergamon, New York, 1978), p. 339.
5. T. Saito and I. Ohdomari, Phil. Mag. B 43, 673 (1981).
6. L.C. Feldman, in Proc. 1984 Intern. Conf. Physics of VLSI (Springer-Verlag, Berlin, 1985), in press.
7. A.C. Wright, G.A.N. Connell, and J.W. Allen, J. Non-Cryst. Solids 42, 69 (1980).
8. P. Chaudhari, J.F. Graczyk, D. Henderson and P. Steinhardt, Phil. Mag. 31, 727 (1975). For a more recent discussion, see: K. Doi, J. Non-Cryst. Solids 68, 17 (1984), particularly Fig. 2.
9. Ph. Lemaire and J.P. Gaspard, unpublished. The present authors are grateful to Ph.L. and J.P.G. for helpful conversations.
10. P. Lambin and F. Herman, in Proc. 1984 Intern. Conf. Physics of Semiconductors (Springer-Verlag, Berlin, 1985), in press.
11. P. Lambin and F. Herman, to be published.
12. F.F Abraham, Rep. Prog. Phys. 45, 1113 (1982).
13. J.P. Hansen and L. Verlet, Phys. Rev 184, 151 (1969).
14. F.W. Smith, Can. J. Phys. 42, 304 (1964).
15. P.N. Keating, Phys. Rev. 145, 637 (1966).
16. Yuan Li and P.J. Lin-Chung, Phys. Rev. B 27, 3465 (1983).
17. P. Lambin and F. Herman, to be published.
18. F. Wooten and D. Weaire, J. Non-Cryst. Solids, in press; Proc. 1984 Intern. Conf. Physics of Semiconductors (Springer-Verlag, Berlin, 1985), in press; see also this volume.
19. J.A. Van Vechten, J. Phys. C: Solid State Phys. 17, L933 (1984).
20. N.F. Mott, Adv. Phys. 26, 363 (1977); J. Non-Cryst. Solids 40, 1 (1980); J. Phys. C: Solid State Phys. 13, 5433 (1980).
21. F. Herman, J. Vac. Sci. Technol. 16, 1101 (1979); F. Herman, J. Vac. Sci. Technol. 21, 643 (1982); F. Herman, J. Phys. (Paris) 45 Colloque C5-375 (1984).
22. J.D. Weeks and J.H. Gilmer, Adv. Chem. Phys. 40, 157 (1979); G.H. Gilmer, Science 208, 355 (1980); G.H. Gilmer and J.Q. Broughton, J. Vac. Sci.Technol. B 1, 298 (1983).
23. F.F. Abraham, Repts. Prog. Phys. 45, 1113 (1982); J. Vac. Sci. Technol. B 2, 534 (1984).
24. F.F. Abraham, private communication.

AMORPHOUS-CRYSTALLINE HETEROJUNCTIONS

Václav Šmíd, Jiří J. Mares, Ladislav Štourač and
Jozef Krištofik

Institute of Physics, Czechoslovak Academy of Sciences
180 40 Prague 8, Czechoslovakia

INTRODUCTION

Investigation of amorphous-crystalline (a/c) heterojunctions is useful both for the understanding of the fundamental device physics and for
applications. In some cases it can also give important information about
physical parameters of amorphous semiconductors.

A pioneering work by Grigorovici[1] on a-Ge/c-Ge junctions was followed
by a number of others[2-13]. The physics of a/c heterojunctions is far from
being completely understood at present, nevertheless these structures show
potentiality for practical application. Besides the study of a/c heterojunctions consisting of chalcogenide semiconductors, such as As_2Se_3[4,5],
As_xSe_y, $As_xSe_yTe_z$[2], $GaTe_3$[6], which were motivated by improvement in TV pick-
up tubes or switching devices, a considerable effort is made to investigate
a/c heterojunctions with a-Ge[7,8] and a-Si:H[9-12]. The a/c heterojunctions
with a-Si:H were used for constructing solar cells[11] and for applications
to a silicon vidicon target without a diode array[12].

The existence of space-charge-limited currents (SCLC) has been proved
in several types of a/c heterojunctions[3,8,13,14]. We have shown for the
first time the existence of this type of transport mechanism in the heterojunctions with a-Si:H[9,10]. It has opened a new possibility of evaluating
one of the most important physical quantities of a-Si:H - the energy distribution of localized states in the gap. On improving the analysis of the
SCLC experimental data[14-18], the SCLC method is complementary to the field
effect[19], deep level transient spectroscopy (DLTS) and other methods[21]. The
a/c heterojunctions are also used as suitable structures for DLTS experiments[22].

In this paper we present results of the study of electrical properties
of a/c heterojunctions with a-Si:H and a-Ge on various types of crystalline
substrates. The aim of this paper is to shed new light on the transport
mechanisms in these structures. We construct an energy band diagram and
demonstrate the application of these structures to the evaluation of some
basic physical parameters of amorphous semiconductors.

A MODEL OF AMORPHOUS-CRYSTALLINE HETEROJUNCTION

There exists no consistent microscopic theory of a/c heterojunctions, but for many purposes a phenomenological theory which takes some parameters from measurements is more useful. The first step in building such a theory is the construction of a heterostructure band scheme, supposing that the band structure of both semiconductors is known. This procedure worked out by Anderson[23] for an abrupt crystalline heterostructure may be used analogously for the amorphous one.

After making a contact of two different semiconductors, a contact potential due to the difference in their work functions appears between them. A detailed shape of electrostatic potential (vacuum energy level) is given by the screening ability of both parts forming the junction. In crystalline semiconductors where the screening charge is placed originally at a relatively sharp level, a parabolic solution to the Poisson equation may be a good approximation. We can write it as

$$V(x) = (n_c e/2\varepsilon)(x - x_0)^2 ,$$ (1)

where n_c is the concentration of the screening charge homogeneously spread over a layer of thickness x_0 and $\varepsilon = \varepsilon_0 \varepsilon_r$ is the permittivity. If the potential at the crystalline semiconductor surface is V_c, the induced charge will be

$$\sigma_c = -(2\varepsilon n_c e V_c)^{1/2} .$$ (2)

If we assume that there is no extra charge present at the junction, the induced charges on both sides of the junction must be of equal magnitudes and opposite signs, or in other words, the electric displacement vector must be continuous. From this condition we can obtain a simple, well-known relation which describes how the total potential drop $V_{c1} + V_{c2}$ (regardless of its origin) is divided between the members of the junction:

$$\frac{V_{c1}}{V_{c2}} = \frac{n_{c1}}{n_{c2}} \frac{\varepsilon_1}{\varepsilon_2} ,$$ (3)

where subscripts 1 and 2 denote the quantities in the parts 1 and 2 of the heterojunction, respectively.

In the case that one part of the junction is an amorphous semiconductor this formula cannot be used generally. In amorphous semiconductors with a continuous spectrum of states in the gap the charge is not bound to any sharp energy level, but the redistribution of the charge during screening takes place, and the approximations (1) and (2) are no longer valid. A simultaneous shift of all energy levels by an electrostatic potential results in the fact that the levels originally neutral are transferred above the Fermi level and get electric charge. Thus in any place where the potential is $V(x)$ and the density of states $N(E)$, the density of charge $\rho(x)$ is given by

$$\rho(x) = -e \int_0^{eV(x)} f(E, T) \cdot N(E) \, dE ,$$ (4)

where $f(E, T)$ is the distribution function.

Solution of the Poisson equation with such a right-hand side is almost impossible in a real case. Therefore let us treat only a simplified equation in which $N(E)$ is a constant equal to N_f. In this case we have:

$$\rho(x) = -e^2 N_f V(x). \tag{5}$$

A solution of the Poisson equation with the boundary condition $V_a = V/_{x=0}$ yields:

$$V(x) = V_a \exp(-x/\lambda). \tag{6}$$

Here and in the following formulas the subscripts a and c denote the quantities corresponding to the amorphous and crystalline parts of the heterojunction, respectively, λ is a characteristic length in the amorphous semiconductor. Using this equation and the Gauss theorem, we can obtain the induced charge

$$\sigma_a = \varepsilon_a \left(\frac{dV}{dx}\right)_{x=0} = -\left(N_f e^2 \varepsilon_a\right)^{1/2} V_a. \tag{7}$$

Assuming that the surface states are absent, we may use the above-mentioned condition

$$\left|\sigma_a\right| = \left|\sigma_c\right| \tag{8}$$

and obtain

$$\frac{V_c}{V_a} = \frac{eN_f \varepsilon_a V_a}{2 \varepsilon_c n_c}. \tag{9}$$

This formula has the same significance for the a/c heterojunctions as formula (3) for the crystalline ones and enables us to construct the energy band scheme (see Fig. 1). In spite of the simplification (5) used, this relation conserves an important property, namely that the right-hand side of (9) is a function of V_a. This property, which is obviously a consequence of the redistribution of screening charge in the gap states, is one of the reasons for a complicated behaviour of the a/c heterojunctions. The dependence of the right-hand side of (9) on V_a would be more complicated for a general form of the density of states.

A special case of a/c heterojunctions where a dominant transport mechanism in amorphous semiconductor is the variable range hopping near the Fermi level[24] has been treated by Döhler and Brodsky[7]. They assumed that the behaviour of the a/c heterojunction is to be similar to that of the Schottky barrier where the role of the metal is played by the amorphous semiconductor owing to its high density of states near the Fermi level. The difference between the metal and this type of amorphous semiconductor can be expressed by an effective Richardson constant, which is much lower for the semiconductor, and by the fact that the band bending in the amorphous semiconductor must be taken into account. For the calculation of the band bending the authors used practically the same formulas as (9) which was derived, however, for another boundary condition. According to this model, reverse-biased I-V characteristics do not saturate as for an ideal rectifier, but increase with increasing bias. This behaviour is given by the dependence of the barrier height on the applied voltage which is again a consequence of the above-mentioned properties of (9).

For systems such as c-Ge/a-Ge[25] and c-Si/a-Si[26] where the amorphous semiconductor is prepared by evaporation in high vacuum, an excellent quantitative agreement between this simple theory and experimental results has been established. On the other hand, for many systems[2,4,8,13,27,53] composed of a variety of amorphous and crystalline materials, quite a different behaviour was observed. To explain these results it is assumed that the potential profile in the vicinity of the junction creates suitable boundary conditions for various types of conduction mechanisms in homogeneous amorphous or crystalline semiconductors. The transport is then no longer controlled by the junction itself but is bulk limited.

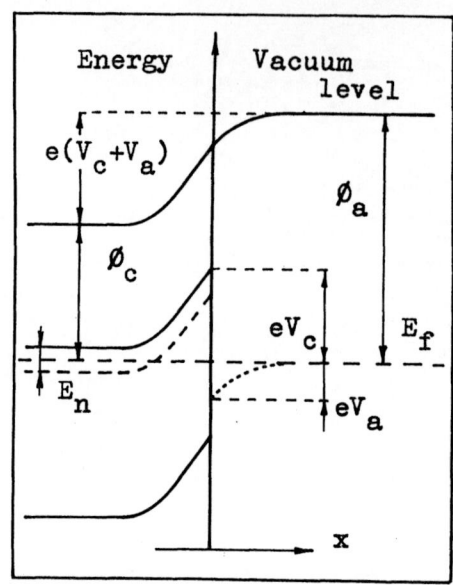

Fig. 1. Scheme of the band structure of an a/c heterojunction.

EXPERIMENTAL PART AND DISCUSSION

Sample Preparation

Table 1 summarizes the types of samples used in this paper. As substrate material we have used crystalline Si and GaAs with different dopings. As the amorphous semiconductor for the a/c heterojunctions we have employed beam-evaporated a-Ge and a-Si:H prepared by reactive sputtering in an argon-hydrogen atmosphere and by glow discharge decomposition of SiH_4.

The crystalline part of the heterojunction was supplied with an ohmic contact. Prior to the deposition of a-Si:H, the crystalline wafer was etched. To minimize the role of the interface between a-Ge and GaAs, the surface of GaAs was cleaned by heating up to 550° C in high vacuum (10^{-8} torr)[28]. At so high temperature the thermal desorption of native oxide takes place[28]. The corresponding structure was prepared in situ.

For the plasma decomposition of silane we used an rf power capacitively coupled system. The heterojunctions were prepared on the anode at a substrate temperature of 570 K from a gas mixture of 1 at % SiH_4 in Ar at a pressure of 4×10^{-2} torr, using high flow rate and a low 13.56 MHz power. The deposition rate was typically 1 - 2 Å/sec.

The sputtering of a-Si:H was performed in the dc sputtering apparatus supplied with an ionization oil trap where the continuous purifying of $Ar:H_2$ mixture at the inlet, as well as the trapping of the back diffused oil vapours at the outlet, was possible[29]. The samples were prepared at a pressure of 6×10^{-2} torr in an argon-hydrogen atmosphere (~10 at %) at 570 K and a power density of 6.4 mW/cm². The deposition rate was ~1 Å/s. As a target we employed N-type polycrystalline Si with a resistivity of 300 Ωcm.

As the second ohmic contact to the heterojunctions with a-Si:H, thermally evaporated aluminium was used. Measurements of thermoelectric power have shown that the a-Si:H films produced by sputtering and by glow discharge are of N-type. In the case of the sputtered a-Si:H films we have found that the type of the conductivity is the same as that of the target used for the sputtering. The electrical conductivity of a-Si:H films deposited by glow discharge was typically 10^{-9} $\bar{\Omega}^{-1}$ cm^{-1} at room temperature with a well de-

Table 1. The composition of a/c heterojunctions

sample number	material	crystal doping (cm^{-3})	crystal resistivity at room temperature $(\Omega\,cm)$	amorphous semiconductor
1	Si:B	8×10^{13}	60	a-Si:H (SPUTTERING)
2	Si:B	8×10^{13}	60	a-Si:H (GD)
3	GaAs:Te	1×10^{18}	10^{-3}	a-Si:H (GD)
4	GaAs:Te	1×10^{18}	3×10^{-3}	a-Ge
5	GaAs:V	5×10^{15}	3	a-Ge
6	GaAs:Zn	1×10^{16}	2	a-Ge
7	GaAs pure	10^{15}	2×10^{8}	a-Ge
8	GaAs:Cr	1×10^{16}	4×10^{8}	a-Ge

fined activation energy of 0.7 eV in the temperature range from 220 to 400 K. The electrical conductivity of a-Si:H films prepared by sputtering depends on the substrate temperature T_S. It increases at room temperature from 10^{-8} Ω^{-1} cm^{-1} to 10^{-5} Ω^{-1} cm^{-1} and the activation energy decreases from 0.6 to 0.3 eV with T_S increasing from 500 to 600 K[30]. The IR spectra of a-Si:H films exhibiting a typical doublet at 1990 and 2090 cm^{-1} confirm that the films contain built-in hydrogen similarly to the a-Si:H films prepared by decomposition of SiH_4. The optical gap obtained from the absorption coefficient measurements does not change markedly for T_S above 450 K. In our case E_g^{opt} is equal to ~ 1.60 eV[31]. We did not measure independently the hydrogen content. Using the knowledge of the dependence of E_g^{opt} on hydrogen content c_H[32]

$$E_g^{opt} = 1.48 + 0.019\,c_H \quad ,$$

we can estimate $c_H \sim 5$ at %. Our results show that c_H does not change with T_S.

The crystalline part of the heterojunction a-Ge/c-GaAs was formed by a monocrystalline, (100) - oriented GaAs wafer. One face of the wafer was supplied by an Au-Ni alloyed contact. The a-Ge layer was beam-evaporated in a VT 2 Varian UHV apparatus at a pressure of $\sim 10^{-7}$ torr and a substrate temperature of ~ 300 K. The Au contact to a-Ge was prepared in situ[33,34].

The temperature dependence of the conductivity of a-Ge films satisfies the Mott's law for variable range hopping[24] practically to 300 K; σ_{RT} is 7.5×10^{-3} Ω^{-1} cm^{-1} (see Fig. 2). From the slope of the straight lines in the log σ vs $T^{-1/4}$ coordinates we have estimated the density of localized states at the Fermi level to be about 7×10^{19} eV^{-1} cm^{-3}. The gap of a-Ge is ~1 eV, the thermopower measurements have shown that the a-Ge films are of N-type[33,35].

Transport Mechanism

Fig. 3 shows an example of the current-voltage characteristic of an a/c heterojunction with a-Si:H deposited on c-Si by sputtering (sample No 1) measured at different temperatures. It is a typical example of a heterojunction with highly conductive substrate. For a voltage $V \lesssim 0.1$ volt the characteristics exhibit an ohmic dependence, which at higher voltages is followed by the dependence $I \sim V^m$, where m ranges between 2 and 7. For $V > 1$ volt, there is a decrease in the exponent m. This type of dependence is typical for SCLC[36]. In the reverse direction the ohmic dependence is satisfied to ~ 1 volt. We have not usually observed a saturation of the current in the reverse direction, although there is a decrease in the slope of the I-V dependence. From the temperature dependence of the I-V curves we can evaluate

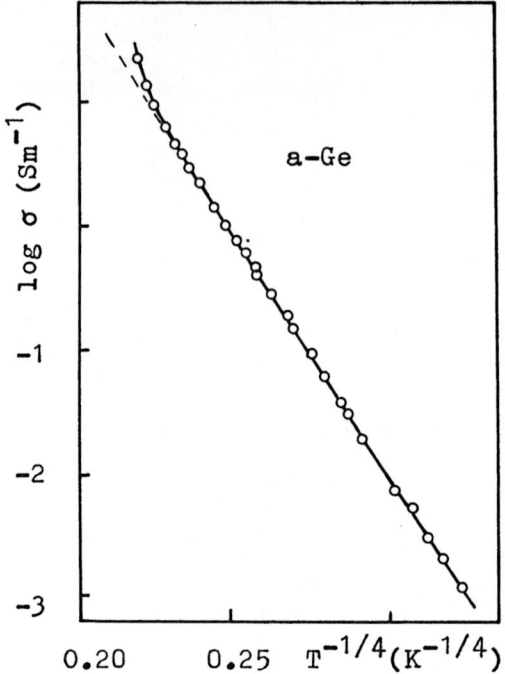

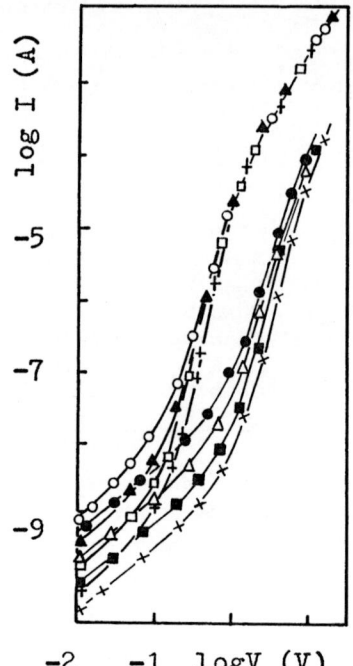

Fig. 2. Temperature dependence of a-Ge film used for a/c heterojunctions Nos 3 to 9.

Fig. 3. Current-voltage characteristics measured at different temperatures for a-Si:H/c-Si heterojunction (sample No 1).

the activation energy of the electrical conductivity at constant voltages. In the forward direction we observe a strong decrease in the activation energy with increasing voltage in the heterojunctions with low-resistive substrates. This is demonstrated in Fig. 4 for heterojunctions Nos 3 and 4. The activation energy decreases below 0.1 eV. This experimental result indicates that in the heterojunctions where the transport is limited by the amorphous constituent there is a remarkable shift of the Fermi level due to the applied voltage. In other words, the Fermi level shifts owing to an efficient injection of carriers and the Fermi level is probably not strongly pinned in the equilibrium position as in materials with negative correlation energy[37,38].

An exact proof of the existence of SCLC in the bulk of the amorphous component of the heterojunction is demonstrated in Fig. 5, where we have summarized the results of measurements of sample No 4 for different a-Ge films thicknesses. According to the theory of the SCLC[36], one should obtain a plot independent of thickness in the I/d vs V/d^2 coordinates[33]. It is the case of the forward direction in the sample of Fig. 5. The deviation from the dependence is caused probably by a metallic contact which might limit the current flow in a relatively low-resistivity amorphous semiconductor, especially at small a-Ge film thickness. Although the shape of the I-V curve in the reverse direction is not very different from that in the forward direction, Fig. 5 indicates that the criterion for SCLC is not fulfilled in the reverse direction. This is supported also by our investigation of the changes of the shape of the I-V characteristics due to high hydrostatic pressure[8,10].

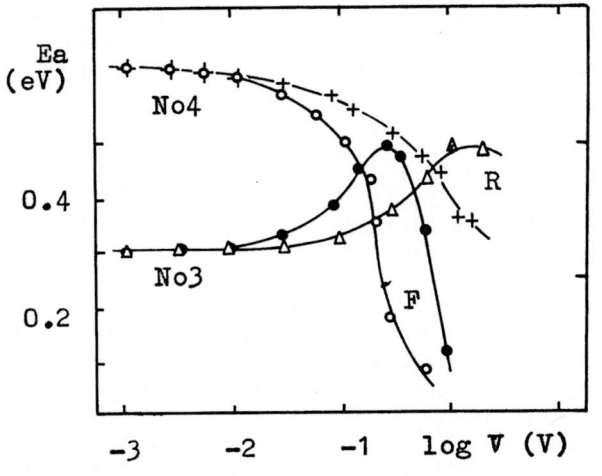

Fig. 4. Dependence of the
activation energy E_a
on applied voltage for
heterojunctions Nos 3
and 4.

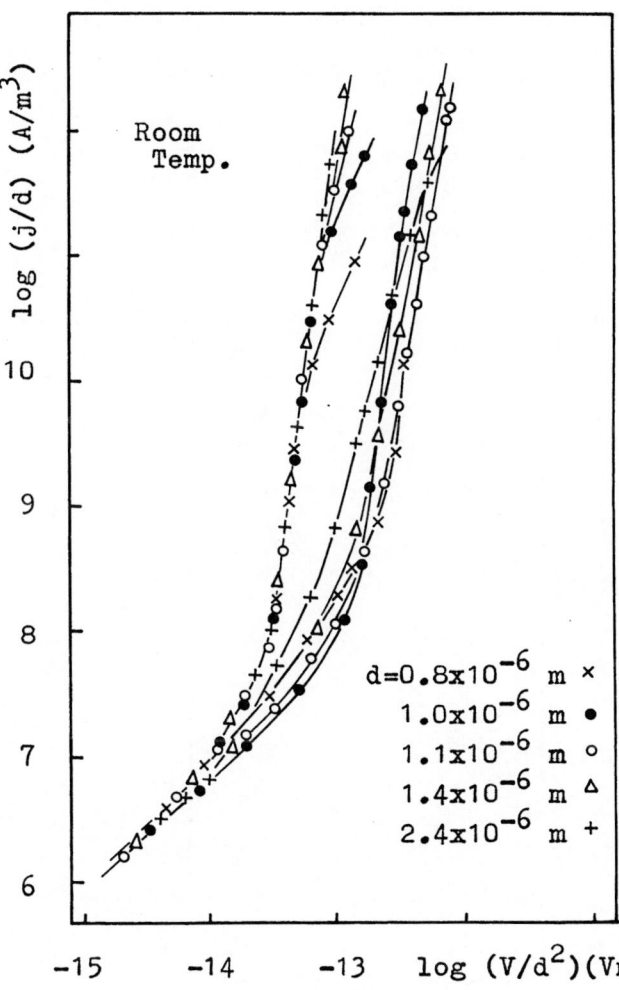

Fig. 5. Normalized plot of
I-V curves for
a-Ge/c-GaAs (sample
No 4)with different
thicknesses of a-Ge.

Fig. 6 shows the influence of hydrostatic pressure on the I-V characteristics of the a-Ge/c-GaAs heterojunction - sample No 4. The decrease in conductivity of the heterojunction with increasing pressure cannot be explained only by a relative increase in resistivity of a-Ge[39] or GaAs[40] because this increase for the pressure up to 1 GPa is relatively small(0.1 - 0.4). With the help of changes in conductivity of a-Ge we can explain only the shift of the ohmic part with the pressure. The stronger change of conductivity in the non-linear part of the characteristics is due to the change in the heterojunction interface or a change in the transport mechanism in a-Ge. The influence of the hydrostatic pressure on the transport properties or on the injection in the heterojunction can be illustrated by Fig. 7. The latter figure shows the pressure dependence of the characteristic voltage V_0 which is determined by the intersection of the extrapolated ohmic and non-linear part of the I-V characteristic. At this voltage the concentration of injected carriers is approximately equal to that of thermally activated carriers. V_0 increases with hydrostatic pressure in both directions. The dependence is steeper in the reverse direction which corresponds to the change of the heterojunction interface. In the forward direction small changes in V_0 may be explained by a change in the density of localized states at the band edges or by a shift of the mobility edges with hydrostatic pressure. Fig. 8 shows the influence of the hydrostatic pressure on the current in the heterojunction with sputtered a-Si:H (sample No 1) at three characteristic voltages. The heterojunction with a-Si:H prepared by glow discharge decomposition of silane also shows an increase in current due to high hydrostatic pressure. In the ohmic part there is no difference in the behaviour between the reverse and forward

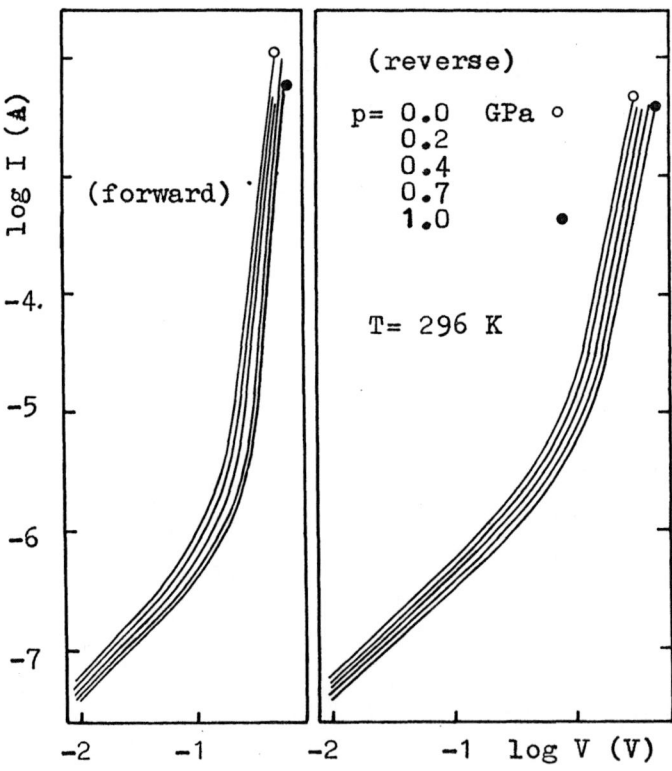

Fig. 6. Variation of the I-V curve measured at room temperature with high hydrostatic pressure (sample No 4). F and R indicate forward and reverse bias, respectively.

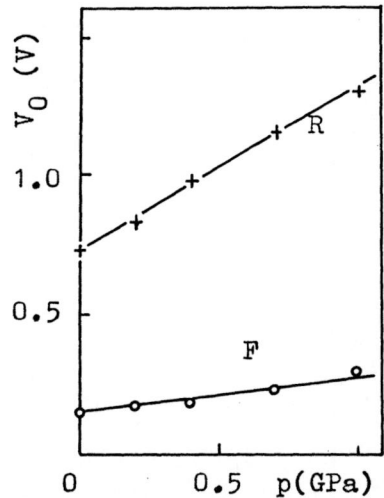

Fig. 7. Dependence of the characteristic voltage V_0 on
hydrostatic pressure. V_0
is the intersection of the
extrapolated ohmic and non-
linear part of the I-V
curves in the log I vs
log V coordinates.

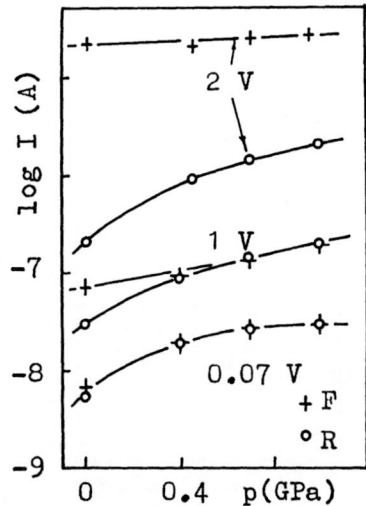

Fig. 8. Dependence of current in
a-Si:H/c-Si heterojunction
(sample No 1) on high
hydrostatic pressure for
three different voltages
corresponding to ohmic,
$I \sim V^m$ and $I \sim V^2$ parts.

directions, which indicates the influence of pressure on the bulk of a-Si:H.
When the current deviates from the ohmic dependence (at a voltage of ~ 1 V),
we observe a steeper change in the current with pressure for the reverse
direction. The third voltage depicted in Fig. 8 corresponds to the $I \sim V^2$
dependence. In the forward direction the current practically does not depend
on the pressure. This observation indicates that the transport mechanism in
the reverse direction is determined by a potential barrier which is modified
by the hydrostatic pressure. Therefore the rectification ratio decreases
with increasing pressure.

    Roberts and Schmidlin[41] have shown that a well-defined activation energy
can be expected only if one specific energy is responsible for the contribution to injected charge. A different situation is in the case of the
a-Ge/c-GaAs heterojunction. Dominant transport mechanism in the ohmic part
is the hopping near the Fermi level. Analysing the SCLC in this structure
we lack direct information about the transport path position. As can be seen
from Fig. 9, the transport in extended states predominates when the applied
voltage increases. In this region the activation energy continuously decreases (see Fig. 4). Nevertheless our experimental results are the first
observations of the SCLC mechanism in the system with the hopping conduction
in the ohmic part of the I-V curves.

    In evaluating the transport mechanism, it is necessary to analyse I-V
characteristics over wide voltage and temperature ranges and to take into
account the scaling rules corresponding to a particular transport mechanism.
If we take only the shape of the I-V characteristic as a proof of the mechanism, we can make serious mistakes. To distinguish between $I \sim V^m$ and
$I \sim \exp (\alpha V)$ dependences it is necessary that I-V characteristics expand
over several decades of current and be linear in the corresponding coordinates.

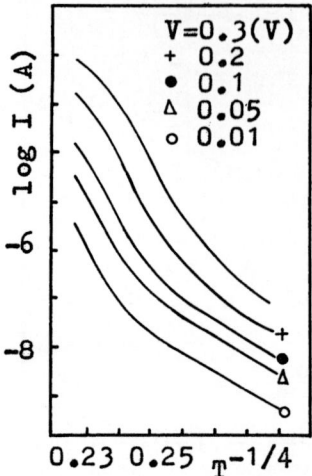

Fig. 9. Temperature dependence of current in a/c heterojunction No 4 for different voltages, which indicates a transition from the hopping conduction at low voltage to the transport in the extended states.

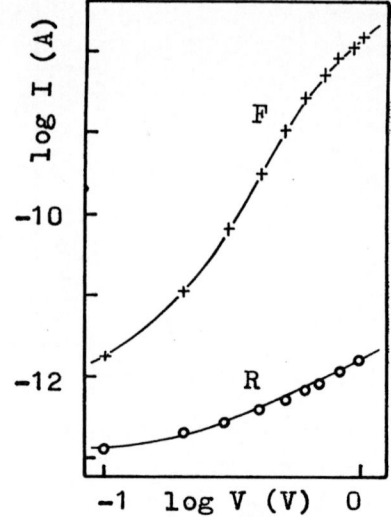

Fig. 10. I-V characteristic taken from Fig. 1b in the paper by Matsuura et al.[42] plotted in the log I vs log V coordinates.

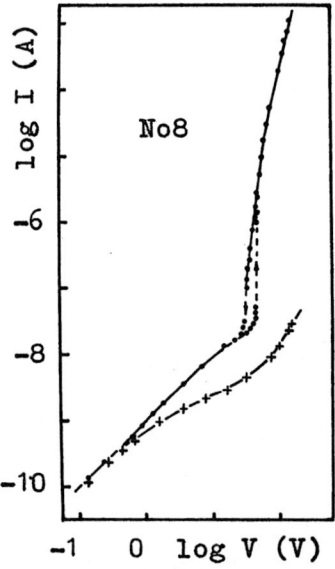

Fig. 11. Current-voltage characteristic of a/c heterojunction with a high-resistivity substrate (sample No 8) taken at room temperature.

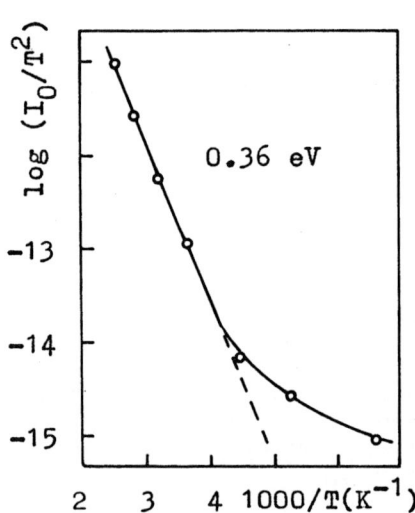

Fig. 12. Temperature dependence of the saturation current in the reverse direction for a-Ge/c-GaAs heterojunction No 4.

If one takes, for example, the experimental results of I-V measurements on a-Si:H/c-Si heterojunctions obtained by Matsuura et al.[42] in a limited range of voltages, one can plot the data both in log I vs V coordinates (as the authors[42] did) and in the coordinates corresponding to SCLC (see Fig.10) and obtain good agreement in both cases. Without a more exact proof concerning the transport mechanism, a sophisticated interpretation of the results may be misleading[43].

We agree that in the limit of small voltages, the transport in the forward direction in the a/c heterojunctions may be barrier limited, but at higher voltages SCLC take place. A sensitive tool for distinguishing between these two parts of the I-V curve is a simultaneous evaluation of the activation energy of the transport mechanism as it will be mentioned in the next section. This is illustrated in Fig. 4 for a heterojunction No 3 in which the activation energy increases at small voltages. This is the case where the native oxide on GaAs was not thermally desorbed in the vacuum chamber as it was done in the case of a-Ge/c-GaAs structures.

Summarizing the experimental data, we believe that the transport mechanism in the forward direction in the heterojunctions with the low-resistivity substrates(samples 1 to 6) is determined, at higher voltages, by the SCLC. In the reverse direction the transport is limited by the heterojunction interface.

There is a different situation in samples Nos 7 and 8 where the amorphous layer was deposited on a substrate having high resistivity. In the forward direction the I-V characteristics exhibit sublinear parts and in some cases also a well-pronounced reversible hysteresis. This shape of I-V characteristics may be attributed to a dominant role of the substrate which behaves like a relaxation semiconductor. A similar behaviour has also been reported for a structure formed by semiinsulating GaAs and an injection contact[45]. Fig. 11 shows an example of the I-V curve taken at room temperature for sample No 8.

Energy Band Model

In constructing the energy band model of a-Ge/c-GaAs heterojunction we have used as a phenomenological parameter the height of a potential barrier deduced from the temperature dependence of the saturation current for the reverse bias (Fig. 12). The calculation of the potential difference at the heterojunction interface has been carried out according to formula (9) de-

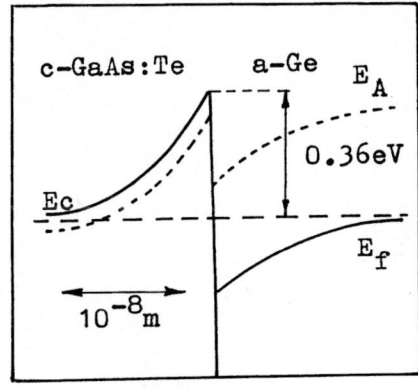

Fig. 13. Energy band model of a-Ge/c-GaAs heterojunction No 4.

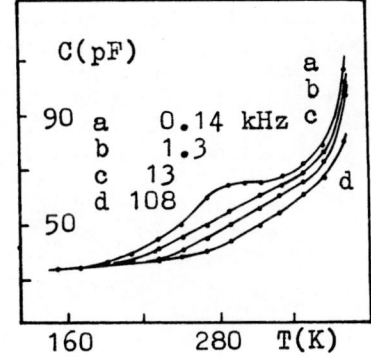

Fig. 14. Temperature dependence of capacitance of a/c heterojunction No 1 measured at different frequencies.

493

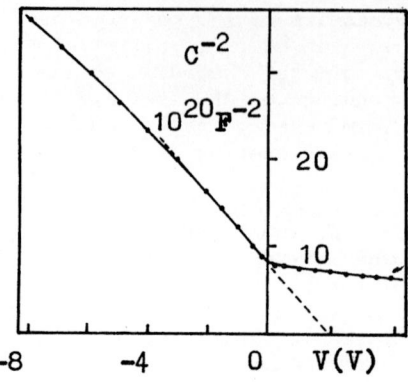

Fig. 15. Capacitance-voltage
characteristic of a/c
heterojunction No 1
taken at 1 MHz and a
temperature of 192 K.

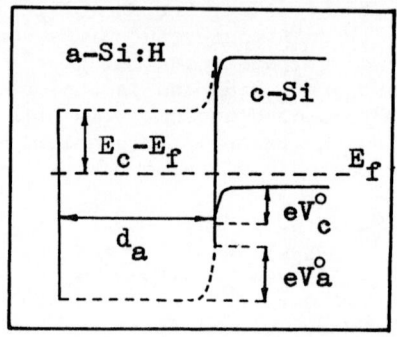

Fig. 16. Energy band model of
a-Si:H/c-Si heterojunc-
tion (sample No 1, see
Table 1).

rived in the framework of the model of a/c heterojunctions where the distribution of the localized states in the amorphous part of the heterojunction
can be approximated by a constant value ($N(E) = 7x10^{19}$ eV^{-1} cm^{-3}). The band
bending at the interface is calculated from the screening length corresponding to the concentration of localized states. As an input parameter for this procedure we have employed a known position of the Fermi level in both parts of
the heterojunction. Fig. 13 shows the band model of the heterojunctions with
a-Ge. In the forward direction the transport is determined by the injection
of electrons from the conduction band of c-GaAs into the gap states of a-Ge.
The current flow is determined by a-Ge. In contrast to this example, in an
a/c heterojunction with a semiinsulating substrate the GaAs does not inject
a sufficient carrier density, and the transport mechanism is substrate
limited.

It is well known that in a-Si:H the density of localized states changes
by many orders of magnitude from the band edges to the Fermi level. Therefore the above-mentioned procedure is no longer correct. It is not possible
either to use the difference in the electron affinities because this value
is not known with a sufficient accuracy for a-Si:H. For the purpose of the
construction of the energy band model of the heterojunctions with a-Si:H
we have employed C-V measurements at high enough frequencies in order that
the MIS model[46-48] be acceptable. If we measure a differential capacitance
at a frequency above the cut-off frequency, the diferential capacitance of
the amorphous layer C_a does not depend on the applied voltage since the
ionization of the gap states cannot respond to such a high frequency.

The measured capacitance C corresponds to a series combination of C_a
and the capacitance of c-Si, C_c; $C = C_a \cdot C_c (V)/(C_a + C_c (V))$, $C_a = \epsilon_r \epsilon_0 \cdot A/d$,
where A and d are the sample area and thickness, respectively. For the he-
terojunction with sputtered a-Si:H, sample No 1, C_a is equal to 35 pF, see
Fig. 14. The voltage variation of C can be accounted for by a voltage variation of C_c. The C-V curve of the heterojunction is depicted in Fig. 15. The
voltage variation of C_c can be calculated theoretically[46]. The space charge
in the crystalline part of the heterojunction, Q_c, is given by[46,47].

$$Q_c = - \frac{2V_c}{e|V_c|} \frac{kT \epsilon_r \epsilon_0}{L_D} F (V_c) ,$$

494

where k is the Boltzmann constant and L_D the Debye length. This relation also gives the value of the space charge in the amorphous component because in this case $Q_a + Q_c = 0$. The charge in the amorphous layer is taken as an effective interface charge of the MIS structure, the charge at the a/c interface is neglected at so high frequencies.

Comparing the theoretical C vs V curve with the experimental curve of Fig. 15, we obtain a value of the flat band voltage $V_{FB} = 0.79$ V and the dependence of Q_a on V_a. If we assume that the applied voltage is divided only in the space charge region of the heterojunction, we obtain $V = V_a + V_c + V_{FB}$. At zero bias $eV_a^0 = 0.47$ eV and $eV_c^0 = 0.32$ eV. In addition to these values we have used, for the construction of the band model shown in Fig. 16, the values of the activation energy in the ohmic part of the I-V curve and the value of the optical gap. It is seen that in the c-Si there is a depleted region near the interface.

Density of Localized States and the SCLC in a/c Heterojunctions

If the contact acts as an infinite reservoir of free carriers, the contact does not influence the distribution of the electric field in the sample, the conduction mechanism is bulk limited and SCLC may occur. The method of SCLC has been traditionally used for the crystalline semiconductors, but the I-V characteristics typical for SCLC have been observed in several types of amorphous-crystalline heterostructures[49].

In analysing the experimental data of the SCLC, one usually assumes a typical energy distribution of localized states, N(E), and then finds agreement with experimental data by choosing an appropriate value of the fitting parameters. In the case of the exponential distribution of states:

$$N(E) = \frac{N_t}{kT_c} \cdot \exp\left((E - E_c)/kT_c \right) .$$

The value of the parameters T_c and N_t, characterizing the distribution of states is found from the current-voltage characteristics

$$I \sim N_t^{-1} \cdot \frac{V^{1+1}}{d^{21+1}} ,$$

where $1 = T_c/T$.

Fig. 17 shows the temperature dependence of the exponent m=1+1, which gives the value of T_c. The higher T_c, the steeper is the decrease in N(E) from the conduction band edge. When using an arbitrary distribution of states, it is possible to calculate the I-V curves in the SCLC region[50]. A comparison of the results obtained for different N(E) distributions is not simple, especially when looking for details in the N(E) function.

The SCLC can be used as a spectroscopic method for evaluationg N(E). This results from the fact that the Fermi level changes its position owing to high injection. By evaluating $N(E_f(V))$ and $E_f(V)$ we can reconstruct N(E). In this procedure we start from the equation

$$\frac{dn}{dE_f} = \int N(E) \frac{df\ (E-E_f)}{d\ (E-E_f)} dE ,$$

where f is the distribution function. The latter equation describes the shift

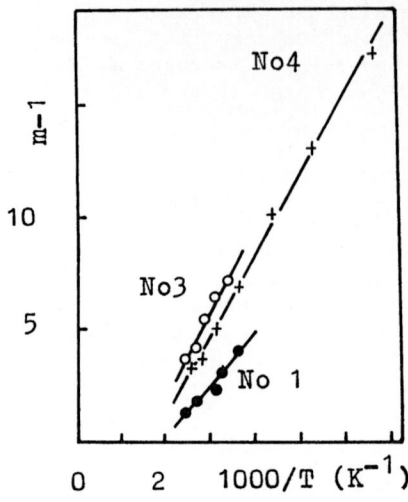 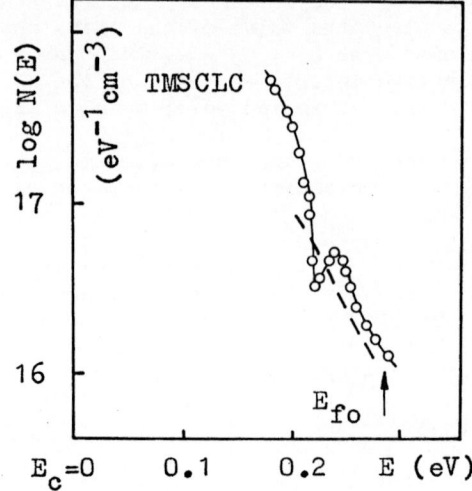

Fig. 17. Temperature dependence of the exponent m of the relation $I \sim V^m$ for three heterojunctions, samples Nos 1, 2,4, see Table 1.

Fig. 18. Energy distribution of localized states calculated according to SCLC method using TMSCLC (full line) and a simplified approach (dashed curve) for sample 1.

of the Fermi level due to the increase in the concentration by dn. The deconvolution of the integral on the right-hand side gives us N(E).

In a simplified case with zero temperature statistics, a simple approach may be used[10,51,52]:

$$E_f = kT . \ln \left(\frac{I_2 V_1}{I_1 V_2} \right) \quad \text{and} \quad \overline{N} (E_{f_o} + \Delta E_f) = \frac{a . \mathcal{E} . \Delta V}{ed^2 \Delta E_f}$$

where ΔE_f is the shift of the Fermi level caused by an increase in the voltage ΔV in the region of SCLC between the points (V_1, I_1) and (V_2, I_2) of the I-V curve. A parameter $\underline{a}$ lying between 1 and 2 describes non-uniformity of the internal space-charge field. We have taken $a=2$[10]. The value of the concentration of localized states N is the average value over the interval having a width ΔE_f. This analysis allows us to reconstruct the function N(E) without any presumption of the shape. Nevertheless, because of the simplification in the calculation it is not possible to evaluate the detail structure of N(E). To establish the energy axis, the activation energy in the ohmic region is usually interpreted as a distance of E_f from E_c. This is correct if there is no influence of the contacts in the ohmic part of I-V curve[44]. Even if this is not the case, in the energy gap there may be an energy level E_d, generally not identical with E_f,[41] which determined the position of the Fermi level. Roberts and Schmidlin[41] have introduced the so-called dominant energy E_d

$$E_d = -d (\ln \Theta) / d \left(\frac{1}{kT} \right) ,$$

where Θ describes the ratio of free carriers to the whole concentration of carriers. The position of the dominant level E_d can be calculated from the value of the activation energy E_a[14]

$$E_a = -d(\ln I)/d\left(\frac{1}{kT}\right).$$

This concept[16], which elaborates in more detail the approach of Nešpůrek and Sworakowski[16] and Stöckmann[15], was originally used for the a/c heterojunction with a-Si:H[14,17]. The method of thermally modulated space-charge-limited currents (TMSCLC) employs simultaneous measurements of the I-V curves and the activation energy.

Fig. 18 depicts the result of evaluation of N(E) by both described methods for the same sample of a/c heterojunction No 1. It indicates that the more sophisticated method (TMSCLC) can also give information about the details in N(E), see Ref. 17 too. Nevertheless the simplified approach may be used as a routine control of the properties of a-Si:H films. We have utilized it for a comparison of properties of a-Si:H prepared by glow discharge and by sputtering[18]. As we have explained in previous sections the evaluation of N(E) function from the SCLC may be employed only for the forward bias. In the reverse direction the C-V characteristics can be used. We have applied a model of MIS structure to the a/c heterojunction No 1 and found that the results of N(E) obtained from the SCLC and MIS follow each other; there is, however, inaccuracy in the determination of the energy axis (~0.06 eV)[48]. This may be taken as a good agreement of the results of the two methods.

In principle the SCLC method utilizable for a/c heterojunctions enables us to measure directly the microscopic mobility. We make use of the part of the I-V curve at high voltages where $I \sim V^2$ is satisfied (see, e.g. Fig. 3). Then according to the theory of SCLC[36] there is direct proportionality between the current density and μ. The measurement must not, however, be influenced by contacts. For example, see Fig. 5 where the deviation from the $I \sim V^m$ dependence occurs, but the dependence at high voltages does not make the calculation of μ possible.

For the a/c heterojunction No 1 we have found that $\mu = 5 \times 10^{-3} cm^2 V^{-1} s^{-1}$ which is at room temperature in close agreement with the results of μ_{FE} obtained on the FET structure[30]. In the latter case we have a higher activation energy, $\Delta E_\mu \approx 0.3$ eV, which is in contradiction with the results from the SCLC where $\Delta E_\mu < 0.1$ eV. These results are to be analysed in more detail.

CONCLUSIONS

The results of the investigation of a/c heterojunctions can be summarized as follows:

(i) In describing the behaviour of a/c heterojunctions, it is necessary to take into account specific properties of amorphous material, especially the continuous distribution of localized states. By comparison with the crystalline heterojunctions this difference is to be taken into consideration in the construction of the energy band model.

(ii) There exist a number of a/c heterojunctions in which the predominant transport mechanism is limited by the bulk of amorphous semiconductor, especially when a low-resistivity substrate is used. Then SCLC can be observed even if the transport mechanism in the ohmic region of amorphous semiconductor is the hopping. This has been proven in a-Ge/c-GaAs heterojunctions.

(iii) The existence of SCLC in a/c heterojunctions makes it possible to evaluate the density of the localized states distribution in the amorphous semiconductor. Special attention has to be paid to the influence of the contacts. In this respect, simultaneous evaluation of the activation energy of the transport mechanism is a recommended procedure.

(iv) Although a more sophisticated method (TMSCLC) yields a more complete information on the structure of N(E), a simplified approach discussed in this paper may be well used for routine measurements.

(v) We have found a fairly good agreement of the calculated N(E) function from both the SCLC and C-V measurements. Some uncertainties may arise in the determination of the appropriate energy axis. The SCLC method makes it also possible to calculate the value of microscopic mobility.

REFERENCES

1. R. Grigorovici, N. Croitoru, A. Devenyi and E. Teleman, Band structure and electrical conductivity in amorphous germanium, in: "Proc. Int. Conf. Physics of Semiconductors, Paris 1964," Academic Press, New York and London (1964), p. 423.

2. D. Ležal, I. Srb, F. Šrobár, V. Šmíd and J. Mišek, Photoelectric properties of chalcogenide glass thin films heterostructures, Thin Solid Films 34:51 (1976).

3. Tran Dam and L. Štourač, Amorphous As_2Te_3 monocrystalline GaAs heterojunctions, in: "Proc. Int. Conf. Amorphous Semiconductors ´78, Pardubice 1978, "Inst. Solid State Physics, Prague (1978), vol. 2, p. 613.

4. Z. Cimpl, M. Jedlička, F. Kosek and F. Schauer, Electrical properties of CdSe - amorphous chalcogenide heterostructure, in: Ref. 3, p. 635.

5. A. M. Andriesh, V. V. Bivol, V. I. Verlan, M. S. Iovu and E. V. Russu, Investigation of isotype InP - chalcogenide glass heterostructures, in:"Proc. Int. Conf. Amorphous Semiconductors ´84, Gabrovo," Inst. Solid State Physics, Sofia (1984), vol. 2, p. 65. (in Russian)

6. A. Cesnis, A. Oginskas, V. Lisauskas, N. Shiktorov and E. Butina-vishute, Two differences in transport mechanism in heterostructures amorphous ($GeTe_3$, $GaTe_3$ or $SiTe_3$) - crystalline (Si), in: Ref. 5, p. 16. (in Russian)

7. G. Döhler and M. H. Brodsky, Amorphous-crystalline heterojunctions, in: "Tetrahedrally Bonded Amorphous Semiconductors," M. H. Brodsky, S. Kirkpatrick adn D. Weaire, eds., American Institute of Physics, New York (1974), p. 361.

8. J. J. Mareš, J. Krištofik and V. Šmíd, Transport properties of a-Ge/c-GaAs heterojunctions, in: "Proc. Int. Conf. Amorphous Semi-conductors ´82, Bucharest 1982," Central Institute of Physics, Bucharest (1982), p. 219.

9. V. Šmíd and M. Libra, Transport properties of a-Si:H/c-Si heterojunctions, in: "Proc. 7th Conf. Czech. Physicists, Prague 1981," JČSMF Prague, p. 05-19. (in Czech)

10. V. Šmíd, J. Krištofik and J. J. Mareš, Comparison of properties of a-Si:H/c-Si heterojunctions prepared by reactive sputtering and plasma decomposition of SiH_4, in: Ref. 8, p. 225.

11. M. M. Rahman and S. Furukawa, Amorphous/crystalline heterostructure as a noval approach to fabrication of a solar cell, Electronics Letters 20:57 (1984).

12. H. Mimura and Y. Hatanaka, Optoelectrical properties of amorphous-crystalline silicon heterojunctions, Appl. Phys. Lett. 45:452 (1984).

13. B. Dunn and J. D. Mackenzie, Transport properties of glass-silicon heterojunction, J. Appl. Phys. 47:1010 (1976).

14. F. Schauer, V. Šmíd, O. Zmeškal and L. Štourač, Evaluation of space charge limited currents in a-Si:H/c-Si heterojunctions, Phys. Stat. Sol. (a) 73:K199 (1982).

15. F. Stöckmann, An exact evaluation of steady-state space-charge-limited currents for arbitrary trap distributions, Phys. Stat. Sol. (a) 64:475 (1981).

16. S. Nešpůrek and J. Sworakowski, Use of space-charge-limited current measurements to determine the properties of energetic distributions of bulk traps, J. Appl. Phys. 51:2098 (1980).

17. V. Šmíd, F. Schauer and N. M. Dung, Density of states in a-Si prepared by sputtering in Ne atmosphere, in: Ref. 5, vol. 1, p. 180.

18. F. Schauer, O. Zmeškal and S. Nešpůrek, A new approach to the analyses of steady-state space-charge-limited currents using their activation energies, Phys. Stat. Sol. (a) 75:591 (1983).

19. W. E. Spear and P. G. LeComber, Electronic properties of substitutionally doped amorphous Si and Ge, Phil. Mag. 33:935 (1976).

20. J. D. Cohen and D. V. Lang, Calculation of the dynamic response of Schottky barriers with a continuous distribution of gap states, Phys. Rev. B 25:5321 (1982).

21. R. A. Street and D. K. Biegelsen, The spectroscopy of localized states, in: "The Physics of Hydrogenated Amorphous Silicon II," J. D. Joannopoulos and G. Lucovsky, eds., Topic in Applied Physics vol. 56, Springer-Verlag, Berlin-Heidelberg-New York-Tokyo (1984), p. 194.

22. D. V. Lang, J. D. Cohen and J. P. Harbison, Measurement of the density of gap states in hydrogenated amorphous silicon by space charge spectroscopy, Phys. Rev. B. 25:5285 (1982).

23. R. L. Anderson, Experiments on Ge-GaAs heterojunctions, Solid State Electron. 5:341 (1962).

24. N. F. Mott and E. A. Davis,"Electron Processes in Non-Crystalline Materials," second edition, Clarendon Press, Oxford (1979).

25. A. Yoshida, T. Ido and A. Arizumi, Heterojunctions composed of amorphous and crystalline Ge, in: "Electronic Phenomena in Non-Crystalline Semiconductors," B. T. Kolomiets, ed., Nauka, Leningrad (1976), p. 407.

26. M. H. Brodsky, G. H. Döhler and P. J. Steinhard, On the measurement of the conductivity density of states of evaporated amorphous silicon films, Phys. Stat. Sol. (b) 72:761 (1975).

27. B. T. Kolomiets and V. M. Lyubin, Photoelectric phenomena in amorphous chalcogenide semiconductors, Phys. Stat. Sol. (a) 17:11 (1973).

28. A. Munoz-Yague, J. Piqueras and N. Fabre, Preparation of carbon-free GaAs surfaces, J. Electrochem. Soc.:Solid State Science and Technology 128:149 (1981).

29. J. Zemek, M. Závětová and S. Koc, On the role of hydrogen in a-Si, J. Non-Crystalline Solids 37:15 (1980).

30. V. Šmíd, N. M. Dung, L. Štourač and K. Jurek, Field effect in dc-sputtered a-Si:H in structure using SiN_x prepared in situ. J. Non-Crystalline Solids, in press.

31. M, Závětová, unpublished results.

32. L. Ley, Photoemission and optical properties, in: Ref. 21, p. 61.

33. J. J. Mareš, "Transport Properties of a-Ge/c-GaAs Heterojunctions," Ph.D. thesis, Institute of Physics, Prague 1983. (in Czech)

34. J. J. Mareš, V. Šmíd, L. Štourač and M. Libra, Heterojunctions c-GaAs/amorphous semiconductor, in: "Proc. 5th Czech. Conf. on GaAs and Related Compounds,"Smolenice 1982, Slovak Acad. Sci., Bratislava (1982), p. 41. (in Czech)

35. J. J. Mareš, L. Štourač, J. Krištofik and V. Šmíd, I-V measurements of a-Ge/c-GaAs heterojunctions, in: Ref. 5, Vol. 1, p. 242.

36. M. A. Lampert and P. Mark, "Current Injection In Solids," Academic Press, New York and London (1970).

37. N. F. Mott, Electrons in glass, Contemp. Phys. 18:225 (1977).

38. F. Schauer, V. Šmíd,and V. Žaludek, On contact capacitance and the field effect in chalcogenide glasses, Phys. Stat. Sol. (a) 49:K157 (1978).

39. D. L. Camphausen, G. A. N. Connell and W. Paul, A model for transport in amorphous germanium, J. Non-Crystalline Solids (8-10):223 (1972).
40. A. R. Hutson, A. Jayaraman and A. S. Coriell, Effects of high pressure, uniaxial stress, and temperature on the electrical resistivity of n-GaAs, Phys. Rev. 155:786 (1967).
41. G. G. Roberts and F. W. Schmidlin, Study of localized levels in semi-insulators by combined measurements of thermally activated ohmic and space-charge-limited conduction, Phys. Rev. 180:785 (1969).
42. H. Matsuura, A. Matsuda, H. Okushi, T. Okuno and K. Tanaka, Metal-semiconductor junctions and amorphous-crystalline heterojunctions using B-doped hydrogenated amorphous silicon, Appl. Phys. Lett. 45:433 (1984).
43. H. Matsuura, T. Okuno, H. Okushi and K. Tanaka, Electrical properties of n-amorphous/p-crystalline silicon heterojunctions, J. Appl. Phys. 55:1012 (1984).
44. S. Nešpůrek, O. Zmeškal and F. Schauer, An ohmic contact test using the activation energy of steady-state SCLC, Phys. Stat. Sol. (a) 85:619 (1984).
45. H. J. Quessier, H. C. Casey, Jr. and W. van Roosbroeck, Carrier transport and potential distributions for a semiconductor p-n junction in the relaxation regime, Phys. Rev. Lett. 26:551 (1971).
46. A. S. Grove, B. E. Deal, E. H. Snow and C. T. Sah, Investigation of thermally oxidised silicon surfaces using metal-oxide-semiconductor structures, Solid State Electron. 8:145 (1965).
47. G. Sasaki, S. Fujita and A. Sasaki, Gap-state measurement of chemically vapor-deposited amorphous silicon, J. Appl. Phys. 53:1013 (1982).
48. V. Šmíd, Z. Chaloupka and J. Krištofik, Capacitance-voltage measurements of a-Si/a-Si:H heterojunctions, in: Ref. 5, vol. 1, p. 239.
49. F. Schauer, O. Zmeškal and V. Šmíd, Interpretation of space charge limited currents in amorphous semiconductors, in: Ref. 8, p. 243.
50. J. W. Orton and M. J. Powell, The relationship between space-charge-limited current and density of states in amorphous silicon, Phil. Mag. 50:11 (1984).
51. W. den Boer, Determination of midgap density of states in a-Si:H using space-charge-limited current measurements, Journal de Physique 42:C4-451 (1981).
52. K. D. Mackenzie, P. G. LeComber and W. E. Spear, The density of states in amorphous silicon determined by space-charge-limited current measurements, Phil. Mag. 46:377 (1982)
53. L. Štourač, Amorphous-crystalline heterojunctions, in: Ref. 8, p. 104.

ON THE PROPERTIES OF QUASI ONE DIMENSIONAL HYDROGENATED AMORPHOUS SILICON FILMS

Shoji Nitta, Masato Kawai, Masahiko Sakaida,
Isao Murase and Akitsugu Hatano

Department of Electrical Engineering
Gifu University, Yanaido, Gifu 501-11
Japan

ABSTRACT

Structural and optical properties of quasi-one dimensional hydrogenated amorphous silicon films which contain mainly polysilane $(SiH_2)_n$ are discussed using the gas evolution spectra, the infrared absorption spectra, ESR, the reflectance and transmittance spectra and the photoluminescence spectra.

INTRODUCTION

Hydrogenated amorphous silicon (a-Si:H) has been investigated fundamentally and has been applied to the solar cells, the imaging devices and others. Other tetrahedrally bonded hydrogenated amorphous semiconductors such as $a\text{-}Si_{1-x}C_x\text{:}H$, $a\text{-}Si_{1-x}N_x\text{:}H$, $a\text{-}Si_{1-x}Ge_x\text{:}H$ etc. have been investigated and apllied also. These tetrahedrally bonded amorphous semiconductors have three dimensional structures roughly approximated by three dimensional random network.

Recently the interest on the electric properties of lower dimensional amorphous semiconductors has been increased. In the tetrahedrally bonded disordered semiconductors, the two dimensional system is realized by two methods. Making a potential well or amorphous superlattice, the electronically two dimensional system is realized even though the network of the chemical bonds is three dimensional.[1,2]

The other is siloxene in which one bond in tetrahedral four-sp^3-chemical bonds is terminated by a hydrogen atom H or OH and three other bonds are used to form two dimensional lattice and these layers are stacked by van der Waals force.[3,4,5] Siloxiene has disordered layer structure and is structurally two dimensional. Siloxene is made from $CaSi_2$ by deintercalation and intercalation reaction.

One dimensional system using a potential well is realized by MOSFET using a gate of submicron (0.07~0.1 μm) width for crystalline silicon but not yet for amorphous tetrahedrally bonded semiconductors.

In the films grown on relatively low temperature substrates by glow discharge decomposition of disilane Si_2H_6 and by HOMOCVD of SiH_4, two bonds in four-sp^3 chemical bonds of a Si atom are terminated by hydrogen

Table. 1. Time of deposition and substrate temperature

Sample number	Time of deposition (minuits)	Substrate temperature (°C)
94	5	16
98	5.5	105
99	2	246
103	5	58
108	5	24

atoms then the chain structure is formed using another two bonds. These materials show some charactristic of quasi-one dimensional amorphous semiconductors having the fundamental chain of polysilan$(SiH_2)_n$.

These low temperature deposited films show efficient visible photoluminescence[6~9] The infrared absorption spectra of these materials are analysed by Furukawa and Matsumoto[10] They used the ratio of the peak absorption coefficient α near $850cm^{-1}$ to that near $890cm^{-1}$ $\alpha(850cm^{-1})$/ $\alpha(890cm^{-1})$ as a meaure of a $(SiH_2)_n$ chain formation.

Recently Yoshida et. al. observed[6] a big luminescence fatigue effect in the high energy part of the luminescence spectrum which is due to the creation of dangling bonds at the region containing a large amount of hydrogen in the form of polysilane.[11,12]

In this paper we present the experimental results on the gas evolution spectra, the infrared absorption spectra, ESR, the reflectance and transmittance spectra and on photoluminescence spectra of sample prepared by glow-discharge decomposition of disilane.

EXPERIMENTAL

The samples were prepared by the glow discharge of $\sim$1 Torr 100% disilane[13] using a 13.56 MHz rf-inductively-coupled system. Time of deposition and substrate temperature are shown in table 1. A high accurate apparatus were used to study the gas evolution spectra.[14,15] Conventional apparatus JASCO-A-102 and JEOL-JES-FE1X were used to study the infrared absorption spectra and ESR. The reflectance and transmittance of the samples were measured at near normal incidence. And the optical constants, the film thickness d and the absorption coefficient α were determined by the self-consistent method.[16]

Photoluminescence spectra were taken at 77 K under the excitation by unfocused argon ion laser light 514.5nm using a single prism monochrometer and a photomultipler, and the output of lock-in-amplifier were stored and analysed in the microcomputer system.

EXPERIMENTAL RESULTS AND DISCUSSIONS

The gas evolution rate dp/dT versus temperature T characteristic curve at low substrate temperature T_s=16 C is shown in fig. 1. Size of a sample is 3x10 mm^2 and thickness of films d is 3.75 μm. Heating ratio is constant and 20 C/min. All of the evolved gases are hydrogen as identified by mass spectroscopy. Two peaks of dp/dT observed at $\sim$300 C and

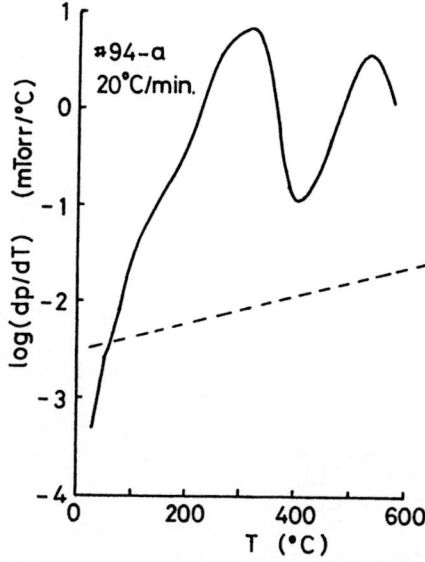

Fig. 1. The gas evolution spectrum dp/dT–T of a quasi-one dimensional hydrogenated amorphous silicon film. A dotted line shows the noise level of the experiment. T is temperature and p is pressure.

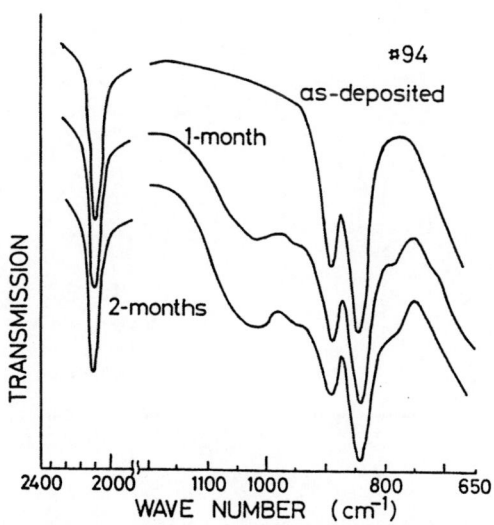

Fig. 2. The infrared absorption spectra of a sample kept in air change with time.

~500 C are similar with conventional three dimensional hydrogenated amorphous silicon (a-Si:H) films. A peak at ~300 C corresponds to the hydrogen concentrated phase mainly of $(SiH_2)_n$, and is larger and clearer compared with conventional a-Si:H. Second peak at ~500 C is from the hydrogen diluted phase of a-Si:H after the films changed to conventional three dimensional a-Si:H by annealing in the process of heating. A shoulder observed at ~150 C is assigned as originated from softly bonded and/or adsorbed gases at surface mainly of hydrogen.[15] This result is consistent with recent results from photoemission.[17]

Properties of films deposited on low temperature substrates change in air with the time scale of 10 days.

In Fig 2, the infrared absorption spectra of films deposited at 16 C are shown. Absorption near 1000 cm^{-1} is showing that films are gradually oxidizing. This result shows the porosity of films. A large absorption peak at 845 cm^{-1} relevant to $(SiH_2)_n$ wagging mode is showing that films contain a large amount of polysilane $(SiH_2)_n$.[10] A peak at 890 cm^{-1} is relevant to $(SiH_2)_n$ and SiH_2 bending modes.

Fig. 3 shows the dependence of absorption spectra of films on the substrate temperature T_s. In films deposited as T_s=16 C and 105 C, the integrated absorption strength at 845 cm^{-1} are large enough to show the existence of polysilane $(SiH_2)_n$. There is no indication of the existence of polysilane in a film deposited at T_s=246 C. From these results, it is clear that the properties of films change drastically between T_s=105 C and 246 C.

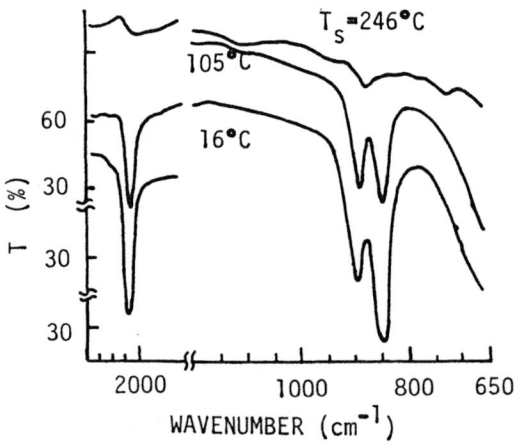

Fig. 3. Dependence of infrared absorption spectra of sample #94, 98 and 99 on substrate temperature Ts. Single crystalline silicon were used for the substrate.

The reflectance spectra of several samples are shown in Fig. 4. The reflectance is larger for the higher substrate temperature. These results are showing that the films deposited at higher substrate temperature are structurally more dense. In a film deposited at 16 C, the dependence of reflectance on the substrate is observed. And the reflectance of a film deposited at 16 C on a Corning 7059 substrate is smaller than that on a sapphire substrate. These results are showing that a film deposited on a sapphire substrate is more dense or has different morphology compared with that on Corning 7059 substrate.

Dependence of optical gap E_0 obtained from Tauc's plot on substrate is shown in Fig. 5. Dependence of E_0 on substrate is observed at $T_s < 100C$.

The real part of refractive indices n and the imaginary part of dielectric constants ε_2 in fig. 6 and 7 are obtained from self-consistent analysis using the Kramers-Kronig transformation of the reflectance and transmittance of samples. N and ε_2 spectra for films deposited at 246 C are similar with conventional three dimensional a-Si:H prepared from SiH_4. Especially ε_2 spectra of films on lower substrate temperature such as 105 C and 16 C are much smaller than that on $T_s=246$ C. These are related with the difference of the electronic density of states with dimension.

Relative luminescence spectra measured at 77K are compared in Fig. 8. A film on 246C shows similar characteristics to convensional a-Si:H. Films on 16 C and 105 C show almost same photoluminescence spectra but different intensity. In these films, the luminescence in visible light region is observed as is consistent with the results by Wolford et al.[7,8] As is shown in Fig. 8, the luminescence intensity in visible light region is stronger in a film deposited at 105 C than that at 16 C. This result is interesting for the application of these films to the light emitting diode.

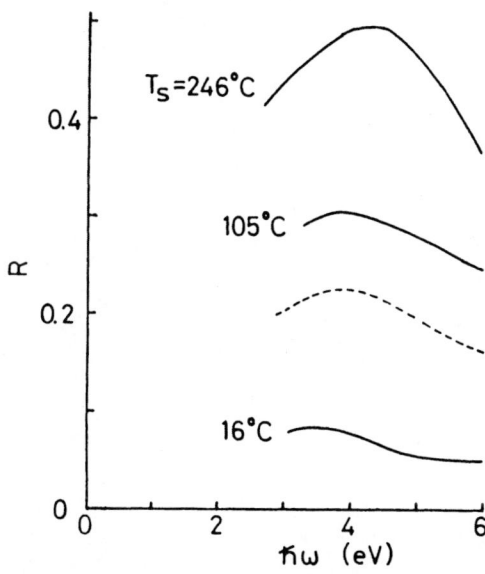

Fig. 4. Dependence of reflectance spectra on substrate temperature of samples deposited on Corning 7059 substrate. A dotted line is the reflectance spectrum of a sample deposited on sapphire of 16 C.

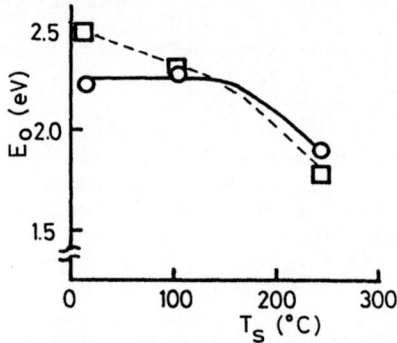

Fig. 5. Dependence of the optical gap on substrate temprature. Circles and a real line are for samples deposited on Corning 7059 substrate and regular squares and a dotted line are for samples deposited on sapphire substrate.

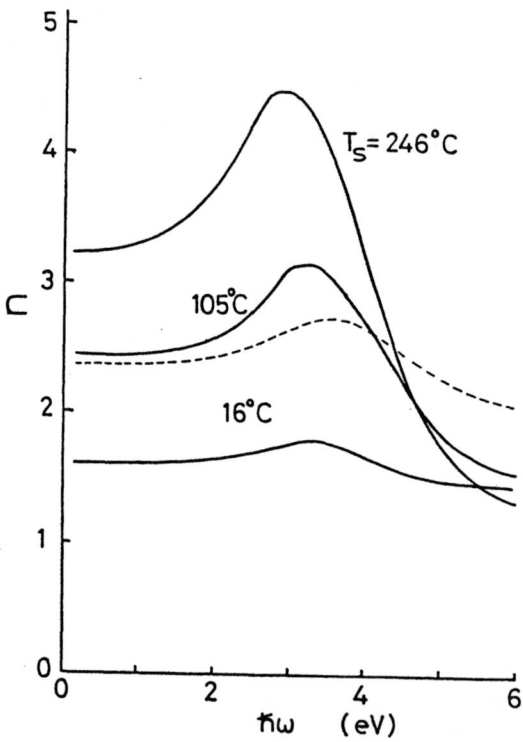

Fig. 6. Dependence of the real part of the refractive index n of samples deposited on Corning 7059 substrate on substrate temperature. A dotted line shows that for a sapphire substrate kept at 16 C.

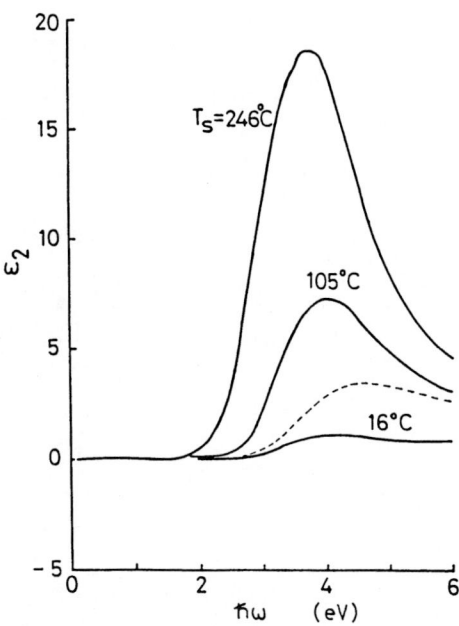

Fig 7. Dependence of the imaginary part of the
dielectric   constants ε_2 of   samples
deposited   on Corning 7059 substrates on
substrate temperature.   A   dotted line
shows that for a sapphire substrate kept
16 C.

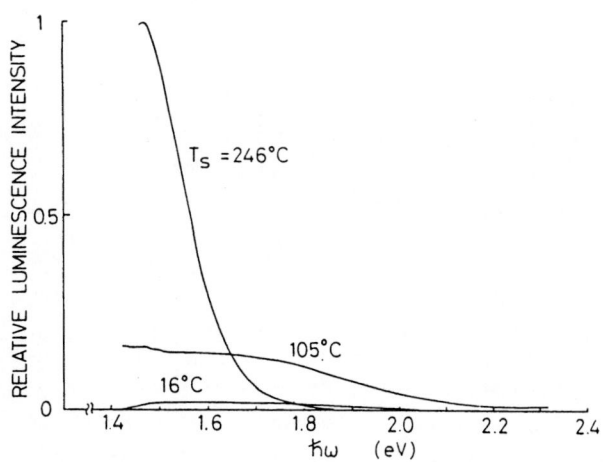

Fig. 8. The photoluminescence intensity are compared
for sample deposited at different temperature.

Annealing experiments are done on films deposited at low substrate temperature in vacuum of $\sim 10^{-6}$ Torr for 30 minutes at each temperature. Dependence of the ratio of infrared absorption coefficients α at 845 cm^{-1} and 890cm^{-1} $\alpha(845cm^{-1})/\alpha(890cm^{-1})$, that is a good measure of a polysilane $(SiH_2)_n$ chain formation in the films[10] is shown in Fig. 9. This figure is showing that the drastical change from quasi-one dimensional to three dimensional occurs between $T_a=250$ C and 300 C. The ratio $\alpha(845cm^{-1})/\alpha(890 cm^{-1})$ do not change so much below $T_a \leq 250$ C. And the effects of substrate temperature T_c on the polysilane chain formation is larger compared with that of annealing temperature T_a below $T_a < 250$ C.

Fig. 10 shows the dependence of electron spin density and g-value on the annealing. Dependense of spin density on T is similar with the results by Wolford et al.[6] The value of g factor about 2.004 that is due to dangling bond center changes to that for conventional a-Si:H between $T_a=300$ C and 350 C. These results also show that films deposited on low temperature substrates change from quasi-one dimensional films to

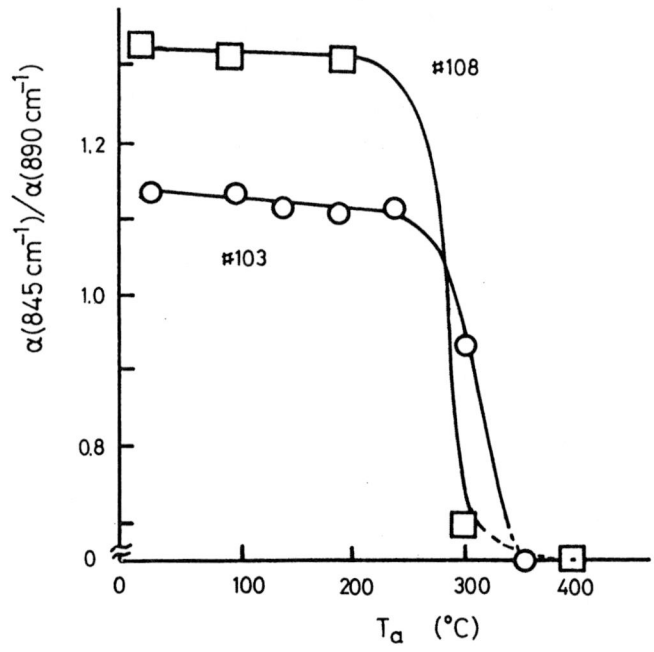

Fig. 9. Dependence of the ratio of absorption coefficient α at 890 cm^{-1} and 845 cm^{-1} on the annealing temperature Ta.

three dimensional by annealing at $T_a \sim 300$ C. Glow discharge film from Si_2H_6 deposited at low substrate temperature such as $T_s=16$ C and 105C show the characteristics of the quasi-one dimesional amorphous semiconductors. The average covalent coordination number m for polysilane $(SiH_2)_n$ is two. But as shown in Fig. 11, the position in J.C.Phillips-classification figure of these low substrate temperature films depends on the substrate and annealing temperature. In the process of annealing, there is no indication that films have two dimensinal structure as in siloxene. The difference of the quasi- one dimensional hydrogenated amorphous silicon and the conventional three dimensional hydrogenated silicon is clear from IR spectra, ε_2 spectra and g-value. These quasi-one dimensional hydrogenated silicon films contain mainly polysilane $(SiH_2)_n$. And the termination and the crosslink of chain as shown in fig.12 depends on the substrate and annealing temperature.

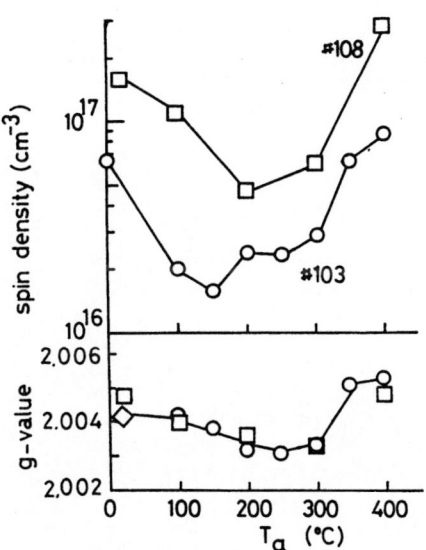

Fig. 10. Dependence of electron spin density and g-value of samples on the annealing temperature.

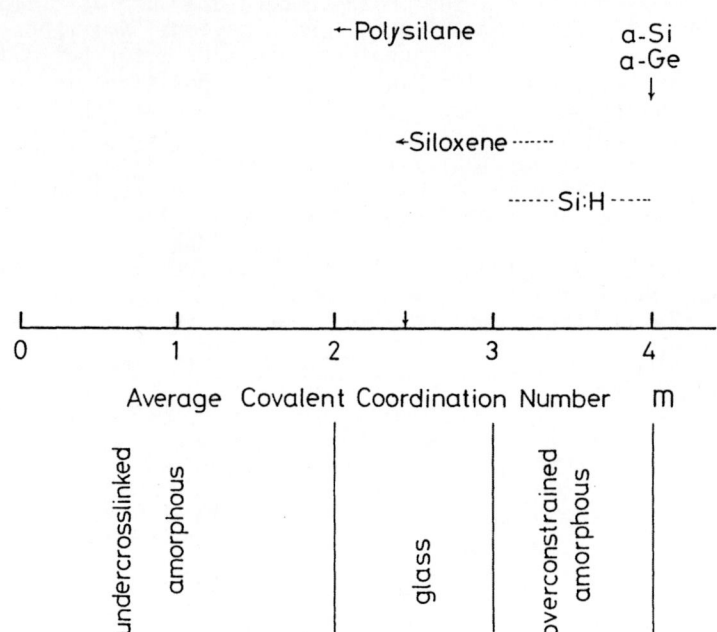

Fig. 11. Quasi-one dimensional hydrogenated amorphous silicon in J.C.Phillips's classification of amorphous semiconductors.

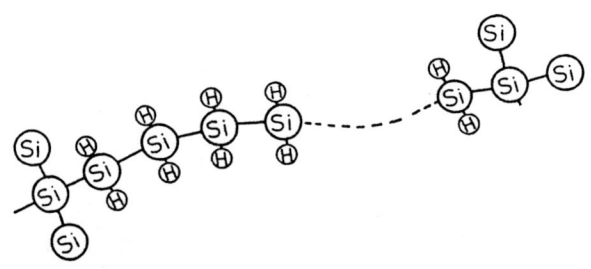

Fig. 12. A model for the quasi–one dimensional hydrogenated amorphous silicon.

CONCLUSION

It is shown that films deposited on low substrate temperature such as 16 C and 105 C show quasi-one dimensional properties. These films are oxidized easily in air at the time scale of 10 days. Amount of polysilane $(SiH_2)_n$ in films depends on substrate and annealing temperature. These low substrate temperature films change to conventional three dimensional a-Si: H by annealing for 30 minuits at $T_a \sim 300$ C.

ACKNOWLEDGEMENTS

We wish to thank for Prof. K.Morigaki, Dr. I.Hirabayashi and Miss. M.Yoshida for helpful discussions, and Mr. M. Watanabe and Mr. N.Matsunami for help for experiments. This work was supported in part from Toagousei Chemical Industry Co. financially and in part by the contract from the Foundation for Promotion on Material Science and Technology.

REFERENCES

1. B. Abeles & T. Tiedje, Phys. Rev. Lett., 51, 2003 (1983)
2. H. Munekata, M. Mizuta & H. Kukimoto, J. Non-Cryst. Solids, 59&60, 1167 (1983)
3. I. Hirabayashi, K. Morigaki & S. Yamanaka, J. Phys. Soc. Japan, 52, 671 (1983)
4. I. Hirabayashi, K. Morigaki & S. Yamanaka, J. Non-Cryst. Solids, 59& 60, 645 (1983)
5. S. Nitta, A. Hatano, I. Hirabayashi & S. Yamanaka, Abstract of 7th International Symposium on Organosilicon Chemistry, (The Chemical Soc. of Japan) p. 143 (1984)
6. D. J. Wolford, J. A. Reimer & B. A. Scott, Appl. Phys. Lett., 42,369 (1983)
7. D. J. Wolford, B. A. Scott, J. A. Reimer & J. A. Bradley, Physica,117B & 118B, 920 (1983)
8. B. A. Scott, R. M. Plecenik & E. E. Simonyi, J. de Phys. C4-42, 635 (1981)
9. B. A. Scott, W. L. Olbricht, B. A. Meyerson, J. A. Reimer & D. J. Wolford, J. Vac. Sci. Technol., A2, 450 (1984)
10. S. Furukawa & N. Matsumoto, Solid State Commun., 48, 539 (1983)
11. M. Yoshida, K. Morigaki & S. Nitta, Solid State Commun., 51, 1 (1984)
12. M. Yoshida, K. Morigaki, I. Hirabayashi, H. Ohta, A. Amanou & S. Nitta, AIP Conf. Proc., No. 120, 141 (1984)
13. 100% Si_2H_6 made at Toagosei Chemical Industry Co., LTD.
14. A. Hatano & S. Nitta, Solid State Physics (in Japanese),19, 283 (1984)
15. S. Nitta & A. Hatano, Tech. Digest of 1st International Photovoltaic S. & E. Conf., 715 (1984)
16. S. Nitta, S. Itoh, M. Tanaka, T. Endo & A. Hatano, Solar Energy Materials, 8, 249 (1982)
17. L. Ley, "The Physics of Hydrogenated Amorphous Silicon II", ed. by J. D. Joannopoulos and G. Lucovsky, p.61 (Springer Verlag, 1984)
18. J. C. Phillips, J. Non-Cryst. Solids, 34, 153 (1979)

RECENT ADVANCES IN AMORPHOUS SILICON AND ITS TECHNOLOGICAL APPLICATIONS

Yoshihiro Hamakawa

Faculty of Engineering Science
Osaka University
Toyonaka, Osaka, Japan 560

INTRODUCTION

In a recent few years, a remarkable progress has been seen in both physics and technologies of a-Si (hydrogenated amorphous silicon) as a new optoelectronic material[1]. Particularly, its significant properties such as excellent photoconductivity with a considerably high optical absorption coefficient for visible light, and ability to produce a non-epitaxial film growth on any foreign substrates match very timely with a strong potential need for the development of a low cost solar cell as a new energy resources[2]. With aids of the national or semi-national project supports for the photovoltaic development, an accelerative promotion has been seen in a wide variety of technologies from the film deposition with valency electron control to device fabrication processing. As the results, more than 12.5% conversion efficiency has been attained in a small area (6x6-10x10 mm^2) laboratory phase cells with the a-Si/poly-Si stacked junction structure[3]. For the large area cells of 10x10 cm^2 module, more than 7% efficiency are presently quite common elsewhere in the industrial productions[4]. These technological achievements have extended to other new device developments, such as photo-sensors, imaging devices, photo-receptors, TFT (Thin Film Transistor), etc.

In this review, firstly, unique advantages of these tetrahedrally bonded amorphous semiconductors are enumerated with remarkable tangible evidences from current technologies. After that, some recent R&D efforts to improve the device performance are introduced, and achieved key technologies in the optoelectronic device fields are demonstrated mainly from progresses in a Japanese domestic activities.

UNIQUE ADVANTAGES OF a-Si ALLOYS AS A NEW ELECTRONIC MATERIAL

First of all, some unique physical properties and remarkable advantages of a-Si alloys as a new optoelectronic material are pointed out from both basic physics and technological view points as following,

a) High Optical Absorption and Large Photoconductivity in Visible Region

Figure 1 shows the optical absorption coefficient spectra of a-Si:H

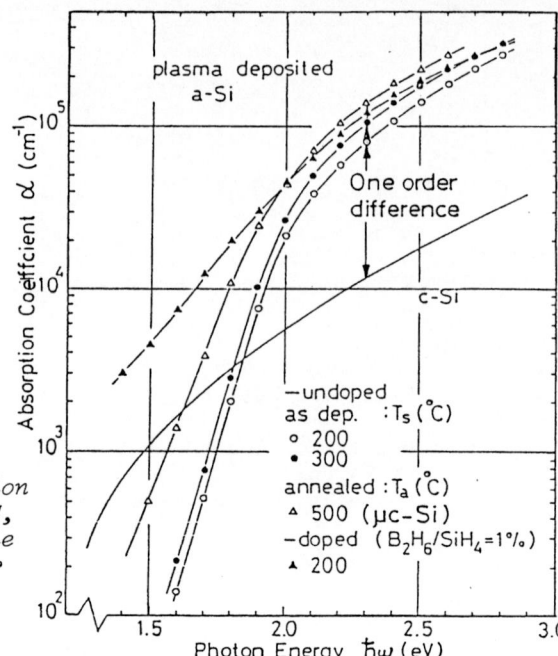

Fig.1 Comparison of absorption
coefficient spectra of a-Si:H,
microcrystalline Si and single
crystalline Si with the solar
irradiation spectrum.

and microsrystalline Si:H (μc–Si:H) compared with that of single crystal silicon. It can be seen that the absorption coefficients of a–Si:H and microcrystalline Si (μc–Si:H) are more than an order of magnitudes larger than that of single crystal silicon near the maximum solar photon energy region near 5000A which is shown by the broken curve in Fig.1[4]. Moreover, a–Si:H has excellent photoconductivity in the visible photon energy region. The ratio of photoconductivity to dark conductivity σ_{ph}/σ_D is 10^5–10^7, and σ_D is of the order of 10^{-9}–10^{-11} Ω^{-1} cm^{-1}.

b) An Existing of Valency Electron Controllability

Another noticeable property in these hydrogenated tetrahedrally bonded amorphous semiconductor is an existence of valency electron controllability by doping of the substitutional impurity atoms. Effects of impurity doping on the conductivity, optical absorption coefficient and optical energy gap have been intensively investigated on a–Si by Okamoto et al.[5], fluorinated amorphous silicon (a–Si:F:H) by Madan et al.[6] and hydrogenated amorphous silicon carbide (a–SiC:H) by Tawada et al.[7], and still now in progress for amorphous silicon germanium (a–SiGe:H) by Yukimoto et al.[8] and amorphous silicon tin (a–SiSn:H) by Kuwano et al.[9]. Figure 2 shows an example of substitutional impurity doping characteristics on the electronic properties of a–Si:H and a–SiC:H[8,9].

c) Low Cost with Energy Saving Material

A systematic calculation based upon the opto–electronic properties of these materials have been made for the optimum thickness of the solar photovoltaic active region for various solar cell materials[9]. The results are shown in Fig.3. The optimum thickness required to obtain the most cost-effective photovoltaic performance for crystalline silicon p/n junction and a–Si p–i–n junction solar cells are notified by point Y[10]. These results indicate that the optimum thickness of active layer in a–Si

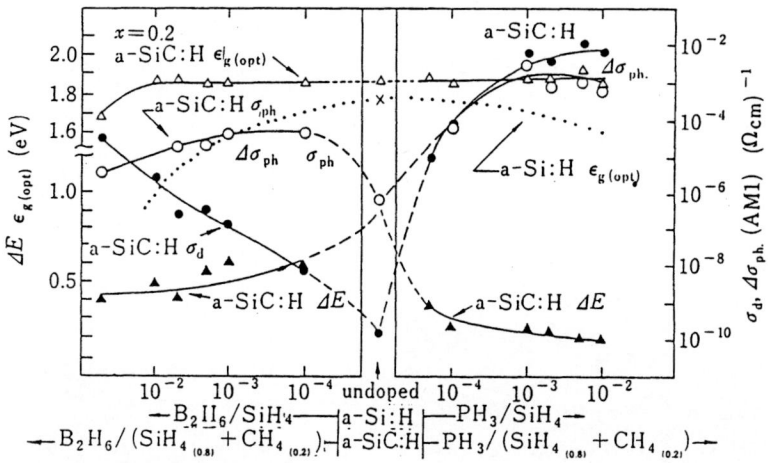

Fig.2 Effect of impurity doping on $E_{g(opt)}$, σ_d, σ_{ph} and activation energies ΔE in a-Si:H and a-SiC:H.

solar cells with backside reflector is only 0.5-0.7 micron depending on the cell structure. This optimum thickness is less than 1/500 to that for single-crystal silicon solar cell. Therefore, a-Si is both an energy saving and a resouce saving material for the large demands in solar photovoltaic applications anticipated in the future.

d) Large Area Non-epitaxial Growth on Any Foreign Substrate Material at Low Temperature

Due to its amorphous structure, a-Si can be deposited on any inexpensive substrate, which needs only to be heated to a relatively low temperature, less than 200-300°C. Moreover, it is possible to form a very wide area solar cell, because it can be deposited directly from a kind of vapor phase growth onto non-crystalline substrates.

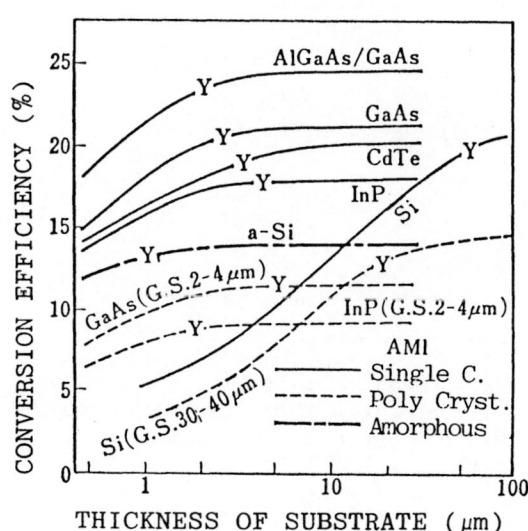

Fig.3 Cell efficiency as a function of cell thickness for p-n junction solar cells made from various semiconductors.

e) Low Balance of System Cost

Utilizing the concept of non-epitaxial deposition technology, it could be possible to reduce BOS (balance of system) costs in photovoltaic arrays by the hybridization of already-built units. Solar tile and sticker from solar cell might be useful to realize this concept. Figure 4 shows an example of various a-Si solar cells deposited on glass[11,12], ceramics[13], Kapton films[14] and stainless steel[15].

Fig.4 a-Si solar cells deposited on various foreign substrate materials; (a) Glass substrate consumer a-Si solar cells by Fuji Electric Co. Ltd., (b) Stainless steel substrate a-Si solar cell produced by role-to-role massproduction line presented by Sharp-ECD Solar Inc.; (c) Ceramic substrate a-Si solar tile by Kyocera Inc., (d) Polymer film substrate flexible a-Si solar cell by Teijin Co. Ltd., (e) deposited on glass roofing tile produced by Sanyo Electric Co. Ltd.

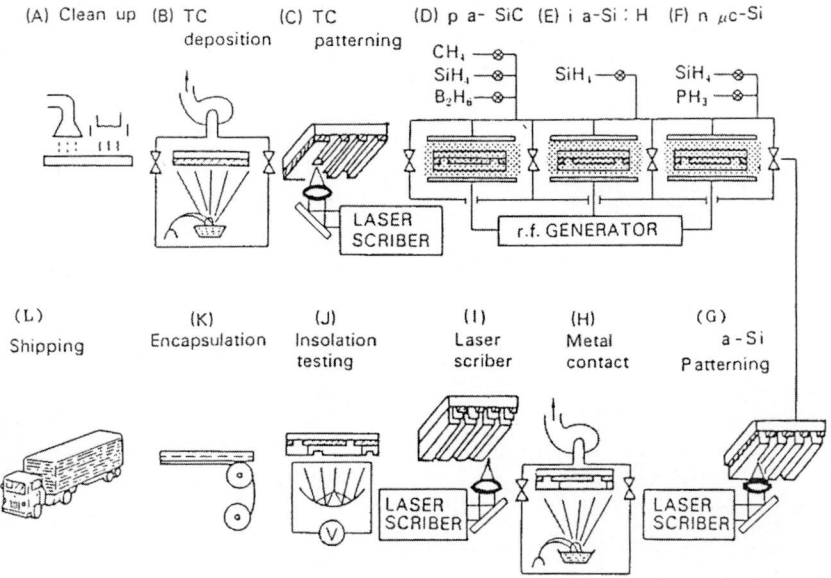

(A) Clean up (B) TC deposition (C) TC patterning (D) p a- SiC (E) i a-Si∶H (F) n μc-Si

CH₄ SiH₄ B₂H₆   SiH₄   SiH₄ PH₃

LASER SCRIBER

r.f. GENERATOR

(L) Shipping (K) Encapsulation (J) Insolation testing (I) Laser scriber (H) Metal contact (G) a-Si Patterning

LASER SCRIBER   LASER SCRIBER   LASER SCRIBER

Fig.5 A schematic representation of the all dry processing of a-Si solar cell massproduction in-line system.

f) Large Scale Merit with Big Massproduceability

As has been pointed out in the early stage of the work, junction formation can be easily made in the same reaction chamber by mixing of substitutional impurity gases into SiH_4 or SiF_4. Moreover, the interconnection of cells can be made in the process of a-Si film deposition with a conventional integrated-circuit photo-mask processing and also laser beam lithography. Combining the mass-production line could be easily accomplished in a all-dry processes as shown in a schematic representation of Fig.5.

RECENT PROGRESS OF THE a-Si SOLAR CELL R&D EFFORTS

In a view of basic physics, there are several peculiar things on the carrier recombination processes and carrier transport phenomena in amorphous semiconductors. The phenomena are mainly based upon an existence of localized states continuously distributed in the mobility gap, and classified as the basic processes of:

a) Dispersive carrier transport[16,17]
b) Geminate recombination[18,19]
c) Ambipolar carrier transport[20]
d) Drift type photovoltaic effect[21,22]

A basic difference in the photovoltaic process of a-Si solar cell to that of single crystal p-n junction is an existence of high electric field in the carrier generation region. Moreover, the internal electric field in the i-layer varies very sharply as functions of both the mid gap state density g_{min} and the induced space charge distributions as shown in Fig.6. Therefore, the geminate recombination process might be taken into account in the analysis of the photocarrier generation in a-Si:H p-i-n junction. This is so called the drift type photovoltaic effect in a-Si

517

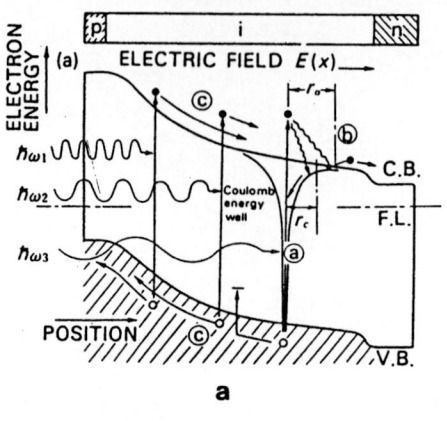

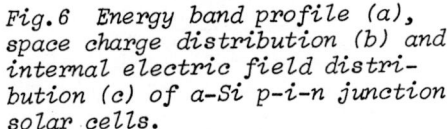

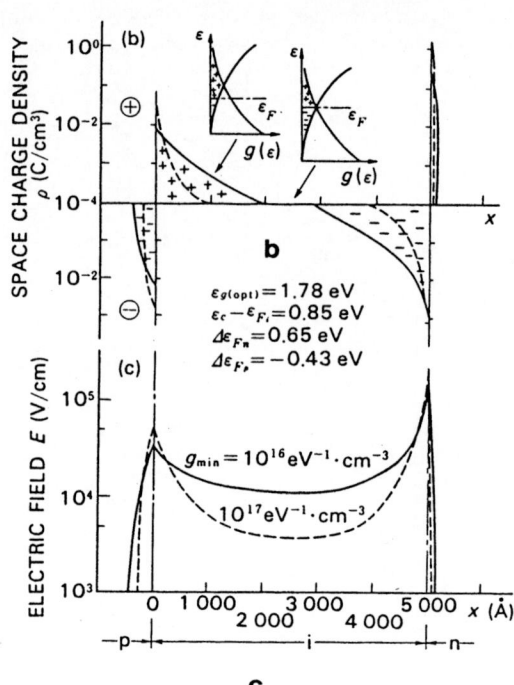

Fig.6 Energy band profile (a), space charge distribution (b) and internal electric field distribution (c) of a-Si p-i-n junction solar cells.

solar cells. Obviously, observable J-V characteristics of the a-Si p-i-n junction are quite different from those of single crystal p-n junction as shown in Fig.7.

The electric field dependence of photocarrier generation probability for a-Si:H has been recently calculated with taking into account of the geminate recombination process by Mort et al.[19] and Okamoto et al.[23] separately. According to their results, both of which are considerably good numerical coincidence, the photocarrier generation probability under the 10^4 V/cm reaches 0.8 or higher with the thermalization distances of more than 200 A. Variations of the photocarrier collection efficiency with the electric field in the a-Si:H p-i-n junction[24] has also been investigated by Osaka University group.

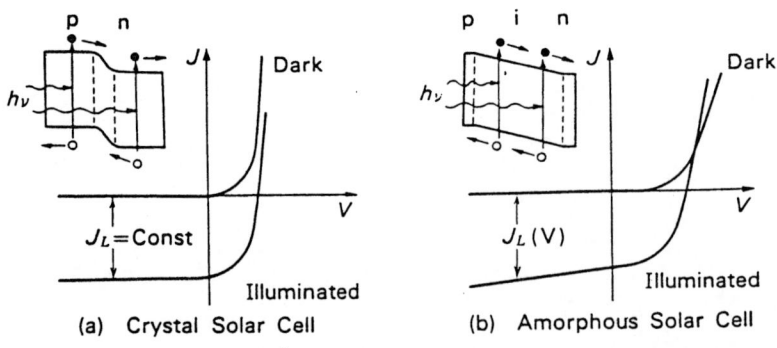

Fig.7 Comparison of J-V characteristics of crystal p-n junction (a) and a-Si p-i-n junction (b) for dark and light illuminated condition.

518

Table I Tetrahedrally bonded amorphous semiconductors and their physical properties.

	a-SiN	a-SiC	μc-Si	a-Si	a-SiGe	a-SiSn
source gases	$SiH_4 + NH_3$	$SiH_4 + CH_4$, C_2H_4 or $Si(CH_3)_4$	$SiH_4 + H_2$	SiH_4	$SiH_4 + GeH_4$	$SiH_4 + Sn(CH_3)_4$
optical energy gap (eV)	1.8 ~ 5.5	1.8 ~ 2.8	$\alpha < 10^4 cm^{-1}$ ($h\nu = 2$ eV)	1.7 ~ 2.0	1.0 ~ 1.7	1.0 ~ 1.7
dark conductivity ($1/\Omega \cdot cm$)	$10^{-10} \sim 10^{-8}$	$< 10^{-10}$	$10^{-5} \sim 10^{-2}$	$10^{-12} \sim 10^{-8}$	$10^{-8} \sim 10^{-5}$	$10^{-10} \sim 10^{-6}$
activation energy (eV)	0.7 ~ 1.0	< 0.9	—	0.7 ~ 0.9	0.5 ~ 0.9	0.3 ~ 0.6
photoconductivity: AM1,100mW/cm2 ($1/\Omega \cdot cm$)	$< 10^{-4}$	$< 10^{-7}$	$< 10^{-4}$	$< 10^{-3}$	$< 10^{-3}$	$< 10^{-6}$
mobility (cm2/Vs)	—	—	< 1 (ambipolar)	< 1 (e), $< 10^{-2}$ (h)	< 0.1 (e), $< 10^{-2}$ (h)	—
lifetime (s)	—	—	$< 10^{-5}$ (ambipolar)	$< 10^{-6}$ (e), $< 10^{-5}$ (h)	—	—
photoluminescence peak energy (eV)	1.4 ~ 1.6	1.5 ~ 2.3	1.2 ~ 1.4, 0.8	1.2 ~ 1.4	0.6 ~ 1.2	—
IR absorption peak position: streching mode (cm^{-1})	2140(SiH), 2340(NH)	2013 ~ 2135 (SiH), 2850, 2910 (CH2)	2089 ~ 2108 (SiH2), 2140, 2158 (SiH3)	2000(SiH), 2100(SiH2)	1850(GeH)	—
Raman shift (cm^{-1})	450	750	480, 520	480	370	—
hydrogen content (at.%)	10 ~ 50	10 ~ 50	< 8	5 ~ 50	4 ~ 40	—
spin density (cm^{-3}) g-value	$> 10^{17}$ 2.002 ~ 2.006	$> 10^{16}$ 2.0028(C)	$> 10^{17}$ 2.0049	$> 10^{15}$ 2.0055	$> 10^{16}$ 2.021 (Ge)	—

Through these basic investigations, electric field dependences of photocarrier generation probability and collection efficiency are parametrically evaluated by taking into account both the geminate and nongeminate recombination processes in the bulk and surface of amorphous silicon[21,24]. On the basis of the theory, the relationship between photovoltaic device performance and design parameters are clarified in the p-i-n basis of amorphous silicon photovoltaic devices[22].

A noticeable progress in the R&D effort in the recent few years is the opening of the amorphous mixed alloy age such as new materials of a-SiC, a-SiGe, a-SiSn etc. While, all these materials have considerably good valency controllability by the hydrogenation passivation of dangling bonds, and doping of substitutional impurity with proper gas mixture technique. At the beginning of introduction of this new material innovation, the optical energy gaps, electrical and optical properties of tetrahedrally bonded mixed alloys are summarized in Table I. Since 1978, a systematic investigations on the valency electron control of amorphous mixed alloy have been made by Hamakawa group at Osaka University. As an application of their results, they developed a-SiC/a-Si heterojunction

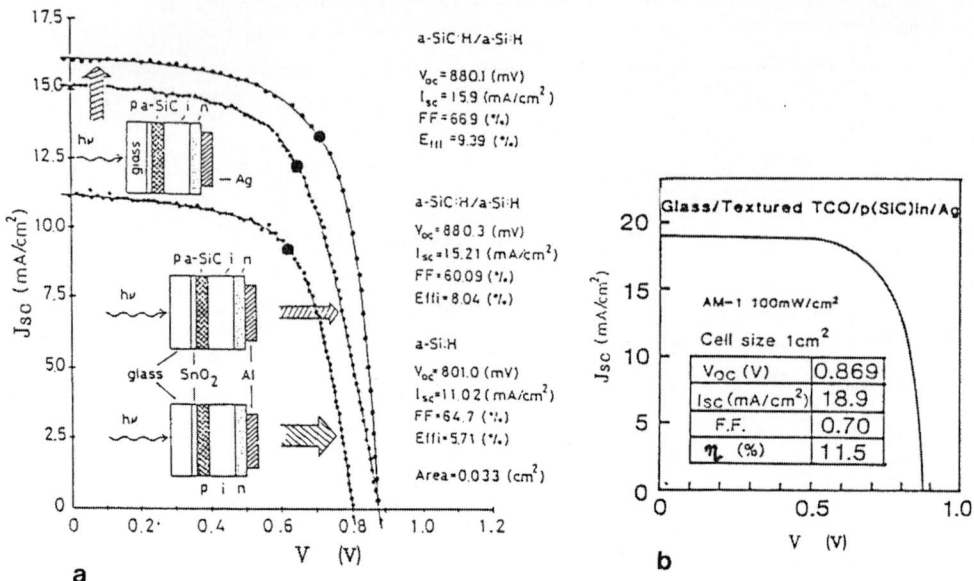

Fig.8 A step-like efficiency increase by p-type a-SiC window junction. (a) A clear evidence of wide gap window effect firstly proved by Hamakawa group in 1980[25], (b) In 1984 technology, the efficiency has been improved up to 11.5%[28].

solar cell having efficiency of more than 8% in 1980[25]. Figure 8 (a) reproduces the memorial record data brokenthrough the 8% efficiency barrier. Comparing with the current density-voltage (J-V) characteristics of an ordinary a-Si solar cell shows that not only the short circuit current density but also the open circuit voltage are clearly improved by utilizing p-type a-SiC:H as the window junction material. Recent developments in material synthesis technology in which the carbon fraction in $a-Si_{1-x}C_x$ is controlled[26] and in optimum design theory[22] based upon the concept of drift type photovoltaic process[20] have made rapid improvemnts in the efficiency of this type of heterojunction solar cell. In 1982, Catalano et al.[27] of the RCA group have obtained an efficiency of 10.1% with an a-SiC/a-Si heterojunction solar cell having an active area of 1.01 cm^2.
The best record of single p-i-n junction for the a-SiC/a-Si heterojunction solar cells having 11.5% efficiency fabricated using a separated three-chamber horizontal plasma furnace is shown in Fig.8 (b)[28].

Figure 9 shows the transitions of cell efficiency for various types of a-Si solar cells since 1976. As can be seen from this figure, a step-like increase of the cell efficiencies is seen in the region -1981, while the slope A before 1981 corresponds to the improvement of the film quality and routine cell fabrication progresses. The key technologies that lead to the steep slope change from A to B at 1981, were due to development of heterojunction solar cells with a-SiC:H[29] and a-SiGe:H[30]. With a full use of the wide gap window effect, an increase of the built-in potential and the minority carrier mirror effect in a-SiC/a-Si heterojunction have been intensively studied by Tawada et al.[26], Nonomura et al.[31] and Okamoto et al.[32]. Due to a recent advance of material synthesis technology by controlling the carbon fraction in $a-Si_{(1-x)}C_x$[33] and of the optimum design theory[34] based upon the concept of the drift type photovoltaic process, this type of heterojunction solar cell

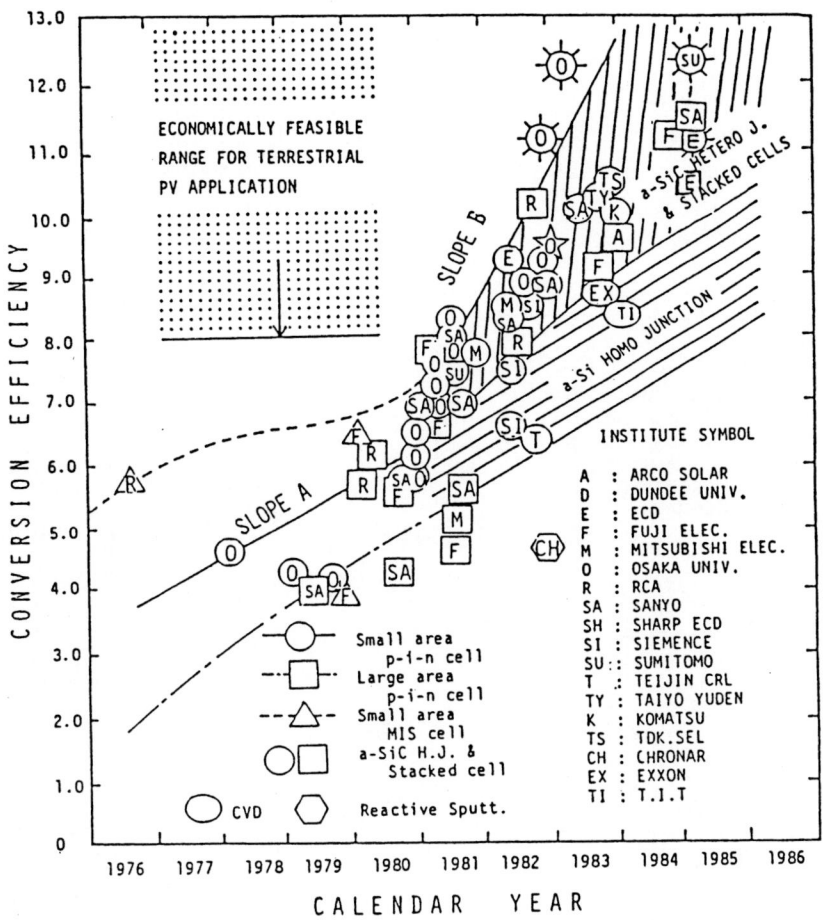

Fig.9 Progress of a-Si solar cell efficiencies for various types junction structures as of Sept. 1983. It is seen a steep slope change with appearance of new amorphous materials around 1980.

efficiency has continued to improve its efficiency with the slope B. The recent world record more than 10% efficiency by RCA[27] and following Sanyo[28], Fuji[35], Komatsu[36], TDK-SEL[37], ARCO[38] and ECD[39] records are also obtained in this type of a-SiC/a-Si heterojunction solar cell.

One important remaining area for a further improvement of a-Si solar cell efficiency is more efficient collection of low energy photons just above the band edge of a-Si, because the penetration depth of 1.8 eV photon, for example, is the order of 5 micron. While, the thickness of a-Si solar cell is only 0.6 micron. A concept of efficient collection of long wavelength photon energy by a highly reflective random surface has been firstly proposed by Boer et al[40] in 1981, and its theoretical basis was established by Exxon group[41]. This concept has been extended to more efficient utilization of optical and carrier confinement in the multilayered heterostructure junction[42]. Recently, Fujimoto et al.[43] have developed a practical technology with the cell structure of ITO/n microcrystalline Si/i-p a-Si/TiO$_2$/Ag plated semi-textured stainless steel having an efficiency of 9.17%. Quite recently, Taiyo-Yuden/ETL groups[44] have reported 10.26 % efficiency with the optical confinement effect employed by MTG (Milky Transparent Glass).

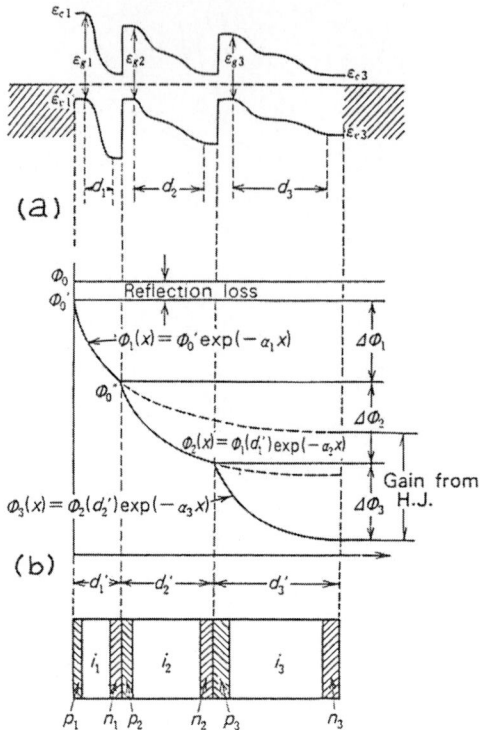

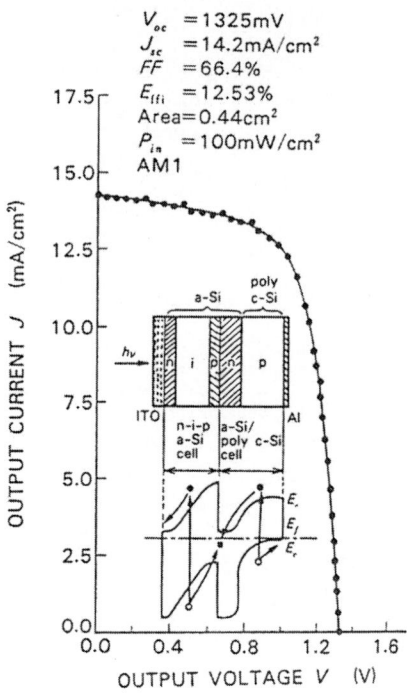

Fig.10 *Band diagram explanation of the multi-band gap staked solar cell (a) and its photon energy collection (b).*

Fig.11 *J-V characteristic of the a-Si/poly Si stacked solar cell and its junction structure.*

Another way to collect the longer wavelength photons is the absorption with the stacked junction of the lower energy gap semiconductor. The concept of efficiency improvement by the heterojunction stacked cell is shown in Fig.10. Because the energy gap in a-Si, 1.7–1.8 eV, is higher than that of crystalline solar cell semiconductor, e.g. 0.66 eV for Ge, 1.1 eV for Si and 1.43 eV for GaAs, the carrier collection efficiency for solar radiation spectrum in a-Si solar cell is considerably low as compared with that of the crystalline basis solar cells. On the other hand, the fabrication of large area polycrystalline thin films have been already established on wide variety of classical semiconductors with CVD, MOCVD, MBE, sputtering and ion plating etc.. By combining these well developed technology with the low temperature a-Si solar cell deposition technology, it is possible to make higher efficiency a-Si basis solar cell with low cost. Quite recently, Osaka University group have developed a new type of the stacked a-Si solar cell deposited on p-type polycrystalline silicon[45]. The cell structure and J-V characteristics are shown in Fig.11. As can be seen in this figure, more than 12.5% efficiency is easily obtained. Although a series of systematic studies on the material selection, and economical feasibility are now in progress, more than 15% efficiency would be obtained with the stacked a-Si solar cell on only one micrometer of polycrystalline GaAs or Ge thin film deposited on stainless steel substrates. In the case of polycrystalline silicon film, about 20–30 micron thickness will be required for the conventional CVD or Photo-CVD deposition method from SiH_4 or $SiHCl_3$.

Fig.12 *Expanding a wide variety of consumer applications (presented by Sanyo Electric Co. Ltd.).*

A wide variety of application systems are developed in a recent few years particularly, consumer electronic applications such as showing in Fig.12 are expanding very rapidly [4]. For instance, about 5.0 Million sets/month of a-Si drived pocketable calculators are fabricated in Japan as of 1984. On the other hand, the field of power application has been still in the experimental phase. Table II shows a list of experimental power plants drived by a-Si solar cells installed in Japan. To demonstrate the present feature in the field, residental a-Si photovoltaic/vacuum tube type photothermal solar house (a) and weather test plant (b) are shown in Fig.13.

INTEGRATED PHOTO-SENSOR AND COLOUR SENSOR

An advantage of large area, uniform film growth in the glow discharge produced a-Si:H is fully utilized on the development of integrated photosensors. Pioneer work has been done by the Musashino ETL group for non-strage type linear sensor in 1980[46]. a-Si strage mode linear photosensors have been developed by Fuji-Xerox [47], Sony [48] and Fujitsu [49]. A 512 element linear photosensor for the facsimile read out element by Kanoh et al. [48] of Sony group. Figure 14 shows the structure of the sensing element which consists of a-Si integrated sensor, light shield plate, two tungsten tubular lamps and guide plate. As can be seen in the figure, 4 MOS IC chips having 4x128 bits shift registor and 4x1278 switching transistors are mounted at the front of the element. The MOS IC shift registor are operated up to 1 MHz clock pulse frequency. This means that a 2048-element a-Si sensors can be operated with a frame time of less than 5 msec.

Fuji Xerox group has also reported 1056 bits, (8 bits/mm) [47], and 210 mm 1728 bits (8 bits/mm) linear sensors [50]. In this device, C-MOS analogue switches, shift resistors and interconnecting circuit are mounted on a single ceramic substrate having the dimension of 240mmx62mm. A new noise reduction technique as high as 25 dB S/N ratio. Recently, a strage mode matrix driven linear photosensor has been

Fig.13(a) a-Si solar photovoltaic with solar heating and
cooling facilities on experimental solar house built in
Osaka City (presented by Sanyo
Electric Co. Ltd.).

Fig.13(b) a-Si solar cell
weathering test plant
installed at Tokyo Electric
Power Co. Ltd. conducting
by a joint research
program with Fuji Electric
Co. Ltd. (presented by
Fuji Electric Co. Ltd.).

Table II a-Si solar cell utility power application demonstration/
experimental plants in Japan (as of December 1983).

Year Installed	Application	Location	Power Size (kW$_p$)	Site Owner	Prime Contractor
1981	Residential	Moriguchi Osaka	2.0	Sanyo Electric Co. Ltd.	Sanyo Electric Co. Ltd.
1981	Test Plant	Chofu Tokyo	0.4	Tokyo Electric Power Co. Ltd.	Fuji Electric Co. Ltd.
1981	Test Plant	Hirakata Osaka	0.4	Sanyo Electic Co. Ltd.	Sanyo Electric Co. Ltd.
1982	Experimental Research	Meguro Tokyo	2.5	TIT	Sanyo Electric Fuji Electric
1984	Sea Water Distillation	Koza Okinawa	0.12	Hitachi Zosen Corp.	Kanegafuchi Chemical Ind.
1984	Demonstration Power Plant	Akihabara Tokyo′	3.5	Dai-ichi Kaden Bldg.	Sanyo Electric Co. Ltd.
1984	Utility Power Experiment	Akagi Gunma	3.0	DRIEPI	Sanyo Electric Fuji Electric
1985	Utility Power Experiment	Saijo Ehime	3.0	NEDO Shikoku Electric Co. Ltd.	Sanyo Electric Fuji Electric
1985	Utility Power Experiment	Sakurajima Osaka	0.56	Hitachi Zosen Corp.	Kanegafuchi Chemical Ind.

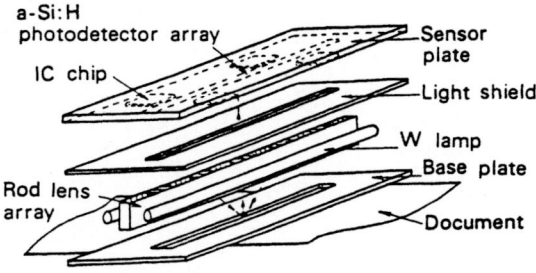

a-Si:H
photodetector array
IC chip
Rod lens array

Sensor plate
Light shield
W lamp
Base plate
Document

Fig.14 Decomposed view of
a-Si:H linear photosensor
developed for the facsimile
read out (after Kohno et al.[48]).

fabricated by Fujitsu group [49]. The active element of this sensors consists of an ITO/a-Si/Al Schottky photodiode and Pt/a-Si/Al blocking diode connected in series which act as an internal connection, and reduce a large number of connecting leads, and switching transistors. Another solution on this bonding problem has been proposed by Matsumura [51] with a-Si TFT integrated to the photosensitive area. The contact type linear photosensors would be expected to reduce the size of device not only of facsimile, but also of Xerographic printers. More large scale intelligent copier will be developed with these a-Si integrated photosensors.

In early 1982, Kuwano group of Sanyo Electric Co. Ltd. has developed an integrated a-Si colour sensor [52]. The structure and out look of the sensor is shown in Fig.15. The typical dimensions and performance are summarized in Table II with comparing that of ordinary visible sensor. Recently, potential needs of this device has been expanded in wide range of application fields such as automatic colour recognition and colour identification in the field of medical science, industrial processes, food science, agricalture, artitecture, etc..

ASPECT OF a-Si IMAGING DEVICE APPLICATIONS

The development of "Saticon", visicon tube using Se-As-Te chalcogenide, is historically very famous invention as the first practical application of amorphous semiconductors in 1973 [53]. The inventor Maruyama's group has extended their activities to a-Si, and

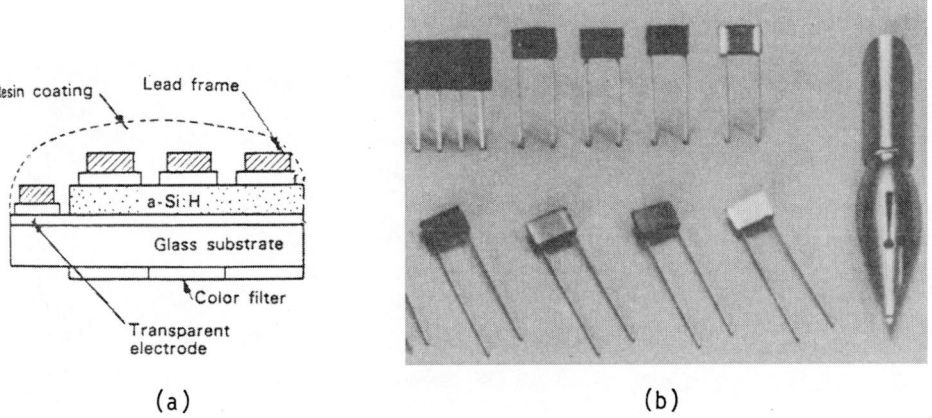

Resin coating
Lead frame
a-Si:H
Glass substrate
Color filter
Transparent electrode

(a)   (b)

Fig.15 Cross sectional view (a) and its outlook of a-Si full colour photosensor (after Nakano et al.[52]).

Table III *Characteristics of a-Si photosensor.*

	Visible light sensor	Monocolor sensor	Integrated color sensor
Size (mm³)	$3.5 \times 6 \times 1.5$	$3.5 \times 6 \times 2$	$6.6 \times 5.1 \times 2$
Effective area (mm²)	$4 \sim 6.8$	$4 \sim 6.8$	4×3
Sensitivity	$0.6 \sim 1.0 \mu A$ (100 lux)	$R = 1.56,\ G = 1.22$ $B = 0.56 \mu A / 0.1 mW$	$R : G : B = 5 : 3 : 2$
Dark current (A) ($V_R = 1$ V)	$3 \sim 20 \times 10^{-11}$	$3 \sim 20 \times 10^{-11}$	$3 \sim 20 \times 10^{-11}$
Terminal capacitance (pF)($V = 0,\ f = 10$ kHz)	$800 \sim 1\,200$	$800 \sim 1\,200$	$800 \sim 1\,200$

developed a-Si vidicon in 1979 [54]. The target structure of R.F.
sputtered a-Si vidicon is shown in Fig. 16(a). The role of thin SiO_2
layer in Fig.16(a) is to suppress the dark current by blocking the
injected holes from the transparent electrode. Comparison of the dark
current with one without SiO_2 blocking layer is shown in the Figure
16(b). Recently, Shimizu et al. have reported a higher sensitivity
vidicon with the glow discharge produced a-Si target having a blocking
layer of a-Si$_x$N$_{1-x}$ [55]. Figure 16(c) shows a picture taken by a-Si target
vidicon single tube color pick-up. A remarkable merit of these size
reduction is low power dissipation of periferal circuit which enable to
utilization of LSI driver circuit, and hardly type application.

A solid state colour imaging device using a-Si photosensitive layer
has also been developed by Hitachi group in 1981 [56]. A cross-sectional
view of the picture element is shown in Fig.17(a). As can be seen in the
figure, a-Si photosensitive layer is directly stacked on MOS signal
processing part. A 485 vertical x 384 horizontal picture element device
has been fabricated. The Figure 17(b) shows an example of picture taken
by the a-Si solid state imaging device. The signal charge stored in the
p-n junction is read through n-MOS transistor, and thin n$^+$ side of the
a-Si layer. With this stacked structure, the effective aperture reaches

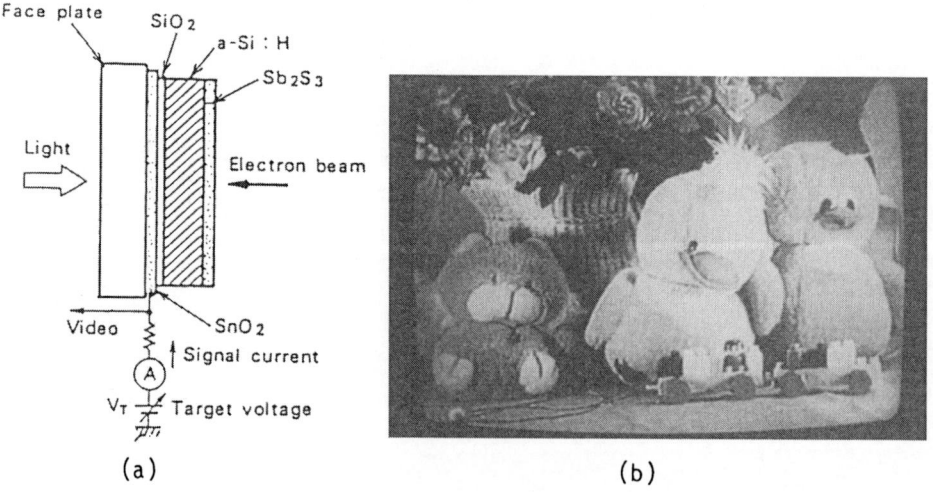

(a)   (b)

Fig.16 *Target structure (a) and an example of picture taken by
the developed a-Si vidicon tube (after Maruyama et al.[53]).*

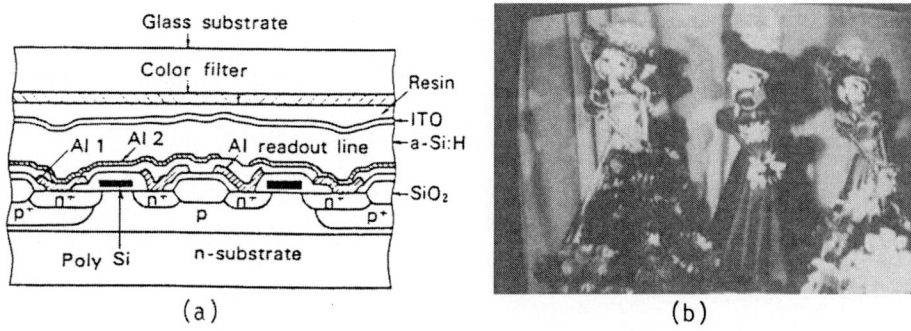

Fig.17 Cross-sectional view of a picture element of an a-Si two-story photosensor (a) and an example of picture (b) (after Baji et al.[56]).

up to 73% of the raster area, while ordinary CCD imaging device has less than 47% aperture [57]. The high-light exposure was about 250 times as intense as the saturation level of photocurrent. Because of the high photoconductivity in a-Si, developed imaging device has much high sensitivity with non blooming performance as compared with ordinary devices [58].

a-Si ELECTROPHOTOGRAPHIC APPLICATIONS

Aiming with the full use of wide area uniform deposition capability in a-Si, the development of electrophotographic receptor with a-Si has been initiated even in 1979, and more than ten groups have worked out in this area in the world [59]. Current technological R&D objectives in this area are focused on (a) a technique for obtaining high deposition rate, (b) optimization of deposition conditions and doping level in a monolayered a-Si photoreceptor, (c) structural improvement of a multilayered a-Si photoreceptor and (d) improvement in long wavelength photosensitivity to cope with problems related to the semiconductor laser diode with the use of a-SiGe alloy.

In principle, the photoreceptor consists of four separated functional layers, that is, i) a surface passivation layer (SPL), ii) a charge generation layer (CGL), iii) a charge transport layer (CTL) and a bottom blocking layer (BL). Full use of electrical and optoelectronic property controllability in a-Si and its alloys with controlling impurity doping and compositional regulation, a wide varieties of material combinations have been recently proposed as shown in Fig.18. For example, Shimizu et al. [60] have developed a-SiN (SPL)/s-Si (CGL-CTL)/a-SiN (BL) multilayer, and Nishikawa et al. [61] have substituted a-SiC layers for SPL-BL combinations as shown in the figure.

For the purpose of small size laser printer and also a new type intelligent copier, there exists a considerable potential need to develop an infrared sensitive high speed photoreceptor. On this objective, Professor Kawamura and his group [67] at University of Osaka Prefecture have recently developed the boron doped a-Si SPL with a-SiGe alloy CTL and CGL. While, a-SiGe layer act as the CTL for short wavelength photons and as CGL for infrared photons as shown in Fig.18 (b). Through these basic investigations in the university institute, Kyocera Co. has announced firstly the successful mass production of "Amorphous Silicon Drum" for electrophotography in November, 1981, developed with the cooperation of Kawamura's group. Stanley Electric Co. Ltd. also has announced same kind of machine in May, 1982. A remarkable advantages of this a-Si drum are i) long life due to the high surface hardness, ii) high sensitivity even

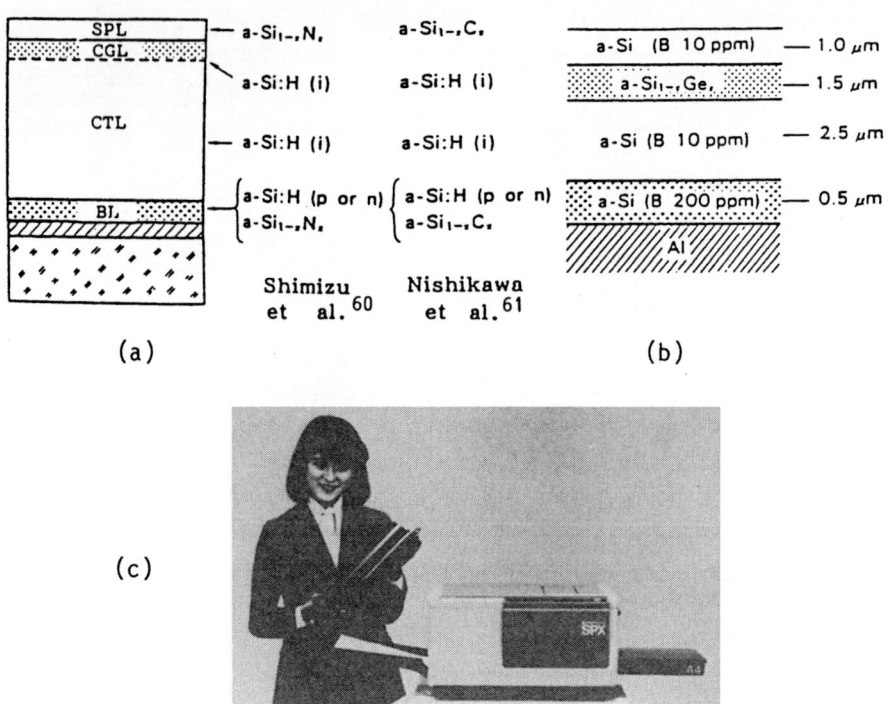

Fig. 18 *Having various options for the spectral sensitivities in the structure of a-Si alloy photoreceptor proposed by Shimidzu et al.[60] (a), by Nishikawa et al.[61] (b) and an example of high speed a-Si copy machine developed by Sanyo Electric Co. Ltd. (c).*

for the infrared region, iii) high thermal stability which makes possible high speed operation with non-tuxity, v) bichorgeable property which can be applied to new functional devices.

Quite recently, Sanyo has also commercialized a-SiC/a-Si drum drived copy machine as shown in Fig. 18(c) in 1984. Efforts have also been made to develop an intelligent copier using an a-Si photoreceptor in combination with a semiconductor laser diode and a microcomputer[68]. In response to the tremendous amount of potential needs for the office automation, new types of information processing systems combining high speed electrophotographic technology; such as laser line printer, with word processor, graphic converter, signal process have come to play a leading role of optoelectronic information sciences. Newly developed tetrahedrally bonded amorphous alloys, such as a-SiC, a-SiGe, a-SiSn could be responsible to this social needs for the new civilization.

ACKNOWLEDGEMENT

The author wishes to express his sincere thanks to all his research friends who kindly supplied the newest informations from all over the world in the field. However, due to the limitted space, he could not cite on details, even many other advanced technologies are being developed, particularly in U.S.A. and Europian countries. The author also gratefully acknowledges to Drs. H. Takakura and H. Okamoto for their advices and assistance in preparation of this reviews.

REFERENCES

1. W. E. Spear and P. G. LeComber, Solid State Comm., 17 (1975) 1193.
2. Y. Hamakawa, Sol. Energy Mater., 8 (1982) 101.
3 K. Okuda, H. Okamoto and Y. Hamakawa, Jpn. J. Appl. Phys. Lett., 35 (1983) L605.
4. For example, Review is given in JARECT Vol.6, edited by Y. Hamakawa, "Amorphous Semiconductor - Technologies & Devices", Ohm-North Holland (1983).
5. H. Okamoto, Y. Nitta, T. Yamaguchi and Y. Hamakawa, Solar Energy Mat., 2 (1980) 313.
6. A. Madan, S. R. Ovshinsky and E. Benn, Phil. Mag., B40 (1979) 259.
7. Y. Tawada, M. Kondo, H. Okamoto and Y. Hamakawa, Appl. Phys. Lett., 39 (1981) 237.
8. Y. Higaki, M. Kato, M. Aiga and Y. Yukimoto, Proc. 4th EC PVSEC, Stresa (1982) 745.
9. S. Tsuda, N. Nakamura, Y. Nakashima, H. Tarui and Y. Kuwano, Jpn. J. Appl. Phys., 21 (1982) suppl.22-2, 251.
10. Y. Hamakawa, Sur. Sci., 86 (1979) 444.
11. Y. Uchida and H. Haruki, The 3rd New Energy Industrial Symposium, NEF, Session 2A-3 (1983) Nov. p.24.
12. M. Ohnishi, H. Nishiwaki, S. Tsuda and Y. Kuwano, Proc. 17th IEEE PV Specialists Conf., (1984).
13. K. Ishibitsu, Y. Nitta and K. Kimura, J.P.S. Spring Meeting (1984).
14. H. Okaniwa, M. Asano, K. Natatani, M. Yano and K. Suzuki, Jpn. J. Appl. Phys. 21 (1982) suppl.21-2, 239.
15. presented by Sharp-ECD Solar Inc.
16. T. Tiedje and A. Rose, Solid State Comm., 37 (1980) 49.
17. J. Orenstein and M. Kastner, Phys. Rev. Lett., 46 (1981) 1421.
18. D. M. Pai and R. C. Enck, Phys. Rev. 11 (1975) 5163.
19. J. Mort, A. Troup , M. Morgan, S. Grammatica, J.C. Knights and T. Lujan, Appl. Phys. Lett., 38 (1981) 277.
20. Y. Hamakawa, in:"Amorphous Semiconductor - Technologies & Devices", ed. by Y. Hamakawa (OHM Sha & North Holland, Tokyo, Amstrdom and New York, 1981) Chap.6.
21. T. Yamaguchi, H. Okamoto and Y. Hamakawa, Jpn. J. Appl. Phys. 20 (1981) suppl. 20-2, 195.
22. H. Okamoto, H. Kida and Y. Hamakawa, Solar cells, 8 (1983) 317.
23. H. Okamoto, T. Yamaguchi and Y. Hamakawa, J. Phys. Soc. Japan 49 (1980) suppl.A, 1213.
24. H. Okamoto, T. Yamaguchi and Y. Hamakawa, J. Phys. 49 (1981) C4-507.
25. Y. Tawada, M. Kondo, H. Okamoto and Y. Hamakawa, Sol. Energy Mat., 6 (1982) 237.
26. Y. Tawada, K. Tsuge, M. Kondo, H. Okamoto and Y. Hamakawa, J. Appl. Phys., 53 (1982) 5273 and Y. Hamakawa, 17th IEEE Photovol. Spec. Conf., Orlando (1984) Late News.
27. T. Catalano, A. Friester and B. Fanghman, Proc. 16th IEEE Photovol. Special. Conf., San Diego, 1982, IEEE, New York, p.1421.
28. S. Nakano, H. Kawada, T. Matsuoka, S. Kiyama, S. Sakai, K. Murata, H. Shibuya, Y. Kishi, I. Nagaoka and Y. Kuwano, Technical Digest of the Intern. PVSEC, Kobe, Japan, (1984) 583.
29. Y. Hamakawa, International J. Solar Energy, 1 (1982) 125.
30. G. Nakamura, M. Kato, H. Kondo, Y. Yukimoto and K. Shirahata: J. Phys. (Orsay, Fr.) 42 C4-483 (1981)
31. S. Nonomura, H. Okamoto, K. Fukumoto and Y. Hamakawa, J.Non-cryst. Solids 59/60 (1983) 1099.
32. H. Okamoto, H. Kida, K. Fukumoto and Y. Hamakawa, J. Non-cryst. Solids 59/60 (1983) 1103.
33. Y. Tawada, K. Tsuge, K. Nishimura, H. Okamoto and Y. Hamakawa, Jpn. J. Appl. Phys. 21 (1982) suppl.21-2, 291.
34. H. Okamoto, H. Kida, S. Nonomura and Y. Hamakawa, J. Appl. Phys. 54 (1983) 3236.

35. Y. Uchida, H. Sakai, M. Nishiura, M. Miyagi and K. Maruyama, Proc. 8th Int. Vacuum Congr., Cannes, 1980, Vol.1, North-Holland, Amsterdom, 1981, p.669.
36. Reported by G. Kagaya, Photovoltaic Insider News, Vo.III, No.1, January (1984) 4.
37. S. Yamazaki, K. Itoh, S. Watanabe et al., Proc. 17th IEEE PV Special. Conf., Florida, May (1984) P1-1B-1.
38. D. L. Morel, J. P. Rumburg, R. R. Gay and G. B. Turner, Proc. 17th IEEE PV Special. Conf., Florida, May (1984) Plemary 2-1.
39. private communication from S. Ovshinsky.
40. W. den Boer and R. M. Van Strijp, Proc. 4th Photovoltaic Solar Energy Conf., Stresa (1982) 764.
41. E. Yablonovitchi and G. D. Cody, IEEE Trans. Electron Devices ED-29 (1982) 300.
42. Y. Hamakawa, Y. Tawada, K. Nishimura, K. Tsuge, M. Kondo, K. Fujimoto, S. Nonomura and H. Okamoto: 16th IEEE Photovol. Special. Conf., San-Diego (1982) 679.
43. K. Fujimoto, H. Kawai, H. Okamoto and Y. Hamakawa: Solar Cells 11 (1984) 357.
44. H. Iida, T. Miyado and Y. Hayashi: Proc. 44th Fall Mtg. Soc. Japan Appl. Phys., Sendai, Sept. (1983) 25p-L-2 & 3.
45. K. Okuda, H. Okamoto and Y. Hamakawa, Jpn. J. Appl. Phys. 22 (1983) L605.
46. T. Kagawa, N. Matsumoto and K. Kumabe, Proc. 13th Conf. on Solid State Devices (Tokyo, 1981) 251.
47. T. Hamano, H. Ito, T. Nakamura, T. Ozawa, M. Fuse and M. Takenouchi, Proc. 13th Conf. on Solid State Devices (Tokyo, 1981) 245.
48. Y. Kanoh, S. Usui, S. Sawada and M. Kikuchi, Digest of 1981 IEDM (1981) 313.
49. K. Ozawa, N. Takagi and K. Hiranaka, Spring Mtg JSAP (1982) 4a-z-5.
50. T. Ozawa, M. Takenouchi, T. Hamano, H. Itoh, M. Fuse and T. Nakamura, Proc. Int. Microelectronics Conf. (1982) 132.
51. M. Matsumura, H. Hayama, Y. Nara and K. Ishibashi, Proc. 13th Conf. on Solid State Devices (Tokyo, 1981) 311.
52. S. Nakano, T. Fukatsu, M. Takeuchi, S. Nakajima, S. Nakayama and Y. Kuwano, Proc. 3rd Sensor Simposium (1983) 97.
53. E. Maruyama, T. Hirai, T. Fujita, N. Goto, Y. Isozaki and K. Sidara, Proc. 6th Conf. on Solid State Devices (Tokyo, 1974) 97.
54. Y. Imamura, S. Ataka, Y. Takasaki, C. Kusano, T. Hirai and E. Maruyama, Appl. Phys. Lett., 35 (1979) 349.
55. I. Shimizu, S. Oda, K. Saito, H. Tomita and E. Inoue, Amorphous Zairyo-Bussei ed. by Sakurai (1980) 202 (in Japanese).
56. T. Baji, Y. Shimomoto, H. Matsumaru, N. Koike, T. Akiyama, S. Sasano and T. Tsukada, Proc. 13th Conf. on Solid State Devices (Tokyo, 1981) 269.
57. T. Tsukada, T. Baji, H. Yamamoto, Y. Takasaki, T. Hirai, E. Maruyama, S. ohta, N. Koike, H. Ando and T. Akiyama, Digest of 1979 IEDM (1979) 134.
58. S. Ishioka, Y. Imamura, Y. Takasaki, C. Kusano, T. Hirai and S. Nobutoki, Proc. 14th Conf. on Solid State Dev. (Tokyo, 1982) 461.
59. for example, T. Kawamura and N. Yamamoto, JARECT vol.2 "Amorphous Semiconductor - Technologies & Devices", ed. by Y. Hamakawa (OHM Sha & North Holland, Tokyo, Amsterdom and New York, 1982) p.311.
60. I. Shimizu, S. Oda, K. Saito, H. Tomita and E. Inoue, J. Phys., Paris, 42 (1981) suppl. C-4, 1123.
61. S. Nishikawa, H. Kakinura, A. Uchiyama, M. Akiyama, T. Watanabe and K. Uenishi, Symp. on Semicond. & Mat. of IECI Japan, S6-8 (1981).
62. Y. Nakayama, T. Natsuhara, M. Nakano, N. Yamamoto and T. Kawamura, Proc. 14th Conf. on Solid State Devices (Tokyo, 1982) 453.
63. T. Kawamura, N. Yamamoto and Y. Nakayama, JARECT vol.6 "Amorphous Semiconductor - Technologies & Devices", ed. by Y. Hamakawa (OHM Sha & North Holland, Tokyo, Amsterdom and New York, 1983) p.325.

AMORPHOUS TETRAHEDRALLY-BONDED MATERIALS FOR MACROELECTRONICS

J. Mort and F. Jansen

Xerox Webster Research Center
Webster, New York 14580

INTRODUCTION

Amorphous materials represent one example of technology stimulating a field of scientific research the results of which, in turn, create the opportunity for new technologies. The original technological exploitation of amorphous selenium for electrophotographic applications in the 1950's - 60's was followed by increased scientific study of amorphous materials. Much experimental and theoretical progress was made regarding the effect of disorder on the materials' properties, such as optical and electrical, described within the framework of the band theory of solids developed for ordered or crystalline materials.[1] This, combined with the production of new materials by a variety of techniques such as plasma assisted CVD or glow discharge deposition, led to the demonstration and production of the first doped extrinsic amorphous semiconductor, viz., hydrogenated amorphous silicon (a-Si:H).[2] Although producible by a number of techniques, the most widely used is the glow discharge process.[3] This process allowing, as it does, for production of amorphous extrinsic semiconductors and related materials from gases, affords useful opportunities in the production of electronic devices in dimensions vastly larger than those of the typically affordable single crystal silicon wafers. It is natural to ask how discrete devices fabricated in this way perform compared with their enormously successful crystalline counterpart. However, a more appropriate and perhaps relevant question is what do these new types of extrinsic semiconductors and insulators technologically enable that would not otherwise be possible?

This contribution in honor of Sir Nevill Mott is intended to take this perspective in reviewing the types of devices currently under investigation, the materials properties required and how these can be met by the glow discharge process. It is a perspective entirely consistent with the prescient view, expounded by Sir Nevill in his opening address to the 1962 Exeter Semiconductor Conference, that the future challenge in solid-state physics was to capture the same excitement, attraction and glamour in finding new ways to use materials as is found in elucidating their laws.[4]

The theme of this contribution, therefore, is to discuss the glow discharge process and the materials producible, in terms of their importance in making feasible a radically new sort of electronics, ie., macroelectronics. This should not be perceived narrowly as competitive with crystalline semiconductors, which it may be in e.g. photovoltaics, but as enabling hitherto unattainable technologies. The opportunities afforded by the glow discharge process in truth are unique. It is capable of producing both doped extrinsic semiconductors and insulators by appropriate choice of gases or gas mixtures by one low temperature process. It is also possible to create sharp or graded interfaces without the complications of ion implantation or high temperature diffusion. The technique is naturally compatible with the fabrication of multilayers[5,6] and vertically integrated or tandem devices[7] and all these possibilities are enveloped by their potential production in large areas at extremely low cost.

DEVICES

The characterizing property of amorphous and polycrystalline materials from a technological perspective is their large area producibility in virtually any geometric configuration. It is this feature, together with the ability to deposit extrinsic semiconductors and insulators by the same process, which enables macroelectronics. As discussed in the next Section, the progress made in device fabrication has drawn heavily on the fundamental understanding amassed in the scientific studies. Here, the generic types of devices and potential applications will be outlined.

As is often the case, new technological opportunities eventually co-exist with technological needs. Less common is the conjunction of these with strong coupling to a potentially vast market. In the case of macroelectronics, in fact, two potential markets exist. The first are the needs of the electronic information industry and the second is the need for alternative energy sources.

The explosive growth of processing, transmission and dissemination of information electronically is creating needs for cheap, reliable, alternative means, capable of being mass produced, for the acquisition and conversion of written information into an electronic bit stream and vice versa. This can be facilitated by linear if not areal optical input scanning devices of documents. At the output end the information needs to be produced either in permanent hard copy or temporally on a display. Thus there are growing needs for alternative marking technologies of either full page width or area dimensional capability in a range of quality and cost. These include laser scanned photoreceptors to full page width image bars. Additionally there is increasing demand for low cost, portable displays as alternatives to conventional CRT displays used in word processors or consumer television receivers. Such devices based on amorphous tetrahedrally-bonded materials are being extensively studied for a wide range of applications relevant to the electronics information age. The dominant device type is the optical to electrical transducers in applications such as electrophotographic photoreceptors or photodiode arrays. Successful production of a-Si:H-based photoreceptors has been reported which have excellent photosensitivity throughout the visible spectrum and extension into the infrared using Si/Ge sensitizing layers.[8-11] Significant extension of photoreceptor life over existing chalcogenide or polymer-based photoreceptors is claimed. The different and advantageous mechanical properties of amorphous tetrahedral photoreceptors promise increased latitude for design engineers. Photosensors which vertically integrate a-Si:H photoconductors or photodiodes with a-Si:H thin film transistors have also been investigated for potential application in large one-dimensional arrays for input scanners for facsimile applications.[12] The switching of liquid crystal panels with two-dimensional arrays of a-Si:H thin film transistors are also being developed as potential replacement of cathode ray tubes in video terminals. Already color television sets with 5" diagonal screens with 250 x 666 picture elements have been reported.[13]

The vulnerability of industrialized societies to dependence on uncertain and declining oil supplies has stimulated heightened activity in the development of cost-effective photovoltaic power sources.[14-15] A significant candidate in this search is a-Si:H since, in addition to the ease of manufacturing large areas, the absorption properties of this material are such that films only 1 μm thick are sufficient to ensure an excellent match to the solar spectrum. By comparison, the indirect absorption edge in crystalline silicon requires that crystalline cells must be several hundred microns thick to achieve comparable absorption.

Dramatic improvement has been made in recent years in improving the conversion efficiency of a-Si:H and in the design and fabrication of multilayered cells. Conversion efficiencies of > 10% have been reported by several groups and a 3.5 kW power generating system has been constructed.[13] Already a-Si:H photovoltaic cells have been commercially used in a wide range of consumer applications such as calculators, watches, radios and battery chargers.[13]

MATERIALS REQUIREMENTS

a-Si:H is the outstanding amorphous extrinsic semiconductor, i.e., one in which the Fermi level can be moved either by doping with phosphorus or boron, to produce n or p-type material, or by applied electric fields, as in the field effect. The efficacy with which these phenomena may be induced in a-Si:H is due to the extent that the density of states within the bandgap may be reduced. The critical role of hydrogen in removing silicon dangling bond states was initially uncovered by studies of the field effect which, in turn, now constitutes the basis of the thin film transistor.[16] These properties are innate features of a-Si:H. The challenge to the solid state physicist, materials scientist and technologist has been and continues to be to understand, control and manipulate these properties so as to engineer from a materials perspective useful devices.

Naturally the most important material requirement for macroelectronic applications is that the material can be fabricated in large areas and this will be dealt with in more detail in the next Section. This must be combined, however, with a range of photoelectronic and electrical properties which meet some minimum specifications if useful photoelectrically or electrically active devices are to be realised. First and foremost, the carrier transport parameters of one carrier (for majority carrier devices such as thin film transistors or electrophotographic photoreceptors) or both (for minority carrier devices such as photovoltaic cells) need to be sufficiently large. Since carrier transport properties are the greatest casualties of disorder, this aspect of amorphous materials causes the greatest concern. The requirements concerning carrier transport parameters relate only to active components in devices, although for successful device performance it is often necessary to use highly insulating films, which though vital, are in a sense electrically inactive. Here, therefore, concern centers on the electrical and photoelectronic requirements demanded of the active component, viz., a-Si:H, in these amorphous technological devices.

Based on experimental studies, the room temperature electronic drift mobilities (i.e., modulated by shallow trapping) are ~ 1 cm^2/Vsec and $\sim 10^{-3}$ cm^2/Vsec for electronic and holes respectively.[17] Evidence suggests that these values are dependent on the quality or purity of the material but constitute the highest achievable values to date. Major improvements present a considerable challenge, since the microscopic mobilities of carriers thermalized at the band edges are estimated to be ~ 10 cm^2/Vsec; substantially reduced from the values in crystalline material. The mobility value by itself is actually only important where speed of response or absolute current values are critical, as in the on-current of thin film transistors or diodes. Where electronic information is reproduced visually, either on paper or a display, the overall speed is often determined by secondary steps such as paper handling or the switching speed of liquid crystal elements. So, although the mobilities in a-Si:H are too low to represent a challenge to crystalline devices for information processing in the electronic domain, at the interface to the visual world they are sufficient for a wide range of applications. In other devices, such as photovoltaic cells or electrophotographic photoreceptors which are photogeneration-controlled devices, the more significant transport parameter is the mobility-lifetime product, $\mu\tau$. Here, the currents are controlled by generation rates and successful device functioning depends on how far the carriers diffuse ($\propto\sqrt{\mu\tau}$) or drift ($\propto\mu\tau$). Values of $\mu\tau$ for electrons and holes are $\sim 10^{-6}$ cm^2/V in high quality material and meet mimimum requirements for most photoelectronically active devices.[10,17] Parenthetically, the diffusion lengths of carriers in a-Si:H are several orders of magnitude smaller than in crystalline silicon. Fortuitously, however, the absorption coefficients for visible light are commensurately larger thus offsetting these otherwise unacceptably low values. For photodevices, the efficiency of carrier photogeneration, collection efficiency and the panchromatic spectral response become key issues. Here again, a-Si:H is found to possess excellent properties since photogeneration and collection efficiencies at modest fields are essentially unity and the spectral response encompasses the entire visible spectrum. Studies have shown that these properties are producible in materials ranging in thickness from ~ 1 μm to ~ 50 μm by either r.f. or d.c. glow discharge.

For macroelectronic applications, these electrical and photoelectronic properties must be reproducible in a cost effective way over macroscopic dimensions. Experience with both photovoltaic arrays and electrophotographic photoreceptors has demonstrated that this is possible. Image transducers in general and electrophotographic photoreceptors in particular are acutely sensitive to the uniformity of electrical and photoelectronic properties over large areas. The requirement to produce or reproduce images of the highest quality over areal dimensions > 0.3 m^2 places stringent demands ($\sim$ a few percent) on the variability of such parameters.[18] Based on achieved xerographic print quality, such demands can be met. In image transducer applications, the ability to cyclically recreate the same image or sequentially different images at relatively high rates (up to 200 per minute) requires high cyclic stability against image fatigue or ghosting. These requirements are specifically related to the trapping/detrapping rates associated with defect or impurity associated traps. Considerable evidence now exists to show that these effects, while occurring, are not limiting. By contrast, integrating devices such as solar photovoltaic cells are adversely affected by the inherent prolonged exposure to high intensity illumination, as evidenced by the Staebler-Wronski effect, for which solutions are actively being sought.[19]

As alluded to earlier, all anticipated device applications depend on multilayered structures involving layers of different materials, in addition to the a-Si:H, e.g., insulators like a-Si:N_x, a-Si:C_x or different material compositions such as doped a-Si:H or a-Si:Ge:H alloys. As a consequence, there is the additional requirement for the producibility and reproducibility of the necessary quality of ubiquitous interfaces. For some applications, notably the production of arrays of thin film transistors, technological applications require the successful definition of high resolution patterns. The compatibility of these materials with existing silicon photolithographic technology is a significant factor in their favor. Other properties such as thermal stability, chemical non-toxicity and mechanical toughness are of generic value to all applications but are sine qua non for successful exploitation in others. In electrophotography, for example, the copy quality produced by an amorphous tetrahedral photoreceptor can inherently be no better than in existing materials and parity, although required, is all that can be be expected. However, the superior mechanical properties of a-Si:H and related amorphous tetrahedral materials hold the promise of quantum improvements in life of a device that is unavoidably subjected to considerable mechanical impact during usage.

FABRICATION

The property that distinguishes macroelectronic devices from their monolithic counterparts is their size. The foremost requirement of the fabrication process is therefore the ability to deposit films with uniform properties over large areas. Most thin film deposition processes can be configured, although some easier than others, in such a way that thickness uniformity over large areas is obtained. For amorphous tetrahedrally-bonded semiconductors, the incorporation of hydrogen or fluorine into the material is, as discussed before, essential in order to obtain dopable material with the requisite low density of gap states. This can be accomplished in two fundamentally different ways. First there is the possibility to incorporate reactive hydrogen into the deposition process and to combine it with the semiconductor material either at the source or the substrate. Because of the high melting point of silicon, this reactive deposition process is most easily accomplished using one of a variety of sputtering processes rather than an evaporative process. Although it has been demonstrated[20] that it is possible to obtain semiconductor grade amorphous silicon in this way, reactive sputtering processes are not commonly used for device fabrication. Rather, the reactive process is primarily important as a diagnostic fabrication tool for materials research e.g. to deconvolute the effects of hydrogenation on a material.[21] Historically, dopable amorphous semiconductor films have been fabricated in a different way and this process remains the most practical to date, certainly for the fabrication of macroelectronic devices. In this process, gaseous precursors are used which consist of the semiconductor material saturated with the univalent passivation element, commonly hydrogen or fluorine. Condensable radicals which combine to constitute the film, are formed from the gas by its decomposition in a plasma or glow discharge, from which the process derives its name: plasma or glow discharge deposition process.

The physical and chemical processes in the plasma and the mechanism for film formation are rather complicated. To illustrate some of these complexities the most important steps in the process are briefly discussed for the case of amorphous silicon deposition. Predominantly neutral radicals are created in the plasma by electron impact dissociation of the precursor molecule, generally silane or disilane. A variety of primary active species is formed of which SiH_2, SiH_3 and H are the most prevalent for silane.[22] The relative abundance of the various radicals depends on the electron energy distribution which, in turn, is not only a function of the primary process variables such as power density and pressure but also of the plasma potential and hence reactor design.[3] These active species are transported to the surface of the growing film by convection and diffusion[23] and before condensing, these neutrals can react with other gas molecules, radicals or ions to from more complex species in the gas phase. The process of condensation involves surface reactions[24] whose net effect is to release more hydrogen so that the film contains generally between 5 and 20% of hydrogen. The usual complexities which surround the growth of a film in the vicinity of a plasma are encountered in the glow discharge process. Film bombardment with ions, electrons and photons can significantly affect the final structure and the resultant properties of the film.[25]

The point of view is sometimes taken that the plasma deposition process, due to its inherent complexity and the multitude of phenomena in a low pressure plasma, is basically uncontrollable. This might be true in the sense that it would be very difficult to control the pathway of certain reactions or to predict from first principles the proper operating parameters for a certain reactor geometry. However, it is clear from the essentially similar characteristics which have been obtained in different laboratories for materials produced in entirely different reactors that the end result of the complicated deposition process is controllable and reproducible. In addition to this remarkable reproducibility, it is the inherent compatibility of the glow discharge process to large area film deposition which accounts for its emergence as the most practical fabrication process for macroelectronic devices. This large area capability is primarily a consequence of the fact that the precursor material is in gaseous form and not a spatially constrained source as is the case for any of the reactive processes. The diffuseness of the source, combined with the ability to apply uniform electric fields over large areas, allows the deposition of films on substrates with a uniformity which would be difficult to obtain by other methods. The use of gaseous precursor materials has other a priori advantages over the use of solid or liquid sources. The latter material sources are not only spatially but also chemically constrained. This is a serious limitation for the demanding materials requirements of macro-electronic devices which often involve multilayers of doped and alloyed semiconductors and refractory insulators. The optimization of device characteristics may even require the grading of interfaces to reduce internal fields, or the creation of dopant or alloying profiles in the film. This would be very hard to accomplish with sources of material of fixed composition but this problem is obviated by the variable flow control of gaseous precursors which allow considerable materials and device flexibility.

Despite its many positive characteristics, the glow discharge process does suffer from a number of serious shortcomings which might ultimately limit its applicability to certain technologies. First there is the question of achievable materials quality as, e.g., measured by the density of states distribution in the case of amorphous silicon. The plasma environment is likely to cause a certain minimum density of bond imperfections due to energetic bombardment processes. The resultant increase in the density of states affects the carrier transport and the doping efficiency of the material. Plasma-less deposition processes are actively investigated as alternatives for the glow discharge deposition process.[24,25] Second, the glow discharge process is inherently slow and materials inefficient because only a fraction of the precursor gas is decomposed into condensable species and because only a fraction of these radicals eventually condense to contribute to the growing film. Especially for applications where the use of thick films is required, xerographic photoreceptors being one example, these process characteristics translate into throughput limitations and relatively high cost. For the case of amorphous silicon deposition, the use of higher silanes might alleviate some of the difficulties which arise from the low deposition rate but the cost of this material is presently still extremely high. It is expected that the rapidly increasing level of activity in both industry and at universities will result in significant improvement of the deposition technology as it is practiced today.

FUTURE

The technological applications of amorphous tetrahedrally-bonded materials are being developed at a very rapid rate across a broad front. In some instances the difficult transition from the laboratory to the market place has been successfully made. This has required the commercialization of the glow discharge fabrication process with the concomitant addressing of issues such as economics, reproducibilty and safety. With this barrier broken, increased exploitation of these materials in the various technologies described earlier seems assured. Progress continues to be made in improving the quality of current materials together with the exploration of alternative materials such as alloys and phenomena in new material structures such as compositional or doped superlattices.[5,6] In particular, the superlattice concept may offer increased opportunities in materials design which can lead to improved or novel device performance. Although to date the applications of amorphous tetrahedrally-bonded materials have focussed on their use as electronic devices, increased attention should be paid, to the advantageous mechanical properties of some of these materials in terms of hardness, durability and chemical inertness combined with adjustable electrical and/or optical properties. Applications impacting questions of the wear or strength of engineered parts are conceivably of significant technological value.

REFERENCES

1. N.F. Mott and E.A. Davis, "Electronic Processes in Non-Crystalline Materials," Clarendon Press, Oxford (1979).

2. W.E. Spear, Doped amorphous semiconductors, Adv. Phys., 26 : 811 (1977).

3. F. Jansen, "Plasma Deposited Thin Films", J. Mort and F. Jansen, eds., CRC Press, Boca Raton, (1985).

4. N.F. Mott, Opening address in : "Report of International Conference on the Physics of Semiconductors," Exeter (1962), The Institute of Physics and The Physical Society, London (1962).

5. B. Abeles and T. Tiedje, Amorphous semiconductor superlattices, Phys. Rev. Letters, 51:2003 (1983)

6. J. Kakalios and H. Fritzsche, Persistent photoconductivity in doping-modulated amorphous semiconductors, Phys. Rev. Letters, 53:1602 (1984).

7. Y. Hamakawa, Recent advances in amorphous silicon solar cells and their technologies, J. Non-Cryst. Solids, 54-60:1265 (1983).

8. I. Shimizu, T. Komatsu, K. Santo and E. Inoue, a-Si thin film as a photoreceptor for electrophotography, J. Non-Cryst. Solids, 35-36 : 773 (1980).

9. N. Yamamoto, K Wakita, Y. Nakayama and T. Kawamura, Photoelectronic properties of gd a-Si:H monolayer films for electrophotographic applications, J. de Physique, 42 : C4-495 (1981).

10. J. Mort, F. Jansen, S. Grammatica, M. Morgan and I. Chen, Field-effect phenomena in hydrogenated amorphous silicon photoreceptors, J. Appl. Phys., 55 : 3197 (1984).

11. S. Nishikawa, H. Kakinuma, T. Watanabe and K. Kaminishi, Properties of multilayered photoreceptor with amorphous silicon and its alloys and application to optical printer, J. Non-Cryst. Solids, 59-60 : 1235 (1983).

12. P.G. LeComber, A.J. Snell, K.D. Mackenzie and W.E. Spear, Applications of a-Si field effect transistors in liquid crystal displays and integrated logic circuits, J. de Physique, 42:C4-423 (1981).

13. Y. Kuwano, Recent developments in a-Si devices, Meeting Electrochemical Soc., Toronto, May 1985, (unpublished).

14. D.E. Carlson and C.R. Wronski, Amorphous silicon solar cell, Appl. Phys. Letters, 28 : 671 (1976).

15. Y. Hamakawa, Recent advances in amorphous silicon solar cells and their technologies, J. Non-Cryst. Solids, 59-60 : 1265 (1983).

16. W.E. Spear, Localized states in amorphous semiconductors, "Proceedings 5th Int. Conf. on Amorphous and Liquid Semiconductors", Taylor & Francis, London (1974), p.1.

17. W.E. Spear and H. Steemers, The interpretation of drift mobility experiments on amorphous silicon, Phil. Mag., B47:L77 (1983).

18. H. Fritzsche, Noncrystalline semiconductors, Physics Today, 37:34 (1984).

19. D.L. Staebler and C.R. Wronski, Reversible conductivity changes in discharge produced amorphous silicon, Appl. Phys. Letters, 31:292 (1977).

20. W. Paul and D.A. Anderson, Properties of amorphous hydrogenated silicon, with special emphasis on preparation by sputtering, Solar Energy Materials, 5:229 (1981).

21. F. Jansen, M. Machonkin, S. Kaplan and S. Hark, The effects of hydrogenation on the properties of ion beam sputter deposited amorphous carbon, to be published, J. Vac. Sci. Technol. A, 3, no.3 (1985).

22. G. Turban, Y. Catherine and B. Grollea, Ion and radical reactions in the silane glow discharge deposition of a-Si:H films, Plasma Chemistry and Plasma Processing, 2:61 (1981).

23. I. Chen and F. Jansen, Mass-transfer analysis of a-Si:H deposition, J. Non-Cryst. Solids, 59-60:695 (1983).

24. F. Kampas, Chemical reactions in plasma deposition, in "Hydrogenated Amorphous Silicon", Chapter 8, Semiconductors and Semimetals Vol. 21A, J.I. Pankove, ed, Academic Press, New York, 1984.

25. J.C. Knights, Structural and chemical characterization, in "The Physics of Hydrogenated Amorphous Silicon", Chapter 2, Topics in Applied Physics, Vol. 55, J.O. Joannopoulos and G. Lucovsky, eds., Springer-Verlag, Berlin, 1984.

CONTENTS OF COMPANION VOLUME:

PHYSICS OF DISORDERED MATERIALS

Edited by
David Adler, Hellmut Fritzsche,
and Stanford R. Ovshinsky

AUTHOR INDEX

A

Akimchenko, I.P......157

B

Beeby, J.L..........255
Bellissent, R........11
Beyer, W............129
Butcher, P.N........325

C

Cannella, V.........389
Carius, R...........367
Carlson, D.E........165
Chenevas-Paule, A.....11
Chittick, R.C.........1
Clark, A.H...........63
Cohen, J.D..........299
Collins, R.W.........63
Crandall, R.S.......315

D

Dohler, G.H.........415

E

Evangelisti, F......457

F

Ferrier, R.P........325
Fritzsche, H........425
Frova, A............271
Fuhs, W.............367

G

Gartner, P..........51

Gheorghiu, A........213
Grigorovici, R.......51
Guha, S.............233

H

Hamakawa, Y.........513
Hatano, A...........501
Hayes, T.M..........255
Herman, F...........469
Hirabayashi, I......221
Hirose, M...........441
Huang, C.-Y.........63

I

Ishihara, S.........379
Ishii, N............187

J

Jansen, F...........531
John, P.............107
Johnson, R..........425

K

Kawai, M............501
Kikuchi, M..........27
Kristofik, J........483
Kumeda, M...........187

L

Lambin, P...........469
Lang, D.V...........299
Lin, S.Y............197
Long, A.R...........325
Lucovsky, G.........197

553

M

Madan, A..............79
Mares, J.J..........483
Miyazaki, S.........441
Morigaki, K.........221
Mort, J.............531
Moustakas, T.D.......93
Murase, I...........501
Murayama, N.........441

N

Nitta, S............501

O

Oda, S..............379
Okushi, H...........239
Overhof, H..........287

P

Paesler, M.A.........37
Pankove, J.I........117
Persans, P.D........147

R

Robertson, J........177
Roxlo, C.B..........147
Ruppert, A.F........147

S

Sakaida, M..........501
Sayers, D.E..........37
Scher, H............407
Schiff, E.A.........357
Schrimpf, A.........367
Selloni, A..........271
Shimizu, I..........379
Shimizu, T..........187
Silver, M...........389
Smid, V.............483
Sterling, H.F.........1
Stourac, L..........483
Summerfield, S......325

T

Takagi, S...........379
Takenaka, H.........221
Tanaka, K...........239
Tauc, J.............345
Theye, M.-L.........213
Tsu. R..............433

U

Ugur, H.............425

V

Von Roedern, B.......79

W

Wilson, B.A.........397
Wilson, J.I.B.......107

Y

Yamasaki, S.........239
Yoshida, M..........221

Z

Zavetova, M.........157

Hybrid orbitals   54
Hydrogen bonding,   132

Imaging device,
   a-Si,   525
Impurities,   408
Impurity,
   conduction,   408
   states,   168,325,407
In,   167,168
In-P,   215
InP,
   amorphous,   213,215,217-219
   crystalline,   216,217
Infrared,   367-371,373,376,509
   absorption,   8,9,11,94,110,
      111,125,129,132,144,
      157,159,160,166,170,
      226,384,387,501,502,
      503,504
Injection,   309
Interference function,   214
Ion,
   implantation,   121,157,158,
      164
   plating,   271
Isothermal capacitance
      transient spectroscopy
      (ICTS),   172,239,240,
      242,243,252,304,308
Isotopic substitution,   14

K,   235
Kinetics,   407
Kramers-Kronig relation,   328,
      505

Lattice-relaxation,   309
Lennard-Jones potential,   471
Level shift,   434,435
Li,   235
Lifetimes,   407
Light-induced defects,   315
Light induced electron spin
      resonance signal (LESR),
      372-376
Line broadening,   434
Linewidth,   435-437
Liquid crystal,   535
Localized electrons,   407,408
Localized gap states,   389
Localized states,   165,218,
      233,274,280,299,339,
      340,391,401,458,483,
      494,496,497
Lone pairs,   386

Long-lived carriers,   398
Long-range order,   441
Low-level impurities,   152
Luminescence,   410

Majority carrier,   372,374,375
Mass spectroscopy,   33
Mean square deviation,   41,42
Metastable centers,   171-173,
      315,316
Mg,   157,158,161,163
Microcorrosion,   23
Microcrystallinity,   39,46
Microvoids,   23,108,113,127,
      234
Midgap,   305,309,322,351,391
Midgap states,   185,235,241,
      287,517
Mini-band states,   433
Minority carriers,   118
MISFET,   110
Mo,   6
Mobile carriers,   391
Mobility edge,   241,279,288,
      289,293-295,349,359,
      370,407,417,422,423,
      458,464,490
   conduction band,   299
Mobility
   gap,   79,190,233,272,282,
      299,310,312,357,416,
      464
   lifetime,   152,380
Mode softening,   125
Modulation spectroscopy,   345
Monovacancies,   168
Monte Carlo simulation,   470
MOS,   124
Mott,   27,36,48,51,63,104,177,
      236,239,255,269,325,
      329,333,335,357,408,
      431,532
Mott-CFO model,   9
Muffin-tin potential,   257
Multilayer film,   425-427,429,433
Multiphonon emission,   250
Multiple trapping,   354,401,
      412

N,   235,400
N_2,   169,170,171,173,199
Na,   235
NH_3,   6,442
NMR,   11,181,235
Near-gap,   274
Neutron Scattering,   13,45

p-i-n junction   517
  a-Si,   518
  a-Si:H,   518
p-i-p-i superstructures,   423
p-n junctions   118,304,423,517
PL decay,   409
P-O,   182
Polyacetylene,   340
Polyhydrides,   234
Polymorphs,   53
  high pressure,   54
Polysilane   112,501,502,504,
  508,509
Power law decay,   410
Pseudo-gaps,   207,209,216-
  218,272,274,462,466
Pseudo-potential calculations,
  200

Quantum efficiency,   403,407
Quantum well,   433,438,450-
  452,454
Quasi-Fermi levels,   309,351,
  370,392-394,416

Radial distribution function,
  19,38,214
Radiative,
  decay,   408,412
  lifetimes,   408,409
  rates,   401,407
  recombination,   225,367,
    373,410,446,447
  recombination centers,   221
  tunneling,   397,398,401,
    407,408
Radius of gyration,   15,16,24
Raman spectroscopy,   19,45,94,
  275,346,386
Random network,   501
Random walk theory,   340
Rate equations,   326
Reaction rates   408
Recombination,   408,409
  center coefficient,   152
  centers,   122,170,173,222,
    233,320,321,368,390,
    391,393-395,448
  geminate,   397,517
  lifetime,   390
  luminescence,   407
  nongeminate,   397
  rates,   352,397,400
  time,   354
Reflectance,   426,501,502,505
Relaxation,   409,412
Relaxation rates,   407
Resistivity,   157

Resonant bonding,   183
Resonant tunneling,   436,437
Ring statistics,   217

S,   235
Sb,   122
Scaling model,   337,339
Scanning electron microscopy
  (SEM),   15,94,96-98,100
Schottky barrier,   84,119,
  300,304,312,317,318,
  449,463,485
Schottky diode,   109,230,525
Se,   177,182
  amorphous,   531
Secondary ion mass spectros-
  copy (SIMS),   129,141
Self trapping,   402
Shallow acceptors,   168,300
Shallow donors,   169,300
Shallow states,   303,311,345
Shallow trapping,   535
Shallow traps,   349,361,363
Shock crystallization,   28,
  30-36
Short-circuit current,   86
Short-circuit photocurrent,
  321
Short range order,   17,18,257
Si,   20,75,125,199,202,203,
  206,208,209,234,276,
  462,470,471,474
  amorphous,   1,4,6-8,14-17,
    20,21,24,37,46,52,57,
    72,79,100,108,117,129,
    143,185,199,200,202,
    205,209,213,258,277,
    287,325,332,334,337-
    340,347,348,412,415,
    418,460,461,463,469,
    470,480,509,510,513-
    515,517,523,526,527,538
  B-doped,   527
  H-free,   239,251
  photoreceptor,   527
Si,
  crystalline   2,17,20,21,65,
    69,70,73,117,130,132,
    148,158,163,165,167-
    170,226,235,340,442,
    449,469,470,481,495,
    533
  microcrystalline,   93,104,386
  p-doped,   142
  polymorph,   52
Si-B,   181
SiC,   6
  amorphous,   282,519,528,536

Valence band tail, 235,274,
        280,281,290,402
Vibrational band, 181
Vidicon tube, 526
Voids, 23,138,143,479,480

W, 6
Weak bond, 187-189,192,248,
        251
White-tin structure, 55
Wrong bonds, 215,217

XANES, 37,42,44,47,48
XeF_2, 179,182
X-ray,
  absorption, 42
  diffraction, 16,28,94,95,
        97,100,275
  photoelectron spectroscopy,
        384,387,444
  spectroscopy, 275

Zero bias fill, 305,306